CREO Topics Covered by Lesson

Sketching Overview – Sketch Tools & Practice Shapes

Lesson 1 - Extrude Tool

Lesson 2 - Extrudes, Datums, Mirror Tool, Basic Drawing, Color Settings

Lesson 3 - Revolves, Holes, Rounds, Datums, Drawing Centerlines, Cross-Sections

Lesson 4 - Ribs, Patterns, ASME Style Datums, Hole Notes

Lesson 5 - Palette Tool, Detail (Dimension) Drawings, Detail Views, Projection Views

Lesson 6 - Blend Tool, Shell Tool, Pattern Tool, Drawing Dimension Text

Lesson 7 - Sweep Tool, Text Tool, Custom Isometric View Orientations

Lesson 8 - Helical Sweep Tool, Text Along a Curve, Cosmetic Threads, Multiple Drawing Models

Lesson 9 - Rendering, Draft Tool, Re-ordering the Model Tree

Lesson 10 - Assembly Mode, Constraints, Bill of Materials (BOM) Table & Balloon notes

Lesson 11 - Assembly Drawings, Assembly Cross-Sections, Exploded Views, Layer Tree

Lesson 12 - Vendor Supplied Models (McMaster.com), Assembly Repeats & Copies, Part Activation

Lesson 13 - Measurement Tool, Specifying Hardware, Mechanism Constraint, Re-Ordering an Assembly

Appendix

Basic Drawing Steps

Detail Drawing Steps

Assembly Drawing Steps

Drawing "Class Standards" Checklist

Creating a Custom MyIso view orientation

Creating an Auxiliary Drawing View

Recovering from a Crash (Trail files)

Setting the Model Units

Using the Measuring Tool & Mass Analysis

Importing a Points File or X, Y, Z Coordinates

Setting the Colors of the Model and Rendering

Importing Vendor supplied Parts (McMaster & ServoCity)

Creating a DXF file for CNC Milling

Creating a File for 3D Printing (and. STL files)

Table of Contents

Preface

3D Modeling is an important skillset for many Mechanical Engineers. This textbook was developed to be a practical experience to develop your CAD skills. Instead of lengthy explanations of how the CAD features operate, students will instead apply features and tools to build realistic product models while gaining deeper knowledge through repetition on later lessons. Eventually students will be at an "intermediate" level of CAD skill; knowing what features are available, how to use them, and how to troubleshoot their own models. Students will be able to work on their own designs, work for a company at an internship or early career design job, and be capable of producing models and detail drawings suitable for prototyping or manufacturing. At this skill level students will not be exposed to complex equation driven modeling, simulations, mechanisms, or detailed geometric tolerance drawings. Those complex topics are better suited for a more advanced student that has a reason and motivation to learn those skills through special projects, job training, online learning, etc.

One thing this textbook does not do is prepare students to be a Drafter. Professional Drafters work with CAD and detail drawings for the majority of their job and must have a deeper understanding of the operation and standards for engineering drawings, such as ASME Y14.5, needed for manufacturing or production industries. Some engineers may become expert drafters and modelers over time, but most engineers will have their skills focused elsewhere and may only maintain an intermediate level skillset of CAD instead of being a replacement for a Drafter.

The Software: *CREO Parametric 10.0*

This textbook is specifically setup to teach 3D modeling through the use of **CREO Parametric**, a parametric modeling software provided by PTC. This software was previously known as PRO/Engineer or PRO/E Wildfire, and has been a major industry player on 3D modeling and design software for many years. In fact, PTC's software was the first parametric based CAD software on the market. It is a very powerful CAD software with complex features and applications that can be added in for commercial focused users, similar in capability and market use as Solidworks, SolidEdge, etc.

Due to CREO being an advanced commercial software used in large industries it is not designed to be easy to learn and does have a steeper learning curve than some of the other packages. The content should start making more sense as you work through the lessons and start understanding the purpose of each step, and possible errors that can arise.

Software Access & Installation

PTC CREO Parametric installation instructions are provided by the instructor for downloading and installing the Student version of the software on your own computer (*"PTC CREO 10.0 – Student Edition – Quick Installation Guide"*). The software is free, the license lasts for one year before needing to be updated or reinstalled (at no cost). The only caveat is that the files created in the Student version cannot open, or be opened by, users of a Commercial License of the software. Similar to many engineering software programs, CREO is only capable of being installed on a Windows operating system.

Also note that CREO has a few different software's packages within it – this textbook will only cover *CREO Parametric*, not *CREO Direct, Layouts, Simulate, etc.* which are different modeling methods and packages under the CREO name.

About the Author:

Mr. McNally is a Mechanical Engineer and engineering Instructor that has been teaching CAD modeling and engineering design since 2006. He has designed many prototypes and designs for personal use (motorcycle accessories, electric car conversion, remote control snowblower) as well as professionally (hydrogen fueling stations, material testing systems, custom robotic equipment for disabilities, & remote-control targets for military use).

Getting Started

Before getting started on the first lesson you should work on getting access to the software, whether that is your own computer, a computer lab or work computer, or using *UND Labs Anywhere* to operate a virtual computer through a web browser.

Check your course documents for details on installing the software or using alternative methods. The CREO software company, PTC, provides the student version of the software for free. The installation instructions should be posted by your instructor, or you can try this address:

https://apps.ptc.com/schools/references/install_creo10_unistudent_standard.pdf

Provided Required CAD Files:

Some files for these lessons are provided by the author, such as the C-size drawing format, bill of materials table, and some parts for the assembly lessons. If you are currently enrolled in the course all the files are provided to you within Blackboard. **If you are not currently enrolled in the class you can request the files from the instructor at** (dustin.mcnally@und.edu).

Instructional videos for each lesson, along with Walkthrough videos, are provided in the course lesson folders for those enrolled in the course. Eventually you should reduce your reliance on walkthrough videos and make sure you can complete the lessons without using them step-by-step.

Computer Requirements:

PTC does have a hardware requirements document that lists the minimum requirements for a computer to CREO that you can find online. For the level of work being done in these lessons you do not need a specific 3D modeling computer or expensive CAD specific graphics card (common in industry). But you will want a decent "mid-level" computer with good RAM, fast processor, Windows Operating System, and if possible an OpenGL graphics card. It does not need to be a "workstation" or "gaming" computer for the level of work done by most students. Some cheaper computers may struggle to run the software and a Chromebook, iPad, etc. would not be suitable.

Keep in mind the following:

- Download and install the correct version of the software from PTC.
- CREO is only able to be installed on a windows computer. An Apple Mac may be able to run it if Windows is installed as a dual–boot system, but the newer "M" series Apple chips do not allow this.
- Newer versions of CREO can open older versions files, but not vice versa (e.g. 7.0 files cannot be opened in 5.0 software)
- The academic version cannot open Commercial versions of files, and vice versa.
- You will need a **dedicated 3 button mouse** (Left button, middle scroll wheel, right mouse button). The Middle Mouse button and scroll wheel is used a lot.
- You should be using the Student or University edition. Do not use "setup-**schools**.exe" when running the installation file, which installs a simplified K-12 Schools version of the software.
- Some applications, like Weldments, may not be available in the home student version. These will not be used.
- Always open the software, ***CREO Parametric***, first before opening files. Do not open files from Windows, as Windows may choose CREO Direct or CREO Simulate to open them, which are different programs.
- Always **Set Your Working Directory** and save your work often. **CREO does not auto-save**. Also note you cannot save whenever you want, you must complete or cancel any currently open tools, sketches, or features.
- Is CREO acting strange or seem frozen? Check the message display log in the very bottom left corner. Sometimes a tool will be waiting on you to complete an operation and is prompting you about what to do next. Or press Control + A to activate the current window.

Sketching Overview

Before you can start modeling parts in CREO you must first learn how to create **Sketches**. Sketches act as the foundation for creating features and solid volumes which are used for most features in 3D modeling.

Basically, a sketch is a cross-section or area that defines the location, shape, and size of a specific feature of a model. For now, consider a Sketch to be a two-dimensional area that is created using entities such as lines, arcs, circles, and centerlines on a specific sketch plane.

There are 3 main steps for creating most sketches

1 – Create a Closed-Loop shape *(There are some exceptions to this that we will learn in later lessons, like Ribs)*

2 – Add or Remove Constraints

3 – Create Dimensions and set their values

1) Closed-Loop Shapes

When in Sketch mode, use sketch entities from the top toolbar to sketch the desired geometric shape. The sketch "loop" should be closed by the time you finish; **no dead ends, no gaps, and lines cannot cross over each other.** Multiple loops are OK as long as they do not touch each other. Think of a loop as a race car track; a car can start on the loop and will end up back at the start without getting lost or finding gaps. Centerlines and dotted reference lines are not part of the Closed-Loop.

- Sketch entities are found in the Sketch Toolbar, such as tools to create Arcs, Lines, Circles, Rectangles etc.

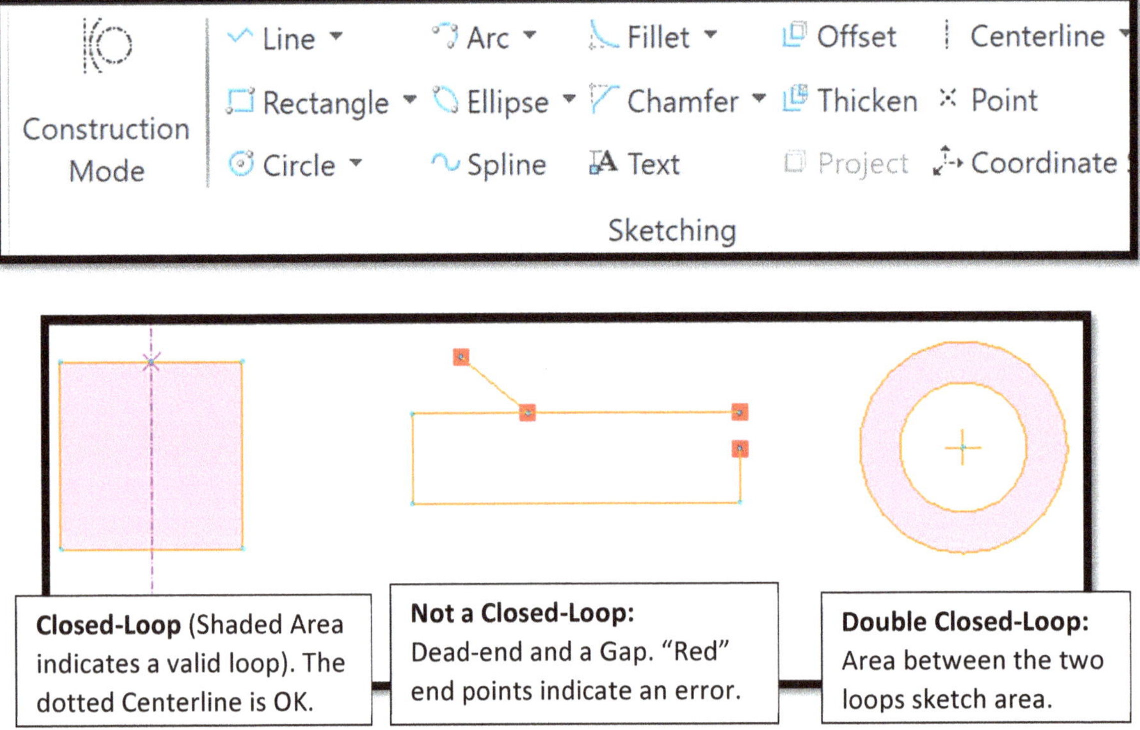

Closed-Loop (Shaded Area indicates a valid loop). The dotted Centerline is OK.

Not a Closed-Loop: Dead-end and a Gap. "Red" end points indicate an error.

Double Closed-Loop: Area between the two loops sketch area.

2) Add or Remove Constraints

As you place entities on the sketch CREO may assume and place some constraints whether you desire them or not. If you sketch a line close to being horizontal, CREO will "snap" the line and force it to always be Horizontal and will show the Horizontal Constraint symbol on that entity. If a line is sketched approximately the same length as another line, CREO may "snap" in an Equal Lengths constraint forcing that line to be the same length as the other line.

Undesirable constraints should be deleted out, while desired constraints can be left in place or added in yourself using the Constraint toolbar.

- **Tip:** The more constraints that are added, the less dimensions that will be needed to fully define the shape, size, and location of the sketch.
- Constraints can be added using the ***Constraint Toolbar*** (i.e. Symmetric, Vertical, Equal Lengths, etc.), and can be deleted from the sketch by clicking on the symbol in the sketch and pressing Delete on the keyboard.

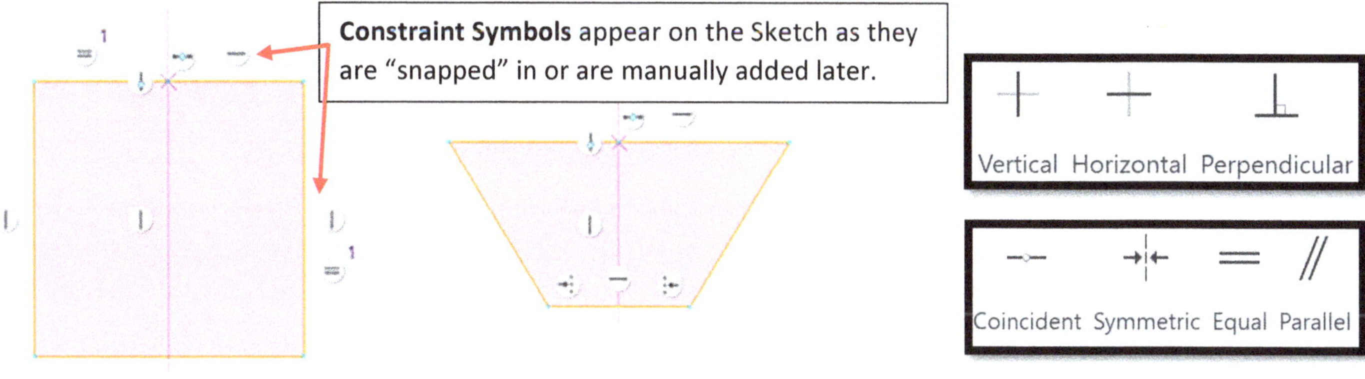

3) Create Dimensions and set their numeric values

After adding sketch entities CREO will shows some measurements, known as **Dimensions**, which define the size or position of the sketch shape. You can create your own dimensions and set the values as needed. Not every single entity needs a dimension, as CREO will "solve the geometry" using the provided dimensions and constraints to calculate and define how all entities are sized. Providing any excess or unnecessary dimensions will result in an Error, as the sketch would then be over-defined and could have competing values.

In the example shown, notice that the left side vertical line is not dimensioned (it is known from an *Equal Length* constraint), and that the slanted lines are also not dimensioned (the vertical component is calculated from the difference between the 0.980" & the 0.520" dimension, while the Horizontal component is calculated as the difference between the 3.250" & 1.500" dimensions. **CREO automatically does the geometry calculation for you.** If you try to add in an additional dimension that is not needed, CREO will either bring up a Resolve Sketch error menu or will remove one of the light blue colored Weak dimensions from the sketch.

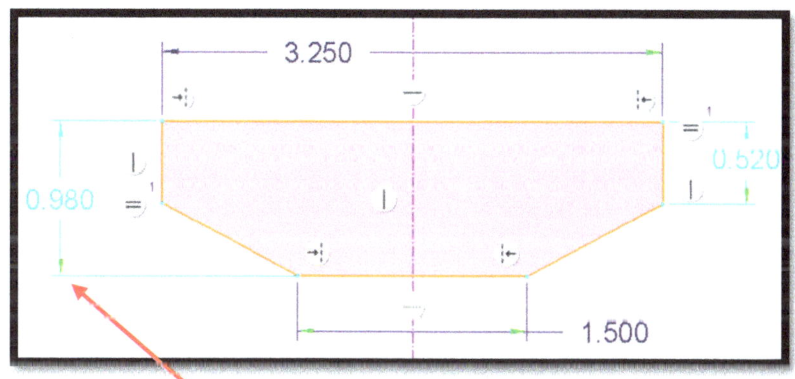

Weak vs Strong Dimensions:

Dimensions on a sketch will either be "Weak", which have a light blue color. These are the dimensions that are added automatically and can be removed or change value on their own. Dimensions will become "Strong" when the user has provided them a specific value, signaled by their darker/purple color. Unlike Weak dimensions, Strong dimensions are not at risk of changing or being deleted by CREO when solving the geometry. **It is standard practice to complete a sketch using strong dimensions with no weak dimensions remaining.**

Resolving Sketcher Errors:

Often you may attempt to provide additional dimensions or constraints that cause the sketch to be over constrained. Anytime CREO has more than the minimum amount of information to solve the geometry of a sketch you will get the "Resolve Sketch" error pop up menu. CREO doesn't want to have possibly conflicting information so it must be fixed.

In the example below, the Dimension Tool was used to define the horizontal measurement between the slanted line start and end points. **This added in too much information causing a conflict!** CREO is asking the user to resolve the sketch by either canceling the new dimension or deleting out one of the conflicting dimensions or constraints.

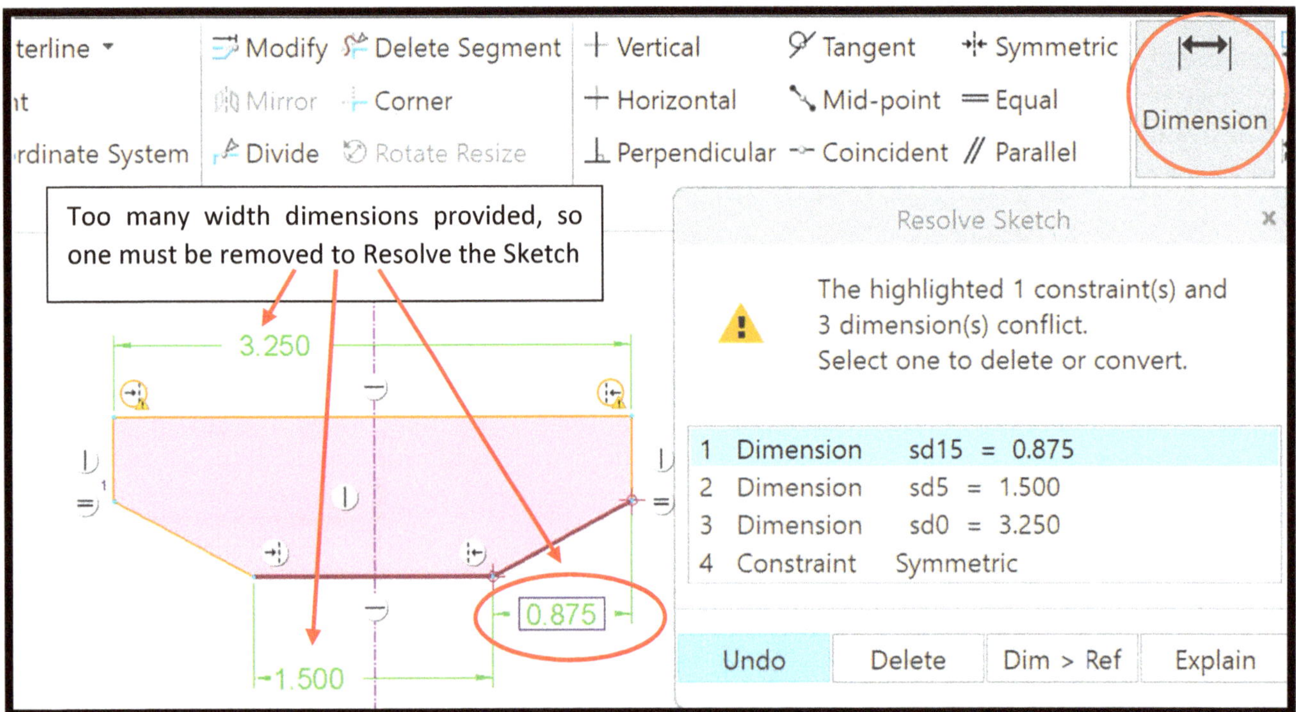

The 1.500" dimension was chosen to "delete" as that line length will now be calculated by CREO using the difference between the top 3.250" length and the new dimension of 0.750" for the slanted line horizontal distance. There are many ways to fully dimension a sketch, but often you will want to use a dimension that is known to you or easily measured.

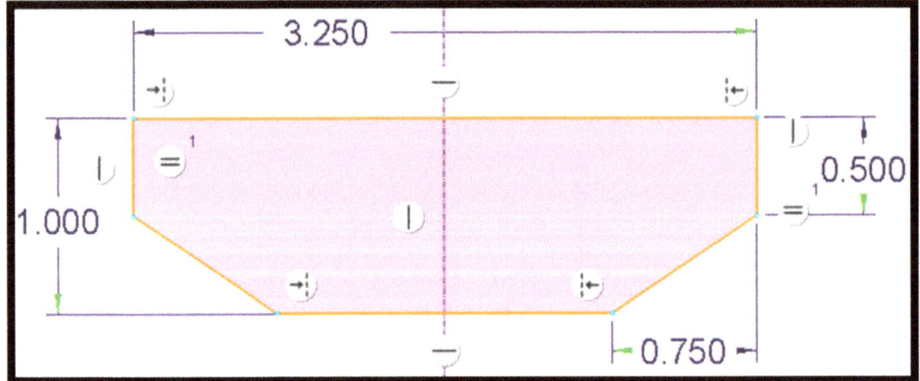

The sketch above is now resolved. Even though not all lines are dimensioned, CREO has all it needs to calculate the size and position of the shape. All dimension values have been set by the user and are now "Strong" (dark colored). Some say at this stage you have become friends with CREO, but don't expect it to act like it while other users are around.

Sketching Practice Exercises

Open up ***CREO Parametric*** and practice making the correct **Closed-loop Shape, Constraints, and Dimensions** of the following sketches using ***CREO Parametric***. You do not need to save or submit any of this work, so start a new Sketch file to move on to the next item after you have created your own sketch that matches that shown in the exercises.

Tip: Start with a brand new, empty sketch for each example. Add in your own **centerlines** which are required for symmetry constraints, and **pay close attention to any extra entities or sketch lines** that you may have accidently placed that will need to be removed in order to match the image shown. If your shape is not working properly you can start a new sketch file to start over.

Step 1 - Start by opening ***CREO Parametric*** and start a new sketch file. The default name "*s2d000X*" or similar is OK. *Detailed Instructions to complete the step are shown below, which is the format for the rest of the textbook.*

-   After opening ***CREO Parametric*** - **close any pop-up windows - then click on File - New – Sketch – Press OK.**
    -   o   If you do not choose *Sketch* as the file type it will default to a *Part* file instead and you will be lost….
-   Repeat this step after completing each of the sketch exercises on the following pages. You do not need to save or submit these Sketches, just create them to match the image and move on to the next one.

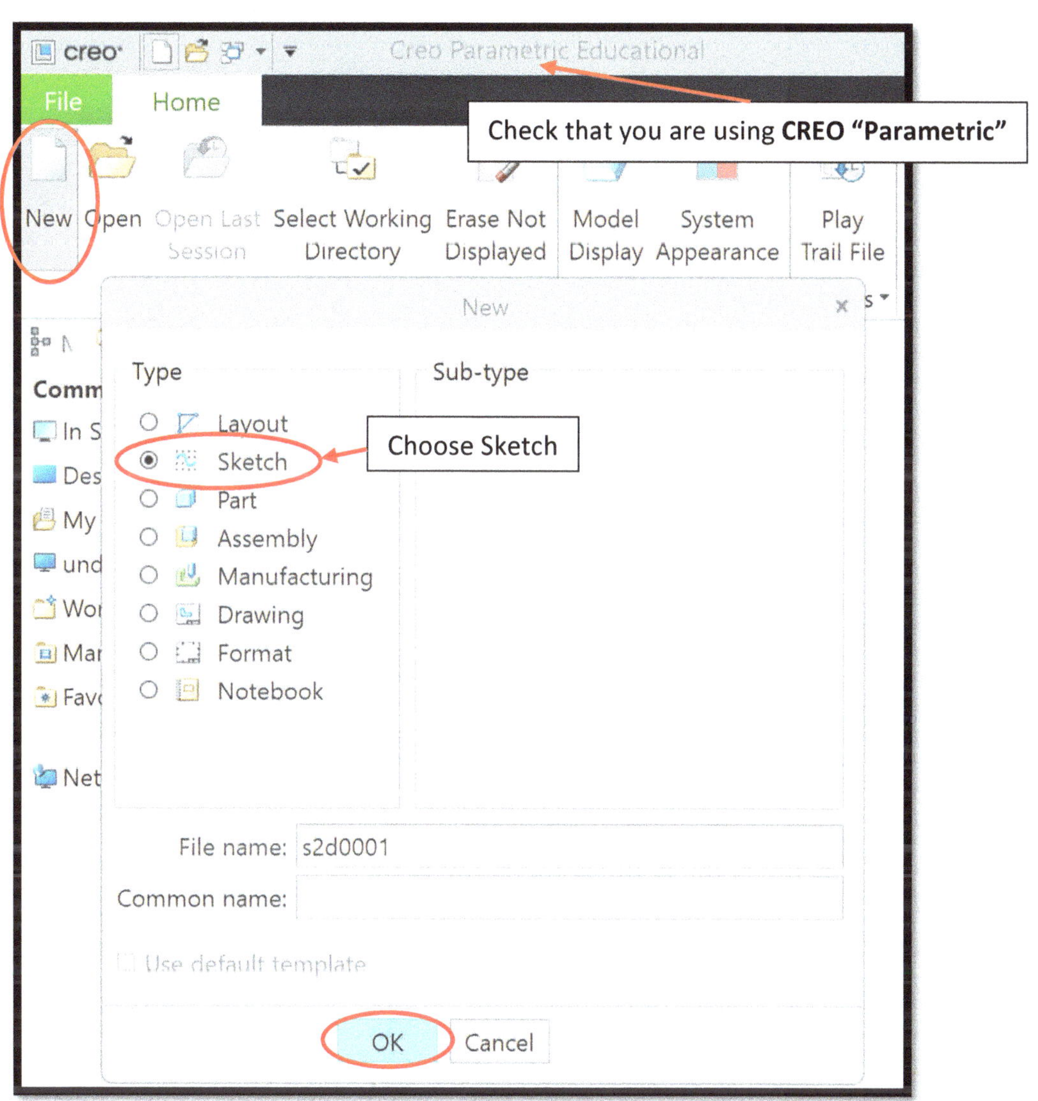

Sketch #1 - Concentric Circles (or Pipe)

Sketch the shape shown below to make your sketch match the image. Use the Sketch tools and details as listed below.

<u>**Closed-Loop Shape**</u>: Add vertical and horizontal **Centerlines**. Then use the **Circle Tool** to create two closed-loop circles with the center points of the circles placed on the intersection of the centerlines.

<u>**Constraints**</u>: The circles should be coincident or centered on the intersection of the centerlines. If they are not aligned properly, you can use the **Coincident constraint** to force the centerlines and circle center points to be coincident by clicking on them.

<u>**Dimensions:**</u> Set the circle dimension values to **5.50" and 4.50"** respectively (double click on the values while on the **Select (Pointer) Tool** and then type in a value and press Enter on the keyboard).

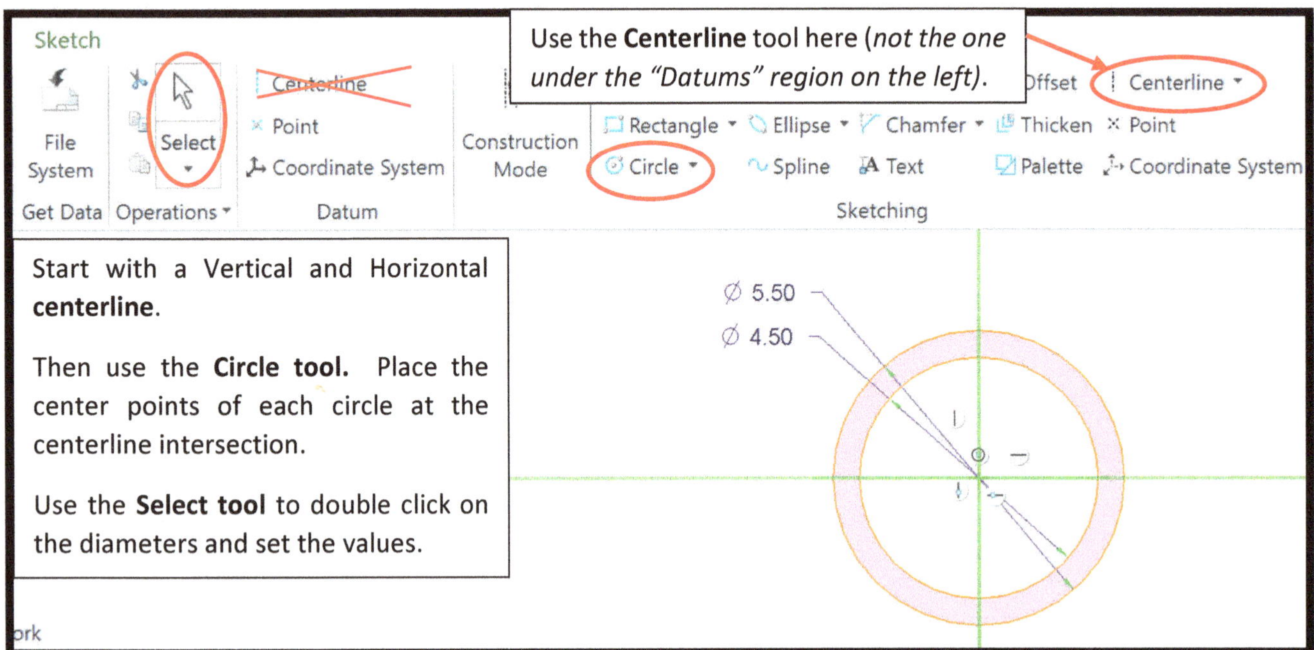

Sketch #2 – Equilateral Triangle

Start a fresh sketch. You do not need to save the previous one. (Click **File – New – "Sketch" – press OK**).

Closed-Loop Shape: use 3 lines to create a triangle with a horizontal bottom line, using the **Line Tool** in the top toolbar.

Constraints: Make sure there is a **Horizontal Constraint** on the bottom line, then use the **Equal Length Constraint** to set all 3 lines to be equal.

- Use the **Equal Constraint Tool** – then **click on each of the 3 lines**. You should notice the $=^1$ symbol on each one, leaving you with a single dimension.

Dimensions: Set the line length to **5.00"** (double click on the dimension value while on the **Select (Pointer) Tool**). Your dimension may appear on any of the lines, it does not need to be the bottom line.

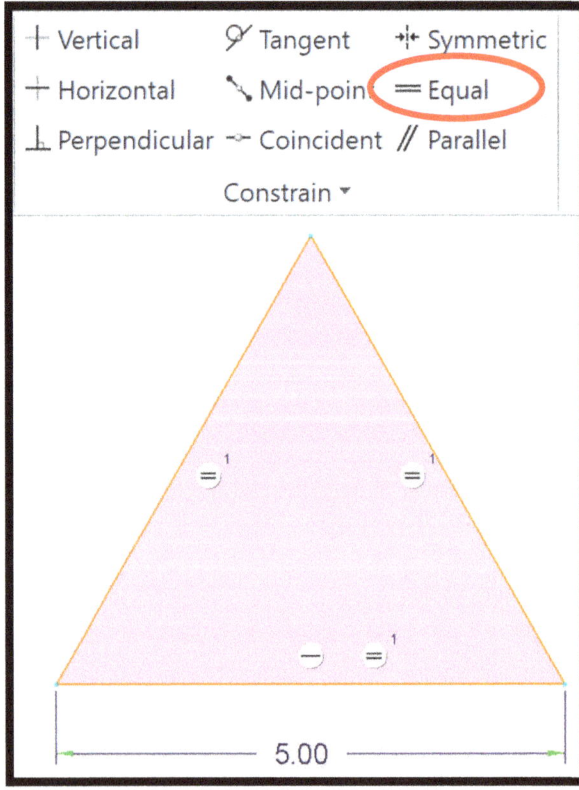

Sketch #3 - Symmetric Double Square

Step 1 - Start a new sketch. Add the **vertical and horizontal sketch centerlines** (use the 'sketch' Centerline on the right middle of the toolbar, not the one in the left 'Datum' menu).

Step 2 - Create the two separate rectangle closed-loop shapes using the **Rectangle Tool.**

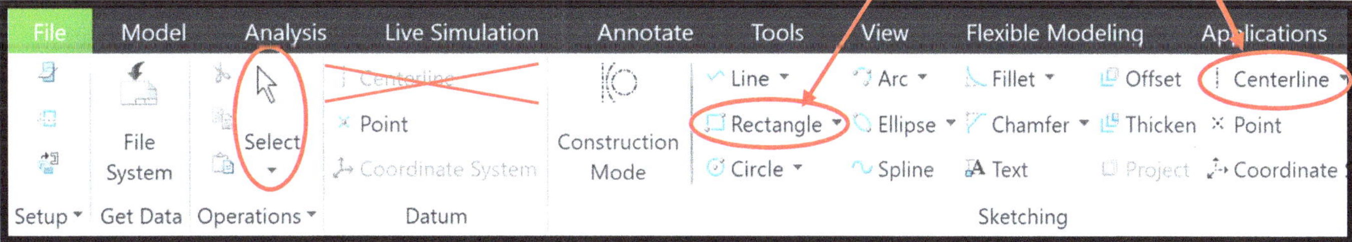

Step 3 - Set the **Vertical & Horizontal <u>Symmetry</u> Constraints** on both Rectangles (*if not automatically added by CREO when you placed the rectangles*). Each rectangle will need its own vertical and horizontal symmetry constraints setup. *Note – do not try to add symmetry to all sides of a rectangle; a top or bottom and a left/right side is all that is needed.*

- Click on the **Symmetry Constraint** – click on the desired **Centerline** - **click on the first Point from a rectangle** that will be symmetric on one side of the Centerline – then **click on a Point on the <u>opposing</u> side** of the centerline. The symmetry arrow symbol should appear. Repeat so that each Rectangle is symmetric vertically and horizontally.

Step 4 - Create **two separate sets** of **<u>Equal Length Constraints</u>**, one for each rectangle.

- Use the **<u>Equal Constraint</u> tool** – click on the **<u>vertical</u> line of the desired rectangle** – then **click on the <u>horizontal</u> line** of the same rectangle. The constraint should be visible with an "=¹" symbol visible on both lines.
- Click on the **Select Tool** to end the current batch of equal lengths.
- Use the **<u>Equal Constraint</u> tool** again to setup the equal lengths on the second rectangle. This symbol should appear as "=²" symbol, signaling it is different than the "=¹" set.

Step 5 - Set the dimension of 6.00" & 3.00": Select the **Select (Pointer) Tool** – then **double-click on the dimension value - type in 6.00"** on the keyboard - **repeat for the smaller rectangle and set it to 3.00".**

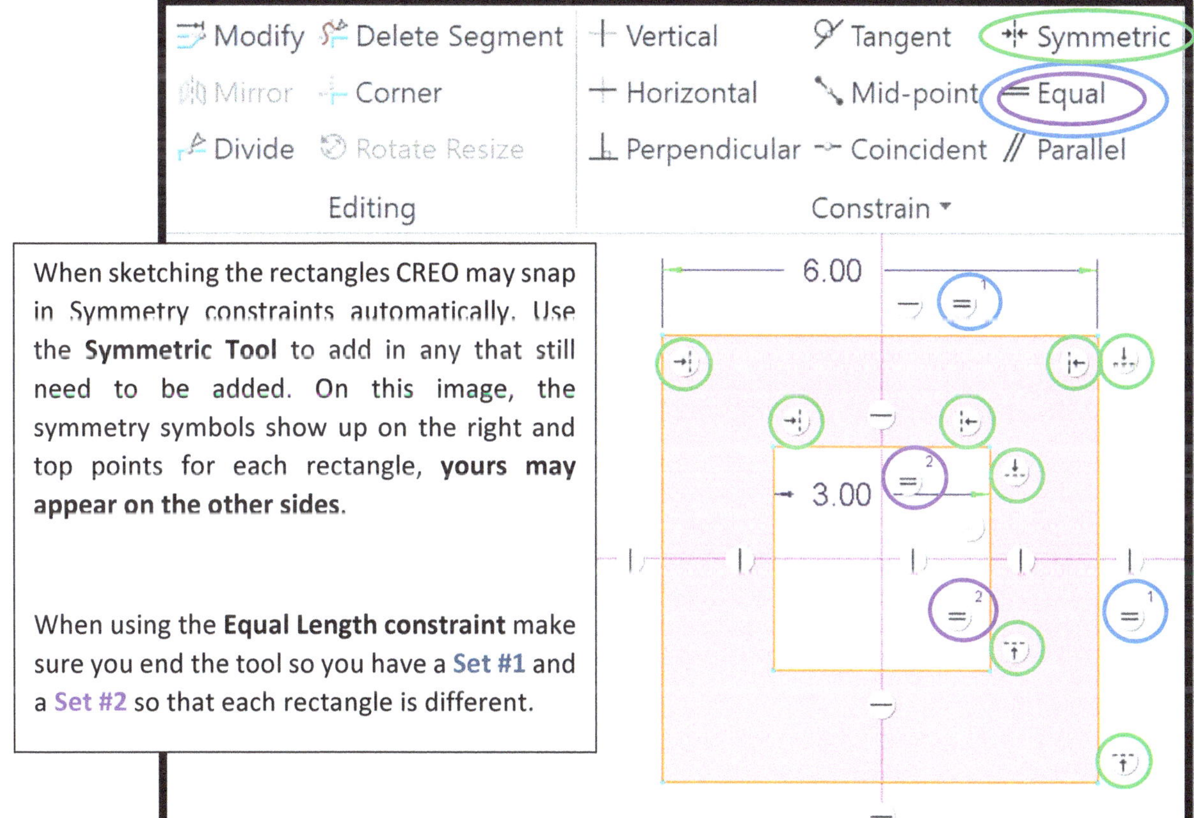

Sketch #4 - Angled Shape with a Fillet (Rounded) Corner:

Again, start a new sketch (**File – New – Sketch – OK**)

Closed-Loop Shape: **Create the shape <u>without the Fillet radius</u> by using the Line Tool**. Then add in the Radius to the top corner using the **Fillet Tool**.

Constraints: Make sure the bottom line is horizontal and the sides are vertical. The Parallel constraint should appear automatically along with the Tangents on the points of the Fillet. You may have slightly different constraints depending on your order of operations (i.e. this image has a perpendicular constraint in the bottom left, you may instead have a horizontal constraint. As long as it ends up as desired you are OK to have slightly different constraints.)

Dimensioning: use the **Dimension Tool** from the top toolbar to place the **angle dimension** between the top angled line and the bottom horizontal line. (- Select the **Dimension Tool - left mouse click the bottom line – left mouse click the slanted top line - then Middle Mouse Button click** to place the numbers somewhere between the two lines.)

Dimensions: 8.00" wide, 2.00" tall on the right side, an **angle of 20 degrees**, and a **radius of 1.00"**. *If CREO will not accept your values, try changing the order that you are adjusting them (i.e. radius last, angle first, and check that you have the same dimensions in place as shown below and not dimensions that reference other entities).*

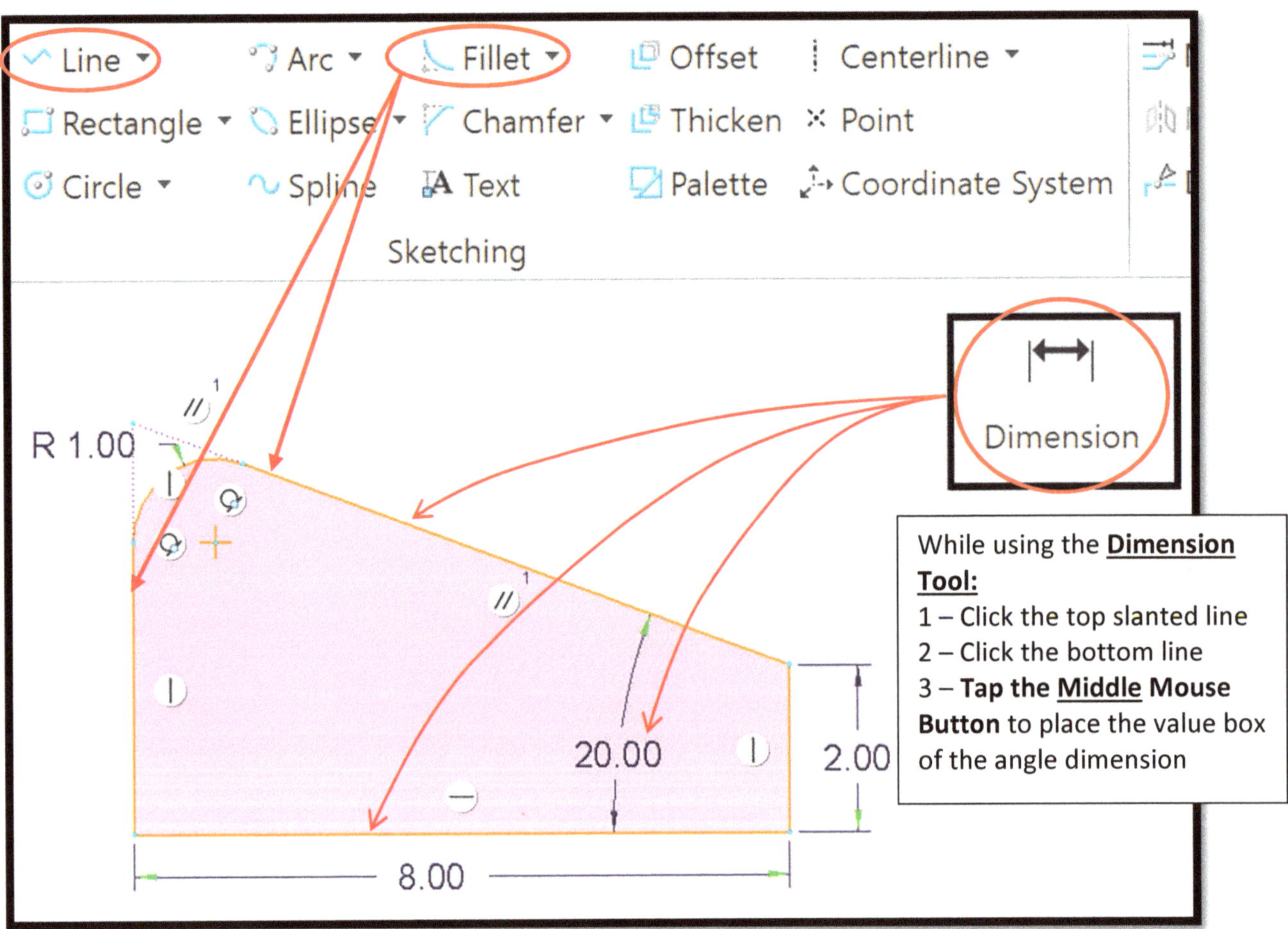

Sketch #5 - Rectangle with Fillet (Rounded) Corners:

Again, start a new sketch (**File – New – Sketch – OK**)

Closed-Loop Shape: Make a **Rectangle**, then use the **Fillet Tool** to convert the intersection between two lines in each corner into a radius. (- Select the **Fillet Tool – click on two intersecting lines** – the radius should appear.)

Constraints: **Add Symmetry** about the Vertical and Horizontal centerlines <u>after</u> **making the fillets***. Also set **Equal Lengths** on all four arcs.

**The fillet tool will remove the corner reference points of the symmetry constraint, so you will need to add symmetry after adding the fillets. Use the start points of the fillet arc as the symmetry reference, not the curve.*

Dimensions: Set the dimensions to be **6.00" x 4.00"** for rectangle, and **1.00" for the Radius**. (Double click on the dimension values while on the **Select Tool** to adjust the values.) If needed, you can use the Dimension Tool to place the correct overall width and height dimensions (- **Dimension Tool – click on the top line – click on the bottom line – tap the Middle Mouse Button** (MMB) to place the dimension. Repeat for the left and right lines if needed.

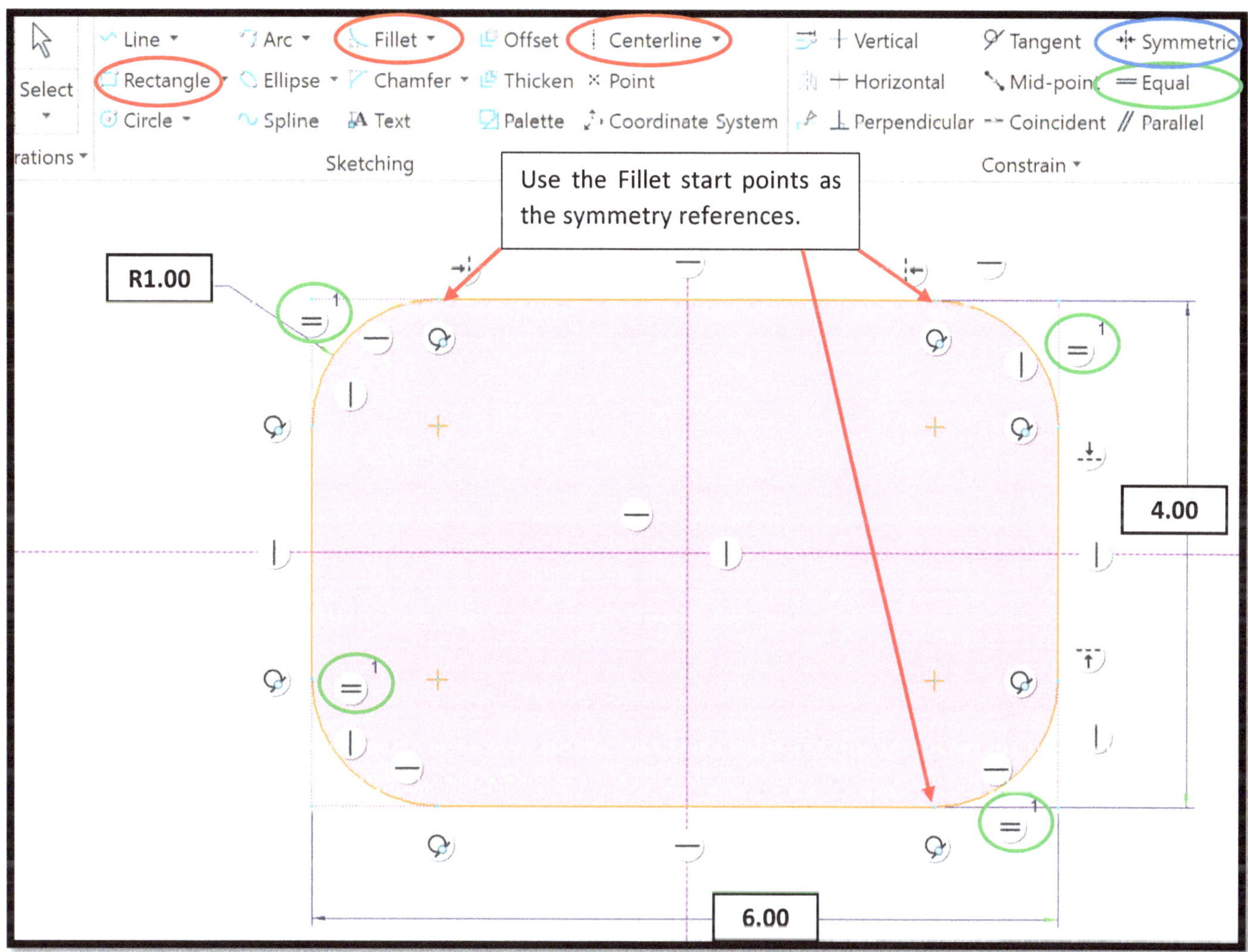

Sketch #6 - Rectangle with Rectangular Cutout:

Closed-Loop Shape: **Add a Vertical Centerline**. Then use the **Rectangle tool** to create the large rectangular shape first. Then **add in the smaller rectangle** coincident with the top horizontal line of the larger rectangle. Then use the **Delete Segment tool** to delete out the top lines of both rectangles that overlap to create the cutout shape.

Note that the cutout is not a square, so if you have an Equal length Constraint it will need to be deleted out. (To remove an equal length constraint, click on the "=" symbol while on the Select Tool and press Delete on the keyboard)

Constraints: **Set a Symmetry Constraint on both rectangles**. Depending on your sketch, you have slightly different constraint possibilities.

Note - If you placed the smaller rectangle with a point **snapped onto the centerline**, you will get a Resolve Sketch error when you try to make it symmetric (Solution: delete out the "point on entity" constraint if you have one). If you placed the centerline so it snaps on a "Mid-Point constraint", you will also get a Resolve Sketch error, and can delete that constraint out from the menu.

Dimensions: use the **Dimension Tool** to place a new dimension for the top horizontal line on one side of the cutout as needed. (Select the **Dimension Tool – click on the top left corner point – click on the left corner point** of the cutout – **then press the Middle Mouse Button** to place the dimension.)

Set the **dimension values** to 8.00" for width, 6.00" for height, 2.50" for the cutout height, and 3.00" for the top horizontal line length on one side of the cutout.

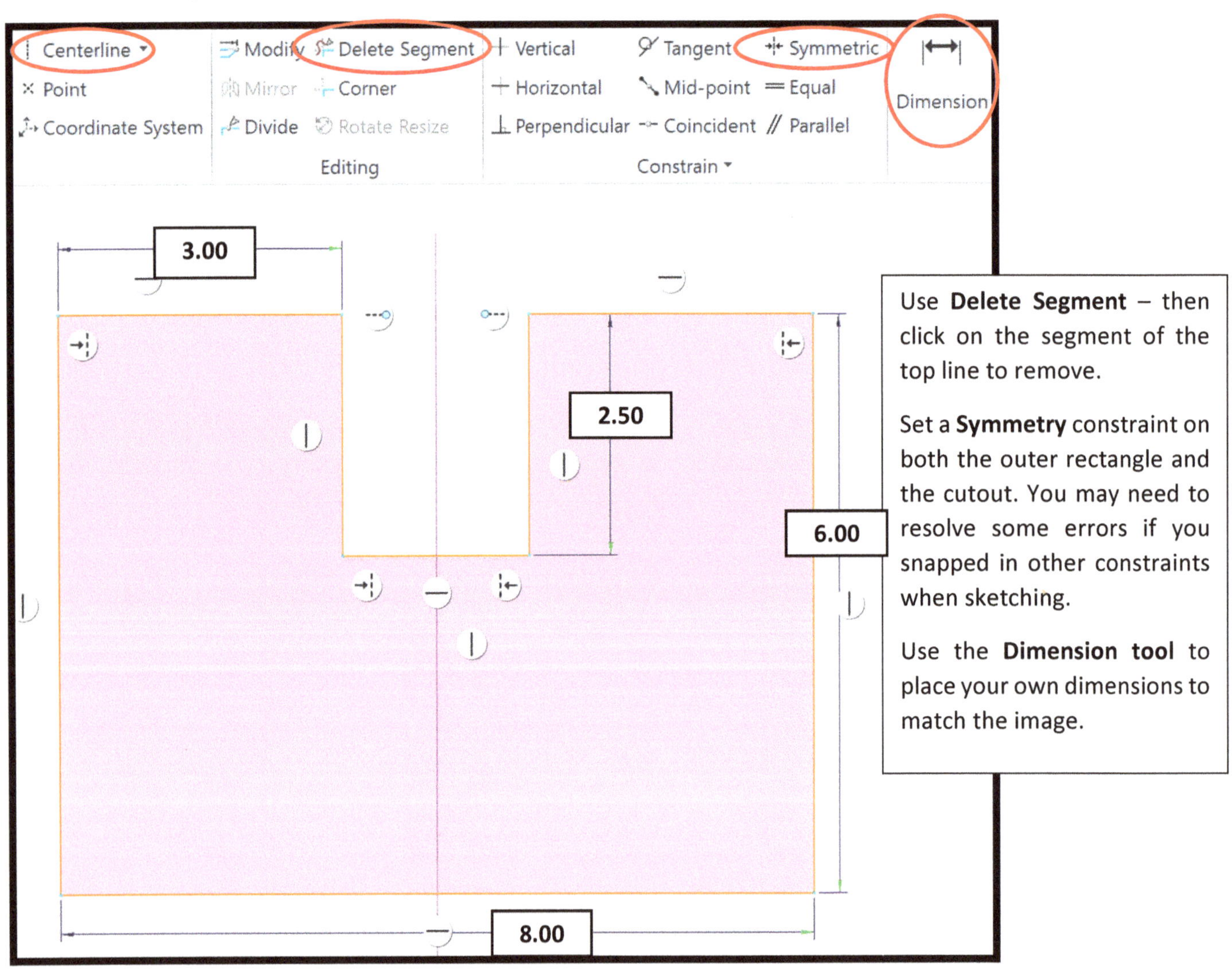

Additional Sketches for Practice:

Below are some additional sketches to practice.

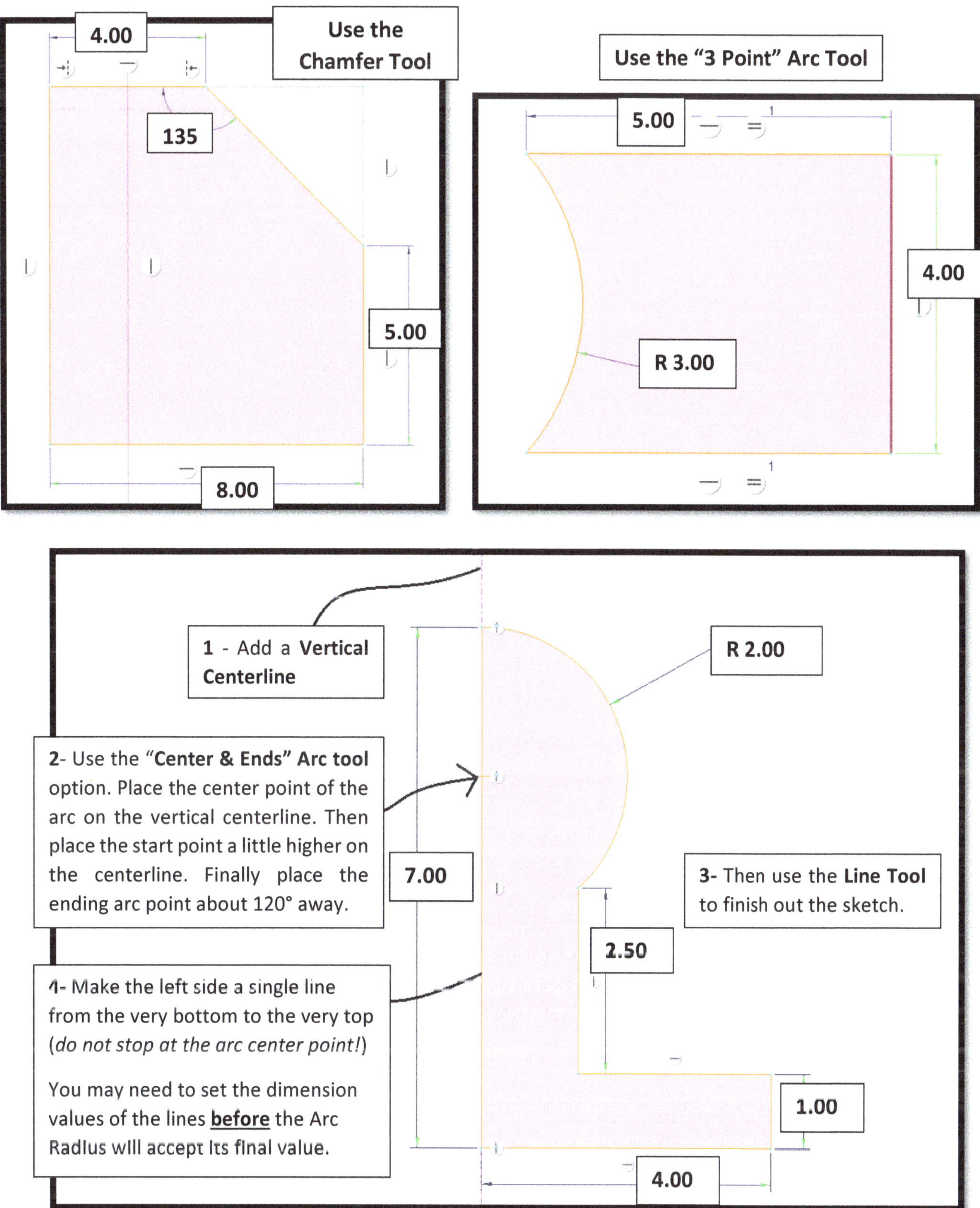

End of the Sketching Practice Exercises

Lesson 1 – Basic Part Modeling (Handle)

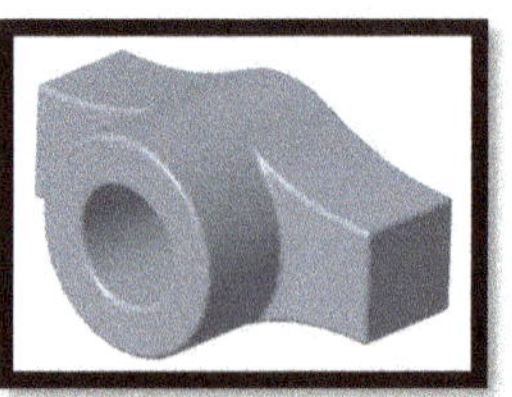

This lesson explores the process to create a three-dimensional **part** in CREO Parametric using a basic feature known as an **Extrude**. Read the steps carefully and refer to the images as you complete each step. Don't worry too much about comprehension as you work through this lesson as the content will be explained and used more in Lesson 2.

Review the command acronyms listed in the box below. When you see "MMB" that means to press the Middle Mouse Button (scroll wheel) on your mouse, which will require you to have a **3-button mouse** to do so.

Basic CREO Commands:

LMB – Left Mouse Button Click

MMB – Middle Mouse Button (Click the Scroll Wheel)

RMB – Right Mouse Button (*Typically you need to **hold** this to get menus to appear*)

Hold MMB and move Mouse – Rotates the 3D space in the graphics window.

Scroll Wheel (MMB) – Scrolling the MMB zooms in/out based on the location of the mouse pointer

Esc (Escape Key) – Will exit out of the current menu or end the use of the current tool.

Quick Toolbar:

At the top of the graphics window you will see the **Quick Toolbar**. Use these tools to quickly refit to screen, adjust the View Orientation and Style, switch to 2D View, turn on/off display filters, etc. as needed.

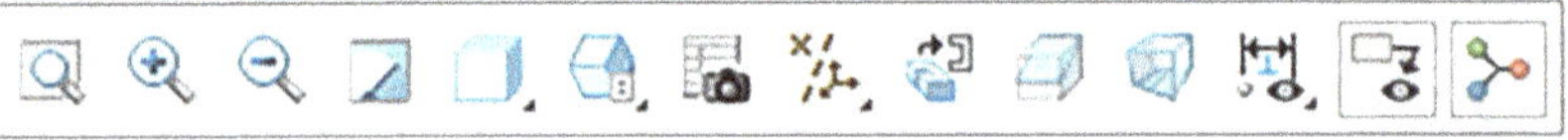

Open up **CREO Parametric** on your computer. If you are working on your own computer, follow the instructions provided by the instructor for installing or getting access to the software.

Step 1 - Set your **Working Directory** to your ME101 Working Directory Folder*.

- Select **File – Manage Session – Set Working Directory – Navigate to your ME101 Directory*** – **press OK**

You may need to create a New Folder in the desired location if you have not done this before, calling it "ME101 Working Directory". You can hold RMB in the desired location, then choose "New Folder" and name the folder appropriately.

- If using your own computer (not in the classroom) create a working directory folder on your C: Drive Desktop.

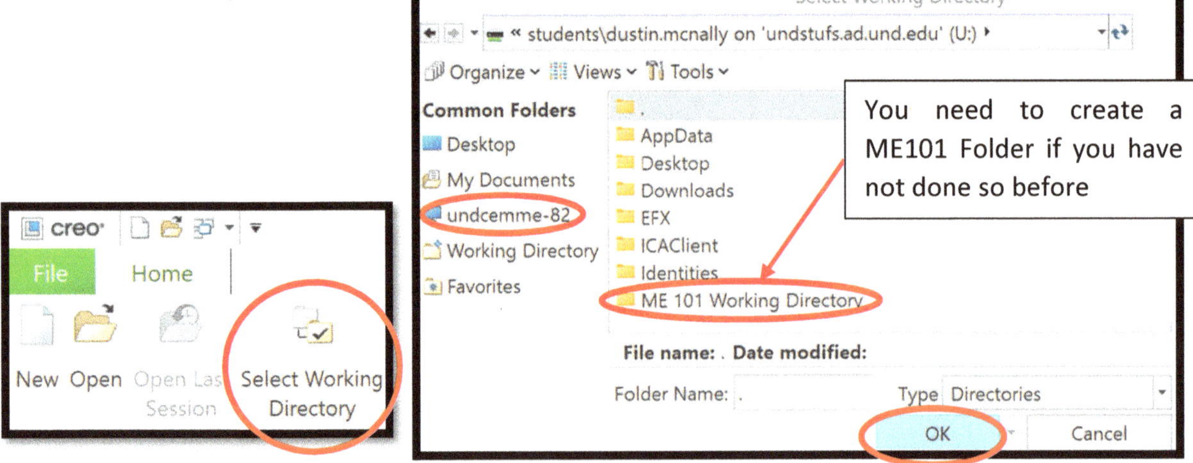

Step 2 - Create a new part called ***"Handle_L1"***. *You cannot use spaces in a filename so use underscores!*

- **File - New – select "Part" as the file type – type in "Handle_L1"- press OK.**

A pop-up menu may* appear asking you to choose a **"Template"** – choose **'inlbs_part_solid_abs'** - press OK.

**The template menu usually will not appear as CREO usually defaults to the desired inlbs_part_solid_abs already, which is what we want to use. This is shown below just in case this menu does appear for you.*

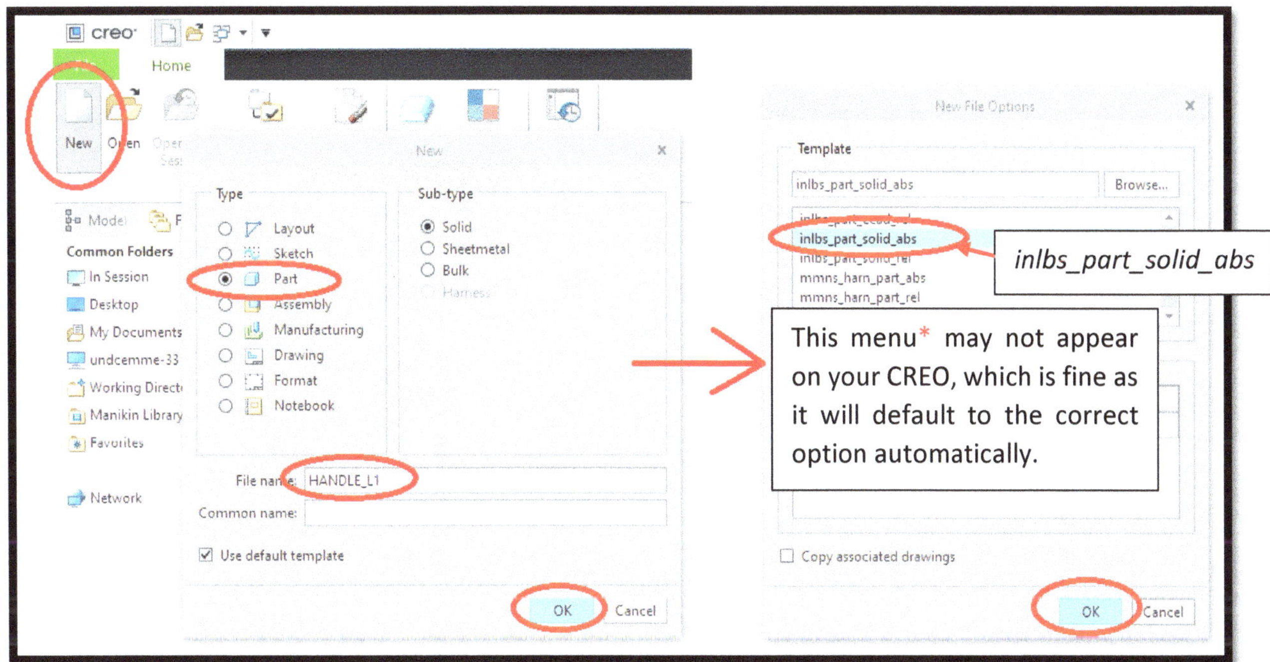

Get Situated in CREO: Notice that your screen is in **"Part"** Modeling Mode and that you are currently on the **Model** tab in the top Toolbar. The Model Tab is the main tab to use for Modeling shapes or features. In the middle of the screen is the **Graphics Window**, which shows the 3D space and any <u>Datums</u> or <u>Features</u> within the part. The **Model Tree** on the left side shows the order of the **Features** that make up the part model. Currently you will only see the Datums on the screen. *If you do not see the datums, you likely chose the wrong type of file when making the new part!*

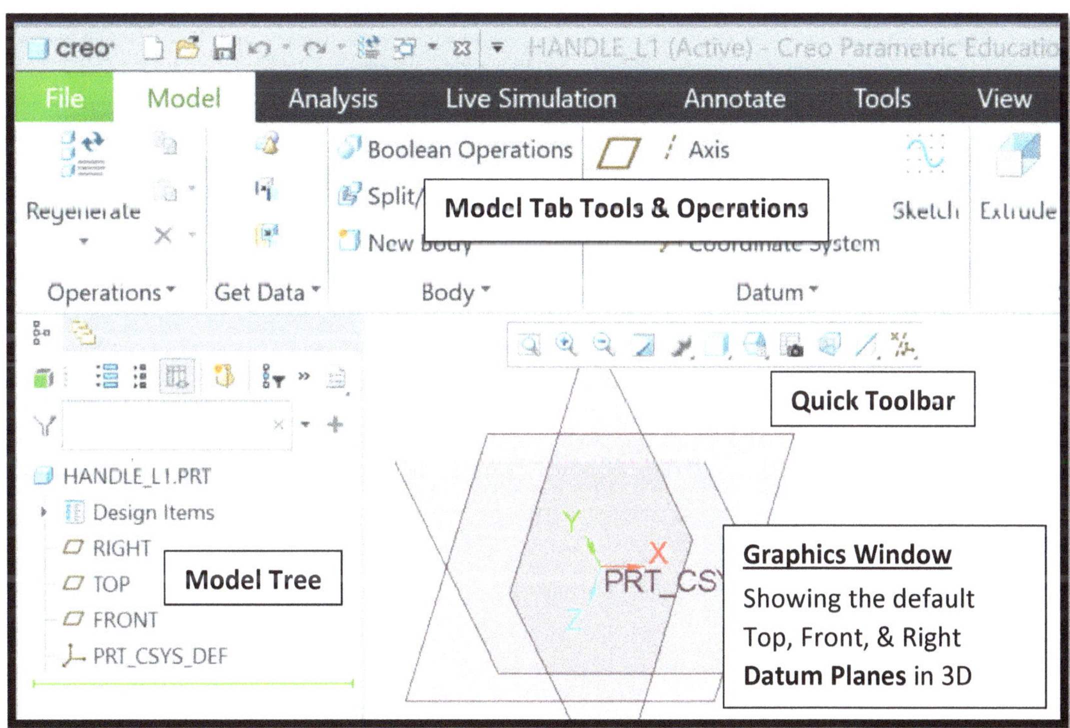

The part model for this lesson will be created with multiple extrusions. The **Extrude** tool will create (or remove) material by <u>sketching</u> a 2D shape and <u>extruding</u> a specific depth to define the <u>3D volume</u> of the feature. The first step is to define the primary sketch plane placement of the Extrude, such as a surface or datum plane.

Step 2 - Create an Extrude feature.

1) Select the **Extrude Tool** in the top toolbar. *(if you don't see this menu, you likely are not in <u>Part</u> mode)*

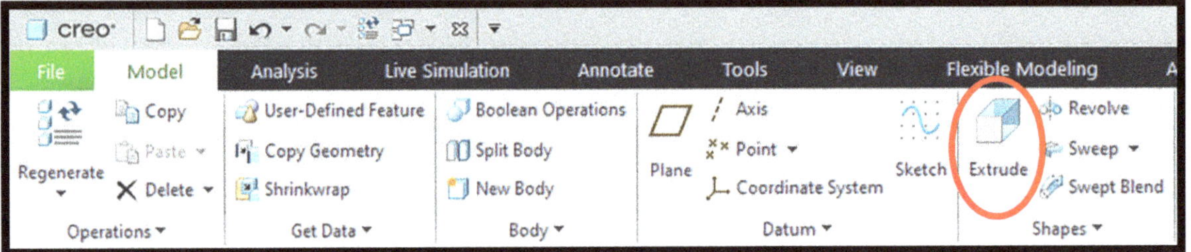

2) Select the **<u>Placement tab</u>** – select **<u>Define</u>** - Click on the **<u>Front</u>** datum to set it as the primary Sketch Plane and leave the Reference and Orientation as the default option (i.e. *Right & Right*).

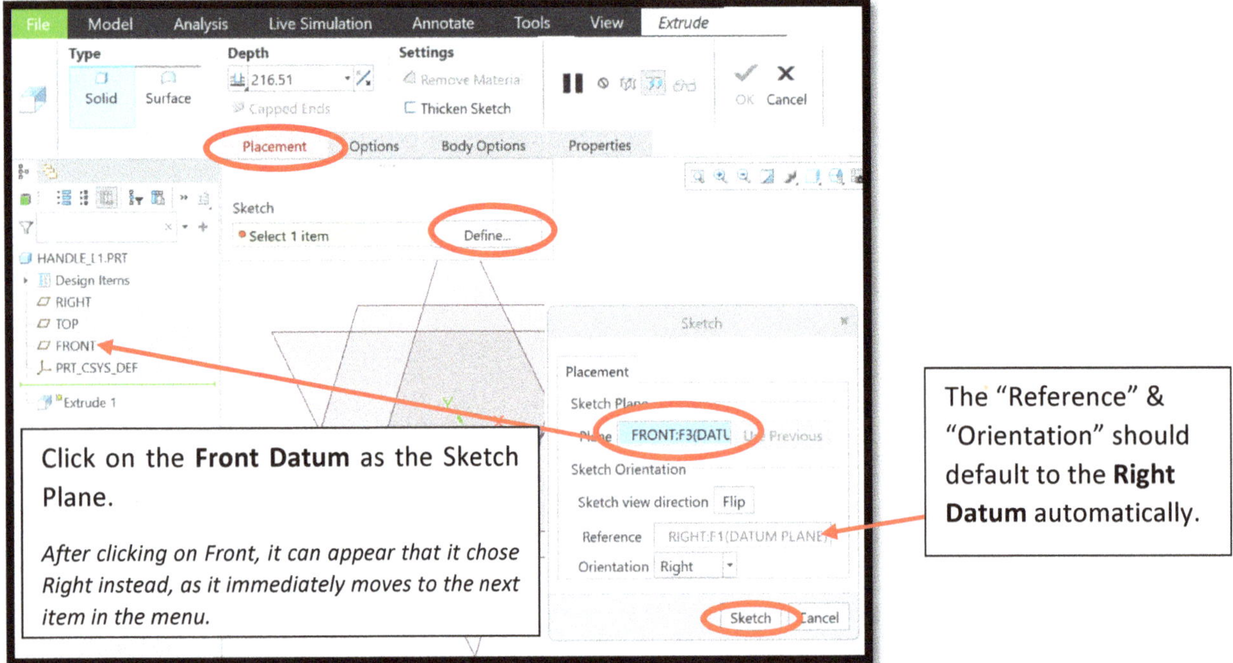

3) **Press Sketch** to continue to the Sketch Mode for this Extrude.

4) Select the "**Sketch View**" icon from the **Quick Toolbar** at the top of the Graphics Window. This will adjust the view to be parallel with the screen.

5) Use the **Centerline tool** to create a **Vertical** and a **Horizontal Centerline** that runs through the origin.

Note that the Blue Dotted lines are datum references, not centerlines, so you must add your own centerlines to the sketch.

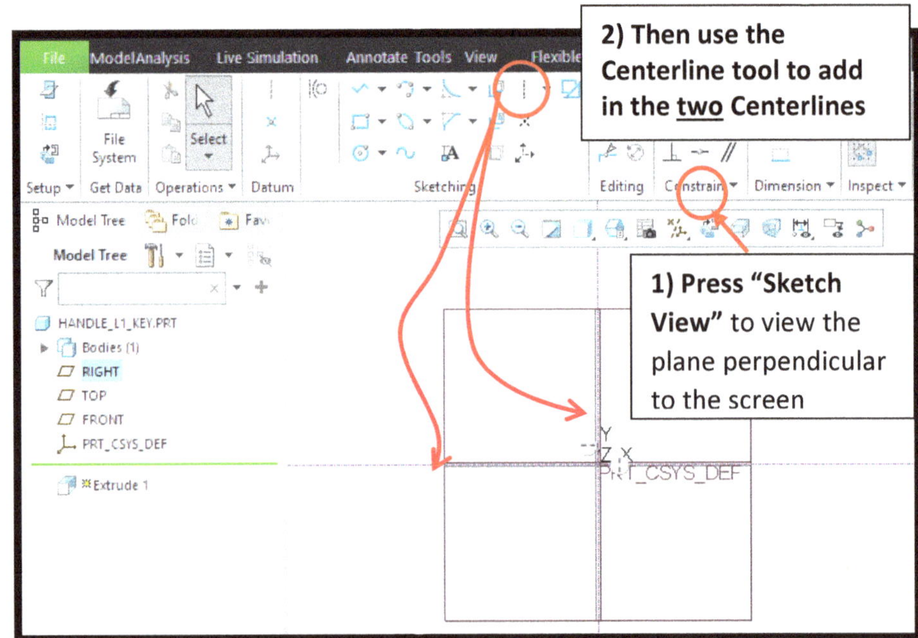

6) Use the **Rectangle** sketch entity tool to create the rectangular Closed-Loop shape below.
7) Use the **Symmetry constraint tool** to ensure the shape is symmetric about both the Horizontal and Vertical centerlines.
    a. Choose the **Symmetric Constraint tool** - **select the desired centerline – select the two points** to make symmetric about the centerline.   **Repeat** for the other symmetry direction.
8) Adjust the **Dimensions** of the Rectangle to be **1.50" wide & 0.40" height**
    a. Use the **Select (Pointer) tool** – double click on the dimension values - change the value.
9) After setting the dimensions your sketch **might become very small on the screen**. Try the following if needed:
    a. Use the **Select Tool** and click and **drag the dimensions closer** to the sketch.
    b. Then select the "**Refit to Screen**" icon in the quick toolbar. Refit will zoom to show all dimensions, so the closer you move the dimensions towards the sketch the more it will zoom in.
10) Press the **Green Checkmark in the top toolbar** to accept the completed sketch.

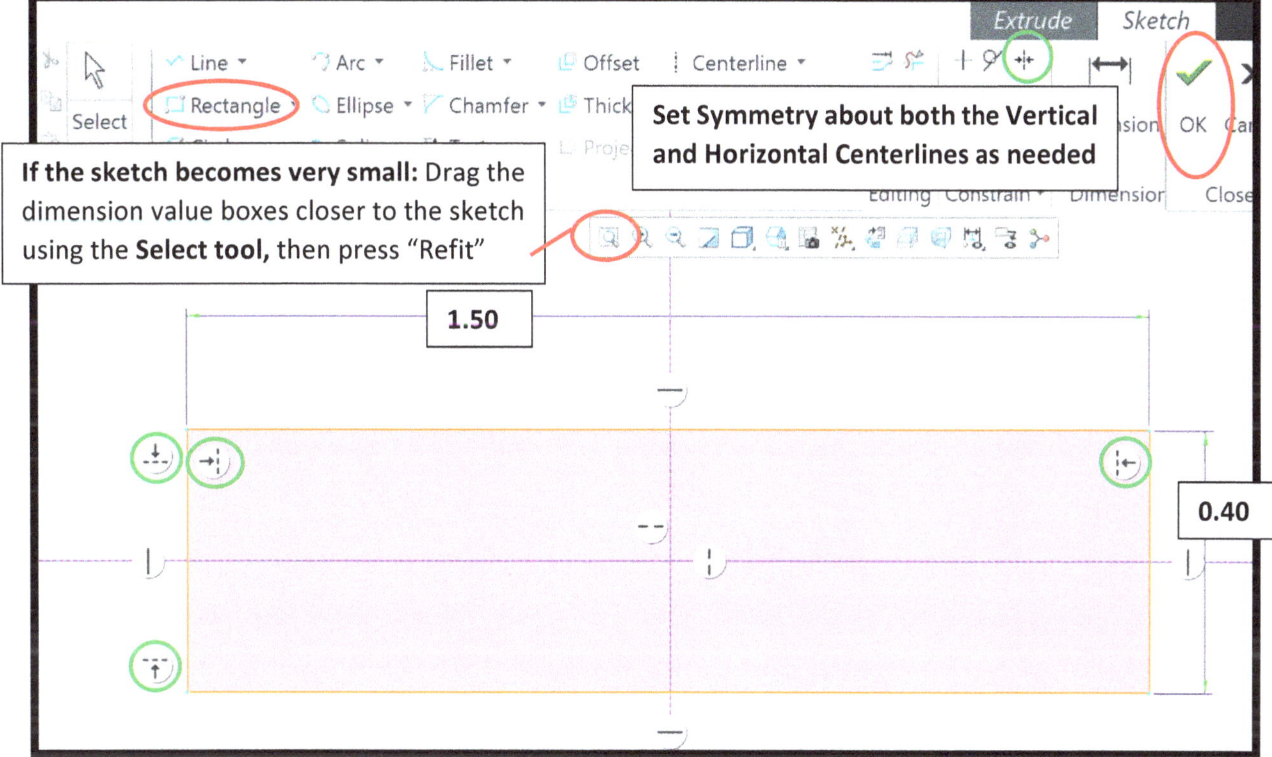

11) **Zoom in on the model** using the mouse scroll wheel (*roll it backwards to zoom in on your pointer*). You can **press the "Refit to Screen" icon** from the quick toolbar to center on the model again if it slides off screen.

12) Set the **Depth** of the extrusion to be **0.40"** in the top toolbar.

13) Press the **Green Check mark** to accept the Extrude.

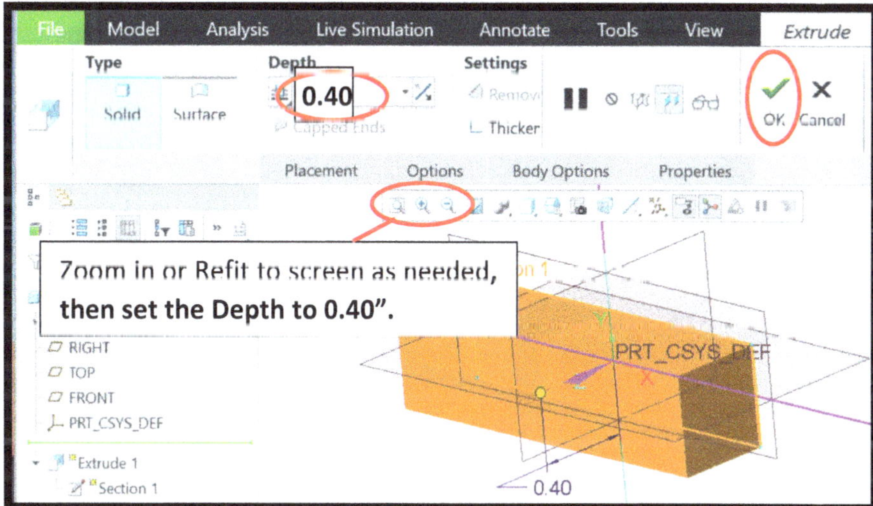

If you accidently accepted the Extrude or made a mistake you can re-enter the feature toolbar using ***Edit Definition*** *shown on last page of Lesson 1.*

14) **After check marking, the model may be too far zoomed in. Press the Refit to Screen icon in the quick toolbar.**

Step 3 – 1 - Use the **Extrude** tool again from the top tool bar to make the next Extrude.

- **Extrude Tool** - click the **Placement** tab option – **Define** - and LMB click on the front planar **surface** of the part as shown below to act as the "Plane Reference". You do not need to adjust the Reference or Orientation input boxes which will likely say Right Datum.

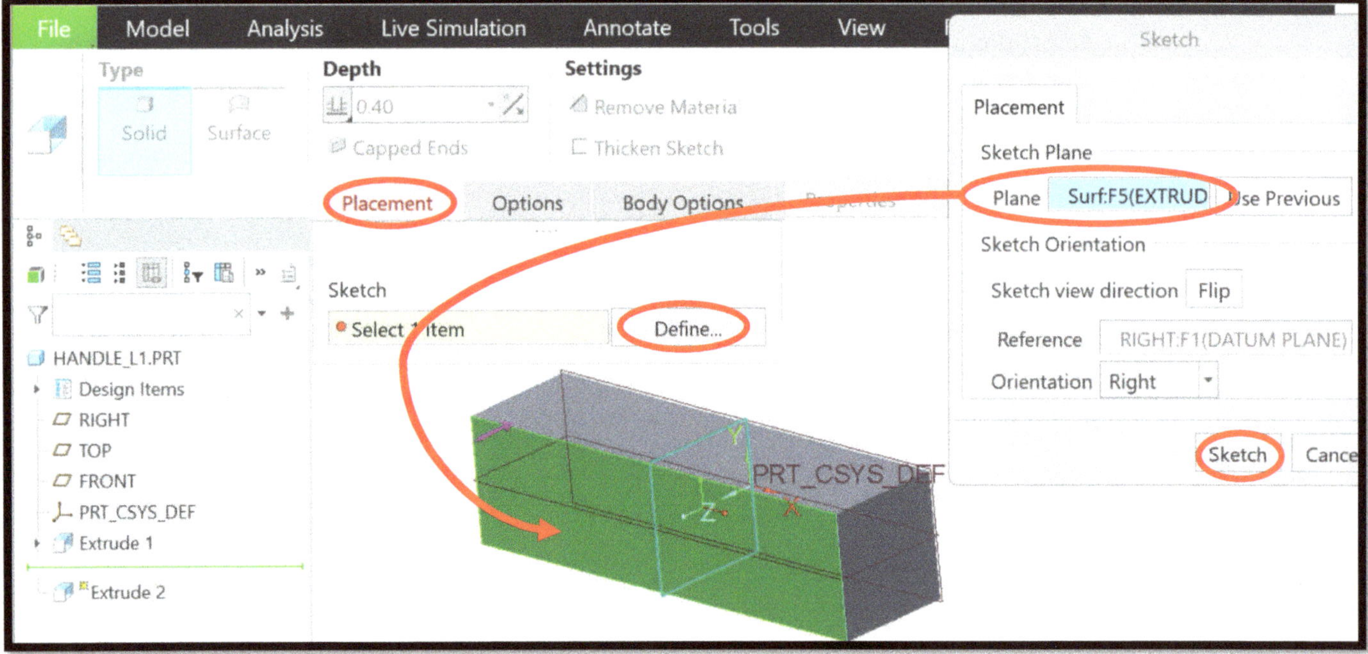

Step 3 – 2 - Click **Sketch** to start sketching on that plane.

- Use the **Circle Tool** to sketch a Closed-Loop circle shape centered on the origin of the rectangle from before.
- Dimension the circle to be **0.75" in Diameter**

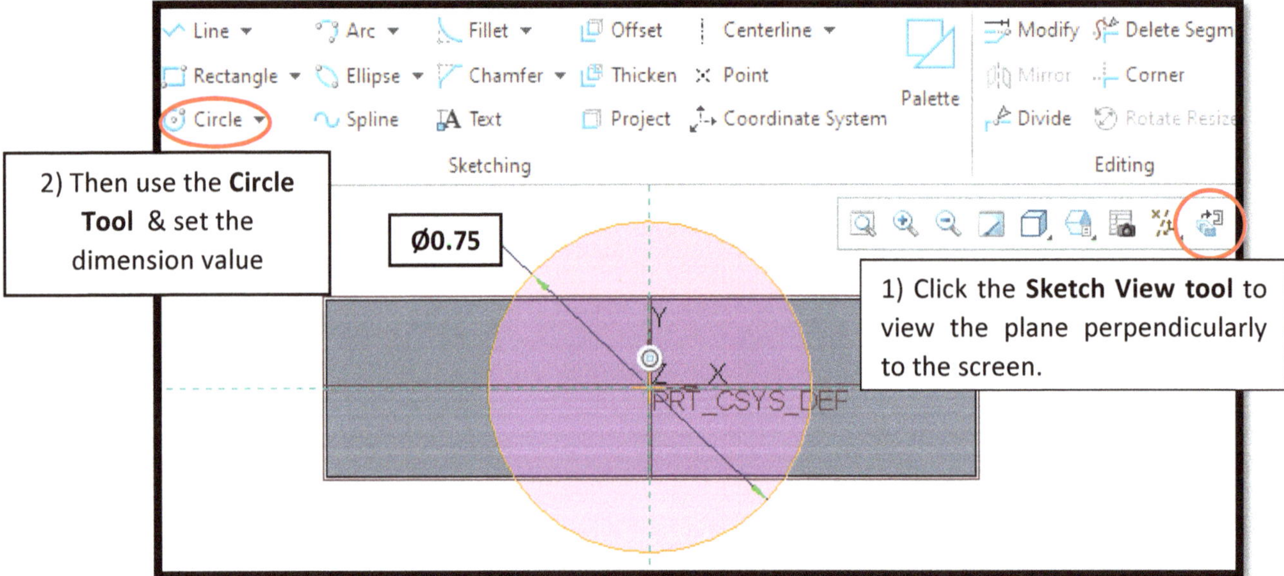

- Press the **green Checkmark** in the top tool bar to accept the completed Sketch.

Step 3 – 3 - Rotate the model and set the extrude Toolbar settings to direct the extrude through the part.

- **Hold the MMB and move the mouse** to rotate the model so you can see the depth of the extrude.
    - o   Warning: if you tap MMB it will press the checkmark and end the tool! Click and Hold MMB to rotate!
- Toggle the **Direction (arrow icon)** icon of the direction setting in the Extrude Toolbar so that the cylinder merges **through** the rectangle as shown below, not away from it.
- Set the depth value of the extrusion **to 0.75"**.

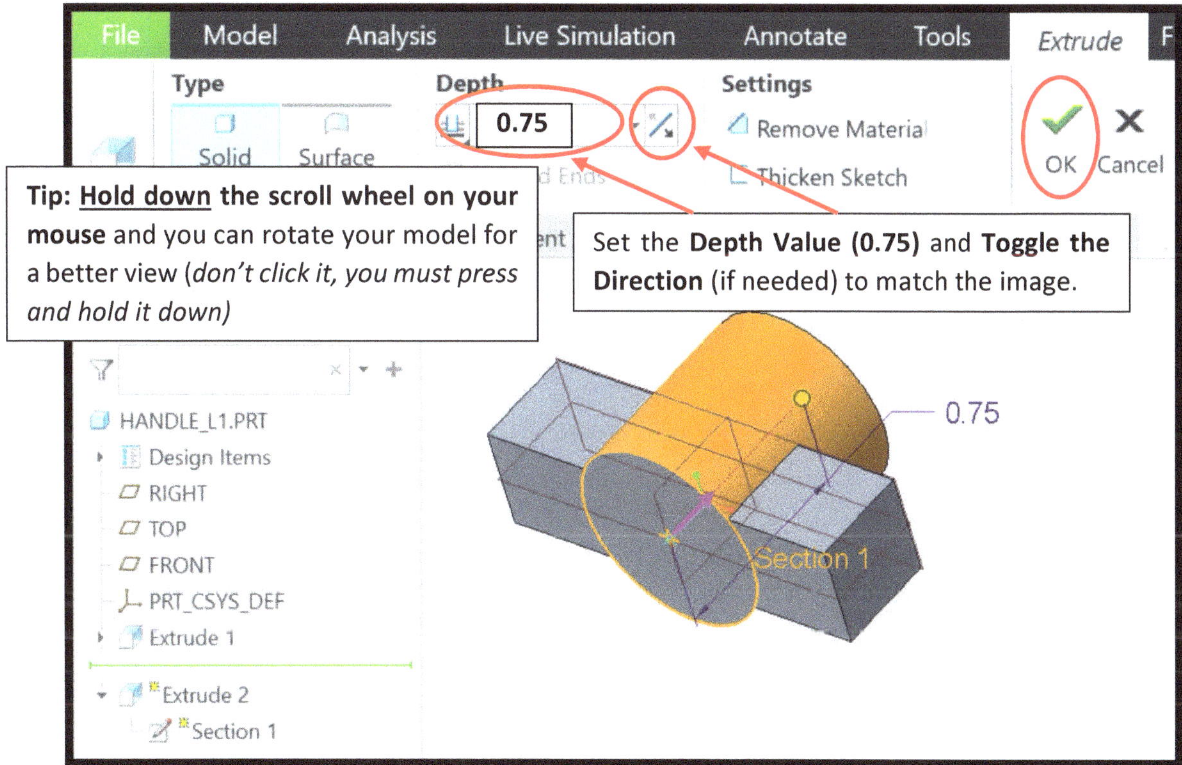

- Then press the **green Checkmark** to accept the Extrude from the top toolbar.   

Step 4 – 1 - Create another **Extrude** to add a hole in the center of the cylinder (on the side that extends past the rectangle)

1) **Rotate the view** of your component by **holding the Middle Mouse Button** or scroll wheel and moving the mouse so you can see the bottom surface of the cylinder *(the end that protrudes away).*

2) Select **Extrude** from the top toolbar.   

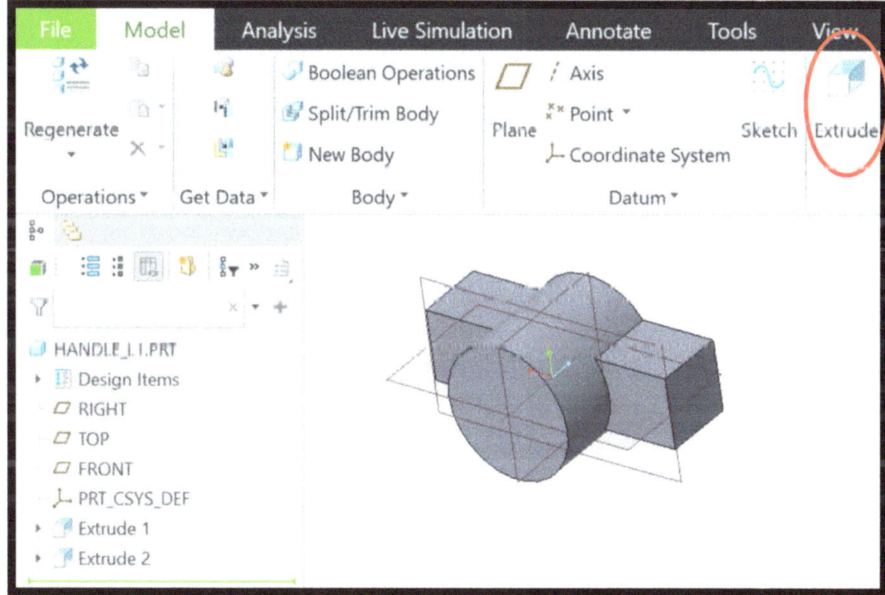

Step 4 – 2 - Click on the **Placement tab** – click **Define** - and select the **flat bottom surface of the Cylinder** as the Sketch Plane. Then click **Sketch** to enter Sketch mode for this feature.

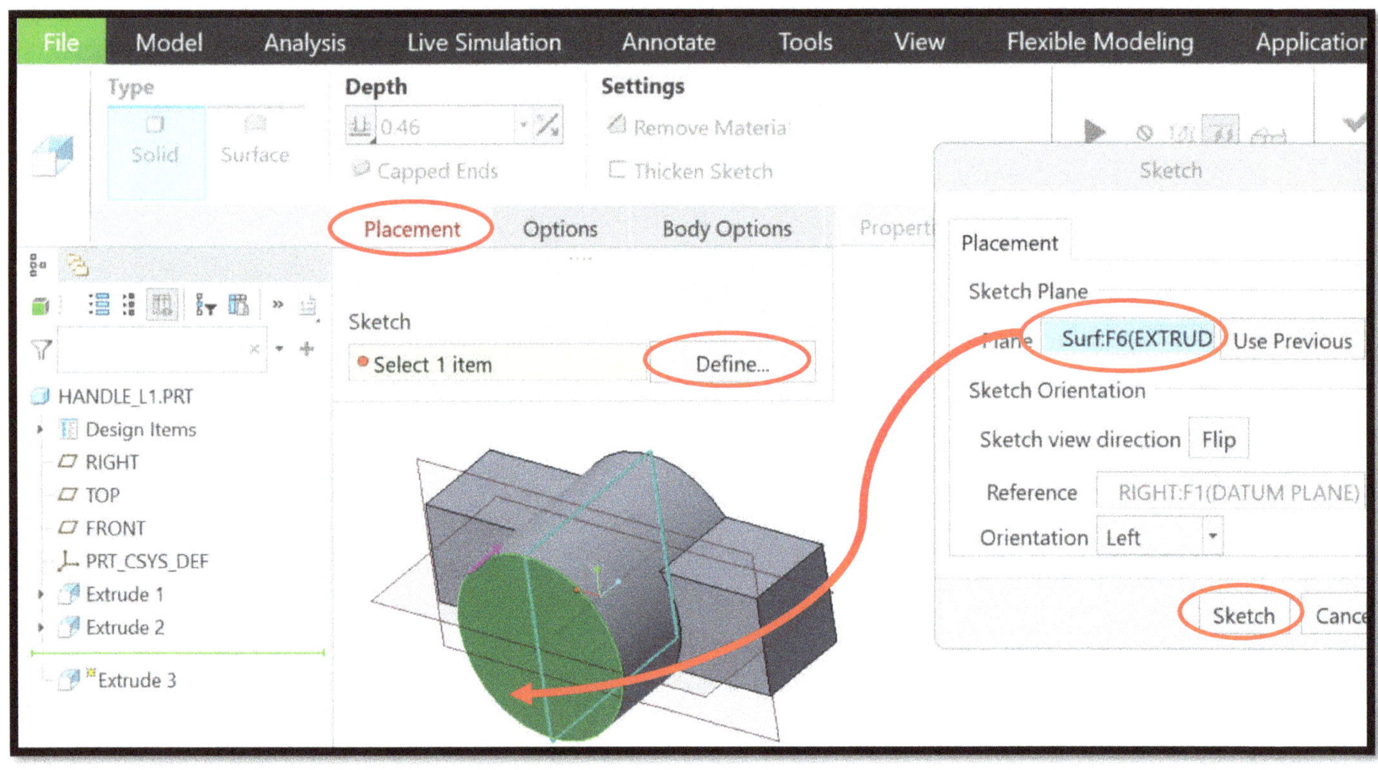

Step 4 - 3 – Sketch a Circle centered at the Origin with a **Diameter of 0.40"**.

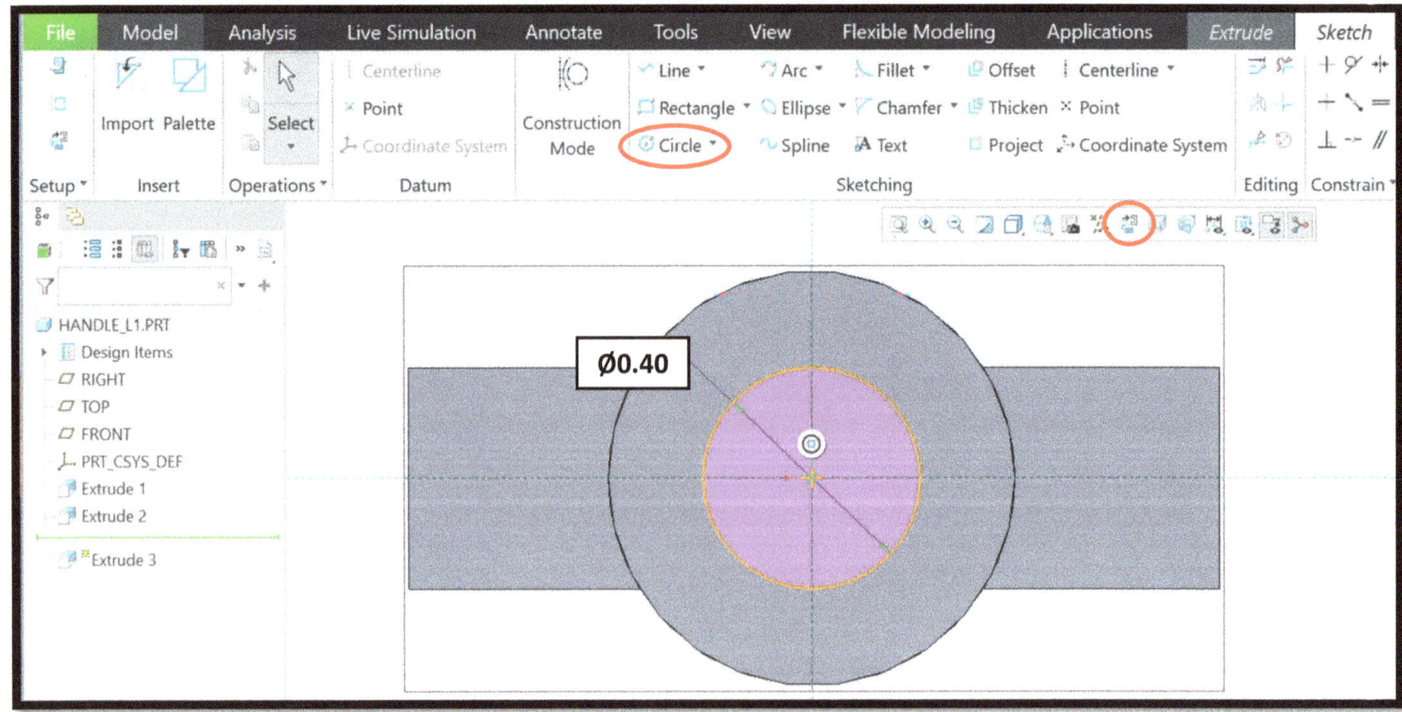

- Click the **green checkmark** to accept the sketch.

Step 4 – 4 - Adjust the Extrude options in the top toolbar:

- Toggle on the **Remove Material** option
- Toggle the extrude **Direction Arrow icon** (click once) to flip the Direction of the Extrude to go into the cylinder
- Adjust **the Depth value to 0.65"**
- **Press the green checkmark** to accept and complete the extrusion.

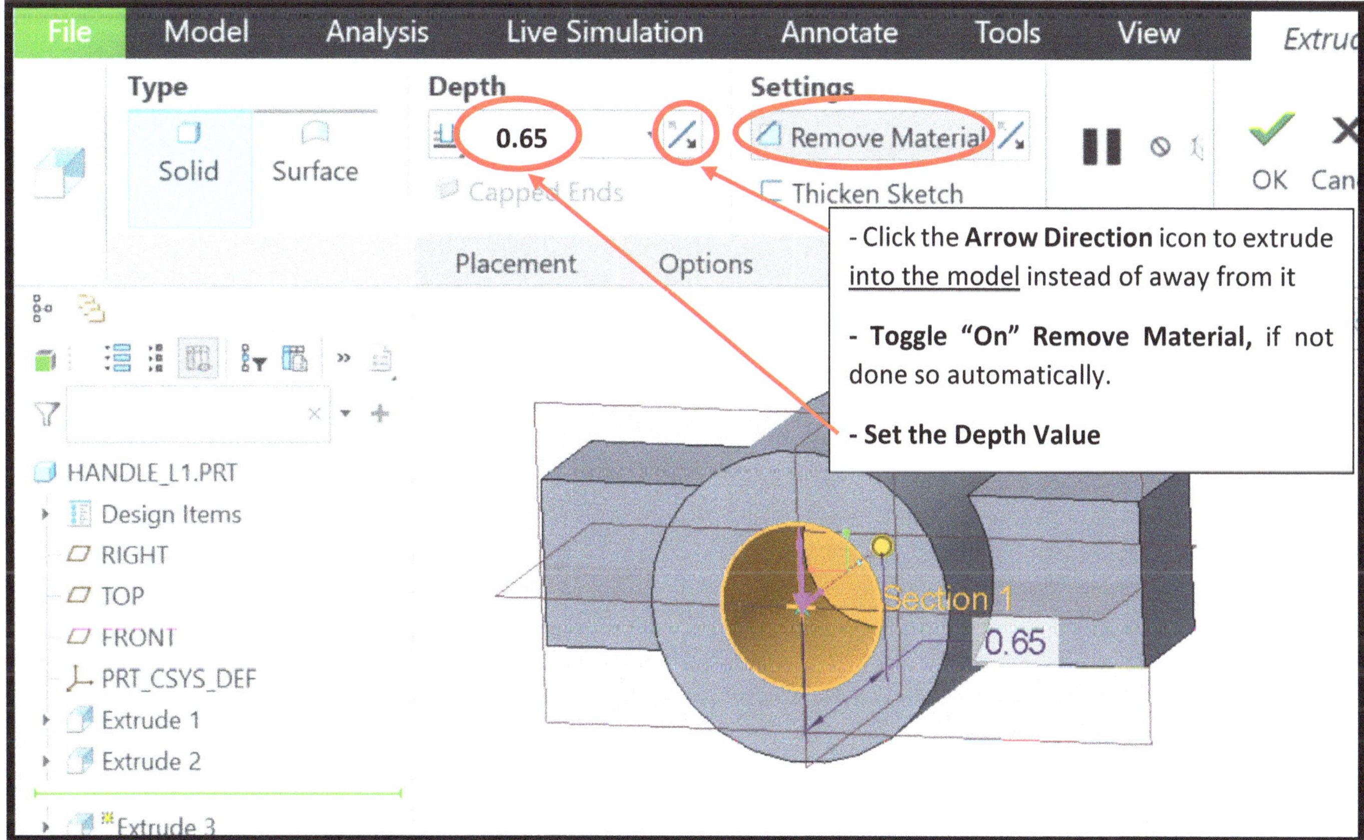

Step 5 – 1 - Now that the Extrude cut is complete, select the **Round** feature tool from the top toolbar. This will add a curve at edges to make a more ergonomic shape to the Handle design.

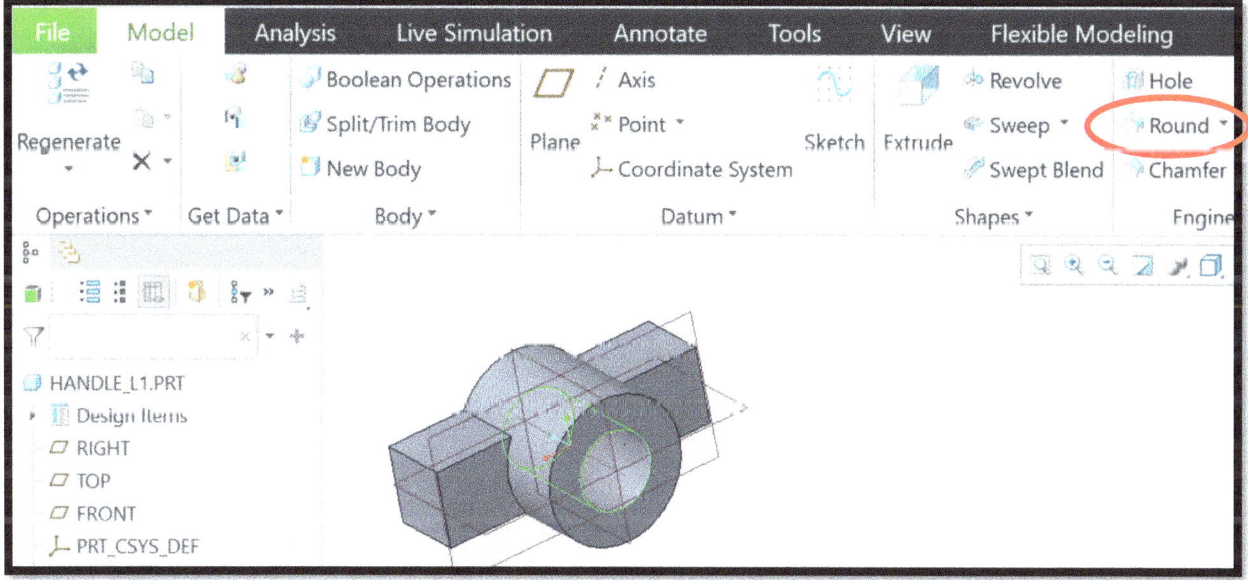

Step 5 – 2 - Set the **Radius** of the Round to **1.00"** in the top toolbar, then click the four edges at the intersection between the cylinder and rectangle as shown below (on both top and bottom sides, 4 in total). You will want to hold MMB and rotate the mouse to rotate the view to select each edge line properly.

Set the Round Radius value before you click the edges otherwise you would need to check and change the value of each Radius value in the "Sets" tab.

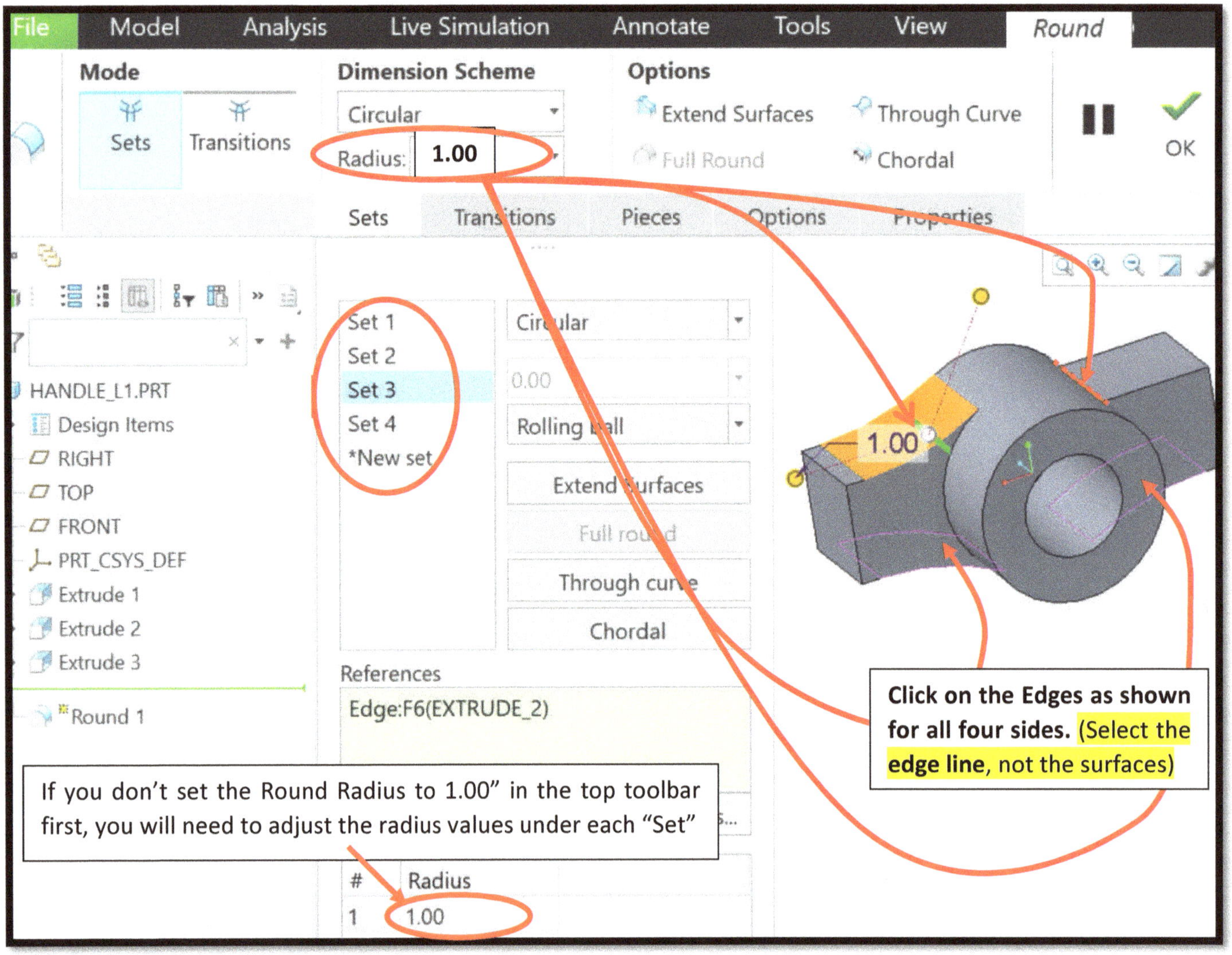

Step 5 – 3 - Press the **green checkmark** to accept the Round feature.

Step 6 – Press Save in the very top toolbar to save your part to your working directory. This is the end of Lesson 1.

FYI about Edit Definition: (modifying a completed feature)

If you need to make any changes to a feature, such as Extrude 1, you must first select the feature in the Model Tree. Clicking on the Feature should bring up a pop-up menu (*or hold RMB on it*) – then select **Edit Definition (***the "Pen & Sphere"* icon**)**. This brings back the **Toolbar** for that feature. You can then modify the Extrude options or return to the **Sketch** for that feature by selecting **Placement – Edit** in top toolbar.

After making any changes you need to press the checkmark to accept your changes, or press cancel to return back to the part model.

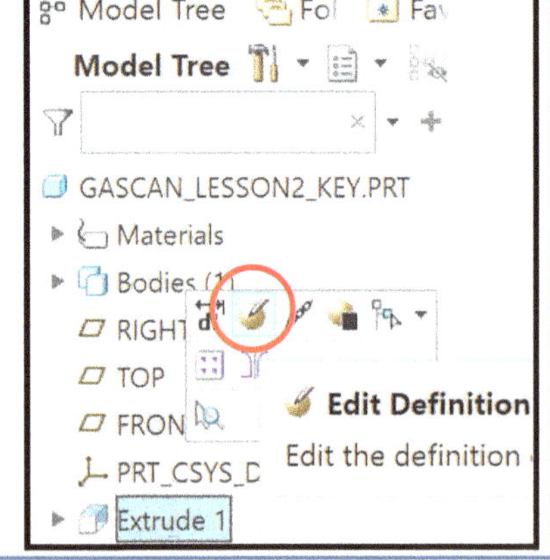

Lesson 2 - Extrudes, Datums, & Mirrors (Gas Can)

Lesson 2 will have you make a semi-realistic model of a "Rotopax" brand off-road style gas can. This part may look complicated but by using small steps to add and remove material it can be accomplished by a beginner. Using proper Datum placement to control the location of an Extrude will add to your skillset to be able to place extrudes away from edges. The Mirror Tool will be used to quickly copy features to opposite sides of the model.

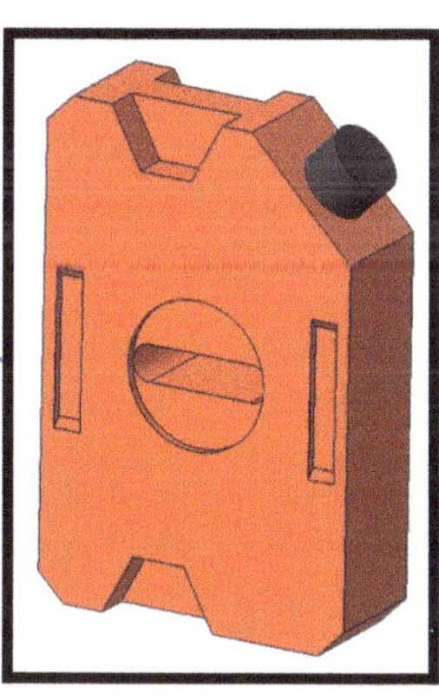

Basic CREO Commands:

LMB – Left Mouse Button Click

MMB – Middle Mouse Button (Click the Scroll Wheel)

RMB – Right Mouse Button (*Typically you need to __hold__ this to get menus to appear*)

Hold MMB and move Mouse – Rotates the 3D space in the graphics window.

Scroll Wheel (MMB) – Scrolling the MMB zooms in/out based on the location of the mouse pointer

Shift + Hold MMB – Slides the 3D space around in the graphics window.

Control key + LMB – To select more than one entity at a time __Hold__ the Control Key and Click the entity

Esc (Escape Key) – Will exit out of the current menu or end the use of the current tool.

Tap RMB - Cycle the selection of entities when holding the pointer over an area that has multiple items

Control + A – Activates the current window you are working on in CREO. Will also exit the current tool

Control + G – Regenerates the Model – forces the model to rebuild with any recent changes

Quick Toolbar:

At the top of the graphics window you will see the Quick Toolbar. Use these icons to quickly refit to screen, adjust the View Orientation and Style, switch to 2D Sketch View, and turn on/off display filters.

Getting Started: Open the **CREO** *__Parametric__* software and follow the lesson steps carefully to complete this lesson. Pay close attention to the instructions and images to ensure your model will work properly with the lesson steps.

Step 1 - Set your working directory to your ME101 Working Directory Folder.

>　　　Select **File – Manage Session – Set Working Directory – Navigate to your ME101 Directory* – OK**

**You may need to create a New Folder in the desired location if you have not done this before, calling it "ME101 Working Directory". You can hold RMB in the desired location, then choose New Folder and name the folder.*

>　　　*- If using your own computer (not in the classroom) create a working directory folder on your C: Drive Desktop.*

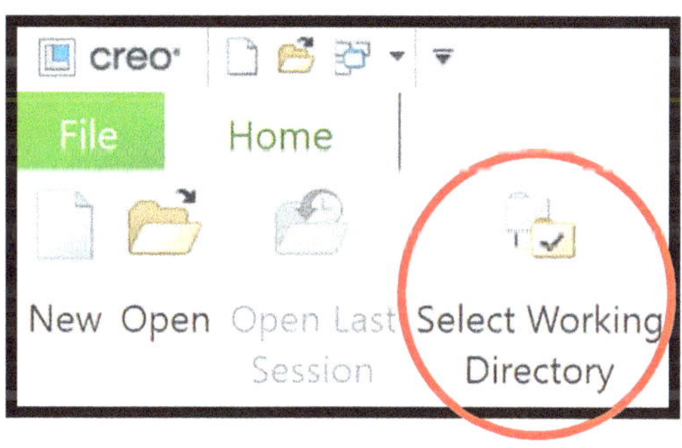

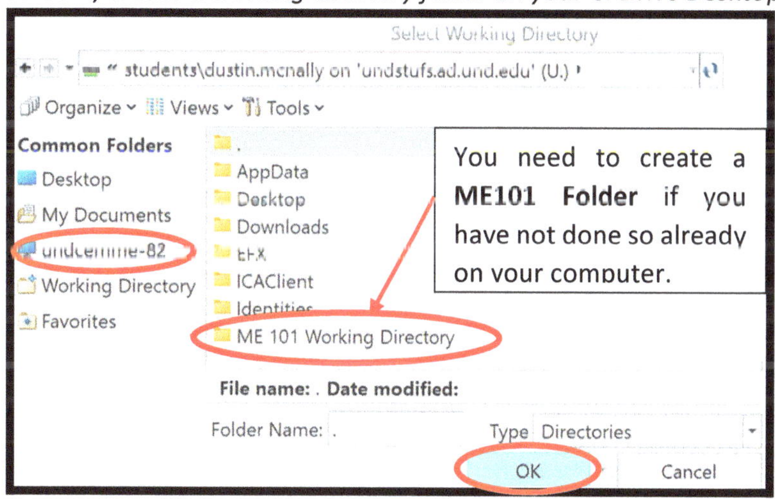

Step 2 - Create a new part file called ***GasCan_L2***. *You cannot use spaces in a filename so use underscores!*

- File - New – select Part as the file type – for the name type in "GasCan_L2"- press OK.

A pop-up menu may* appear asking you to choose a **"Template".** If so, choose **"inlbs_part_solid_abs"** - **press OK.** If not, it will have defaulted to that setting automatically.

**The template menu may not appear in the student home version of CREO, as it defaults to the desired inlbs_part_solid_abs already without needing to be selected.*

Step 3 - Change the Material of the part model to **PVC** *(a common plastic),* and be sure your units are in the CREO Default (*inch-lbm-Second*).

- **File – Prepare - Model Properties -** in the Material option select **change** – select the Legacy Materials folder - **double click** on "PVC.mtl" – press **OK** – press **Close** to close out the Model properties menu.

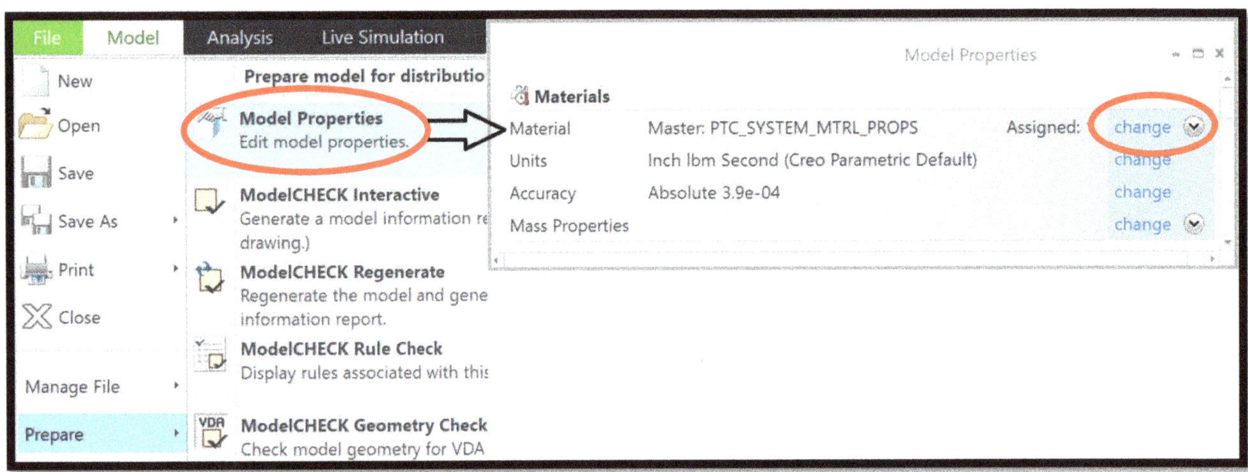

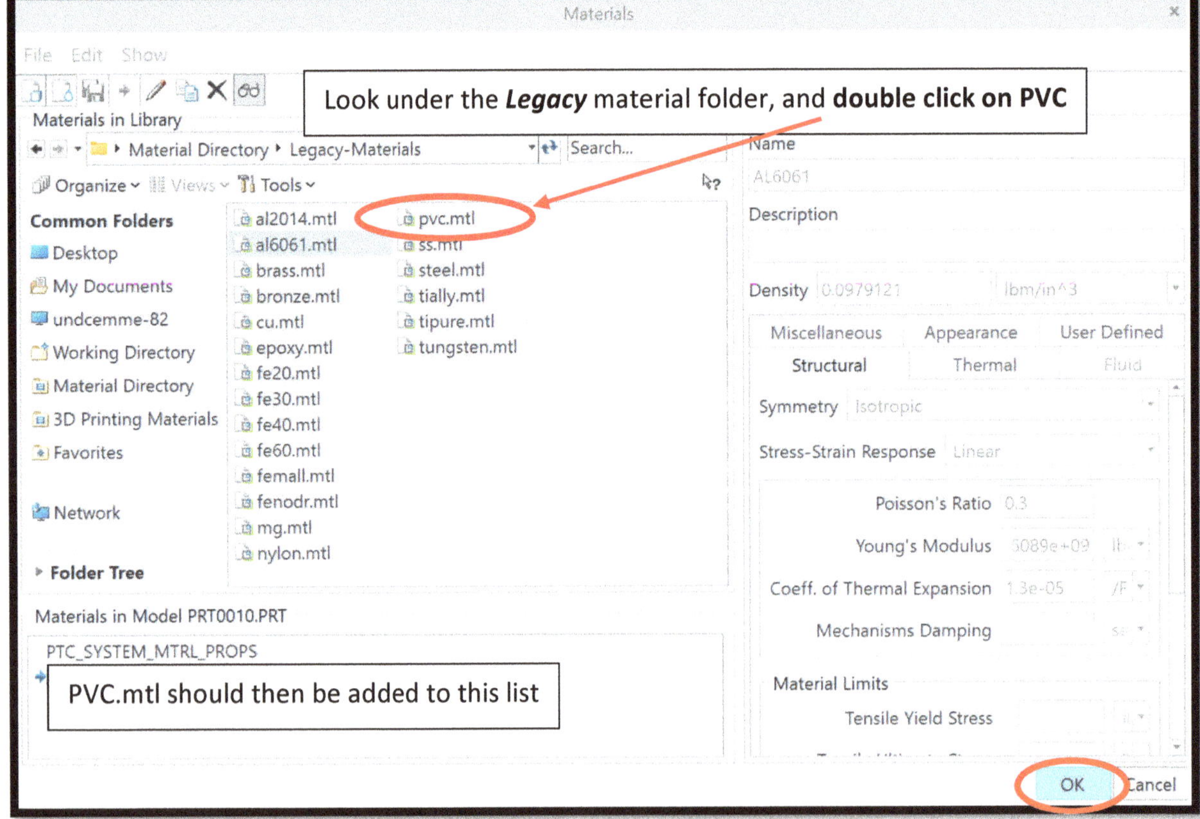

The part model for this lesson will be created with multiple extrusions. The **Extrude** tool will create (or remove) material by sketching a 2D shape and extruding that sketch a specific depth to define the 3D volume of the feature. The first step is to define the primary sketch plane placement of the Extrude, such as a surface, datum, etc.

Step 4 – 1 - Click on the **Extrude Tool** in the top toolbar.

Step 4 – 2 - Select the <u>Placement</u> tab – select **Define** - Click on the **Front*** datum to set it as the primary Sketch Plane and leave the Reference and Orientation as the default option (i.e. Right Datum).

If you choose the wrong sketch plane datum it will affect the orientation and placement of later features. **Double Check that your "Sketch Plane" selection reference says "Front" datum before pressing 'Sketch'. If not, click back in the "Plane" menu box and select the Front datum again.*

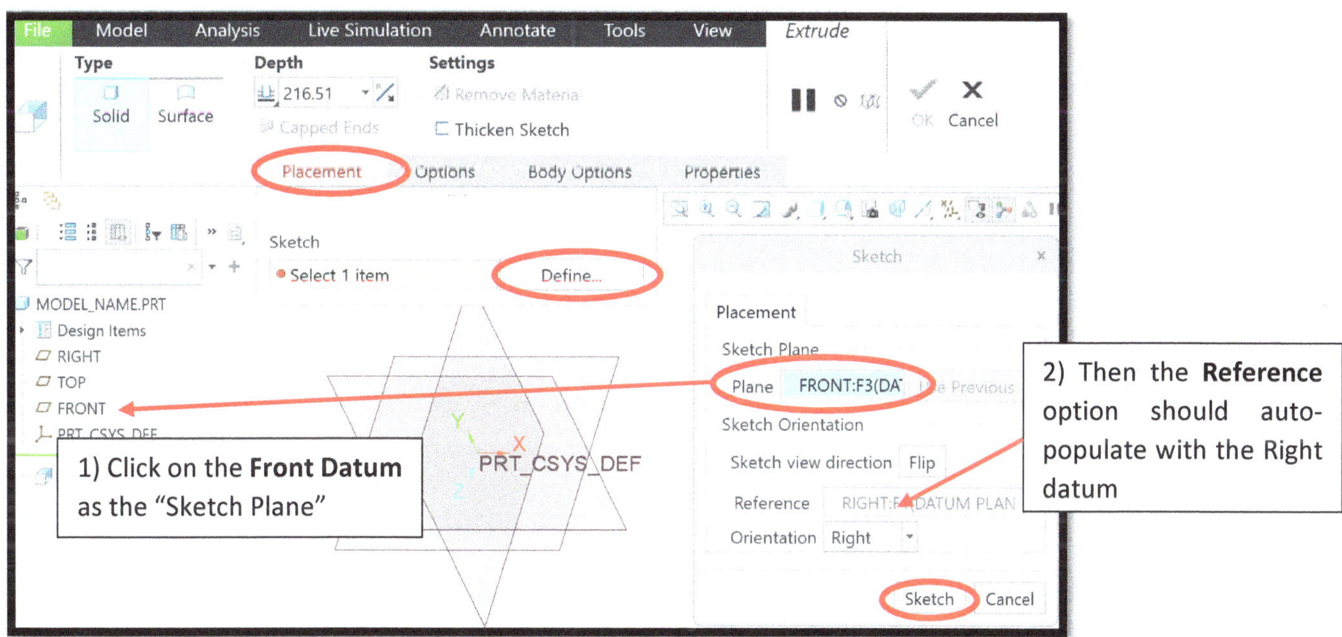

Double Check: Does it say Front in your "Sketch Plane" Plane box? If not, click on that box and select again.

Step 4 – 3 - **Press Sketch** to continue to Sketching mode for this Extrude.

Step 4 – 4 - Select the **2D "Sketch View"** tool from the **Quick Toolbar** at the top of the Graphics Window. This will adjust the graphic window view to be parallel with the screen which is ideal for Sketch mode.

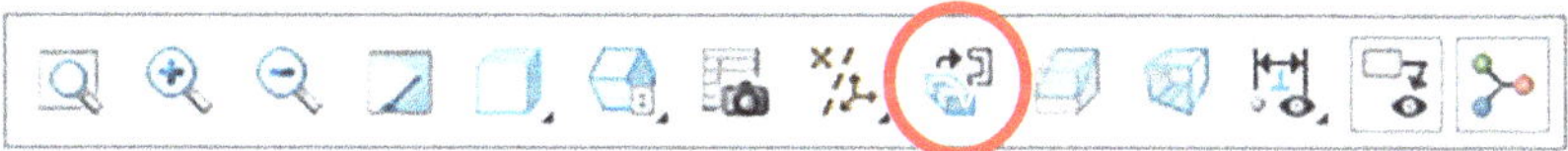

Every time you enter sketch mode you will want to use this tool so you can see your sketch plane clearly, parallel to the screen.

4 – 5 - While in Sketch Mode, place **Vertical** and **Horizontal Centerlines** that run through the Origin. Note that the blue dotted reference lines are not suitable for the Symmetry constraint so you must create your own **Centerlines**.

4 – 6 - Use the **Rectangle** sketch entity tool to create the Closed-Loop rectangle shape as shown about the origin.

4 – 7 - Use the **Symmetry** constraint tool to ensure the shape is symmetric about both the Horizontal and Vertical Centerlines, if CREO did not snap those constraints in already.

- **To add Symmetry:** select the **Symmetric constraint tool** - **select the desired Centerline – select the two opposing points** on the sketch to make symmetric about the centerline. The symmetry icons should appear.

4 – 8 - Adjust the **Dimensions** of the Rectangle to be **9.00"** wide, and **13.50"** inches tall. You can use the **Select tool** to double-click on the values and type in the value.

Note- If you have extra "weak" dimensions you may be missing symmetry or have extra entities that need to be resolved.

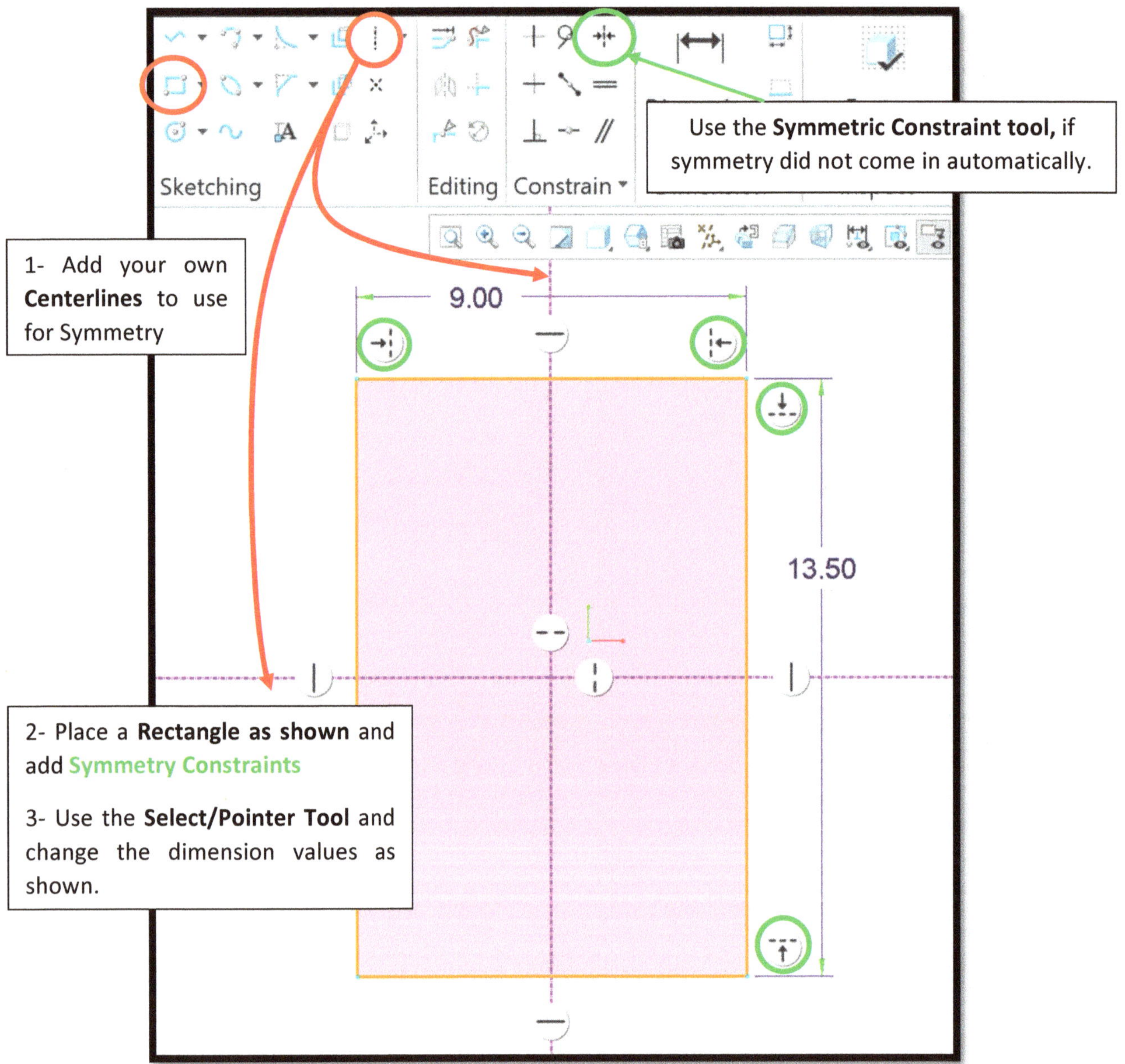

Note - After setting the dimensions your sketch may be very small on the screen. If so, drag the dimensions closer to the sketch and use "Refit to Screen" in the quick toolbar.

Step 4 – 9 - Press the **Green Checkmark** in the sketch toolbar to accept the sketch.

Step 4 – 10 - Set the **Depth** and **Direction** of the Extrude in the Extrude Toolbar so that it extrudes on **both sides** of the sketch plane.

- Rotate the model (**hold** the MMB and move your mouse) to have a better view of the extrude volume.
- Select the **Depth option** to be "**Extrude Symmetrically on Both Sides**"
- Set the **Depth value** to **3.25"**.
- Make sure that the "Type" setting is set to **Solid**. (*A Surface setting would cause problems later on!*)
- Click the **Checkmark** to accept the Extrude.

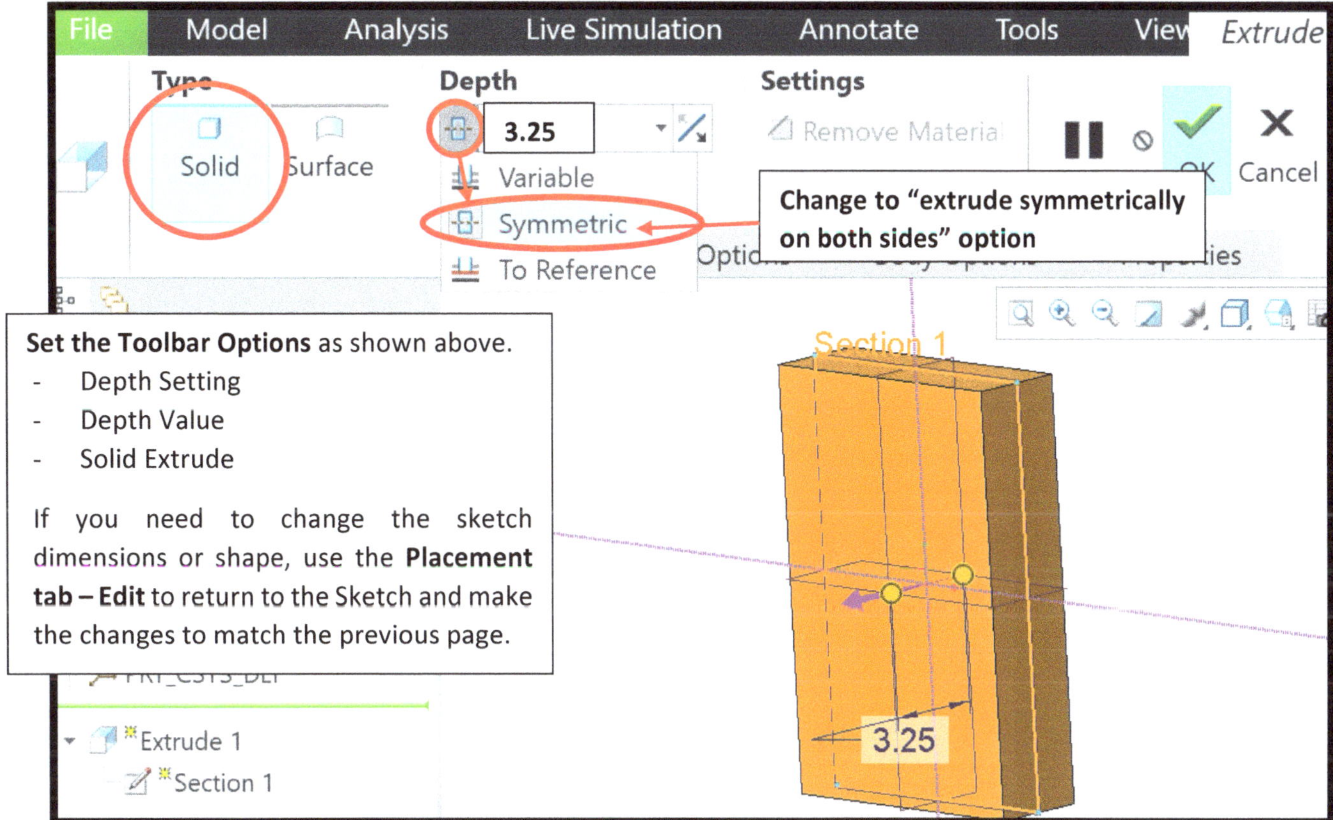

Once you press the checkmark the 3D model will be a Grey color and you will be back in Part Modeling mode. Now is a good time to check if your first extrude appears to be sized & positioned correctly. Rotate the model by holding the **Middle Mouse Button** (MMB) and moving your mouse, or choose the "Standard Orientation" from the Quick toolbar (under the Saved Orientations icon). If you need to modify Extrude 1 use **Edit Definition** (see next page).

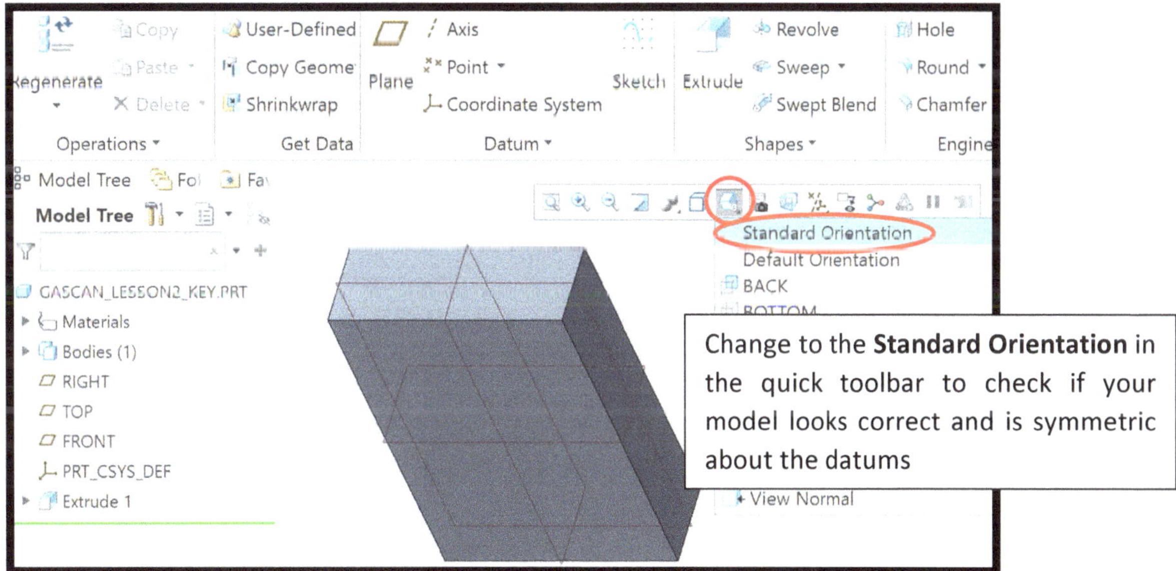

Edit Definition:

If you need to make any changes to a Feature such as Extrude 1, you must first select the feature in the Model Tree. Clicking on the Feature should show the pop-up menu (*or hold RMB on it*) – then select **Edit Definition (***the "Pen & Sphere" icon***)**.

This brings back the **Toolbar** for that feature. You can then modify the Extrude options or return to the **Sketch** for that feature by selecting **Placement – Edit** in top toolbar.

After making any changes you need to press the Checkmark (if you made changes) or Cancel to return back to the part model.

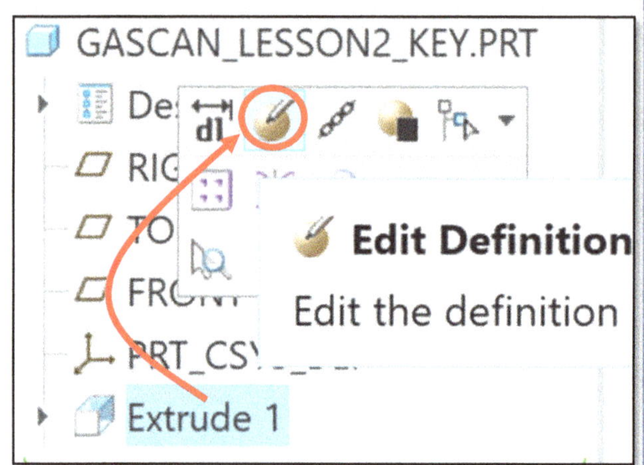

Step 5 – Create a second Extrude to cut away the top corner of the model.

5 – 1 - Click on the **Extrude Tool** in the top toolbar.

5 – 2 - Select the Placement tab – select **Define** - Click on the **Front*** datum to set it as the primary Sketch Plane and leave the *Reference & Orientation* as the default option (Right Datum).

> **If you incorrectly chose the wrong sketch plane in Extrude 1, your Front datum may not match the desired datum shown in the step below. You will need to delete out Extrude 1 and remake it to fix this error.*

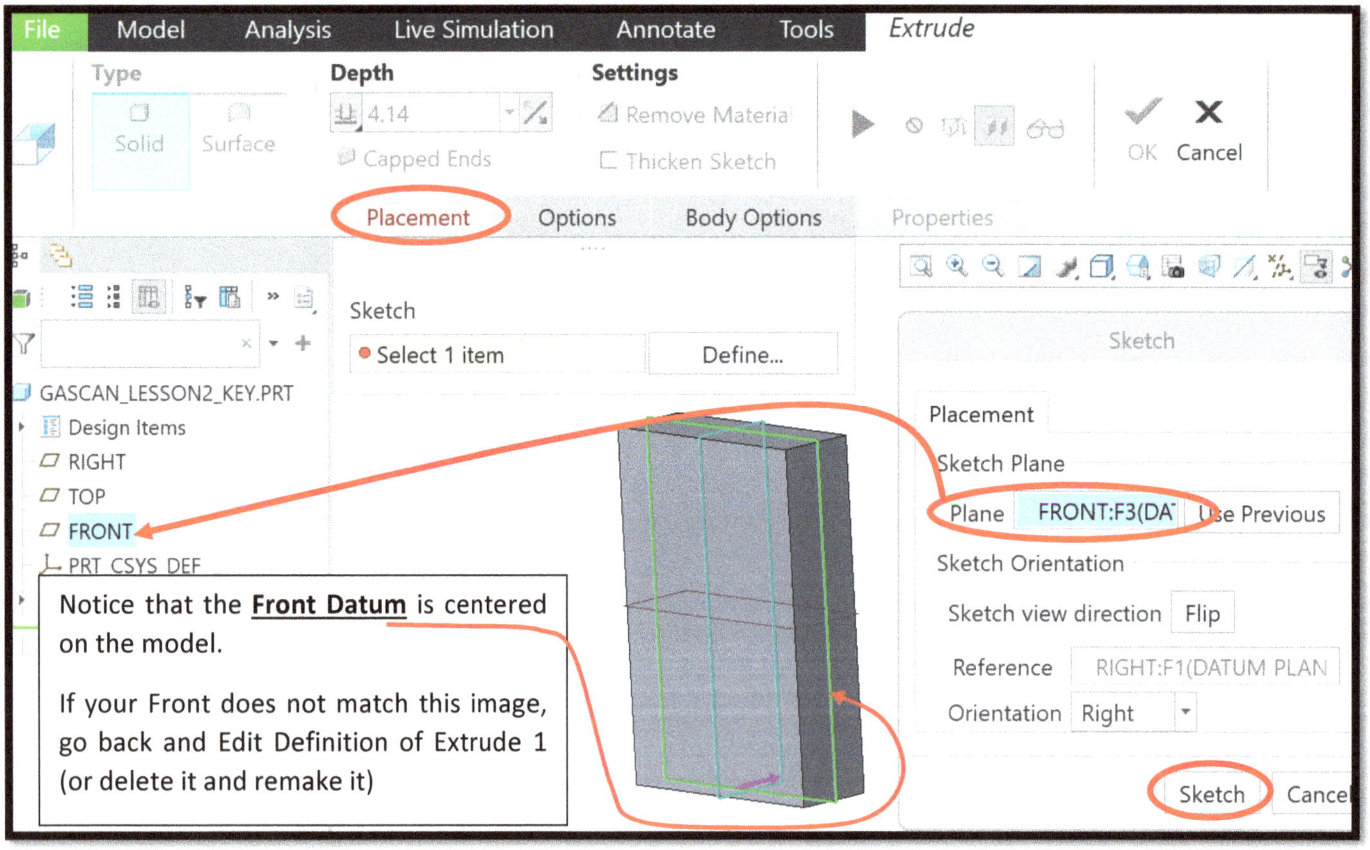

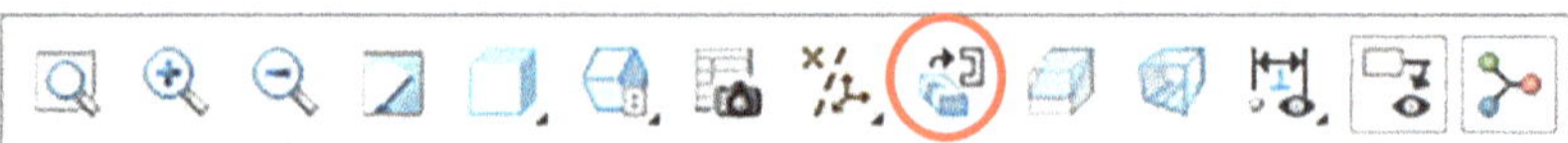

5 – 3 - Press **Sketch** to enter sketch mode for this extrusion. Then select **2D Sketch View** from the **Quick Toolbar** at the top of the Graphics Window.

5 – 4 - Use the **Line Tool** to sketch the **3 Lines** of the triangular closed-loop shape so that the points "snap" directly on the edges of the model.

5 – 5 - Use the **Dimension tool** to create the correct dimensions as shown.

- Select the **Dimension Tool – LMB click the left vertical edge* of the model - LMB the far left corner Point of the triangle** - press the **MMB** to place the dimension value as shown below.
- Repeat to add the vertical dimension: **LMB click on the bottom horizontal edge* of the model - LMB on the bottom Point of the triangle – press the MMB** to place the dimension on the right side of the sketch.

*Tip: Use the **Line**, not the corner **Points** for the edge reference when using the Dimension tool! Clicking on the corner of the model may accidently select the wrong edge and leave you with a 0.00" dimension. Instead measure between the Line of the model and the Point of the triangle.*

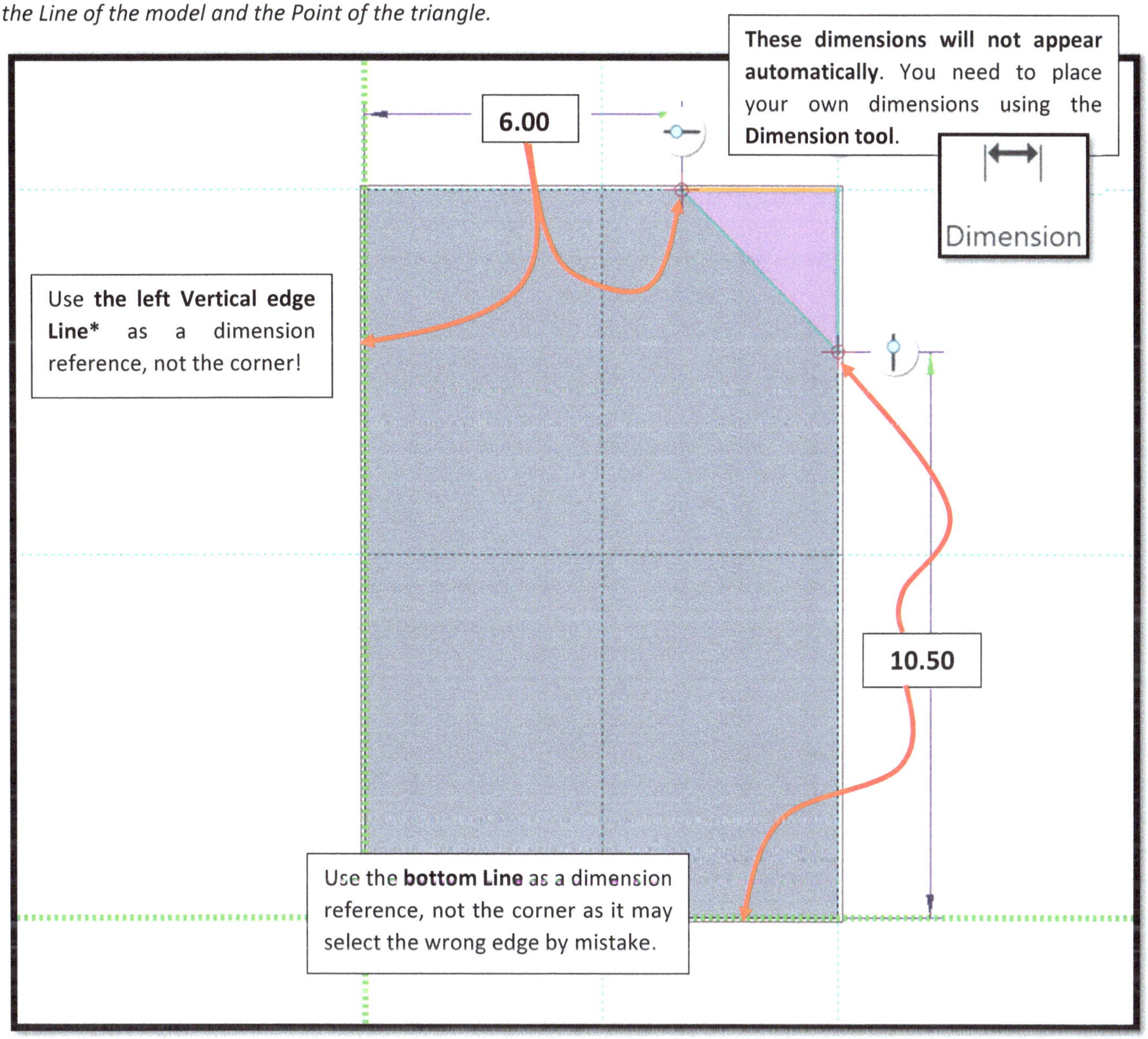

5 – 6 - Set the value of the dimensions to **6.00"** & **10.50"** as shown above. Also, make sure you have a **Closed Loop.**

5 – 7 - Press the **Checkmark** to accept the Sketch.

5 – 8 - Set the **Depth, Direction, and Material Removal options** in the Extrude Toolbar to complete the Extrude options.

- Rotate the Model to view the effects of the Extrude options (hold MMB and rotate the mouse)
- Toggle on **Remove Material** (if it did not do so by default automatically)
- Select the **Depth option** to be "**Extrude Symmetrically on Both Sides**"
- Set the **Depth value** to **10.00"**.
- Check that the "Extrude As" setting is set to **Solid**. (A Surface setting would cause problems later on!)

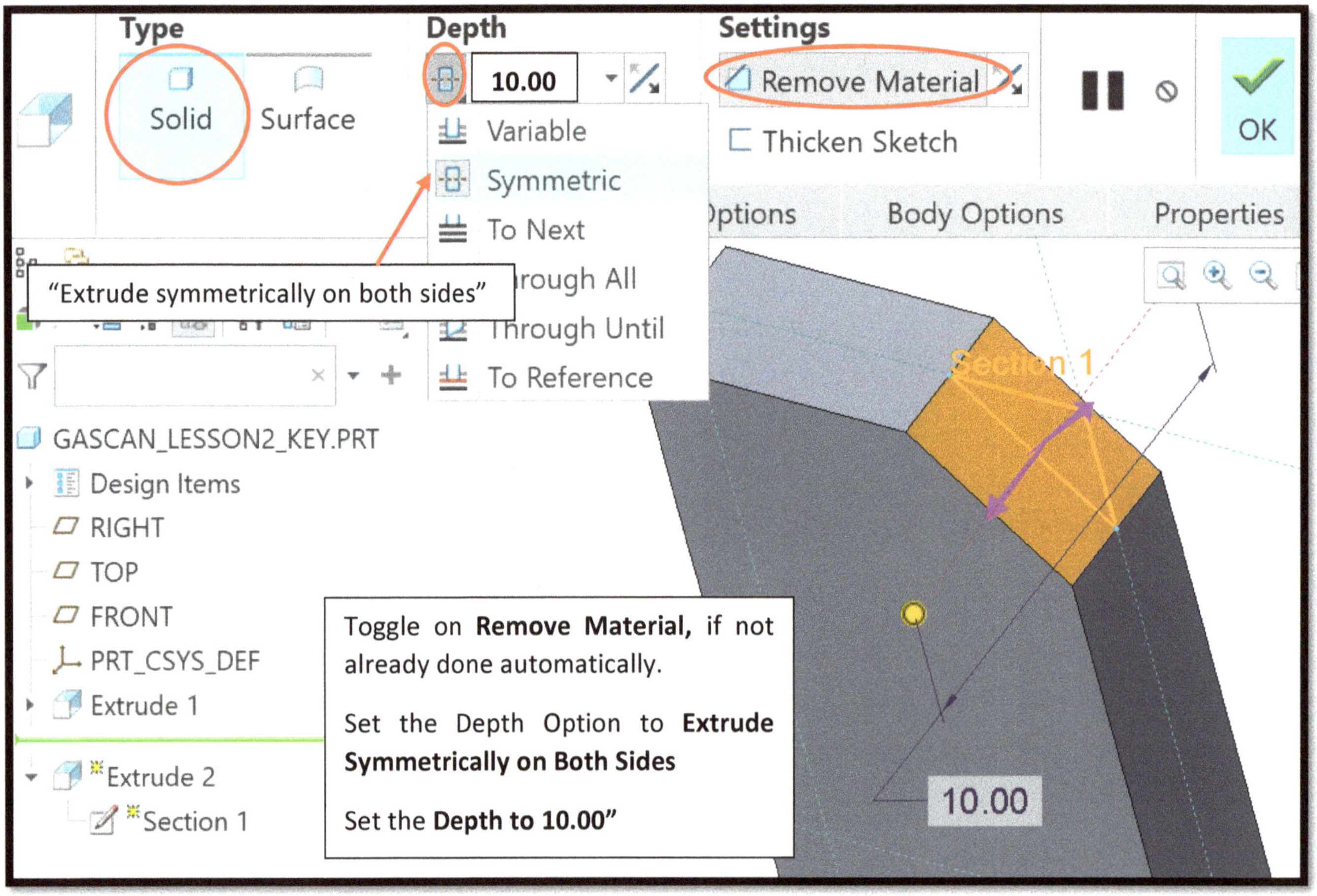

Note – If the part model appears blue/purple colored or thin-walled, you may have accidently clicked on the "Surface" Type option in the toolbar for either Extrude 1 or 2 instead of the Solid setting. If needed, use Edit Definition to bring back the toolbar for each Extrude in the model tree and check for the correct **Solid setting** in the toolbar settings.

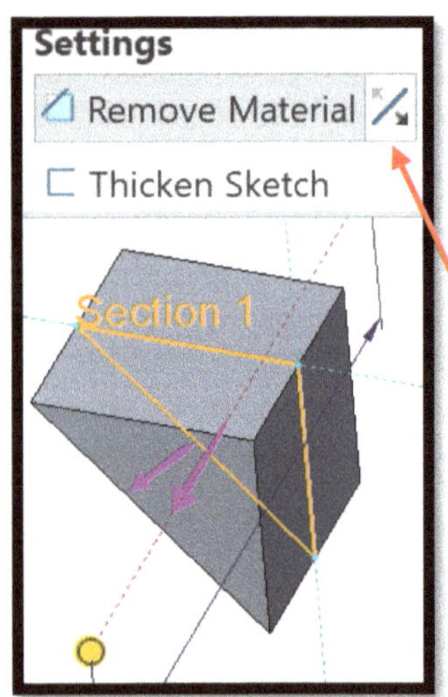

Note 2 - Another potential issue could be if you accidently click on the Arrow option for the Material direction which would then remove material on the outside of the triangular sketch. If that happens, toggle the direction arrow setting icon in the top toolbar of the Extrude.

This will happen if the "Material Direction" arrow icon is clicked accidently......**Don't do this!**

Step 5 – 9 - After assuring that the Extrude is defined correctly select the **Checkmark** to complete the Extrude.

Step 6 – Create a third Extrude to cut a notch into the face surface of the part along the top edge. This extrude will only cut 1.00" into the material, starting from the **outside face** of the model.

Step 6 – 1 - Click in a blank/empty area of the graphics window to deselect any surfaces or datums, then click on the **Extrude tool**.

- o Why? If you are currently selected on any datums or surfaces CREO will use that as the sketch plane as soon as you use the Extrude tool, instead of asking you for the placement.

Step 6 – 2 - Select the **Placement** tab – select **Define**. Select the planar Surface (on the face surface of Extrude 1) to set it as the primary <u>Sketch Plane</u>. Leave the Reference & Orientation as the default option (Right).

Note- your Surface ID# (i.e. Surf: F5 Extrude 1) shown in the image may be named differently in your model.

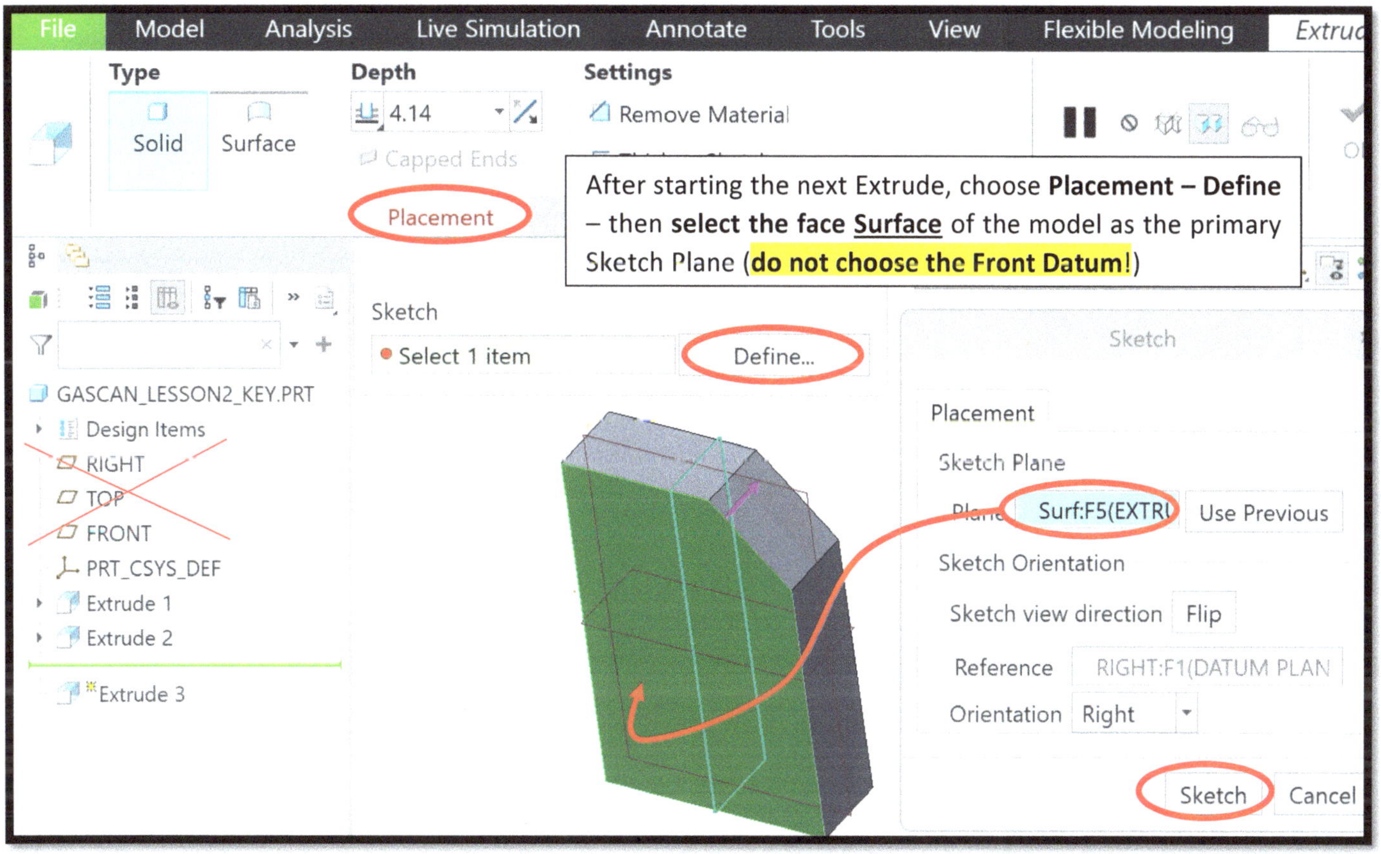

Step 6 – 3- Confirm that your "Plane" option is indeed on the **face surface (*Surf:F5 or similar*)** on not the middle datum, as this Extrude needs to start from the outside face and cut in, and not from the middle datum out.

Step 6 – 4 - Press **Sketch** to enter sketch mode for this extrusion. Then select **2D Sketch View** from the **Quick Toolbar** at the top of the Graphics Window.

Complete the next steps to complete the sketch, being careful of how you are snapping the lines on the model.

Step 6 – 5 - <u>Add a Vertical Centerline</u>

- Place the Centerline such that it is <u>**not**</u> "snapped" onto the ***Mid-Point*** of the top horizontal line nor snapped onto the ***blue vertical datum reference line***. Both of those constraints will interfere with the desired location.

Step 6 – 6 - Dimension the centerline placement

- **Dimension Tool –** click on left vertical <u>edge</u> line as the first reference (*do not select the corner as it may incorrectly select wrong line and place a 0.00" value*) – then click on the centerline – finally click MMB to place the dimension value as desired.

- If you get a Resolve Sketch error, you likely have a constraint snapped in that can be deleted out.

Step 6 – 7 - Set the dimension **value to be 3.50"**

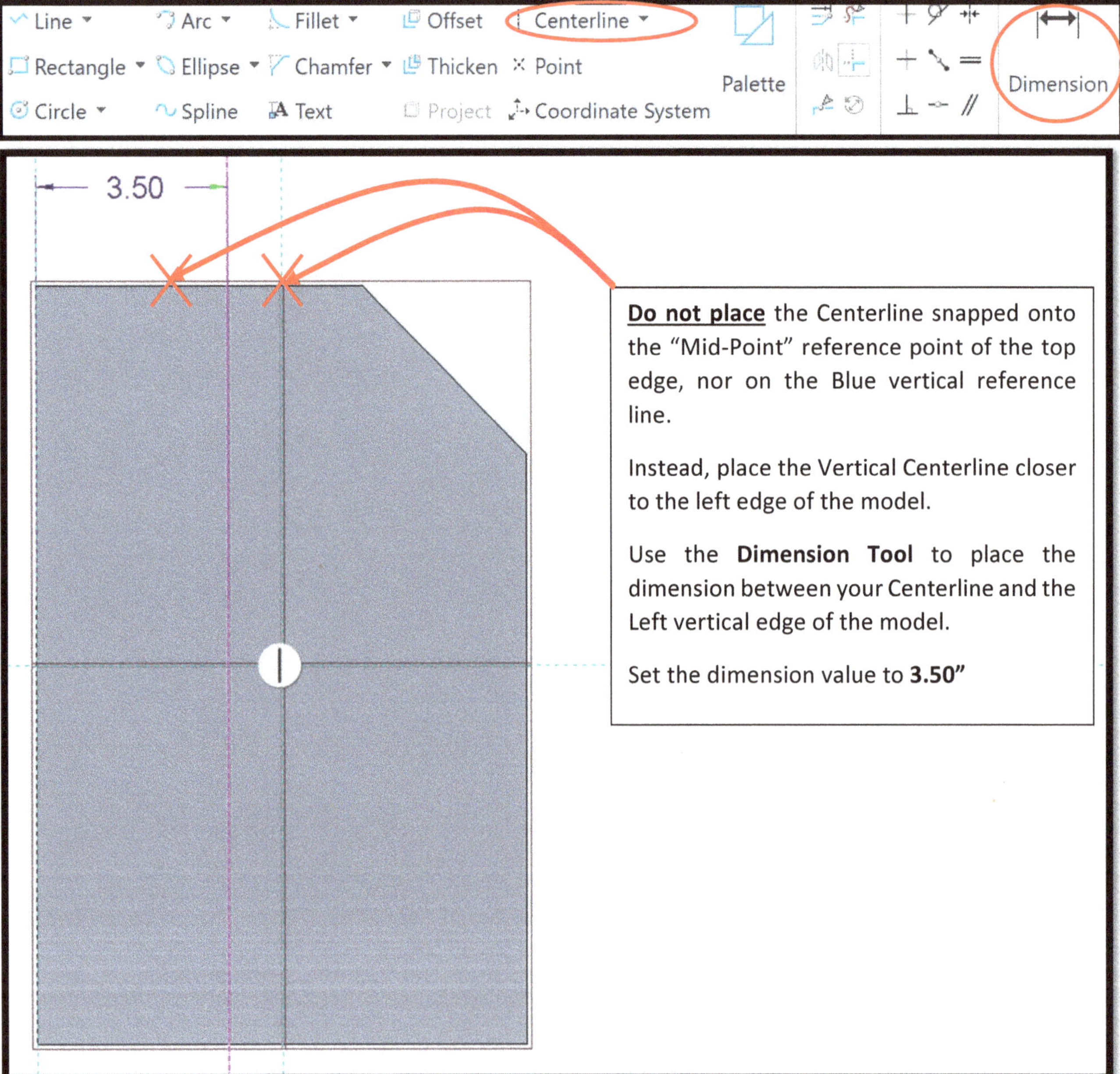

Double Check: Your centerline should not have any constraint symbols on it other than possibly a Vertical Constraint (likely you won't have or need the vertical constraint)

Step 6 – 8 - Sketch the Closed Loop shape: Use the **Line tool** to create the **4 lines** of the trapezoid shown below.

Pay attention to the entities you may be "snapping" the lines onto so you do not add unnecessary constraints.

- o　Not snapped onto the "Mid-Point" of the top edge.
- o　Not snapped onto the blue vertical reference line at either the top or bottom corners.
- o　Not snapped onto the outside corners of the model
- o　Not snapped onto your centerline.

If you do accidently place the sketch lines with those constraints you will need to delete the extra constraints and troubleshoot the Resolve Error menu when you add dimensions, or just delete the lines and re-make them properly without snapping those constraints in place.

Step 6 – 9 - Constrain the Sketch:

- Use the **Symmetry Constraint tool** – click on your Centerline, click on the two opposing points of the upper sketch line to make them symmetric.
- **Repeat the step again** to make the lower horizontal line symmetric about the same centerline.

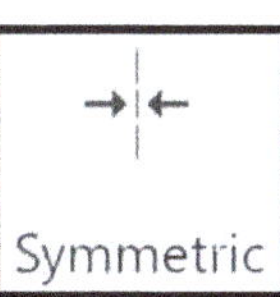

Step 6 – 10 - Dimension the Sketch:

- Use the **Dimension Tool** to create and set the dimension values as shown (**3.25" top length**, **1.50" bottom length**, **1.625" height***). *Note that the 1.625" dimension will truncate to 1.63" on the screen, but the value will remain 1.625" in the model.*

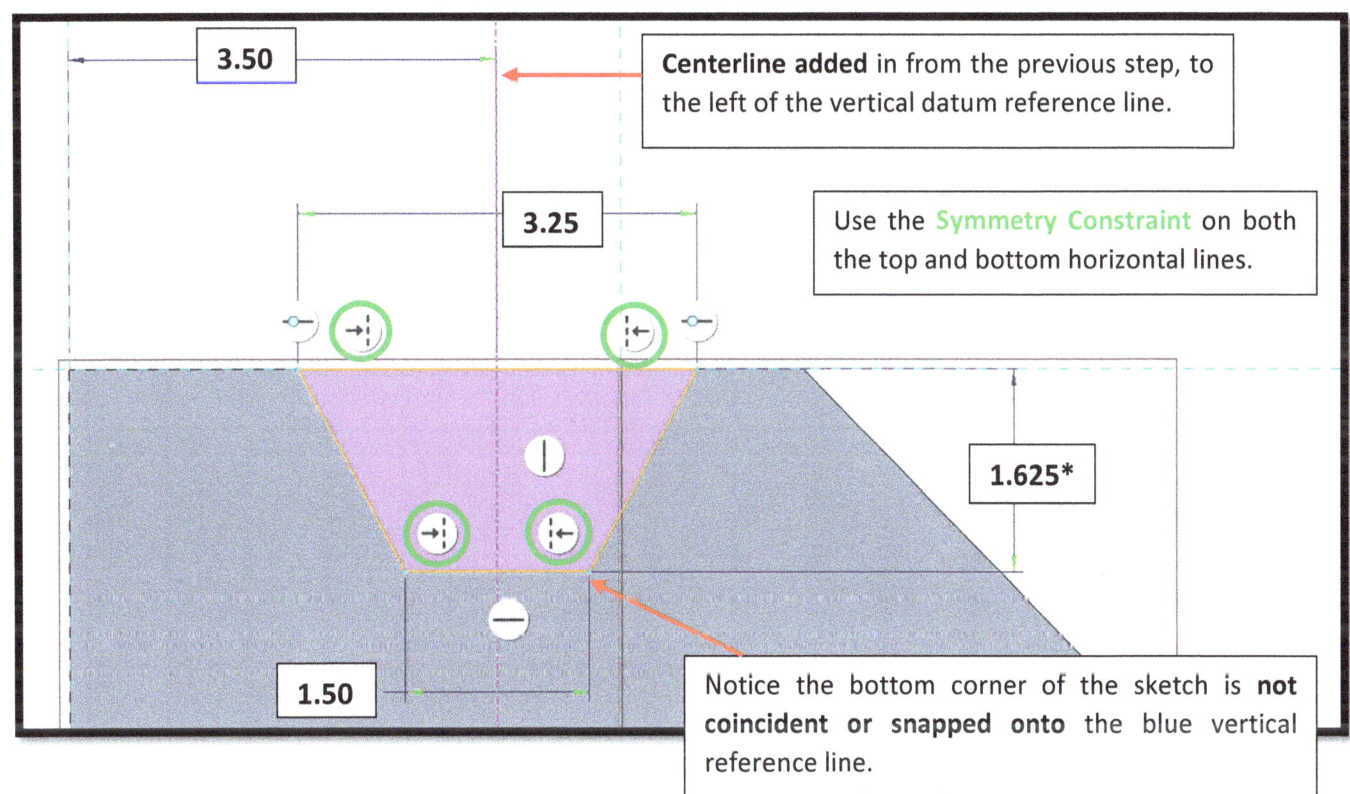

Step 6 – 11 - Accept the Sketch: If the sketch looks correct press the **Checkmark** to accept it.

Step 6 – 12 - Set the **Depth, Direction, and Material Removal options** in the Extrude Toolbar.

- **Rotate the Model** to view the effects of the Extrude options (hold MMB and rotate the mouse)
- Leave the **Depth option** to the default one sided option of *"Extrude from sketch plane by a given value"*
- Toggle the Extrude Depth **Direction Arrow** icon to flip the extrude direction to go into the Model.
- Toggle on **Remove Material** (if it did not do so by default automatically)
- Set the **Depth value** to **1.00"**.
- Check that the Type setting is set to **Solid**.

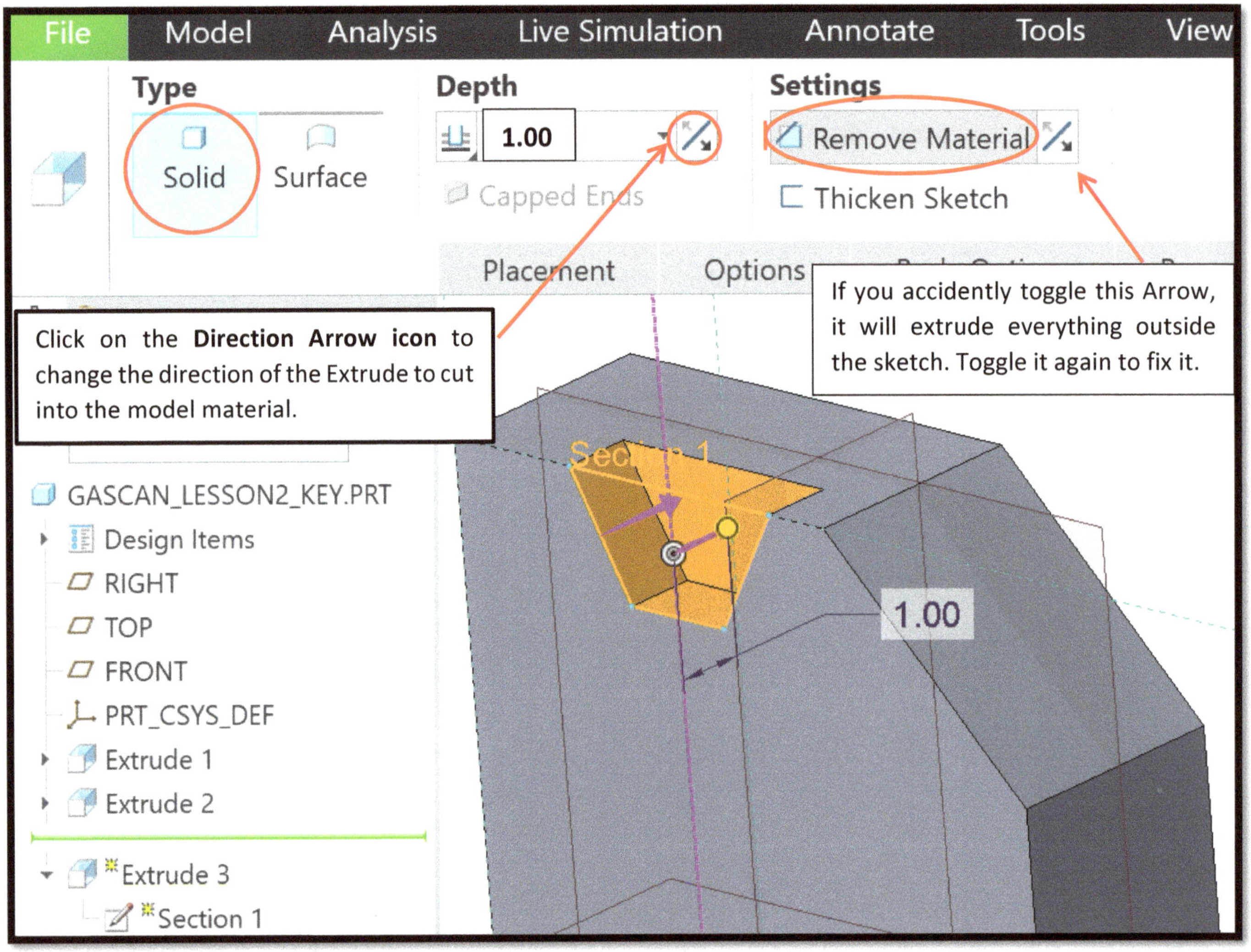

Model Check: If your Extrude did not start on the front "face" surface of the part, you likely chose the incorrect **Sketch Plane**. A common mistake is using the Front Datum instead of the face surface. You will need to cancel/delete this extrude and remake it choosing the correct Sketch Plane as the outer surface.

Step 6 – 13 - Press the **Checkmark** to accept the Extrude.

OK

Step 7 - Go to **Blackboard** and complete **Module 3 (Mirror Tool)** to learn about the Mirror Tool before continuing on to the next step.

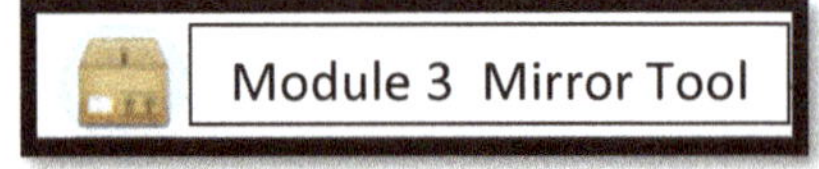

After completing the module, you should see your quiz score reported in the Gradebook by looking at the My Grades tab in Blackboard.

Step 8 - Mirror the previous *Extrude 3* to the other side of the model by using the **Front** datum as the **Mirror Plane**.

- **LMB click** on *Extrude 3* from the Model tree – select the **Mirror Tool** from the top toolbar – **LMB click on the Front** datum from the Model Tree – press the green **Checkmark** to accept the Mirror.

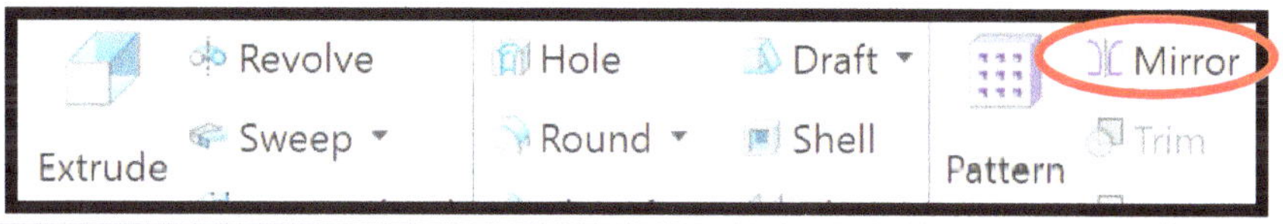

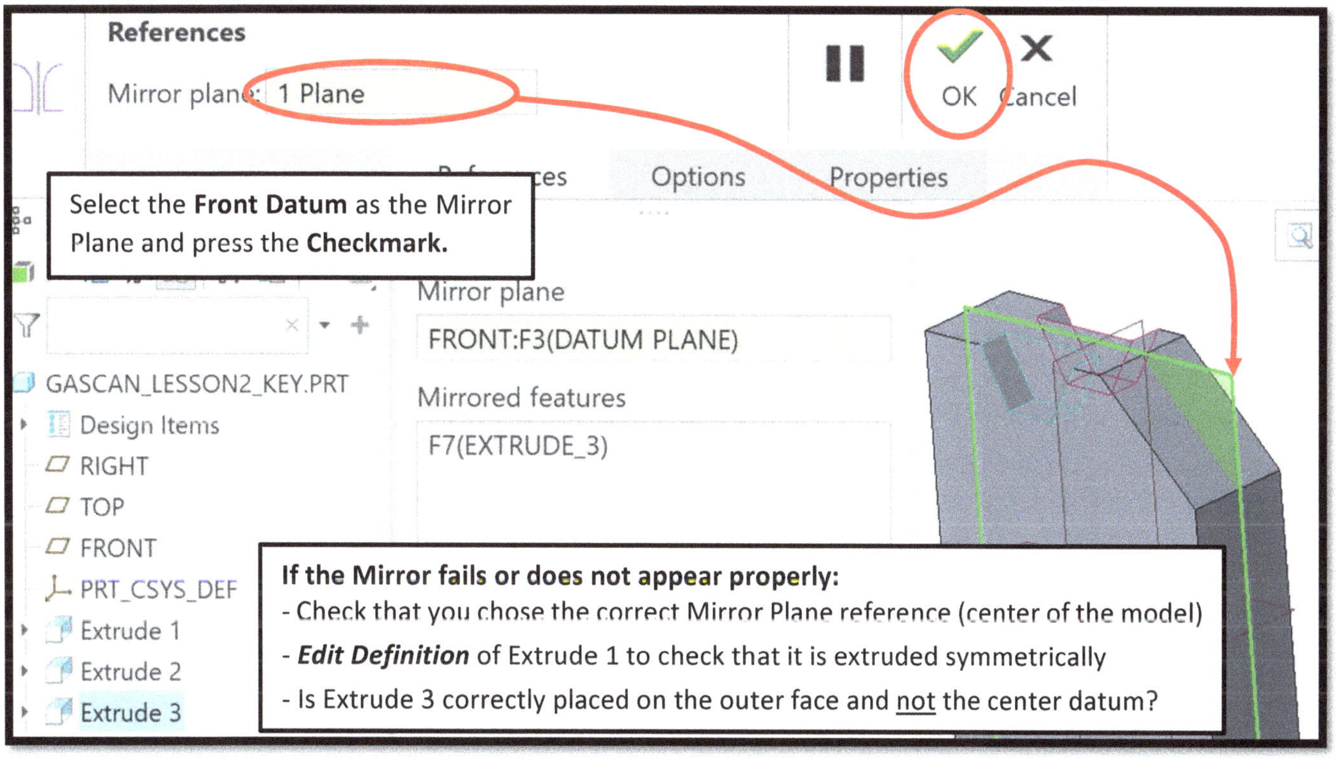

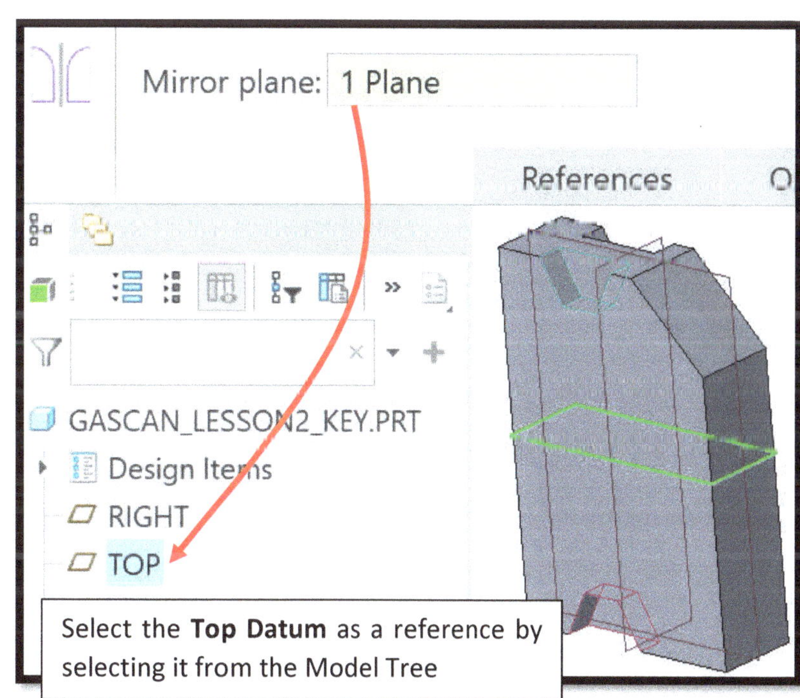

Step 8 – 2 - Use the Mirror tool again to copy Extrude 3 to the bottom of the model.

- **Select** *Extrude 3* in the Model Tree, then choose the **Mirror Tool**. Use the **Top Datum plane** as the Mirror Plane to create a copy on the bottom of the model.

- **Tip:** If this Mirror fails, use *Edit Definition* of Extrude 1 and check that the <u>closed-loop sketch</u> has symmetry about the horizontal reference line.

  Also check if the Extrude 3 sketch has the top line snapped on to the top edge of the model.

Step 8 – 3 - Checkmark to accept the Mirror.

Step 8 – 4 - Repeat the Mirror Tool again to copy *Mirror #2* across the **Front datum** to add a bottom rear cutout. Press the **Checkmark** to accept each Mirror Feature when complete. You should have all 4 cutouts on the top and bottom of both sides.

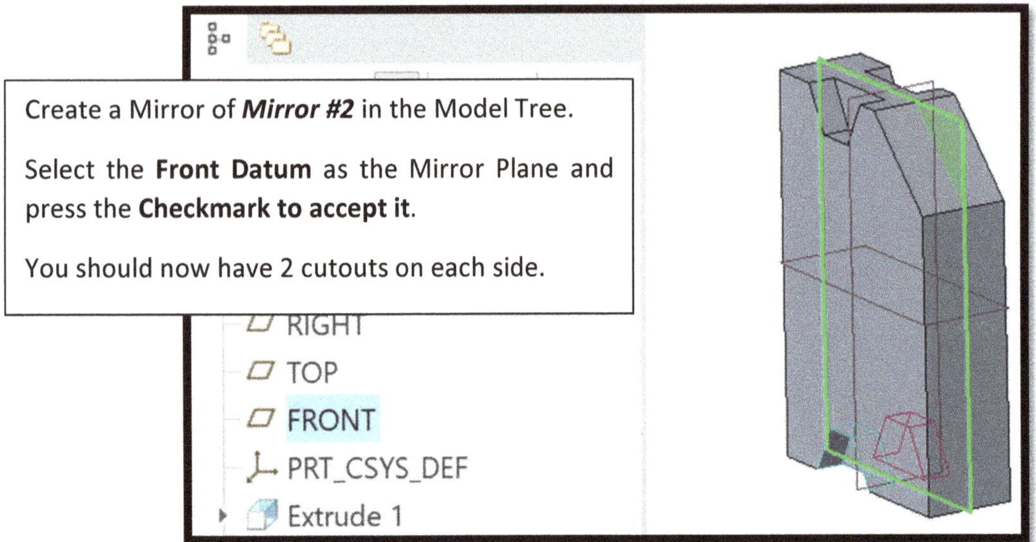

Step 9 – Create a fourth Extrude to form the circular depression in the center of the model's front surface.

Step 9 – 1 – Click in a blank area of the screen to deselect from any entities from the Model Tree, then click on the **Extrude Tool** in the top toolbar.

Step 9 – 2 - Select the **Placement** tab – select **Define**. Choose the **face Surface plane** to set it as the primary <u>Sketch Plane</u> and leave the Reference & Orientation as the default option (Right Datum).

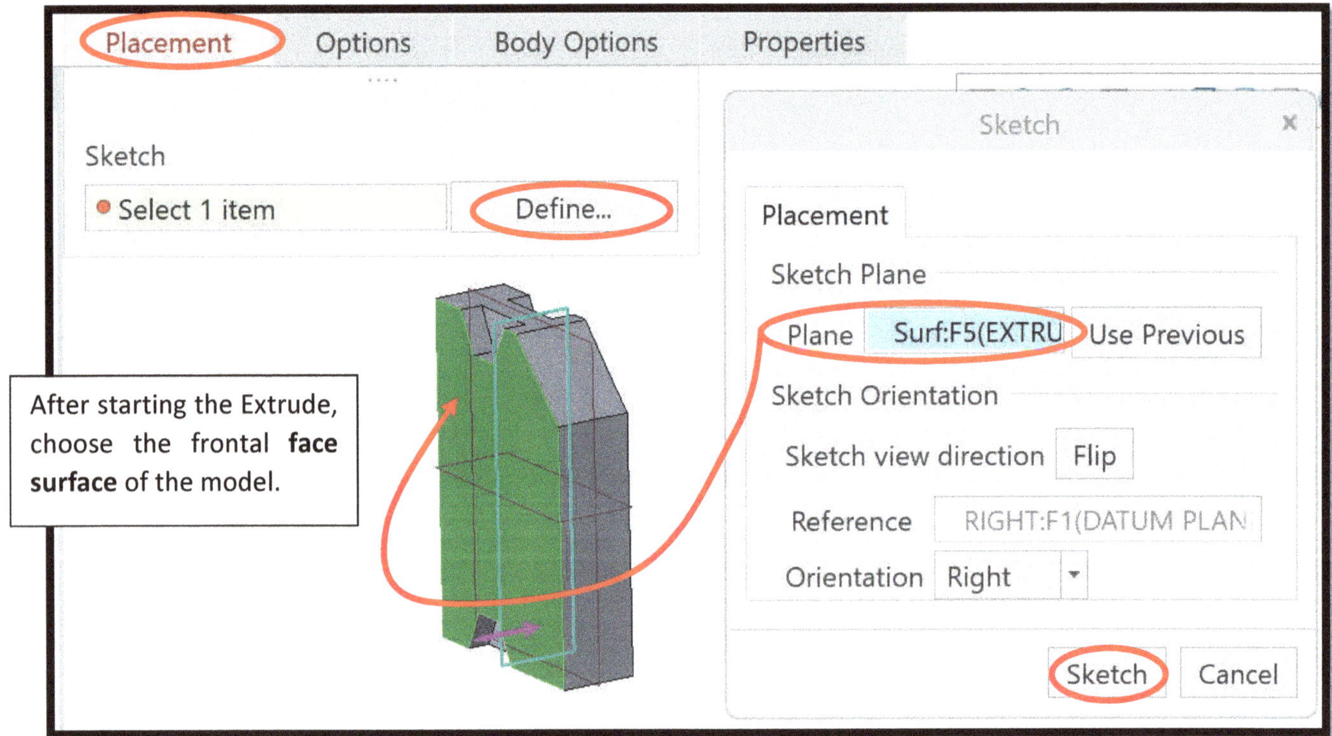

Step 9 – 3 - Press **Sketch** to enter sketch mode for this extrusion.

Step 9 – 4 - Select the **2D View** icon from the **Quick Toolbar** at the top of the Graphics Window.

Step 9 – 5 - Create the closed loop sketch using the **Circle tool** centered on the origin.

Step 9 – 6 - Use the **Select/Pointer Tool** to double click on the dimension value and adjust the diameter of the circle to be **4.00"**.

Step 9 – 7 - Press the **checkmark** to accept the sketch.

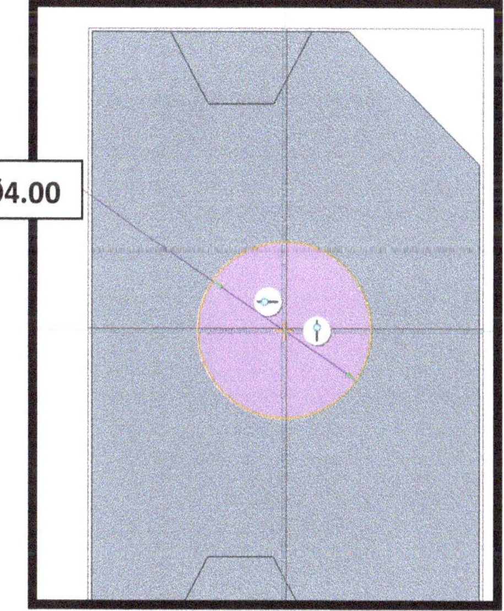

Step 9 – 8 - Set the **Depth, Direction, and Material Removal options** in the Extrude Toolbar.

- **Rotate the Model** to view the effects of the Extrude options (hold MMB and rotate the mouse)
- Leave the **Depth option** to the default one-sided option* of "*Extrude from sketch plane by a given value*"
     Knowledge Tip: This is commonly referred to as the **"Blind" depth option** by most CAD users, as a "*Blind*" depth has a specified value instead of a "*Through*" option that goes through all material in a direction.

- Toggle the Extrude Depth **Direction Arrow** icon to flip the extrude direction to go into the Model.
- Toggle on **Remove Material** (if it did not do so by default automatically)
- Set the **Depth value** to **0.25"**.
- Check that the Type setting is set to **Solid****.
     o   ***this will no longer be stated on future steps as it is the default setting and will always used.*

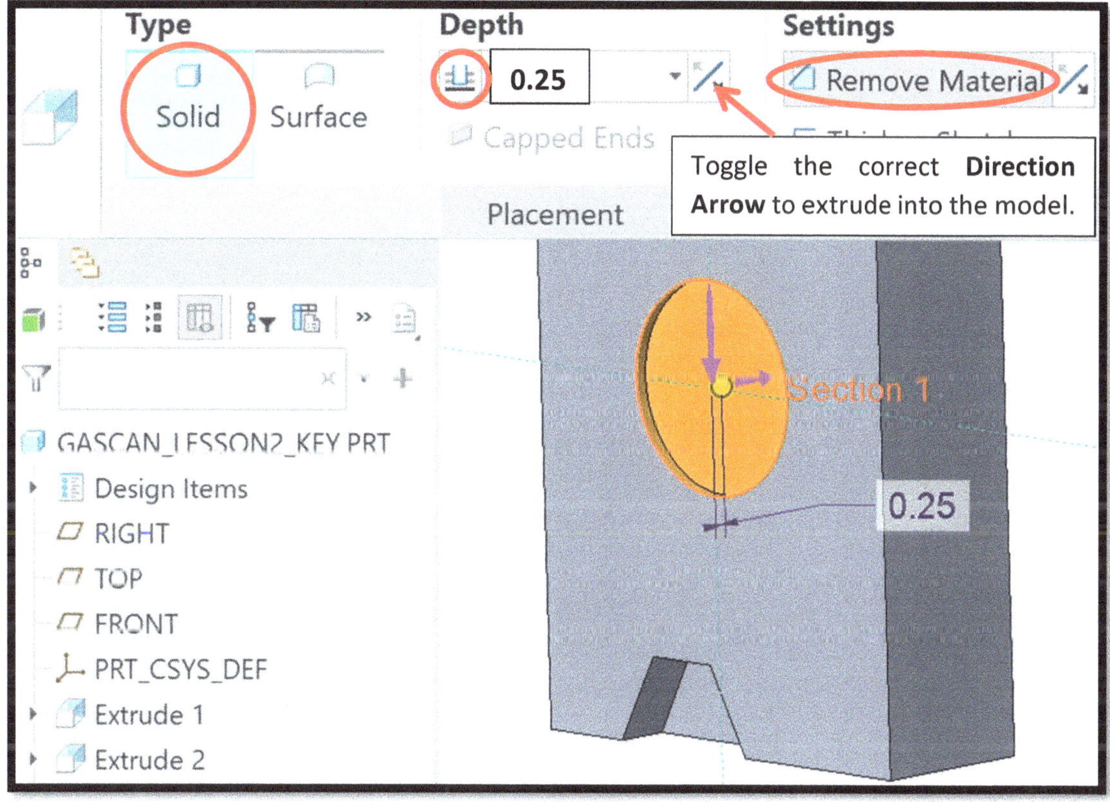

Step 9 – 9 - Press the **Checkmark** to accept the Extrude. 

Step 10 – **Mirror** the previous *Extrude 4* to the rear of the model by using the **Front** datum as the **Mirror Plane**.

- **Select** *Extrude 4* from the Model tree – select the **Mirror Tool** – **select the Front** datum to act as the mirror plane – press the green **Checkmark** to accept the Mirror.

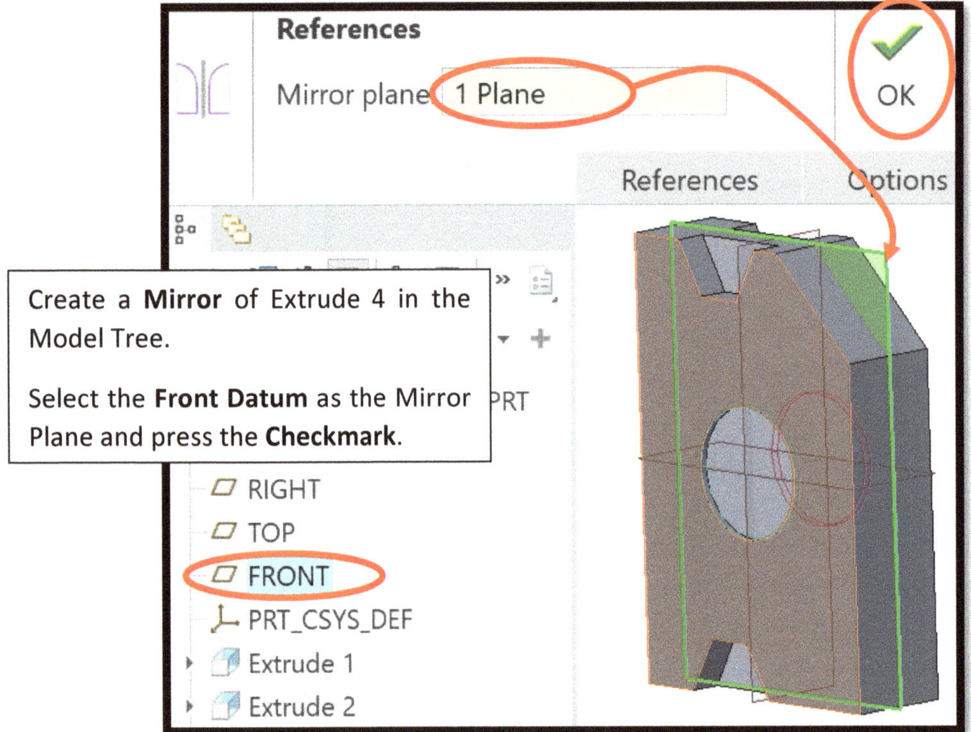

Step 11 – Create a fifth Extrude to cut a slot through the middle of the model using the front surface as the sketch plane.

Step 11 – 1 – Click in a blank area to deselect from any entities, then click on the **Extrude Tool**.

Step 11 – 2 - Select the **Placement** tab – select **Define** - choose the front **face Surface plane** to set it as the primary Sketch Plane and leave the Reference & Orientation as the default option (Right Datum).

Step 11 – 3 - Press **Sketch** to enter sketch mode.

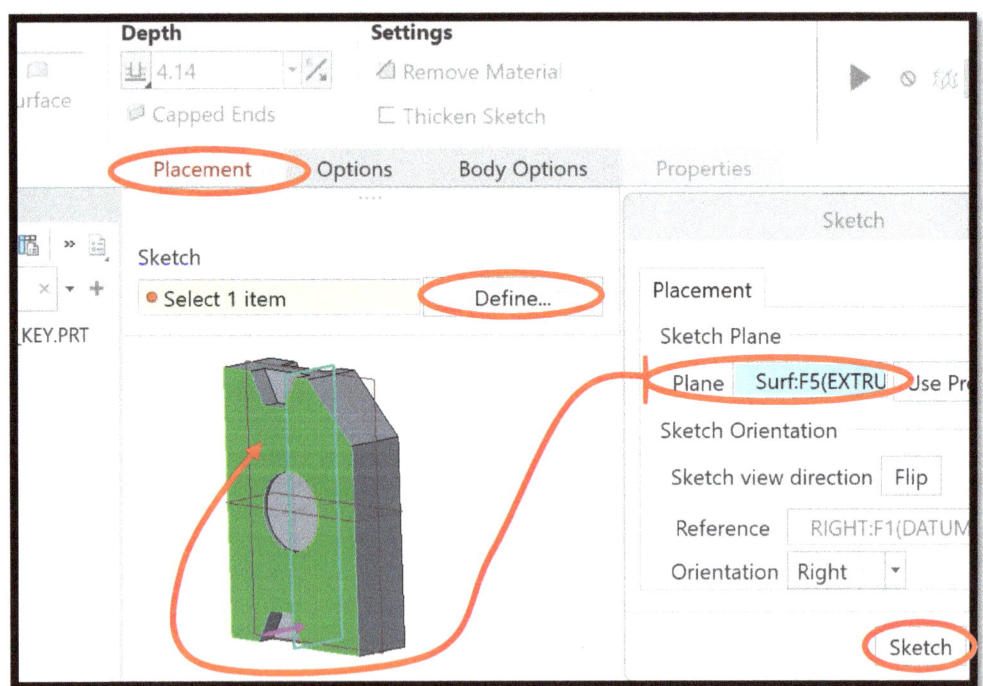

Step 11 – 4 - Select the **2D View** icon from the **Quick Toolbar** at the top of the Graphics Window.

Step 11 – 5 - Complete the steps below to create Sketch.

1) Add Centerlines: Use the **Centerline tool** to place **Vertical & Horizontal centerlines** through the origin.

2) Sketch the Rectangle: Use the **Rectangle Tool** to sketch the shape that crosses over the origin. Make this wider than the model for now. *Do not place the rectangle constrained or "snapped" onto any other edges on the model.*

3) Symmetry Constraints: Use the **Symmetry Constraint tool** about both the vertical and horizontal centerlines. CREO may have already assumed symmetry about either or both centerlines, so add as needed.

4) Center & Ends Arc: Click the **drop-down arrow of the Arc Tool** and select the **Center & Ends** option.

>   - **LMB click to set the center point** of the first arc to be on middle of the left vertical line of the rectangle (snapped onto both the vertical sketch line and the horizontal centerline) - **LMB to select the beginning of the Arc on the top corner** of the rectangle.

5) Delete the lines to make a Closed-Loop: Use the **Delete Segment Tool** to remove the left and right rectangle vertical lines, leaving a closed loop. If you have "red" end points left, that would indicate an open section that needs to be fixed.

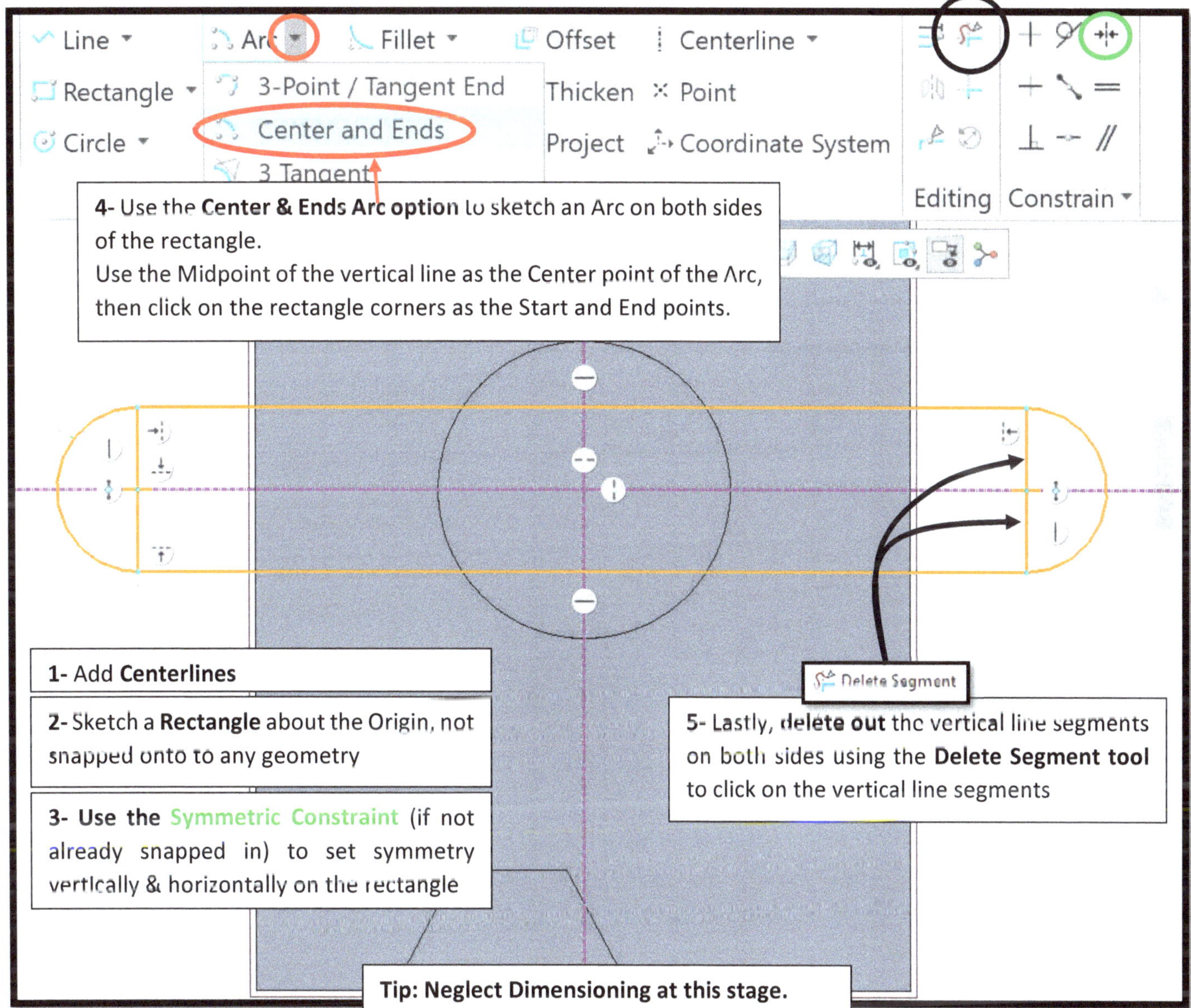

Double Check: Make sure that your Arcs are snapped on to the rectangle corner points & that you have a Closed-Loop. If you leave the Arc short or long it will be a red colored "Open-End" point. If so, fix it by deleting & remaking the Arc, or add a Coincident constraint between the arc end-point and the rectangle corner.

6) Create Dimensions:

- Use the **Dimension Tool** and create a dimension between the outer ends of the arcs (**4.00"**), ==using the arc curved lines as the reference points.==

- Then dimension the height of the rectangle (**1.00"**).

Tip: make sure you have a ==Height dimension from the top horizontal line to the bottom horizontal line.== Some users have a Radius dimension which is not the same!

7) Press the **Checkmark** to accept the sketch.

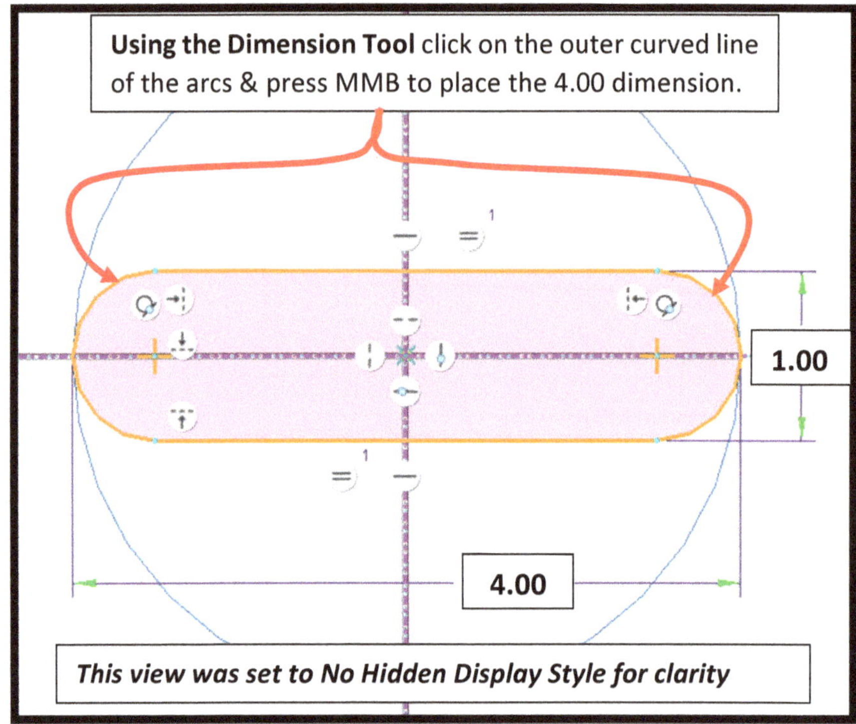

Step 11 – 6 - Set the **Depth, Direction, and Material Removal options** in the Extrude Toolbar.

- Toggle the extrude **Direction Arrow icon** to flip the extrude direction to go into the Model material.
- Toggle on **Remove Material**
- Set the **Depth option** to the **Through All Option** (i.e. "*Extrude to intersect with all surfaces*")

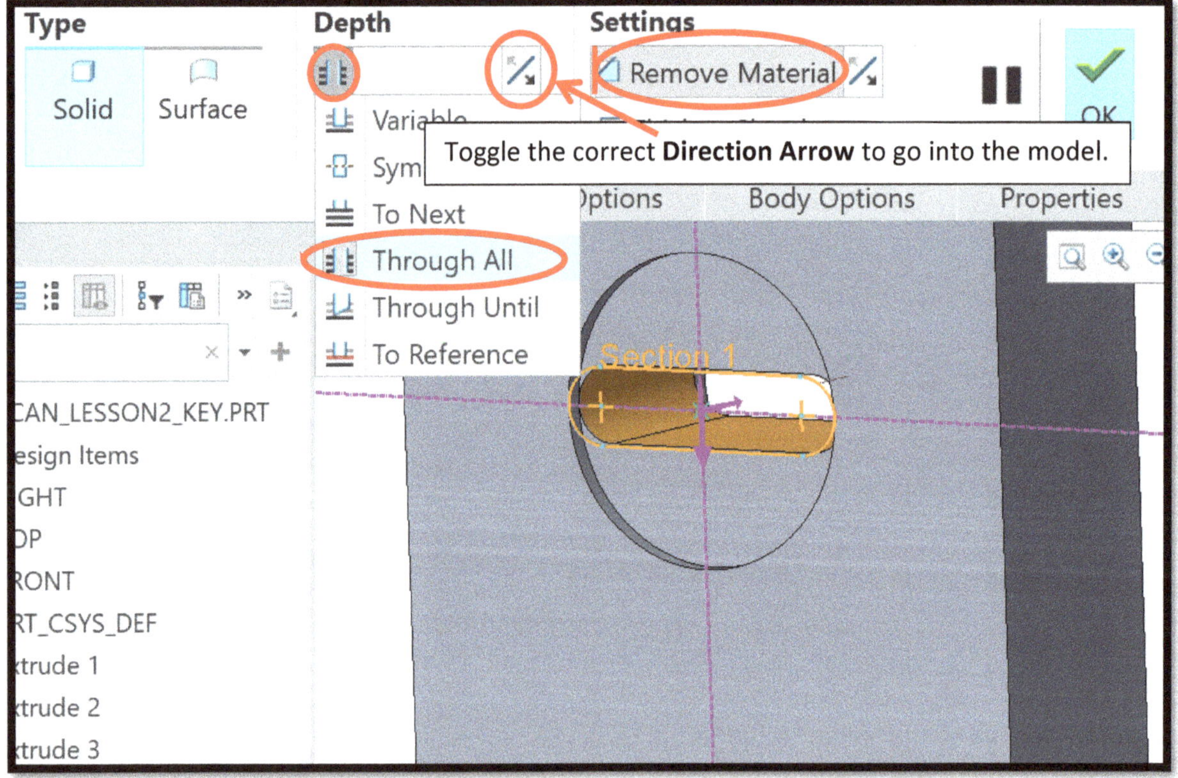

Step 11 – 7 - Checkmark to accept the feature.

Step 12 – Create a new **Datum Plane** offset 3.75" from the Right datum. This new datum plane will be used as a sketch plane for the next extrusion.

- Click in a blank area to **deselect** from any entities - Select the **Plane Tool** at the top toolbar – click on the **Right Datum** (the vertical datum)from the Model Tree or model to add it as a reference – in the **offset box** enter in a **value of 3.75"** – the new datum should be visible on the right side of the model (if it appears on the left side, you can enter in a value of -3.75 to move it to the other side) - **press OK** to accept the Datum. **DTM1** will now be added to the model & Model Tree.

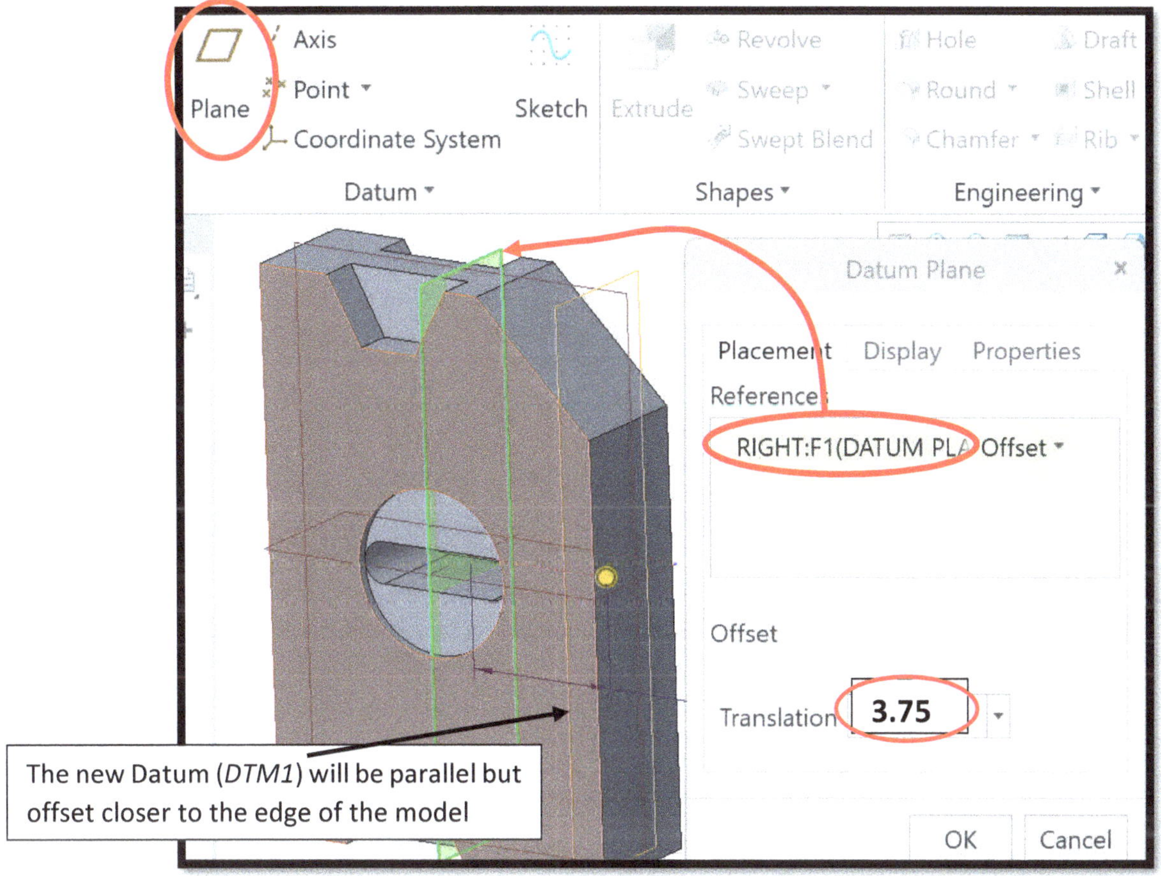

Step 13 – Create a sixth Extrude to cut a slot into the part, using **DTM1** as the sketch plane so the slot will be located away from the edge of the model.

Step 13 – 1 - Click on the **Extrude Tool** in the top toolbar.

Step 13 – 2 - Select the **Placement** tab – select **Define** - select the new datum **DTM1** to set it as the primary Sketch Plane. Leave the Reference & Orientation as the default option (Top Datum).

Step 13 – 3 - Press **Sketch** to enter sketch mode for this extrusion, and press the **2D Sketch View** icon in the quick toolbar.

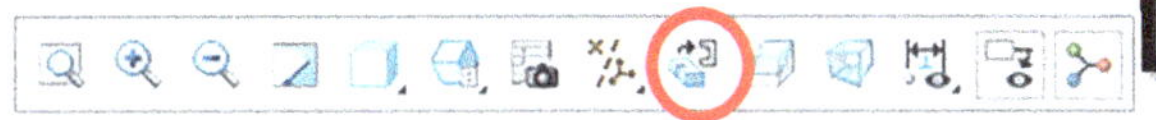

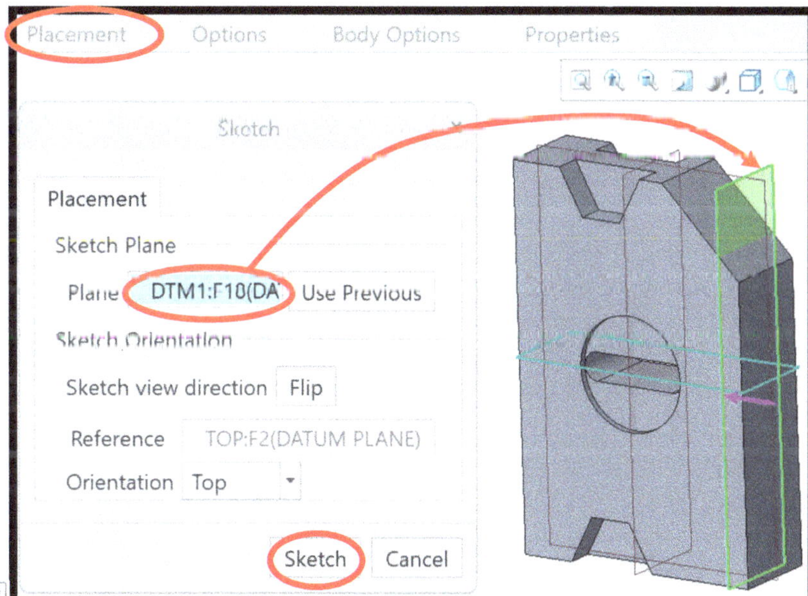

Step 13 – 4 - Create the Sketch shown in the image below, which will be a "side" view of the model looking at DTM1 as the sketch plane. Be careful **not to "snap" onto any hidden line entities** from previous extrusion edges or midpoint of the lines, as those constraints will interfere with the desired sketch.

1) <u>Toggle on a Hidden Line Style</u> under the **View Display** option in the Quick toolbar to better see and avoid the Hidden Geometry of previous features. Also make sure **Perspective View is not toggled on** accidently in the quick toolbar.

2) <u>Add a Horizontal Centerline</u>: Use the **Centerline tool** to place a centerline through the origin.

3) <u>Sketch the Closed-Loop:</u> Use the **Line Tool** to sketch the shape* that crosses over the horizontal centerline, with the left vertical line on the left edge of the model.

> *Be careful that your lines do not accidently "snap" onto unwanted hidden lines or Mid-Points of past feature edges (grey lines).

4) <u>Symmetry Constraints</u>: Use the **Symmetry Constraint tool** about the horizontal centerline on both the **left and then the right vertical lines** of the sketch. Note that CREO may have already assumed symmetry about either or both centerlines.

5) <u>Create Dimensions</u>: use the **Dimension Tool** and create dimensions for the width of the shape (**0.25"**) and the lengths of the left side (**4.25"**) and right side (**3.75"**) vertical lines.

6) Press the **Checkmark** to accept the sketch.

OK

7) Switch back to a **"Shading with Edges"** **View Display** using the Quick Toolbar

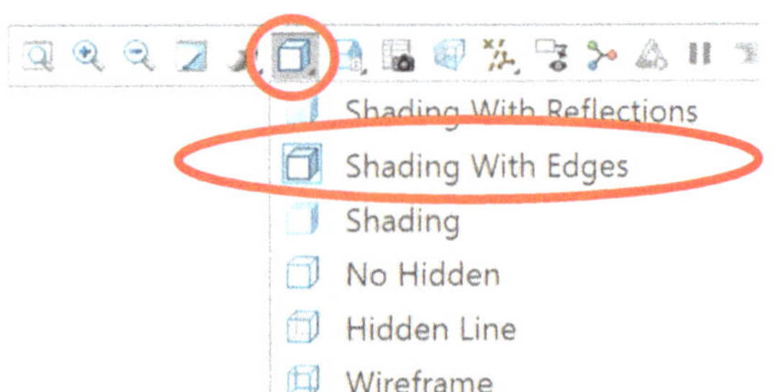

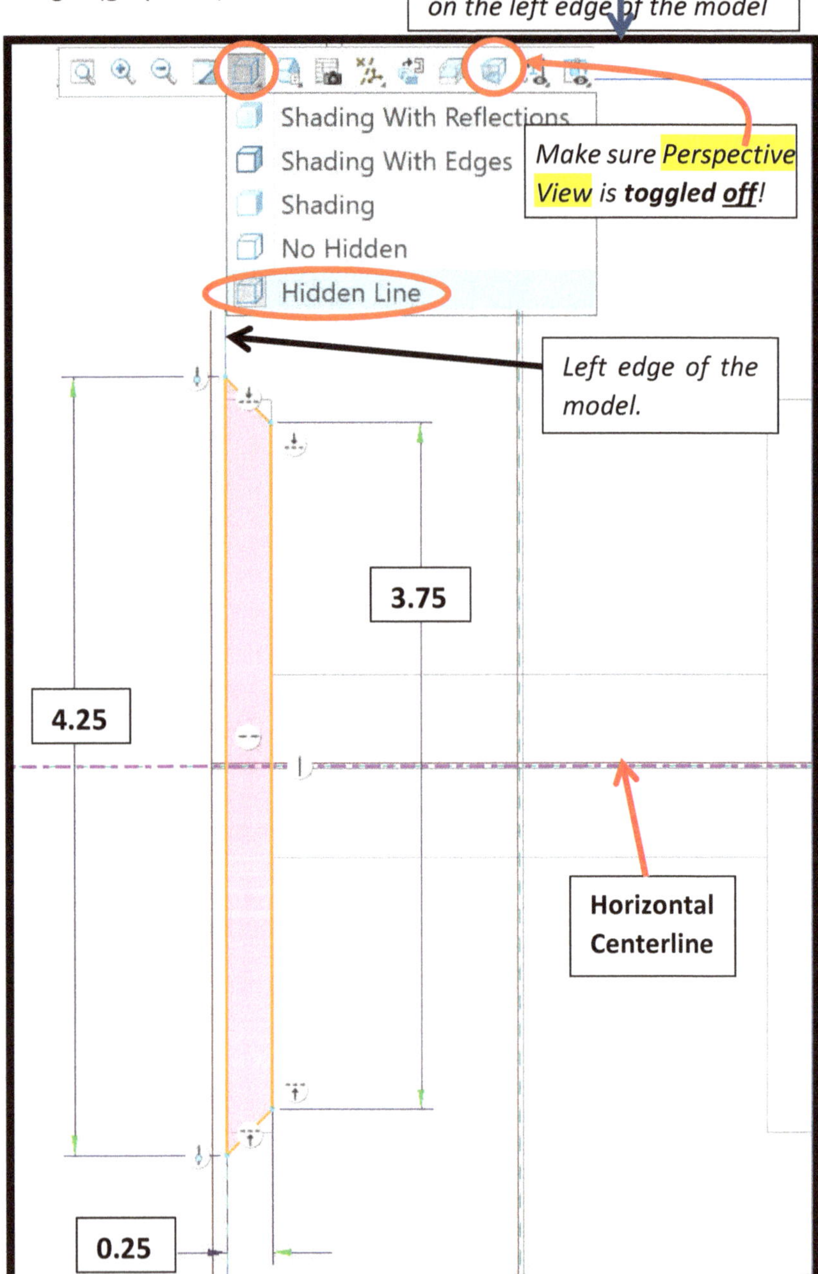

Step 13 – 5 - Set the **Depth, Direction, and Material Removal options** in the Extrude Toolbar.

- Toggle on **Remove Material** (if it did not do so by default automatically)
- Select the **Depth option** to be **Extrude Symmetrically** (i.e. both directions)
- Set the **Depth value** to **0.75"**
- **Press the Checkmark** to accept the Extrude.

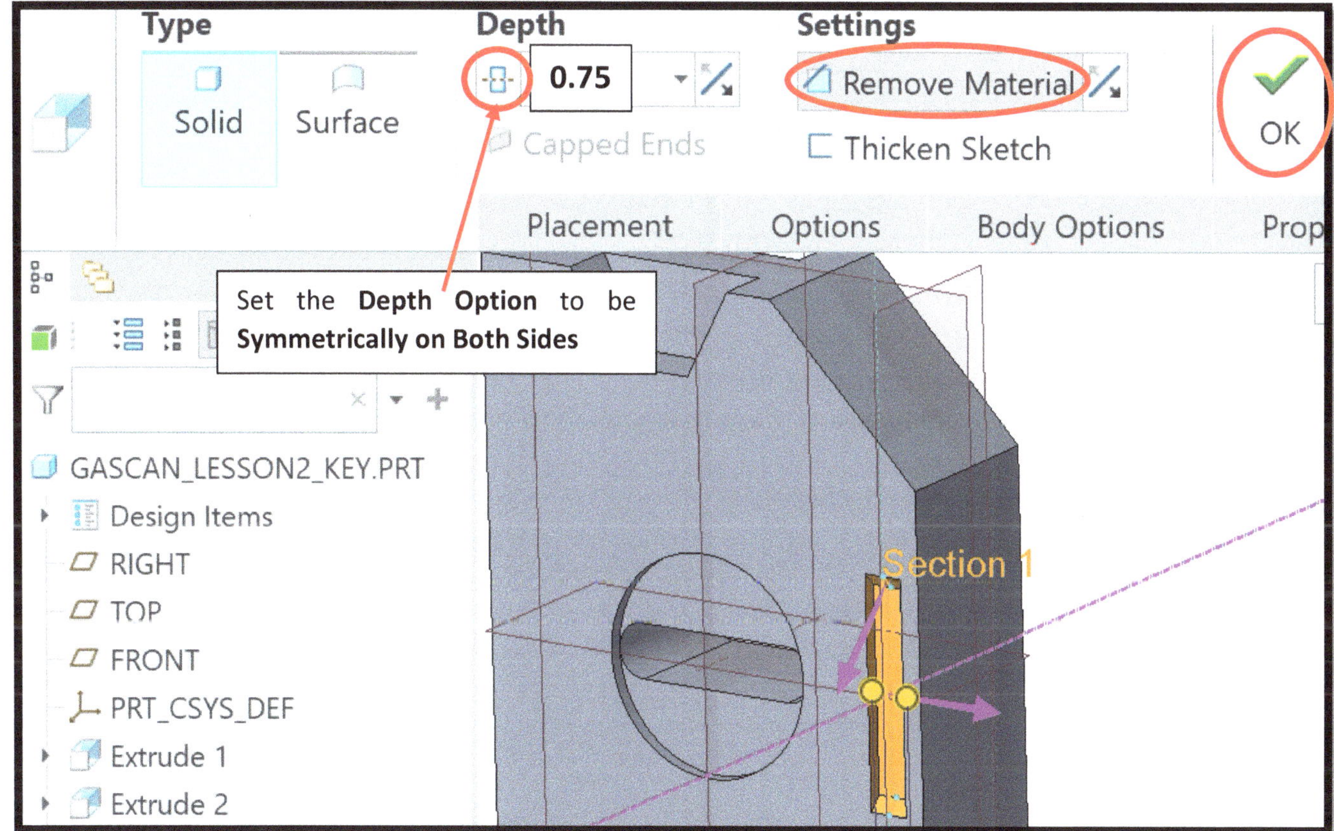

Step 14 – Mirror the previous Extrusion (*Extrude 6*) to the opposite side of the model by using the **Front** datum as the **Mirror Plane**.

- **LMB Click** on *Extrude 6* from the Model tree
- Select the **Mirror Tool** – **LMB click on the Right** datum on the model tree

- Press the green **Checkmark** to accept the Mirror.

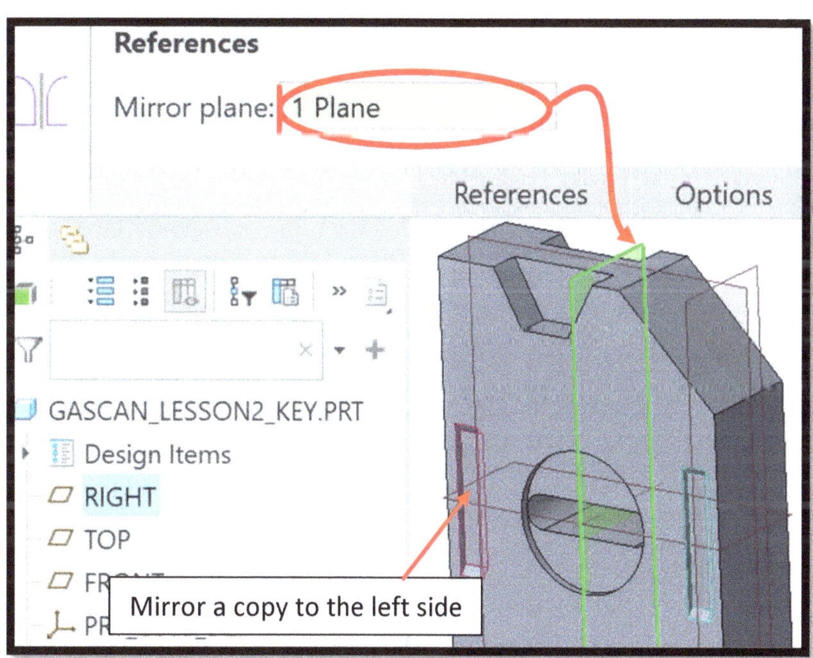

Step 14 – 2- Create another **Mirror of *Extrude 6*** to copy it to the back side of the part.

- Click on **Extrude 6** in the Model Tree

- Select the **Mirror Tool**

- Click on the **Front Datum** as the Mirror Plane.

- Press the green **Checkmark** to accept the Mirror.

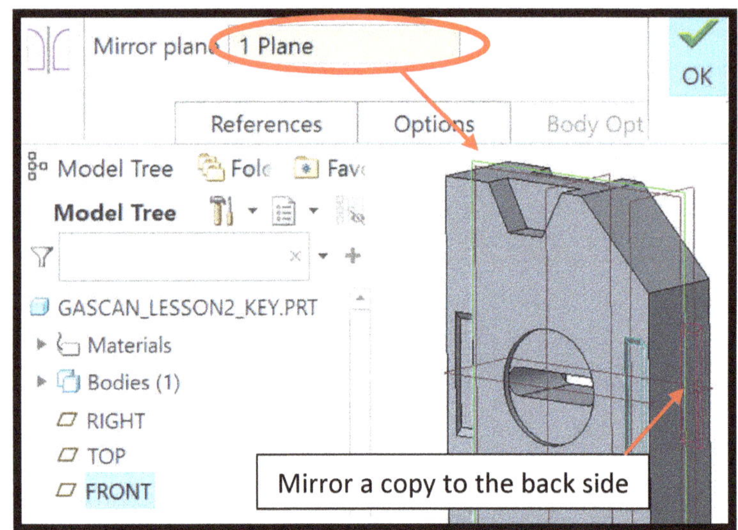

Step 14 – 3 - Create another **Mirror** to create a fourth copy of E*xtrude 6* on the back side of the part. You will Mirror one of the previous Mirrors to create the last copy.

- **LMB Click** on *Mirror 6* (the one you copied to the back side of the part) from the Model tree

- Select the **Mirror Tool – LMB click on the Right** datum on the model tree

- Press the green **Checkmark** to accept the Mirror.

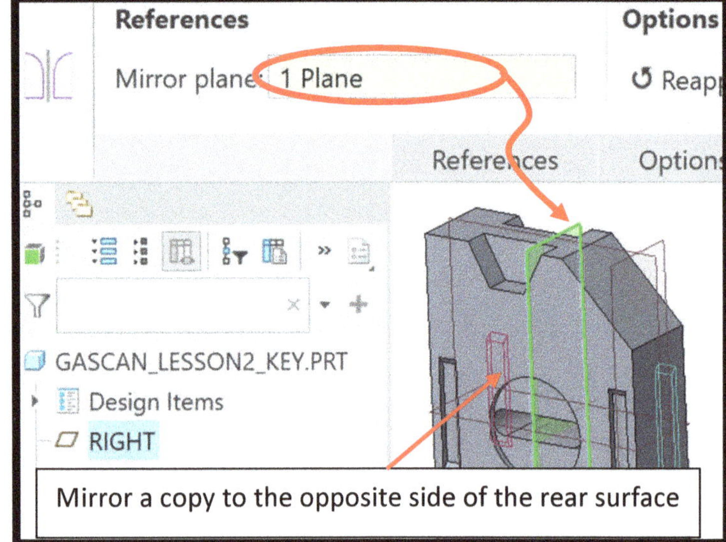

If your model does not match the image shown, use **Edit Definition** to modify the original Extrude 6 to make any needed changes (or use Edit Definition on DTM1 if you need to reposition the datum that the extrude is placed on.)

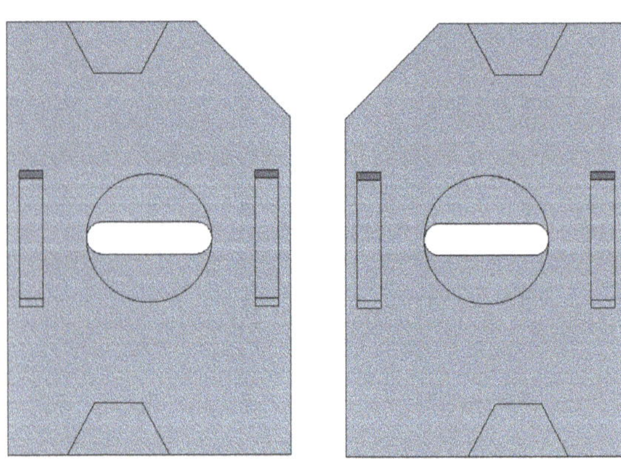

Updating Mirrored Copies: If you do need to make changes to an original feature that a mirror is based on, such as changing the Extrude toolbar direction or depth, you may also need to use **Edit Definition of each mirrored copy** and checkmark the "**re-apply mirror**" option to update the mirrored copies with the new changes.

Front Side   Rear Side

Step 15 – 1 - Create a seventh Extrude to add the "Spout" protrusion on the angled corner of the model.

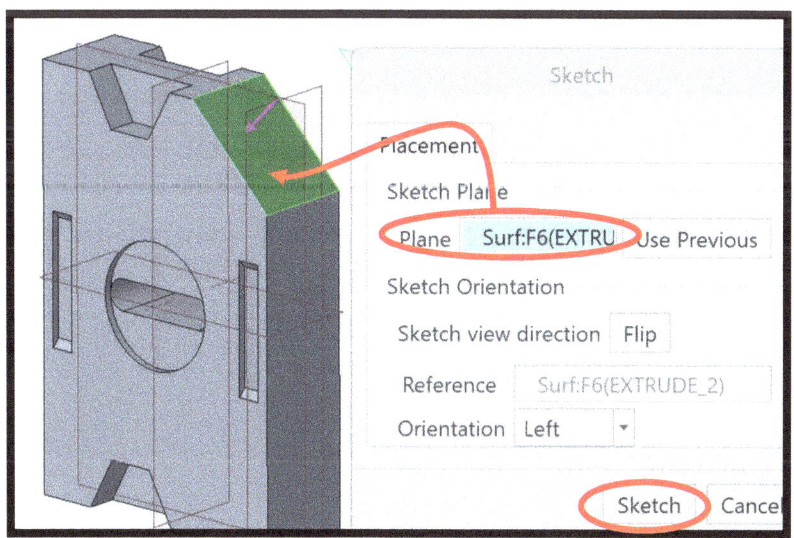

- Click a blank area of the screen to deselect - choose the **Extrude Tool** in the top toolbar - select the **Placement** tab – select **Define**.

- Select the angled **planar Surface** of the model **corner cutout** as the desired **Sketch Plane**.

After selecting the sketch plane, the default Reference & Orientation can be kept as the default values.

Note: your Surface ID# (*Surf:F6*) shown in the image may be different in your model.

Step 15 – 2 - Press **Sketch** to enter sketch mode for this extrusion.

Step 15 – 3 - Click the **2D Sketch View** icon from the **Quick Toolbar**.

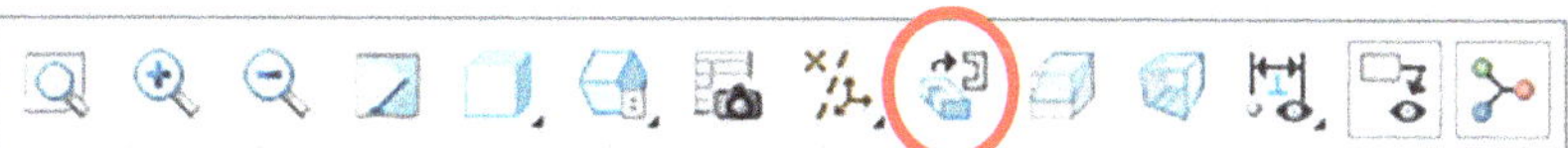

Step 15 – 4 - <u>Sketch References:</u> **Add the bottom edge and middle (Right) datum** as Sketch References using the instructions below. Sketch References add reference lines to allow snapping geometry on those edges.

- **Click in a blank area** of the graphics window – **hold RMB** – select the "**References**" option from the pop-up menu – **LMB click on the bottom edge and the middle datum (Right) as shown in the image below** to add them as sketch references (dotted lines) – press **Close** in the References menu.

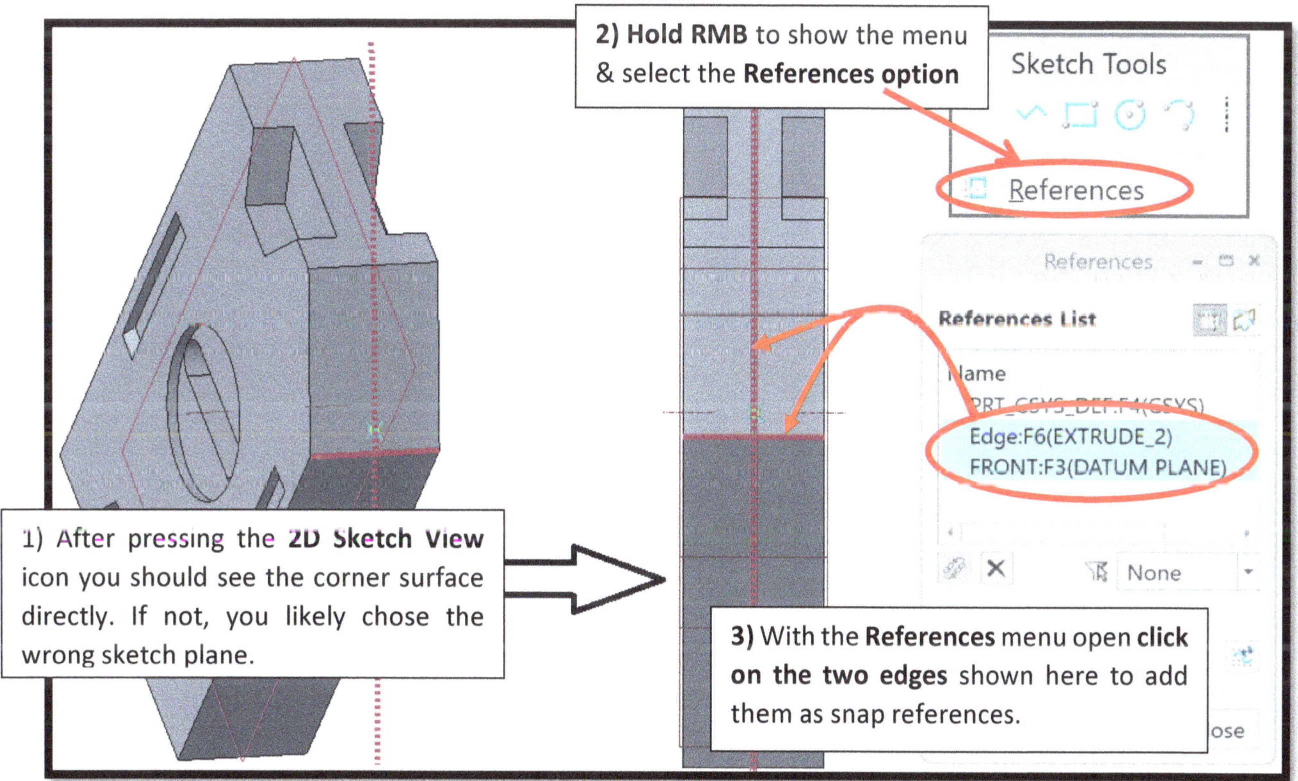

Step 15 – 5 - <u>**Sketch a Circle:**</u> use the **Circle tool** to place a circle centered onto the vertical reference line only. Be careful that your circle placement **does not "snap"** onto the hidden geometry of the previous extrudes or the Mid-Point of any lines, which would then interfere with correct dimensioning and placement of the sketch.

Step 15 – 6 - <u>**Dimension the Sketch**</u>: use the **Dimension Tool** to create a dimension between the bottom reference edge and the center of the circle. Set the value of the circle diameter to be **2.00"** and the distance from the bottom edge of the surface to **2.00"**.

Note: If you get a Resolve Error menu due to over constraining the sketch, you likely snapped the circle onto some hidden line geometry and can try deleting out the correct Point on Entity constraint, or delete the circle and try again.

Step 15 – 7 - Press the **Checkmark** to accept the sketch.

Step 15 – 8 - Set the **Depth** and **Direction options** of the Extrude so it protrudes from the surface a depth of **1.25"**.

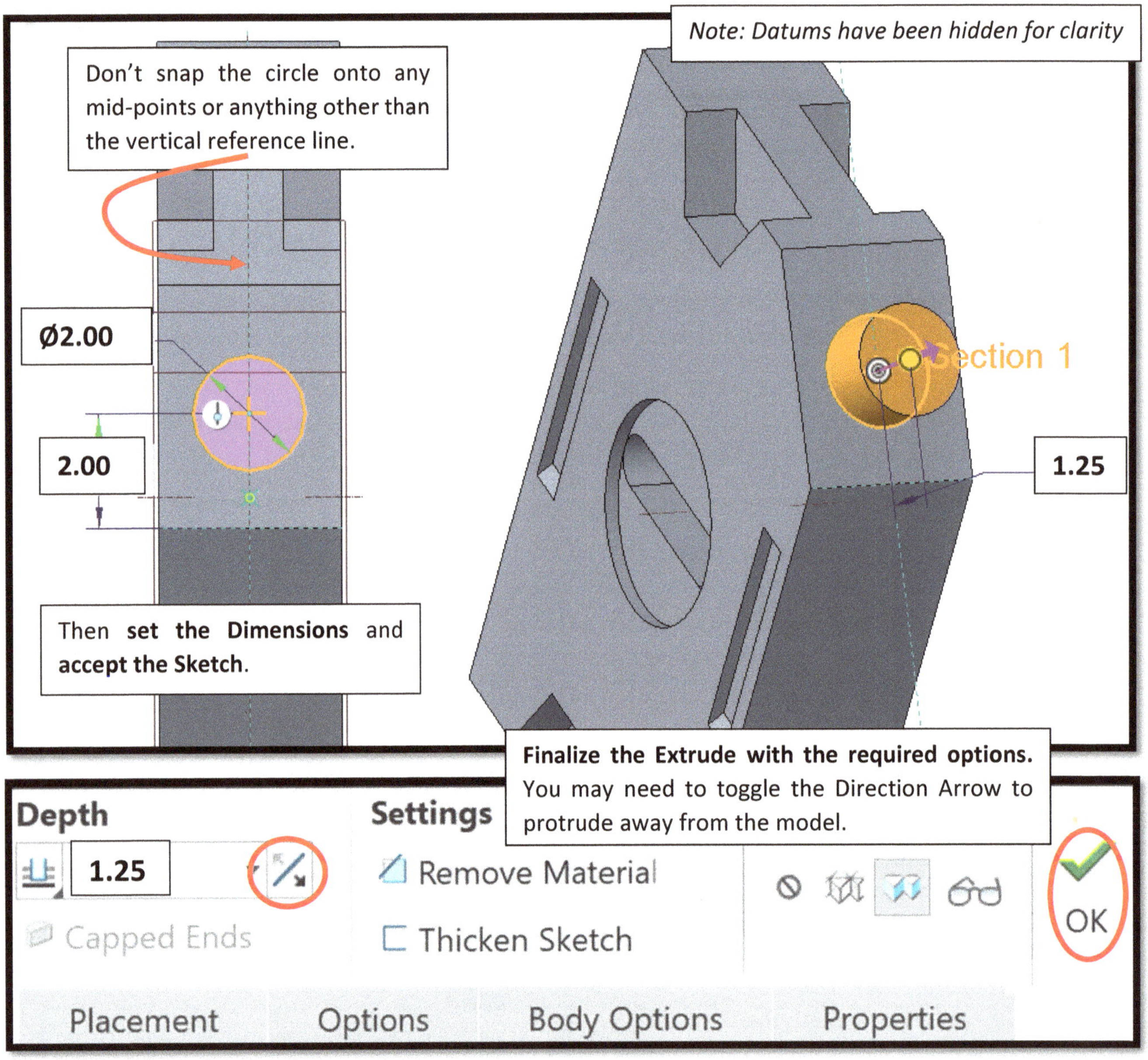

Step 15 – 9 - **Press the Checkmark to accept the Extrude.**

Step 16 – Add **Chamfers** to the three outer corner edges of the model using the **Chamfer Feature**.

Step 16 – 1 - Select the **Chamfer Tool** from the top toolbar.

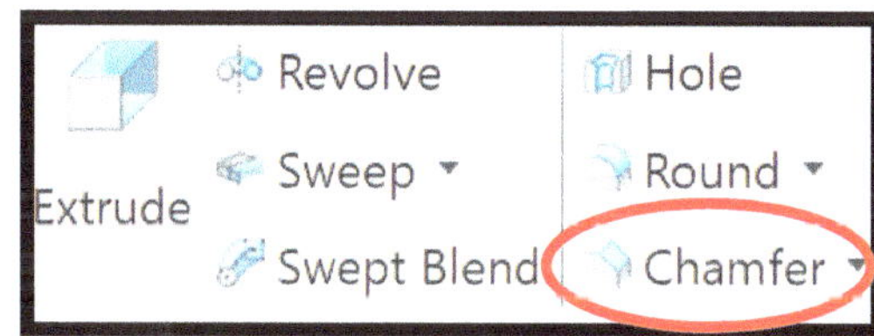

Step 16 – 2 - In the top toolbar set the chamfer edge length dimension (D) **to 0.75".**

Step 16 – 3 - Open the **Sets tab** in the toolbar – use the **LMB** to select the 3 corner <u>edges</u> to be chamfered.

Note - if you need to remove or edit one of the chamfers you can see a list of all of them under the "Sets" option of the Chamfer toolbar and can then select them to edit the values or delete them if needed.

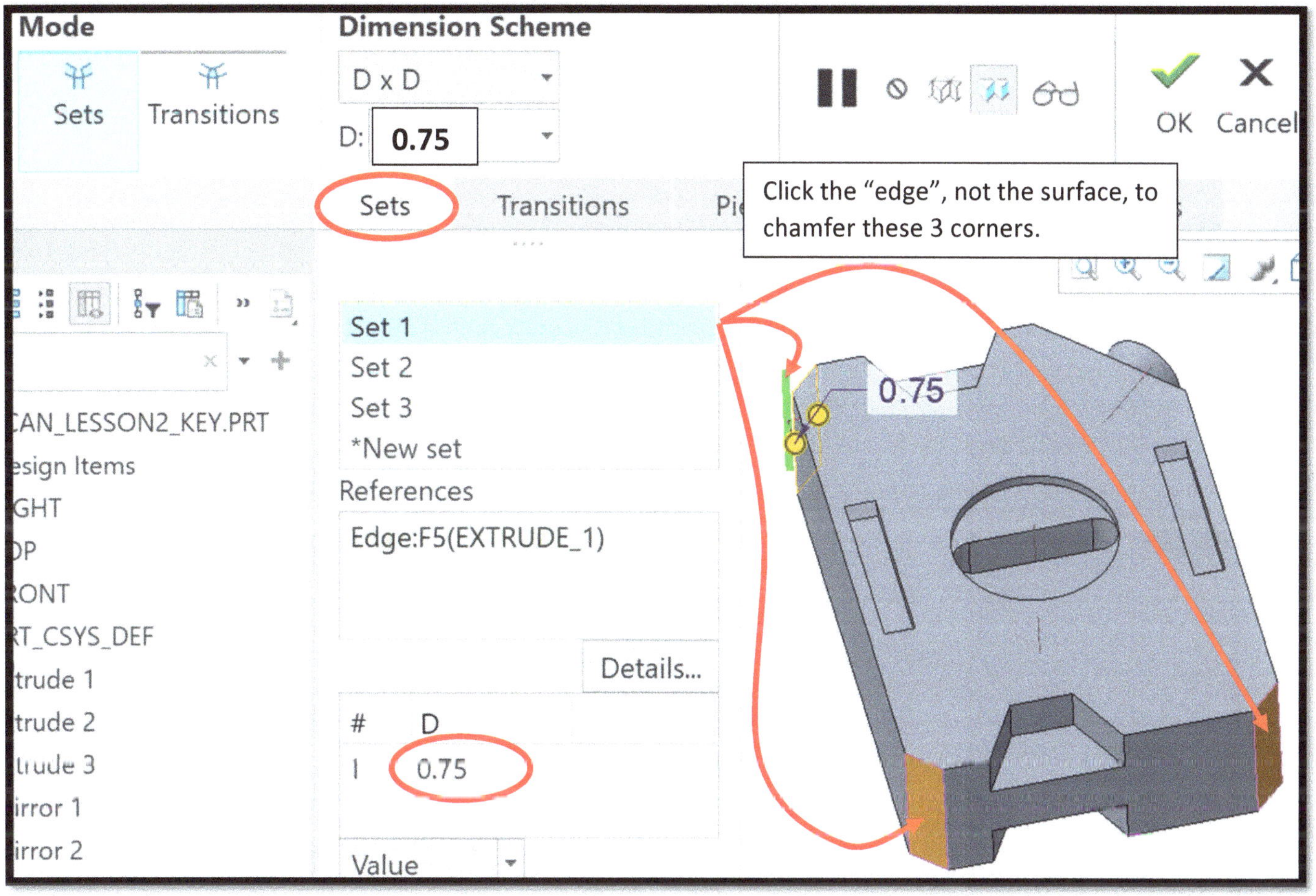

Step 16 – 4 - Press the **Checkmark** to accept the chamfers.

Step 17 – Check your Model – Does every feature look correct? If not, select the desired Feature in the Model Tree and use **Edit Definition** (the *Sphere & Pen icon)* to adjust that feature's toolbar settings or sketch dimensions.

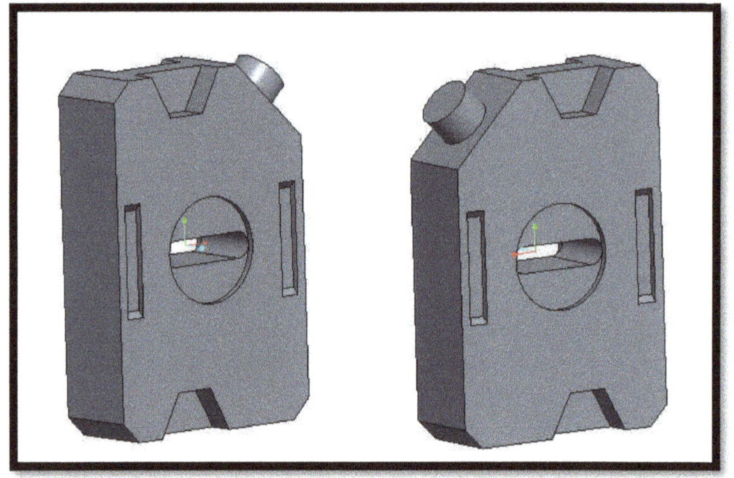

Step 18 – Save your model! Pressing **File - Save** will save the model to the <u>current </u>directory. This could be the default CREO directory (bad!), your correctly specified ME101 Working Directory (good!), or the location that you last used in this session of CREO. You can always look in that folder (through Windows) and see if the Part file (.prt) is saved.

> **Example:** You set your working directory to your ME101 Working Directory folder, but later opened a CREO part file you had located on the Desktop. Since the last location used was different from your set Working Directory, CREO will likely save the file to that "last used" location unless you set the Working Directory again.

<u>Adding Color to the Model (*Optional*)</u>

Shouldn't a gas can be red? You can change the Color of the Model easily with the Appearances tool. You can choose any color your desire for your model, as the grader will not see the color on the drawing anyways.

- Click on the **View** Tab at the top – **Appearances Tool** drop arrow– click on a **red "appearance" color option** (*or browse the libraries for one*) – a small **S**elect Menu will appear in the corner of your screen – with this menu open click on **the Part File name** in the Model tree – **press OK**. This will change all model material to the selected color.

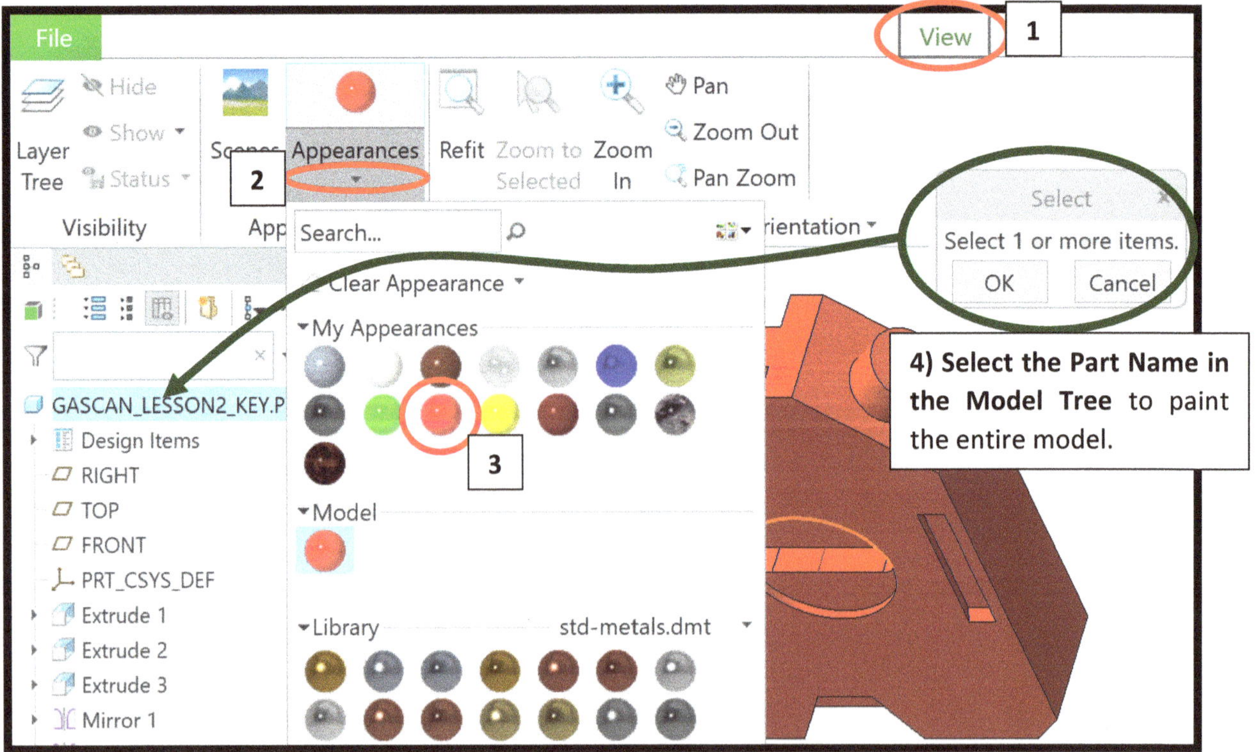

Return to the <u>**Model tab **</u>in the top toolbar. **Press Save** to save the color change.

Step 19 – Create a Drawing of the model that you will submit for grading.

- Select File from the top toolbar **– New – choose Drawing** as the type of file **– uncheck the "Use Drawing Model Name" option – type in "Lesson_2" for the name – OK –** in the pop-up window leave all drawing settings* at the default setting – press OK to get into the drawing.

You can choose Browse if you need to specify the model to use for the "Default Model" if it doesn't come in on its own.

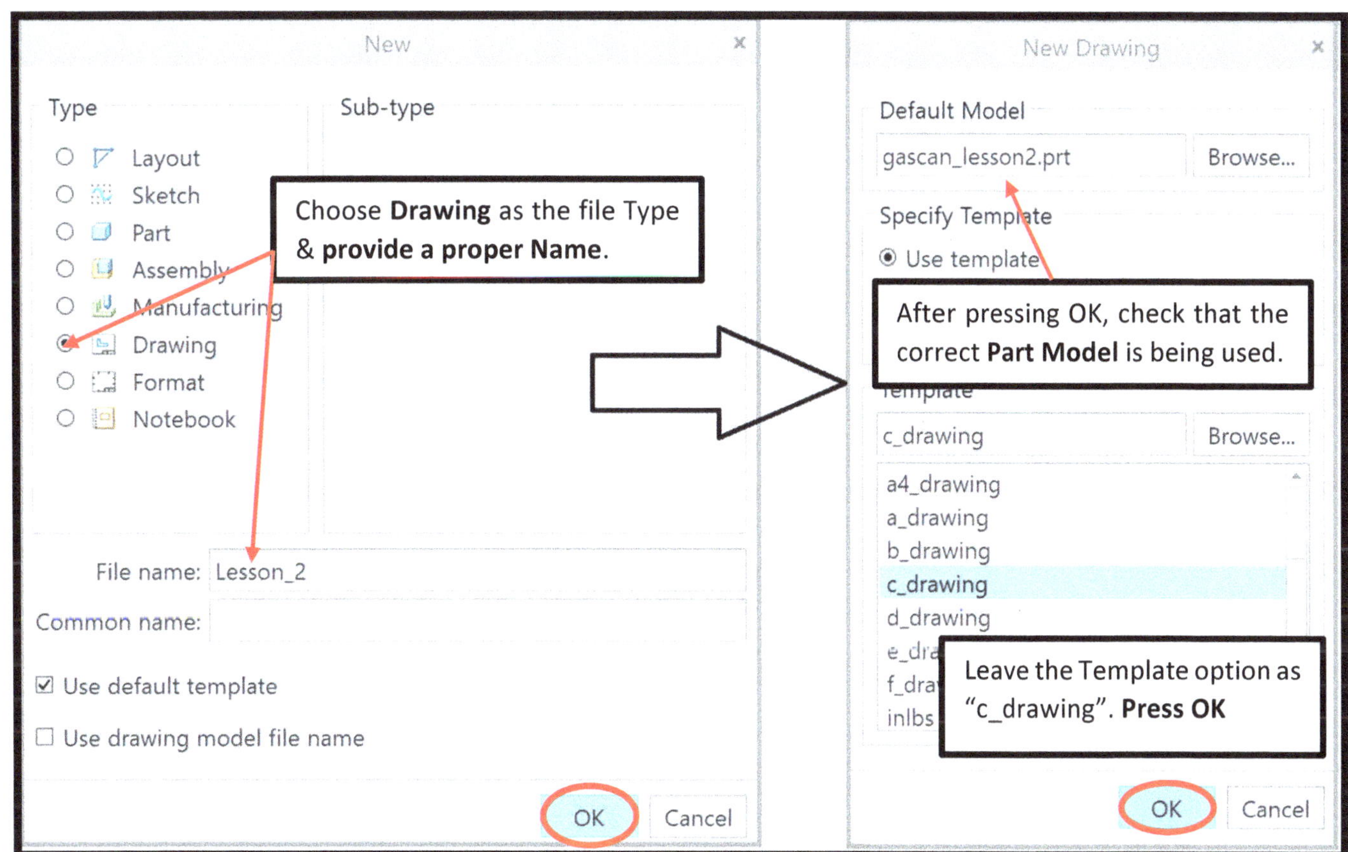

Step 20 - You will need to download the **"C-size_UND" format file** from Blackboard (*Lesson 2 Folder*) and copy it to your ME101 Working Directory on your own computer. This is the UND title block that appears on the drawing key.

-　　Go to the **Lesson 2** folder in Blackboard – click & hold **RMB** on the attached "**C-size_UND.frm**" file – choose **Save As** or **Save Link As** (*depends on your web browser*) – save this to your **ME101 Working Directory***. *You do not want to click on the attachment in Blackboard, as that may try to open the file directly which it cannot do.*

This file must be in the same Directory folder as the Part & Drawing file you are working on for all future lessons.

If all of these files CREO files are not in the same folder you may have an error when trying to re-open a drawing as it will not be able to find the other files it is referencing to build the drawing. This can be resolved by manually moving the files into the same folder or working directory (through Windows File Explorer) and re-opening CREO Parametric.

Step 21 – Complete Module 4 – Basic Drawings in Blackboard to learn how to create a basic two-dimensional (2D) drawing.

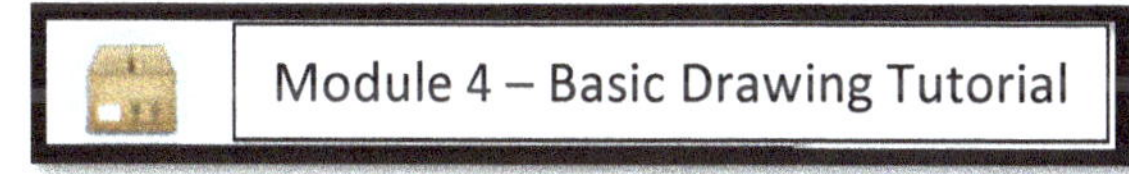

Note: Module 4 shows the **generic** drawing steps. The specific settings you need to match will vary depending on the Lesson Key discussed on the next page.

Step 22 – Follow the steps covered in Module 4 and also explained in the ==**Basic Drawing Steps**== (**in the Appendix**) to create a drawing that matches the drawing Key (posted in Blackboard). Not all steps from the **Basic Drawing Steps** will be required on every lesson, so apply them as needed to match the current lesson key.

Your drawing should generally match the layout and details of the Lesson Key.

- **Do not include the instructional text** shown in red or blue

- **Your Isometric General View** may be rotated different than the key, which is OK. This happens if you accidently choose a different Sketch Plane than the lesson steps.

- **The main View Orientations (Front, Top, Right) should match the key.** Rotate by Angles as needed (refer to appendix).

- Spacing and layout should generally be similar, **but does not need to be exact.**

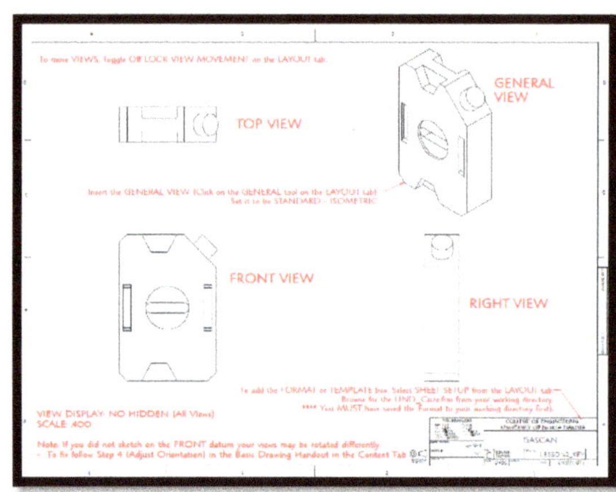

Lesson 2 Key: Posted to the Assignment in Blackboard.
Open it from there so you can read the details!

Step 23 – **Save your drawing** file in CREO (File – Save).

Double Check: Using <u>**Windows File Explorer**</u> look in your directory folder to make sure the files have been saved.

Step 24 – **Save a** ==**PDF version**== of the drawing to submit for grading.

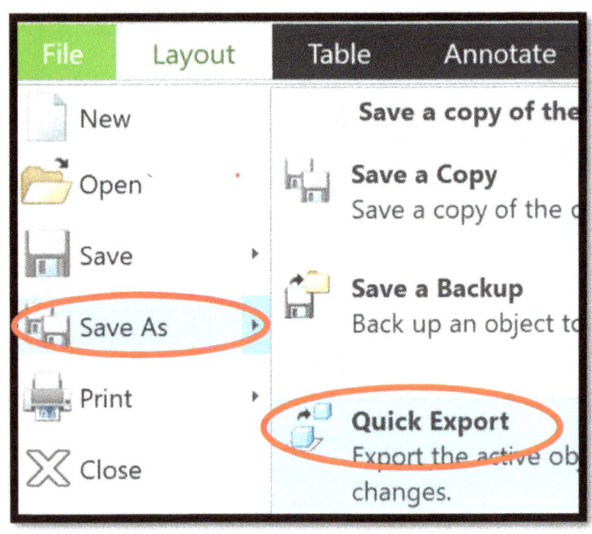

- **Select File** from the top toolbar - <u>**hover**</u> your mouse pointer over the **"Save As" button** – select **"Quick Export (PDF)"** – a PDF of the drawing should open in Adobe Acrobat* (or similar) - **Save that PDF** to your working directory or other known location.

Possible PDF Error Notice: If you get an error in CREO stating that *"the PDF could not be created"* you can instead choose **File –** <u>hover</u> over **Save As** with your mouse – choose <u>**Export**</u> – select the **PDF option** – click **Export.**

Printing a Drawing? If you try to physically print the drawing directly from CREO it will only print a small portion of the drawing due to the C size format (17" x22"). Instead of printing from CREO, instead create the PDF first and then print that so it resizes itself.

Step 25 –**Submit your deliverable for grading** according the instructions on the assignment in Blackboard. Note that you will need to **submit a PDF of the drawing**, not the .DRW file itself. Only PDFs are suitable for grading as it captures exactly how the drawing is setup on your CREO, where as a .DRW file will depend on the user settings that opens it.

==*If you do not submit a PDF it cannot be graded!*==

End of Lesson 2

Lesson 3 – Revolves & Holes (Ram Mount Parts)

Lesson 3 is to create components of a 1" ball "Ram Mount" type system used to carry phones, GPS, or other devices on a handlebar. This lesson will utilize **Revolves, Holes, Extrudes, and Rounds**. The revolve tool is similar to an extrusion but with the addition of "revolved" dimensions and **Axis of Revolutions**. The Hole tool will be used to create holes suitable for fastening parts together and will utilize counterbores and countersinks to allow bolts to sit flush in the surface of the model.

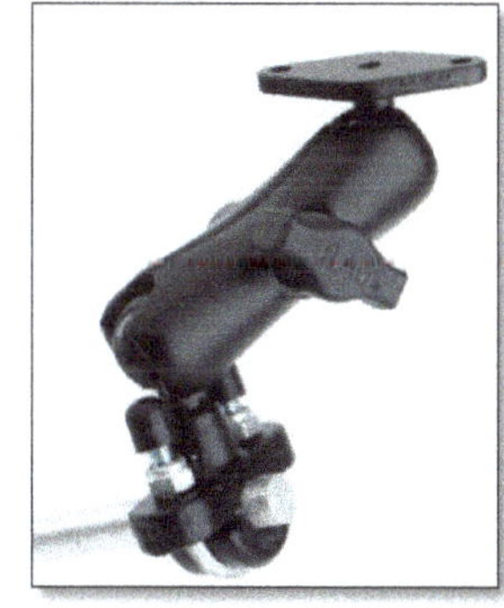

 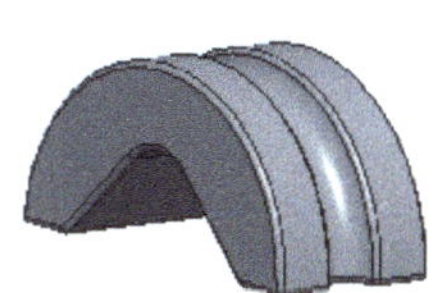 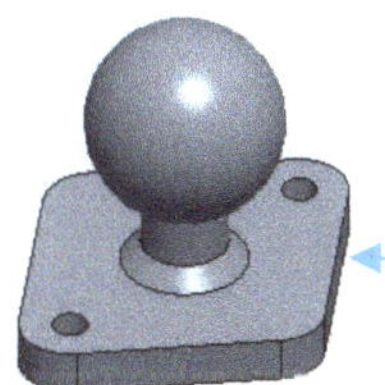

This 3rd part is Optional and not graded!

Axis of Revolution:

When using the Revolve tool the sketch requires an **Axis of Revolution**. This is a specific Axis that has been designated to revolve the sketch about to create the 3D volume. The closed-loop sketch **must not cross** over this Axis of Revolution!

Axis of Revolution: An **Axis of Revolution** must be created while in sketch mode for Revolved features. Without one the feature cannot be dimensioned or completed properly.

Steps: Hold RMB in blank area of the sketch to bring up the **Sketch Tools** menu – select the <u>Axis of Revolution icon</u> in the pop-up menu – **LMB to place the Axis of Revolution.**

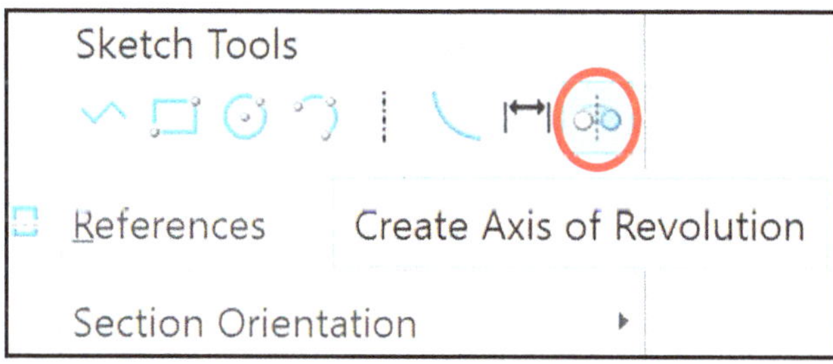

Revolved Dimensions:

Revolving Features will often require the use of a diameter or "revolved" dimensions that set the dimension measurement to the opposite side of the part once it has been revolved about the axis.

To Create a Revolved Diameter Dimension:
Use the Dimension Tool:

1- LMB the Axis of Revolution *(click above/below the sketch lines to ensure you are clicking on the <u>Axis of Revolution</u>).*
2- LMB to select the sketch point or line to dimension from.
3- LMB on the Axis of Revolution again.
4- MMB to place the "revolved diameter" dimension value.

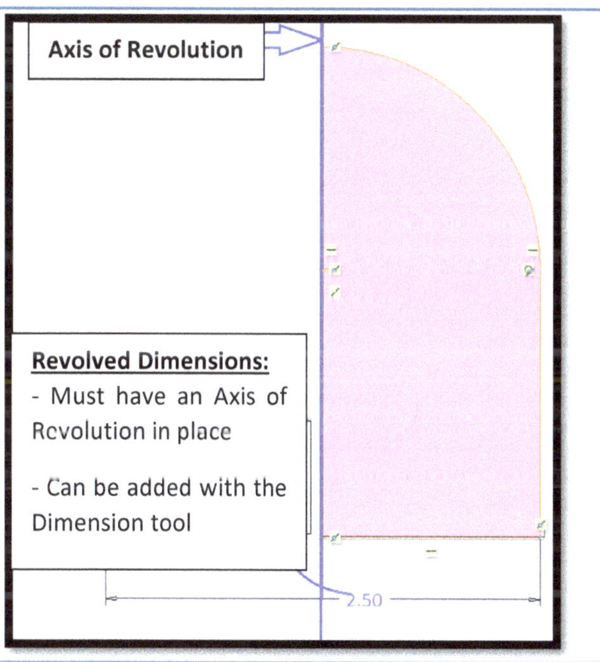

Part #1 - Ram Handlebar Mount

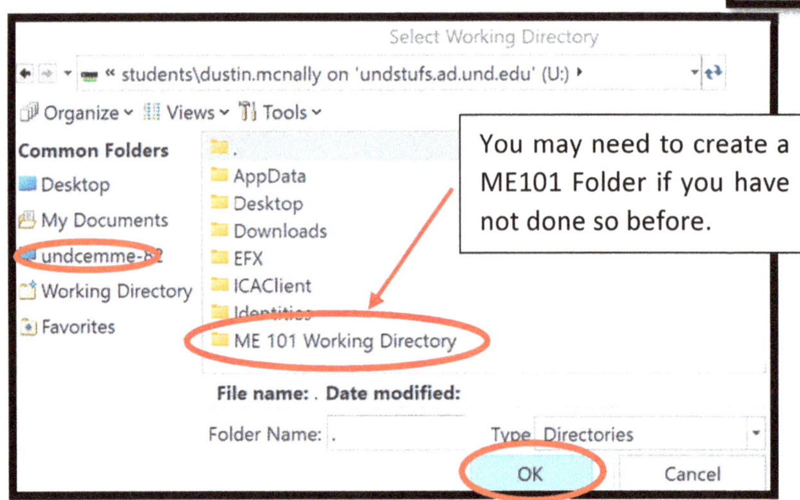

Getting Started: Open the **CREO** *Parametric* software and follow the lesson steps carefully.

Step 1 - Set your working directory to your ME101 Working Directory Folder.

- Select **File – Manage Session – Set Working Directory – Navigate to your ME101 Directory folder – OK**

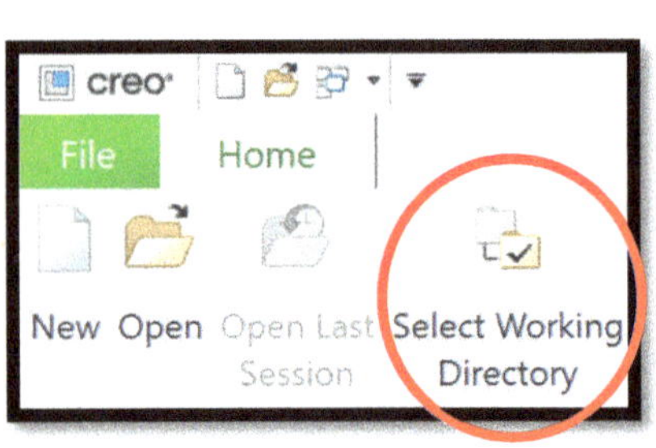

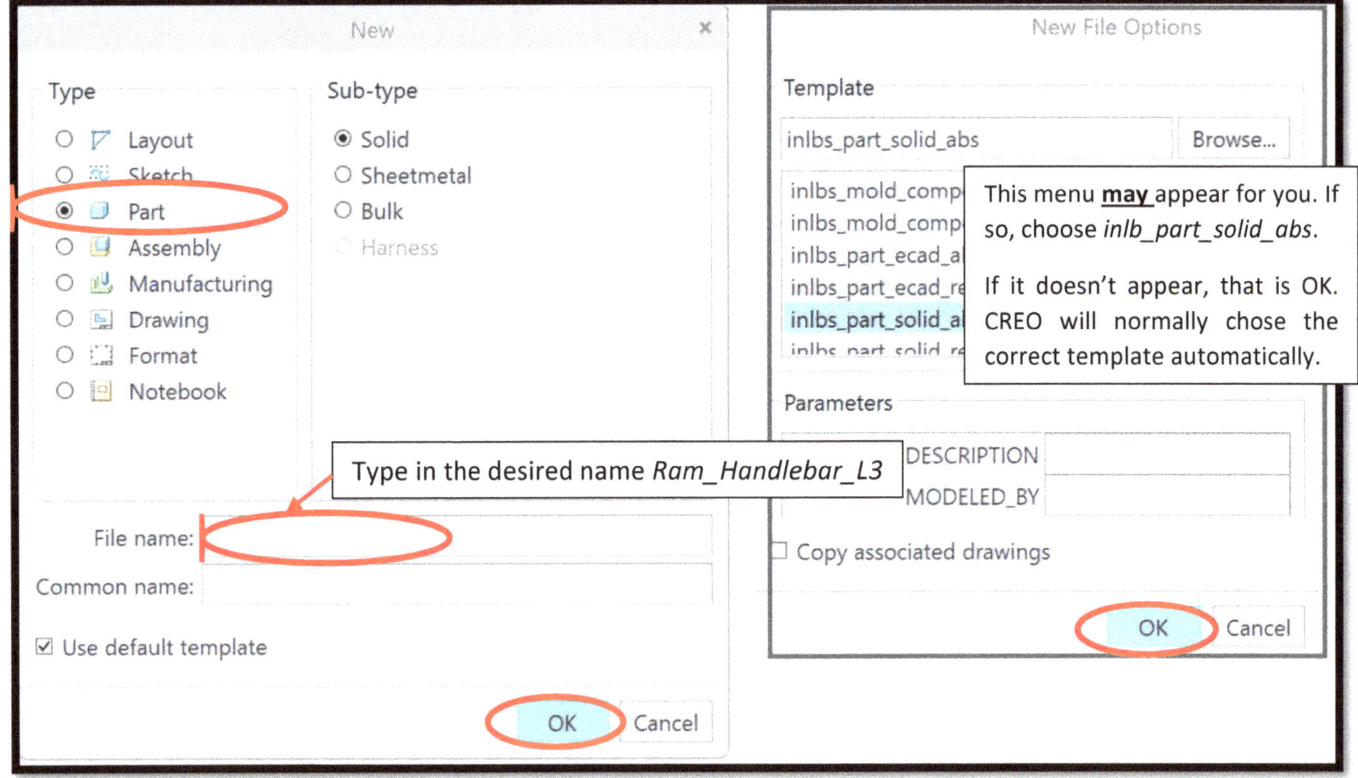

Step 2 - Create a new **Part** called "**Ram_Handlebar_L3**". *You cannot use spaces in a filename!*

- **File - New – select Part as the file type – type in the "Ram_Handlebar" - press OK.**

A pop-up menu may appear asking you to choose a **"Template"** – if so, choose **"inlbs_part_solid_abs"** - **press OK.** *The template menu may not appear in the student home version as it defaults to the desired inlbs_part_solid_abs already without needing to be selected.*

Step 3 - Change the Material of the part to **AL6061** *(Look under Legacy Materials)*, and be sure your units are in the CREO Default (should be inch-lbm-Second).

- **File – Prepare - Model Properties -** in the Material option select **change** – select the Legacy Materials folder - **double click** on "**AL6061.mtl**" – press **OK**

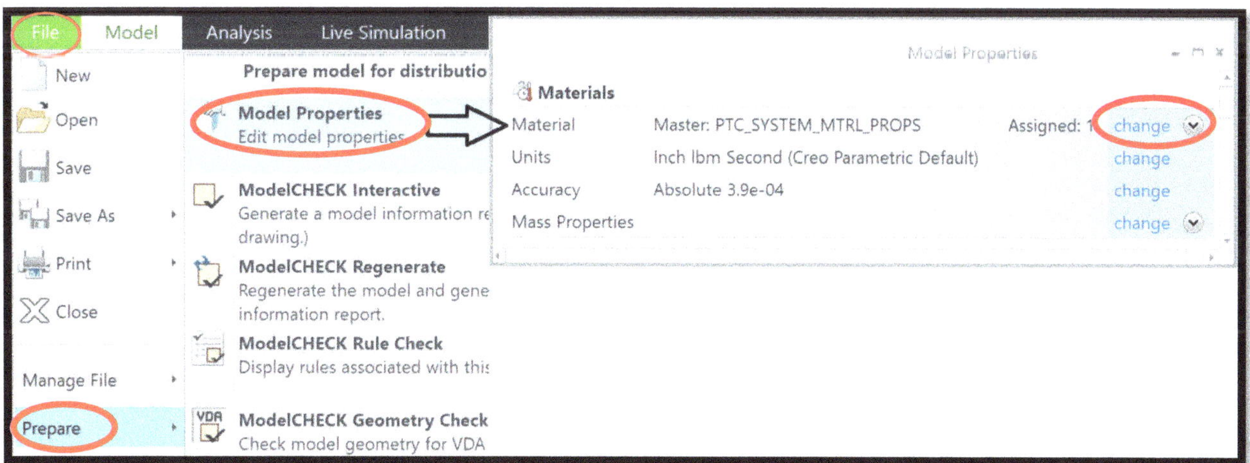

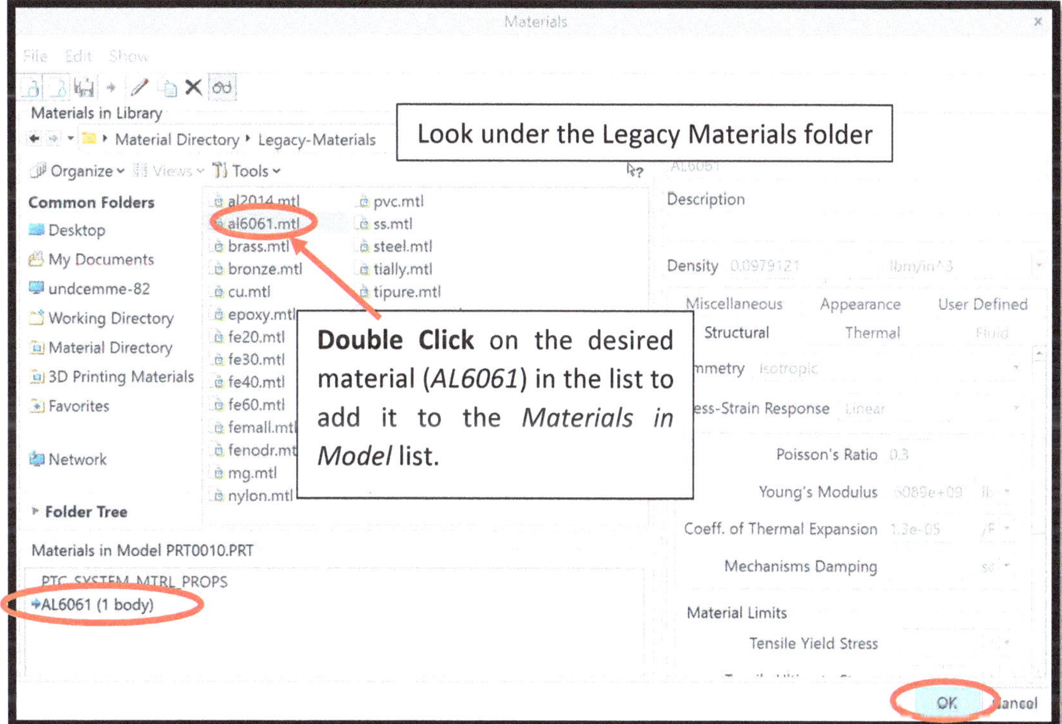

- Press **Close** to exit the Model properties menu.

Step 4 – Adjust the file options so that **3 decimal places** will be shown in sketching mode.

- **File – Options – Sketcher** – set the "**Number of decimal places for dimensions**" to **3** – OK- Press **No*** in the pop-up menu concerning saving the changes to a configuration file.

- **Don't bother saving a Configuration File, as you would need to load it each session to be useful.*

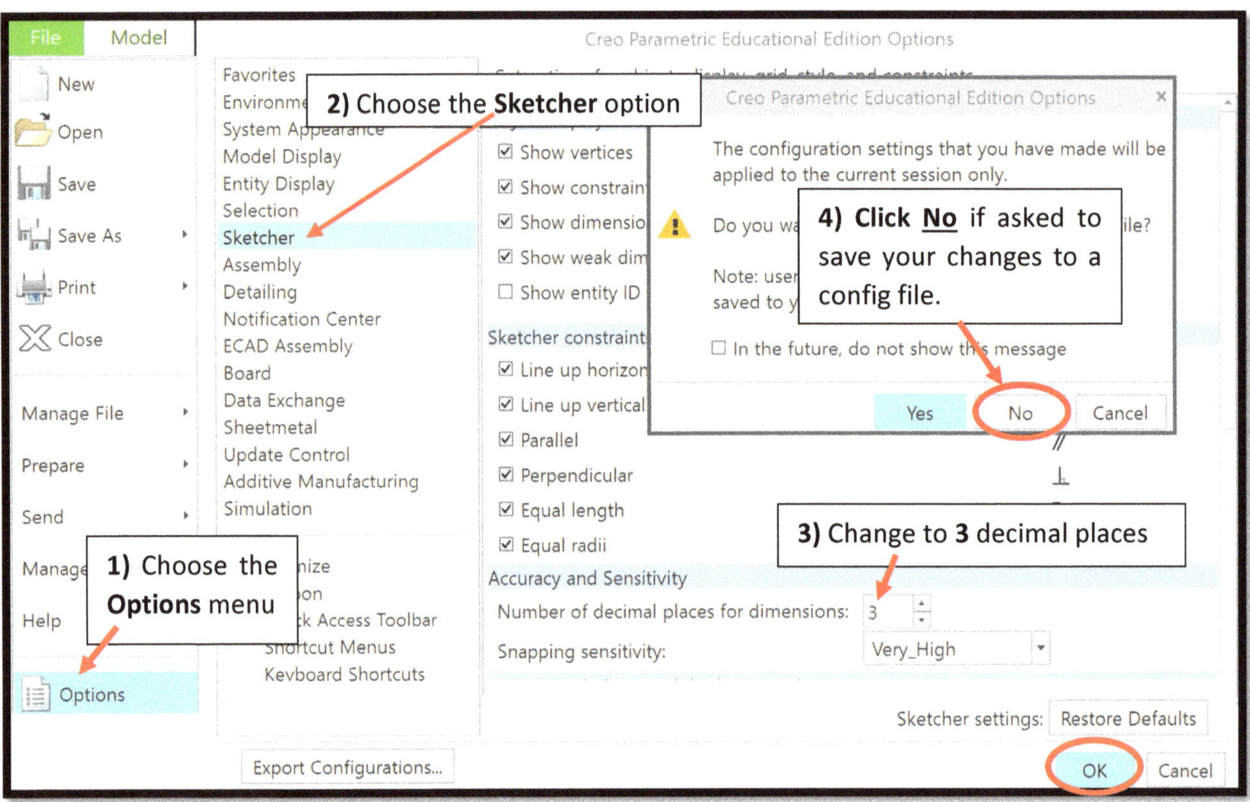

Step 5 - 1 - Create the first Revolve using the Revolve Tool.

Revolve - Click on the **Revolve Tool** in the top toolbar - select the Placement tab – select **Define** - Select the **Front** datum as the Sketch Plane either by clicking the Front Datum in the Model Tree or in the Graphics Window. After selecting **Front** as the sketch plane, the default Reference & Orientation will automatically be added. ***If you choose the wrong datum as the sketch plane you should cancel and redefine the placement***

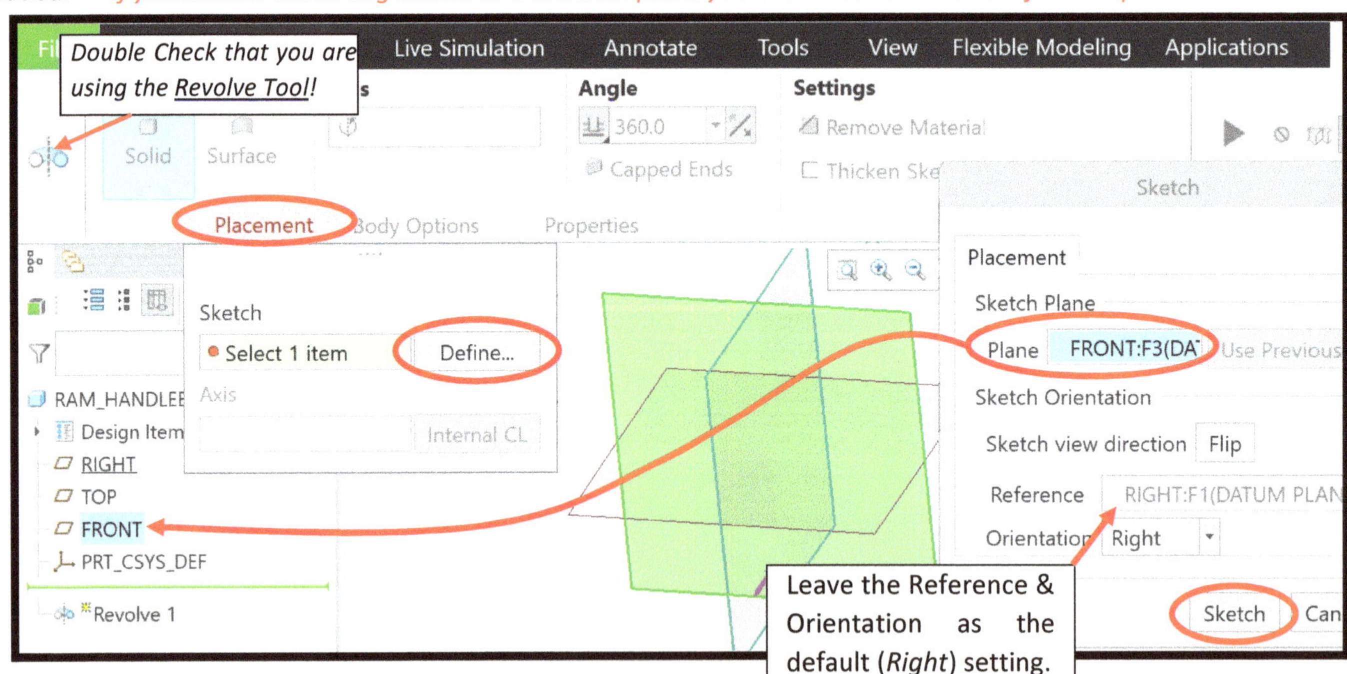

Step 5 – 2 – Press Sketch to enter sketch mode, then select the 2D **Sketch View** icon from the **Quick Toolbar.**

Step 5 - 3 – Place a vertical **Axis of Revolution** that runs through the origin. Then **create the closed-loop shape** shown below, starting with the Arc at the top of the sketch. *Disregard the dimension values in this step.*

1) <u>**Axis of Revolution:**</u> **Hold RMB in a blank area** of the screen – select the **Axis of Revolution tool*** from the pop-up menu (it will either say *Axis of Revolution* or have an icon that looks like a Revolve)- **LMB to place the Axis of Revolution** just like you would a centerline. **Note that you can also place a regular centerline - LMB to select it - hold RMB - select "Designate as Axis of revolution" to set it as the Axis of Revolution.*

2) <u>**Center & Ends Arc:**</u> Click the **drop-down arrow of the Arc Tool** and select the **Center & Ends** option. **LMB click to set the center point** of the first arc to be on the Axis of Revolution – **LMB to select the beginning of the arc** to be on the Axis of Revolution but above the center point – **LMB to select the end of the arc** so that the arc curve is goes past 90 degrees but ends before 180 degrees.

3) <u>**Use the Line Tool to create the rest of the closed-loop**</u>: Note that the right line should be angled and the left line should run all the way from the very bottom of the sketch to the very top of the arc**. Also make sure this sketch is located at the Origin, snapped onto the vertical and horizontal datum reference lines as shown below.

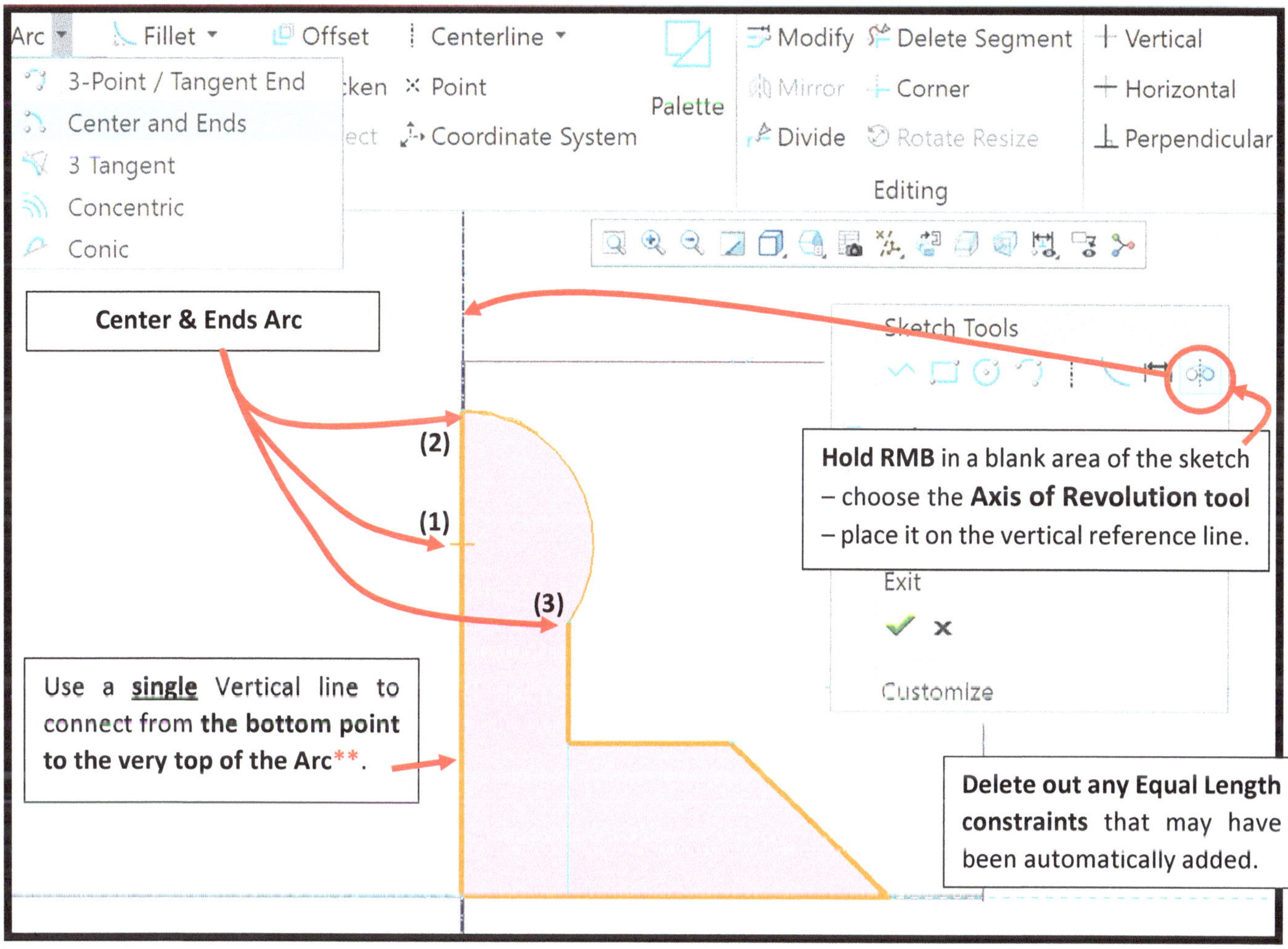

**If the Axis of Revolution Tool will not appear in the Pop-up menu, check that you are creating a Revolve, and not an Extrude Feature!*

***Do not use two lines stopping at the center point of the arc, this can cause errors!*

Step 5 - 4 – Dimension the sketch as shown using "revolved" dimensions for the widths and a Diameter dimension for the arc instead of a Radius.

Note: Do not try to specify the values of the dimensions as you go "one-by-one"; wait until you have placed all of the proper dimensions without setting the numeric values. Then use the **Modify tool** to set the correct values with **regenerate toggled off** to force an update all at once.

1) **Dimension the Arc Diameter:** - Dimension Tool – <u>double click</u> **LMB on the curved line** of the arc- **press MMB** to place the dimension as a Diameter (*a single click creates a Radius, you want a Diameter!*).
2) **Dimension the Revolved Dimensions:** - Dimension Tool – **LMB on the Axis of Revolution** (*click above/below the sketch so you don't click on a sketch line*) – **LMB on the line** or point to dimension to – **LMB again on the Axis of Revolution** – **MMB** to place the dimensions as shown below. *Note- if you do not have an Axis of Revolution this step will not work!*
3) **Dimension the two Height Dimensions:** - Dimension Tool – **LMB to select** the point or line to dimension from - **LMB** to select the point or line to dimension to – **MMB** to place the dimension.
4) **Use Modify (with Regenerate unchecked) to set the dimension values:**
- Use the **Select tool** to make a drag box to select all of the dimensions – click the **Modify Tool** at the top toolbar – check that **all dimensions** are added to the modify box – **toggle off Regenerate** in that menu – **click in the value box** of each dimension and set it to the correct value (the currently selected dimension will highlight on the screen) – **press OK in the modify menu** to update all the values at once*.

Note – if the dimensions will not change or update: You probably have an error in the value, are modifying the wrong dimension for the given value, or the dimension is referencing the wrong entities. A common mistake is to have a Radius (single arrow) instead of a Diameter (double arrows), or the height is not dimensioned to the center of the arc as it should be. Make sure your dimensions are measuring the same entities as shown below.*

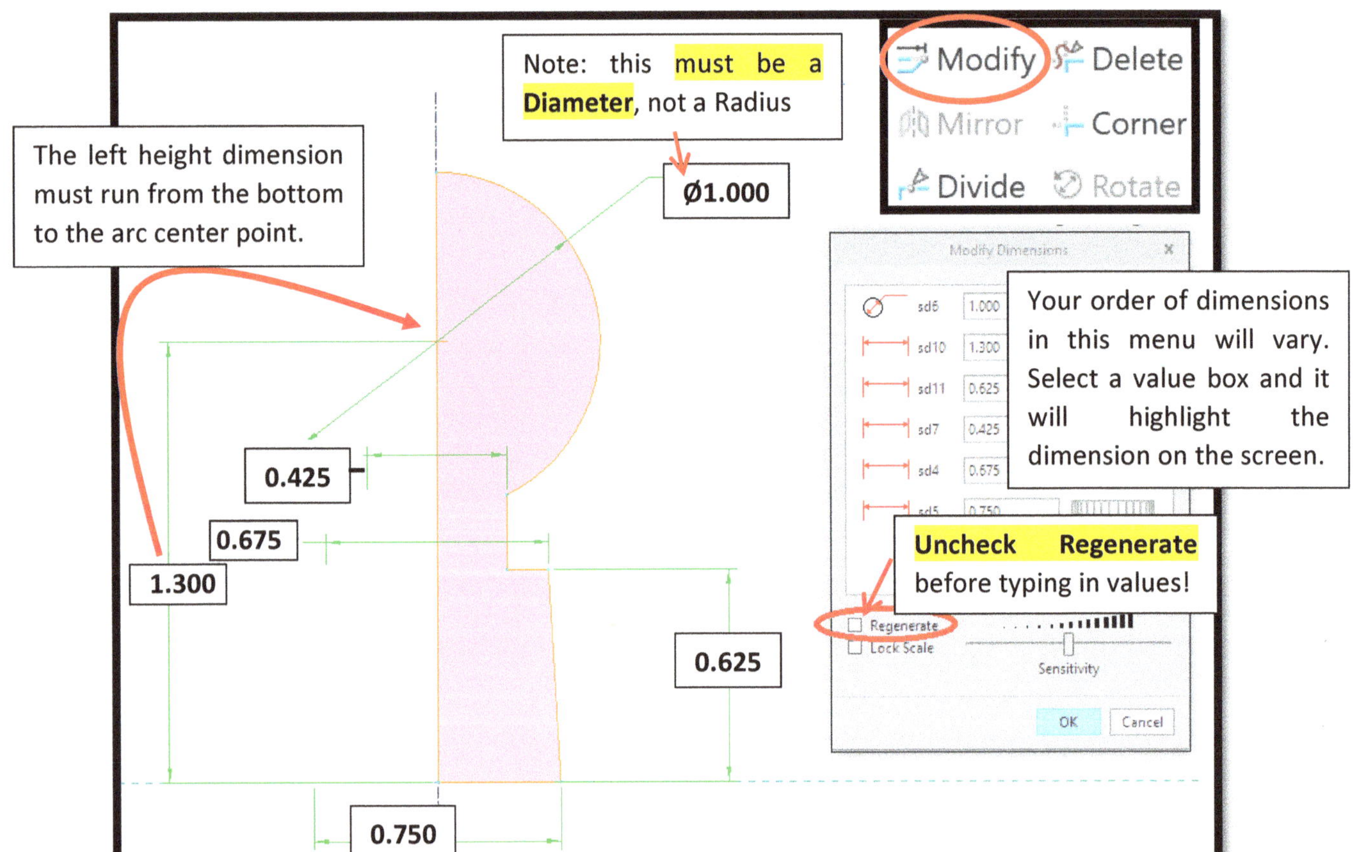

Tip: Double Check that your dimensions measure to the same points as above, as well as make sure to type in the correct numbers for the corresponding dimensions. If the math doesn't work out Modify will not accept it.

Note: **After pressing OK the sketch may become very small on the screen**. This can happen when modifying dimensions to small values, as the dimension value boxes were placed around a large initial sketch size (hundreds of inches). When the sketch is resized to be smaller the dimensions boxes stay in their current position far away from the sketch. Refit to Screen will resize the view around all entities and dimensions, so you need to manually move them closer to the sketch origin in order to refit to screen properly.

-       Use the **Select/Pointer Tool** to click & hold to drag a dimension value box closer to the sketch on each dimension. Then use the **Refit to Screen tool** from the quick toolbar to resize and zoom in on the sketch. **Repeat as needed**.

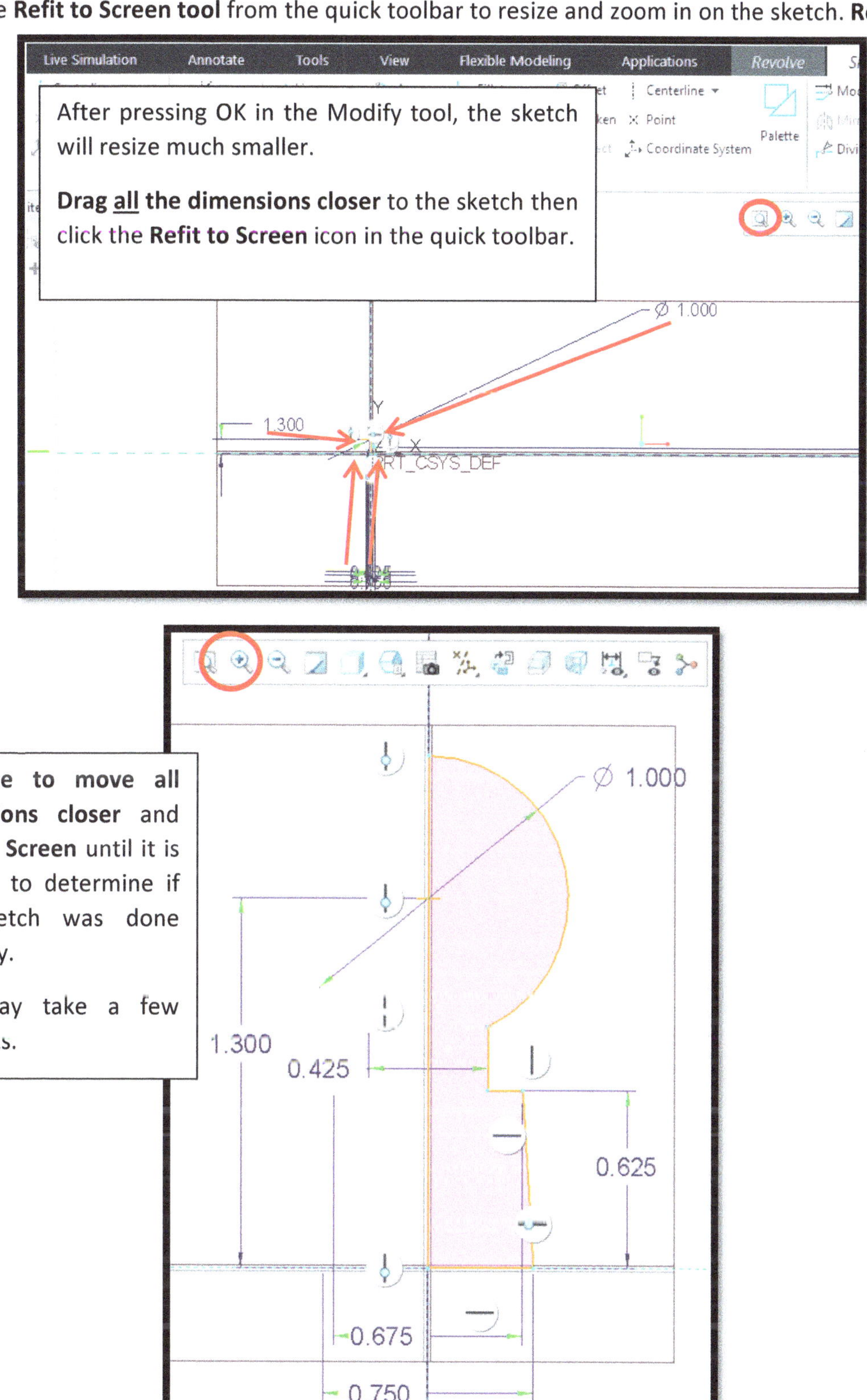

Step 5 - 5 - Press the **Green Checkmark** to accept the sketch. The solid model should default the revolve option to 360 degrees and you should see the Revolved solid material on the screen*. **Press the Green Checkmark to accept it.** Use Refit to Screen if it moves offscreen.

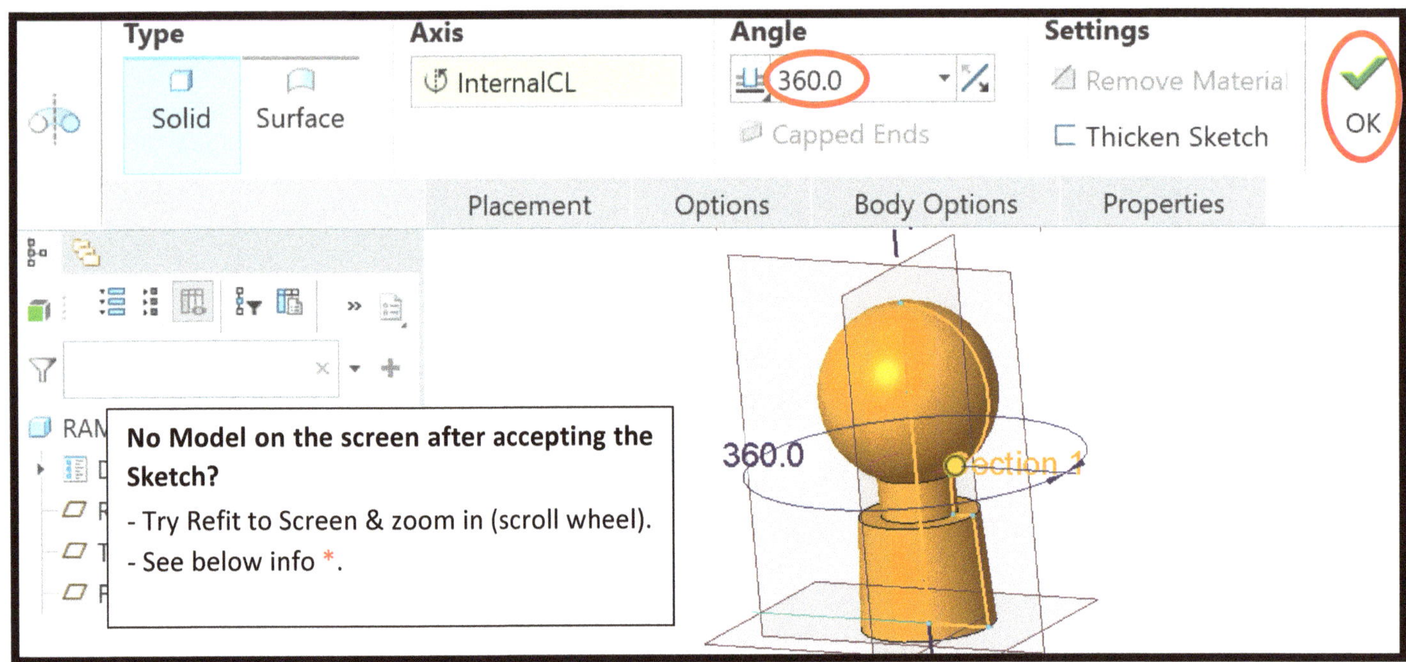

**If your Revolve is not visible on the screen, it is likely that the Sketch is missing an Axis of Revolution (and missing revolved dimensions). Use Placement - Edit to return to sketch mode and fix the sketch to match the previous steps.*

Step 6 - 1 – Create an **Extrude** to add the material on the bottom of the Revolve using the **Top Datum** as the sketch plane (*which should be located on the bottom of the part*).

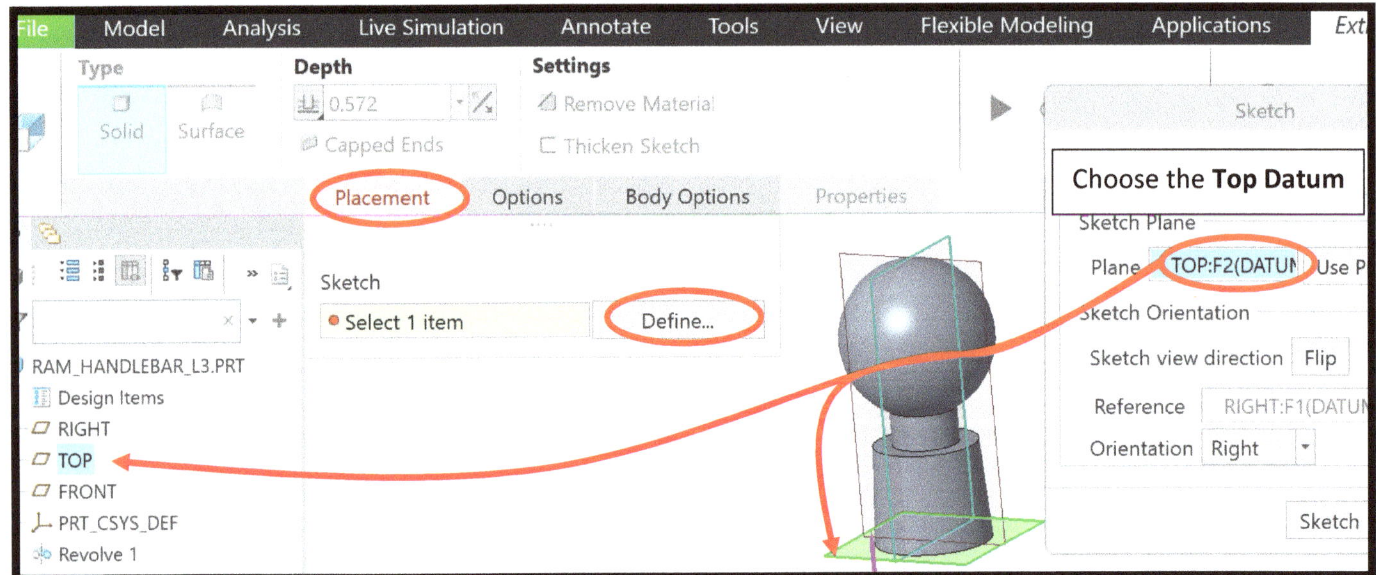

- Click on the **Extrude Tool** in the top toolbar - select the **Placement** tab – select **Define**- select the **Top Datum** and leave the **Reference & Orientation** as default (*Right & Right*).

Step 6 - 2 - Press **Sketch** to enter sketch mode for this extrusion. Then select the **Sketch View** icon from the **Quick Toolbar** to view the sketch plane perpendicular to the screen.

Step 6 - 3 - Sketch the closed loop shape:

1) <u>**Add Centerlines:**</u> Use the **Centerline tool** to place vertical and horizontal centerlines through the origin.

2) <u>**Sketch the Rectangle:**</u> Use the **Rectangle Tool** to sketch the shape that crosses over the origin.

3) <u>**Symmetry Constraints**</u>: Use the **Symmetry Constraint tool** about both the vertical and horizontal centerlines. CREO may have already assumed symmetry about either or both centerlines, so add as needed.

4) <u>**Center & Ends Arc**</u>: Click the **drop-down arrow of the Arc Tool** and select the <u>**Center & Ends**</u> Arc option.

- **LMB click to set the center point** of the first arc to be on middle of the left edge of the rectangle, snapped onto both the vertical sketch line and the horizontal centerline - **LMB to select the beginning of the Arc on the top corner** of the rectangle **– LMB to place the end point** of the arc **leaving it short of the rectangle. Repeat** on the opposite side of the rectangle.

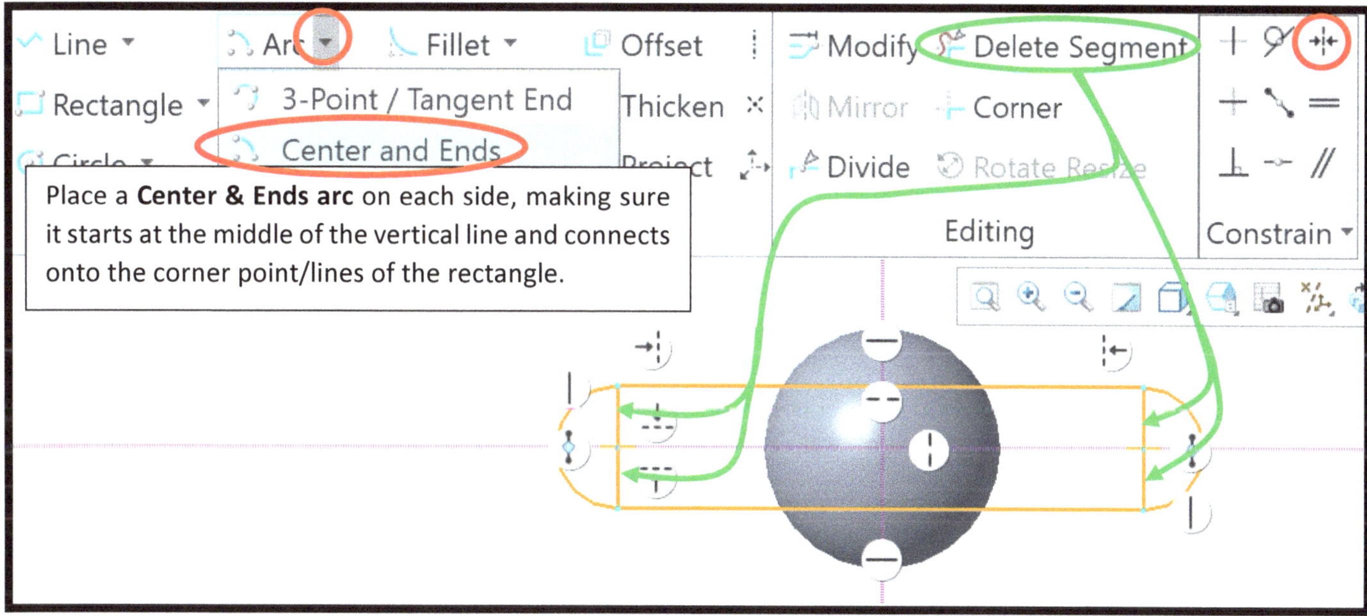

5) **Use the Delete Segment tool** to remove the rectangle vertical lines to create the closed-loop. If it isn't a closed loop, likely the arc is running slightly long/short and will have a Red Point signaling an open loop.

6) **Use the Dimension Tool** to create the dimensions for the height of the rectangle (**0.875"**) and distance between the arcs (**2.330"**) using the outer arc curves as dimension references.

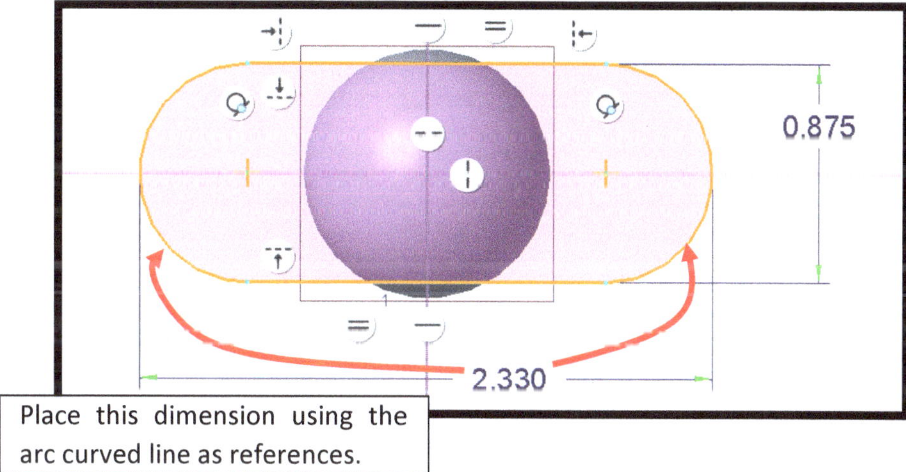

Step 6 - 4 – Press the **checkmark** to accept the sketch.

Step 6 - 5 – Set the **Depth** of the Extrude to **0.500"** and flip the **Direction arrow** to extrude material <u>away</u> from the bottom of the Revolve.

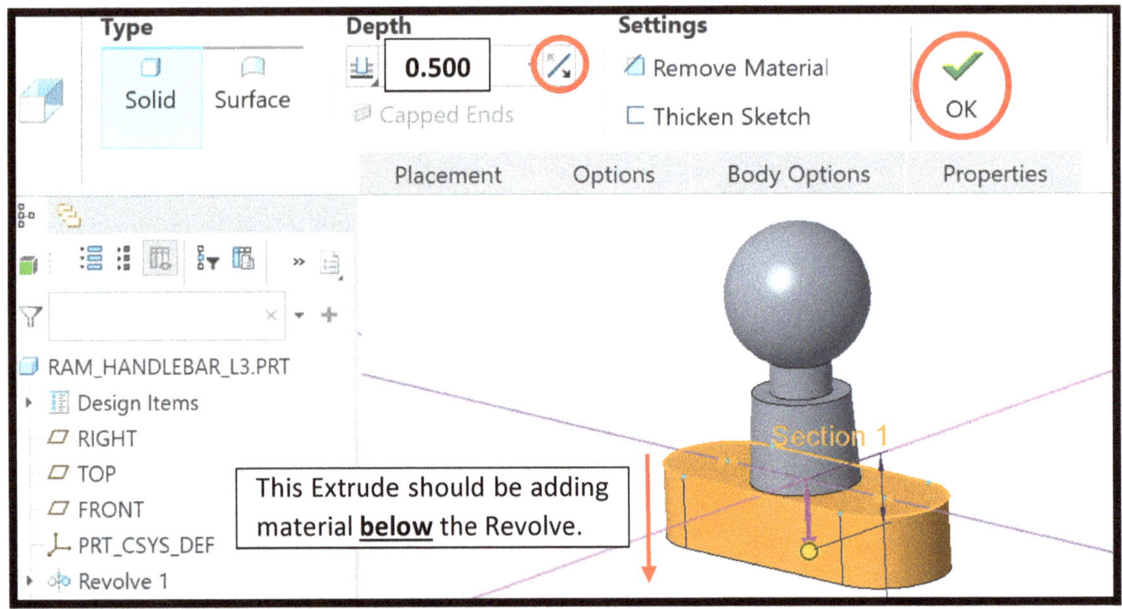

Double Check: Make sure your Extrude is **adding more Height to the model**, and not extruding up into Model.

Step 6 - 6 – Press the **Checkmark** to accept the Extrude. **Press Save** to save your part.   

Step 7 - 1 – Create **Extrude 2** to remove material to create a groove in the bottom of the part to fit against a handlebar. Use the **Front Datum** as the sketch plane (leave the orientation and reference as default.)

- Click on the **Extrude Tool** in the top toolbar - select the **Placement** tab – select **Define –** select the **Front datum** – leave the **Reference & Orientation** as default (*Right & Right*).

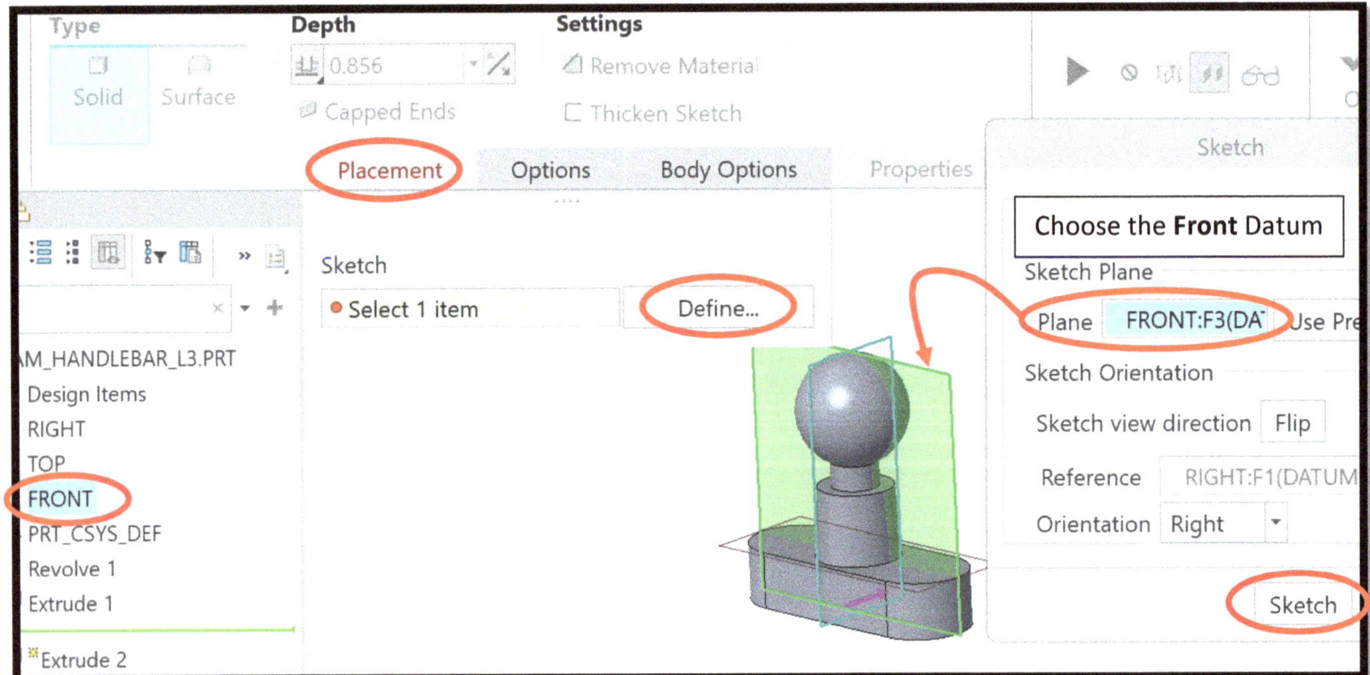

Step 7 -2 - Press **Sketch** to enter sketch mode for this extrusion. Then select the 2D **Sketch View** icon from the **Quick Toolbar** at the top of the Graphics Window.

Step 7 - 3 – Create the closed loop sketch shown below, symmetric about the vertical centerline.

1) **Add a Vertical Centerline:** Use the **Centerline tool** to place a vertical centerline that crosses through the origin to use for symmetry.

2) **Sketch the Triangle:** Use the **Line Tool** to sketch 3 lines of a triangle so that the bottom is "snapped" onto the bottom edge of the model and the peak is snapped on the centerline. *Don't snap onto any midpoints, etc.*

3) **Fillet Tool:** use the **Fillet Tool** – click the intersecting **lines** to add the fillet to the top of the triangle

4) **Symmetry Constraints***: Use the **Symmetry Constraint tool** to add symmetry to the Fillet "end" points.
   **If the symmetry constraints were added before the Fillet was created then you may need to remake them as the Fillet may remove the constraint references.*

5) **Dimension the Sketch:** Use the Dimension tool to create a height dimension from the **top of the fillet curve**** **to the bottom line (0.225")**, the radius of the **fillet (0.125")**, and the bottom **width of the triangle (0.875")**.

   *** You may get a Resolve Sketch notice; **if you have a Perpendicular Constraint** it can be deleted out to resolve it.*

6) **A Coincident Constraint May be needed:** If the **Fillet center point moves off center**, set a **coincident constraint** on the center point of the fillet to the vertical centerline.

Tip: Your constraints may vary on this sketch different than shown, as you may end up with symmetry on the bottom line, a coincident on the "center" point of the fillet, or some other combination of constraints that define the shape properly.

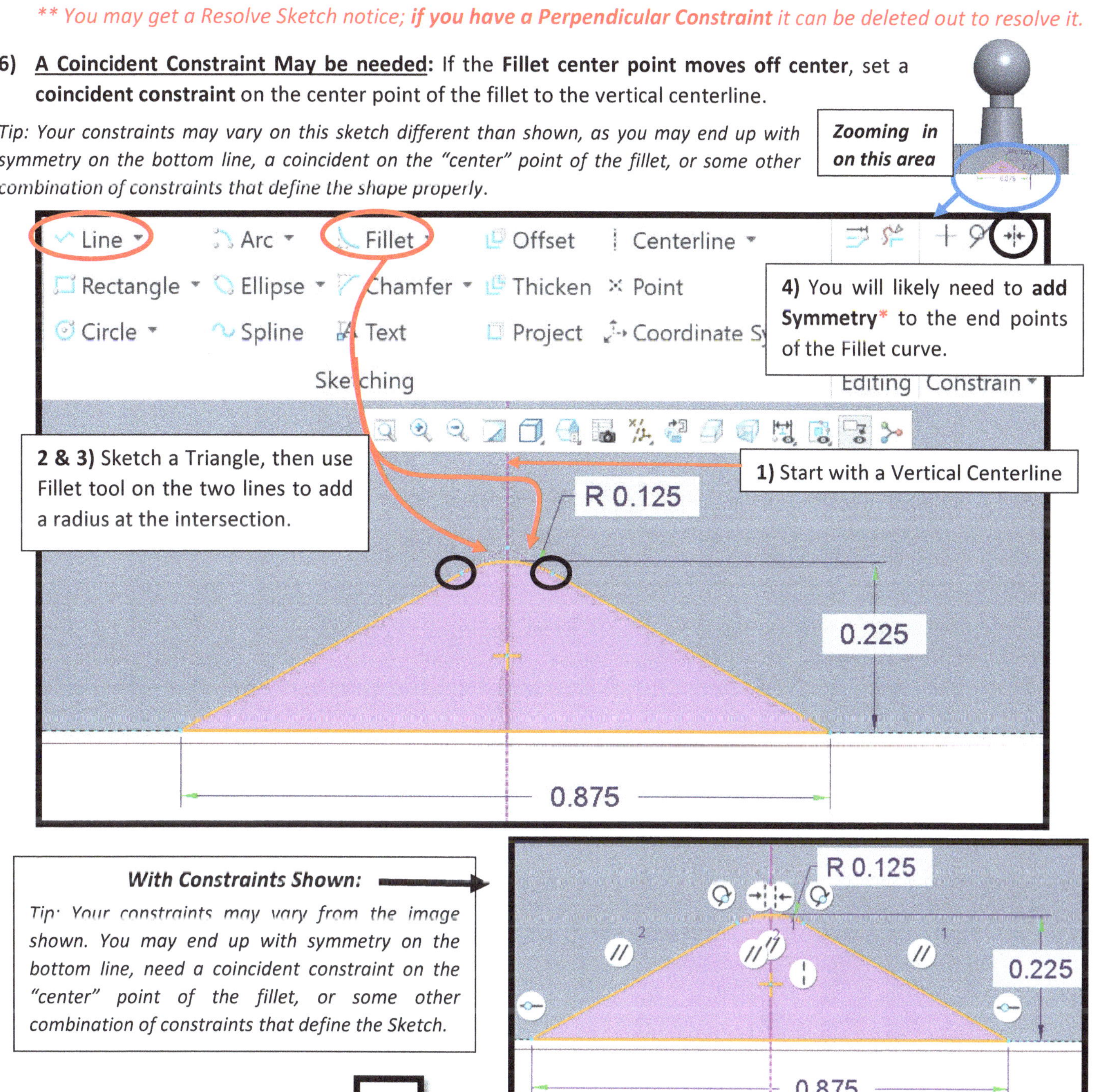

7) **Checkmark** to accept the sketch.

Step 7 – 4 - Set the **Depth, Direction, and Material Removal options** in the Toolbar to complete the Extrude options.

- Toggle on **Remove Material** (if it did not do so by default automatically)
- Select the **Depth option** to be "**Extrude Symmetrically on Both Sides**"
- Set the **Depth Value** to be greater than the width of the part (**10.000"**).

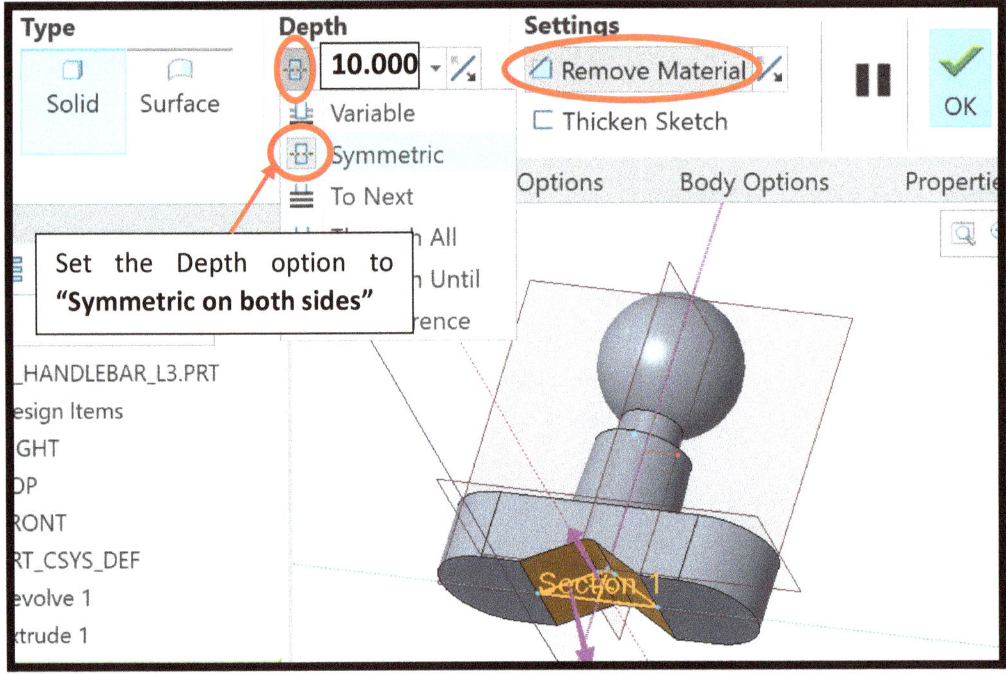

Step 7 - 5 – After assuring that the Extrude is defined correctly select the **Checkmark** to accept it.

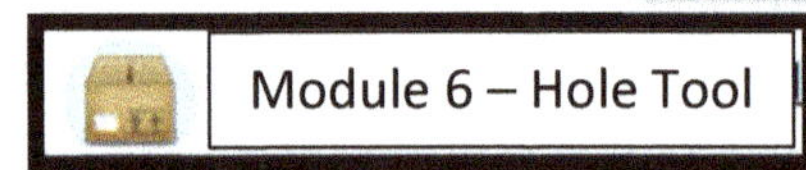

Step 8 – Go into **Blackboard** and complete **Module 6 (Hole Tool)** to learn about the Hole Types, Placement types, and options.

A Brief Review: Hole Placement Types

A hole can be placed in multiple ways (**Linear, Coaxial, and Radial/Diameter**). Below are some Rules of Thumb:

- If the Hole is located along an Axis (i.e. a hole in the center of a cylinder) you should use a **Coaxial** placement.

- If the Hole is located about an axis and you want to control the distance away and the angle about the axis, you should use a **Radial or Diameter** type placement.

- If the Hole is located some distance away from edges (i.e. at the corner of a plate) use a **Linear type** placement to set the X & Y position references and distance values.

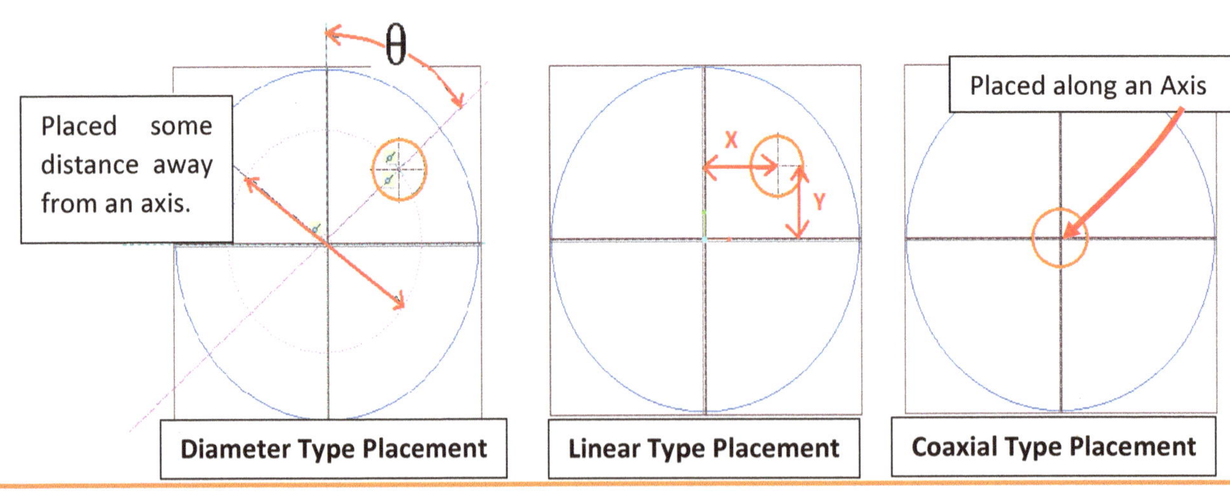

Step 9 - 1 – Use the Hole Tool from the top toolbar to create a Hole in the center of the part. Setup the hole with the settings shown below. *Note: you will need the **View Display** in the quick toolbar to have the "Axis Display" toggled on (should be on by default unless you accidently toggled it off)*

Hole Type: "**Simple**"

Profile Option: Leave as default for the Simple type hole ("***Flat***")

Toolbar Settings: Hole Diameter = **0.250**, Hole Depth = **1.250**

Placement Type: **Coaxial Type** *You must select the Axis <u>first</u> and <u>hold Control</u> on the keyboard to select the second reference otherwise it will just replace the first reference!*

Placement Commands: - Placement Tab – LMB **click on the <u>Axis</u>** (*from Revolve 1 in the center of the part*) – while **holding Ctrl** on the keyboard use **LMB** to select the bottom <u>flat surface </u>of the part as the start surface of the hole.

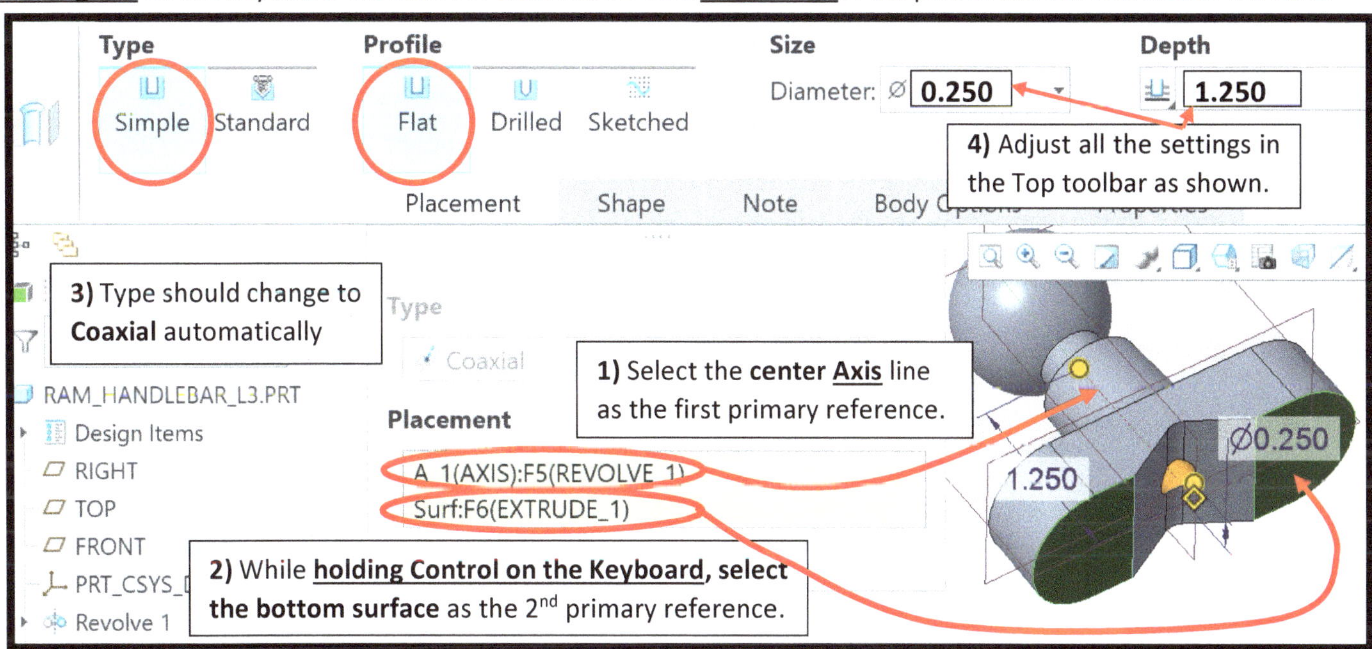

Tip: If you don't **choose the Axis first**, or if you accidently replace it by not **holding CTRL** when selecting the surface, it may not assume a Coaxial Placement. To fix it, RMB in the Placement box – Remove All and try again.

Step 9-2 – Press the green checkmark to accept the hole.

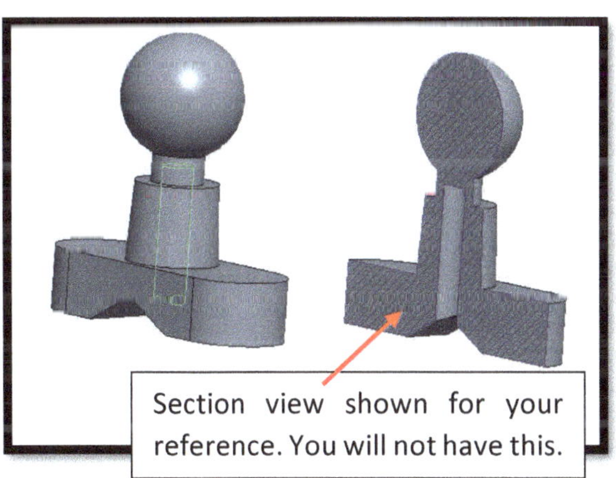

A sectioned view is shown at left for your reference.

The Hole outline (in green) will appear while it is selected in the Model tree which allows you to see the depth and diameter.

If your hole appears to go too far into the Ball, you likely have the wrong Hole Depth, or your Extrude 1 needs to have the Extrude Direction arrow toggled to add material below the Revolve (use Edit Definition on Extrude 1 if needed).

Step 10 - 1 – Use the Hole Tool from the top toolbar again to create a Hole on the top of extruded surface. Setup the hole with the settings shown below. This hole will be some distance away from the center axis so will need to utilize a "Diameter" type placement and have an angle of 0 degrees away from the Front datum.

*Note: you will need the **View Display** in the quick toolbar to show "Axis Display" (should be on by default)*

Hole Type:	Simple
Profile Options:	Toggle on **"Drilled"** and toggle on the "**Counterbore**" option
Toolbar Settings:	Hole Diameter = **0.250***
	Hole Depth = **Thru All** (*Drill to intersect all surfaces*)
Placement Type:	Diameter
Placement References:	*Primary Reference*: top surface of Extrude 1

   Offset References: Axis (of Revolve 1) value of **1.500"** & Front Datum value of **0.0** degrees (*hold Control on the keyboard* while selecting the second reference otherwise it will just replace the first reference!*)

Placement Commands: Placement Tab – click in the Placement box – **LMB** on the top surface of the extrude – change the "Type" to **Diameter – LMB** in the **Offset References** box **–LMB to select the Axis – LMB to select the Front** datum (while **holding Ctrl** on the keyboard) – **set the values to 1.500"** for the Axis reference and **0.0 degrees** for the Front Datum reference.

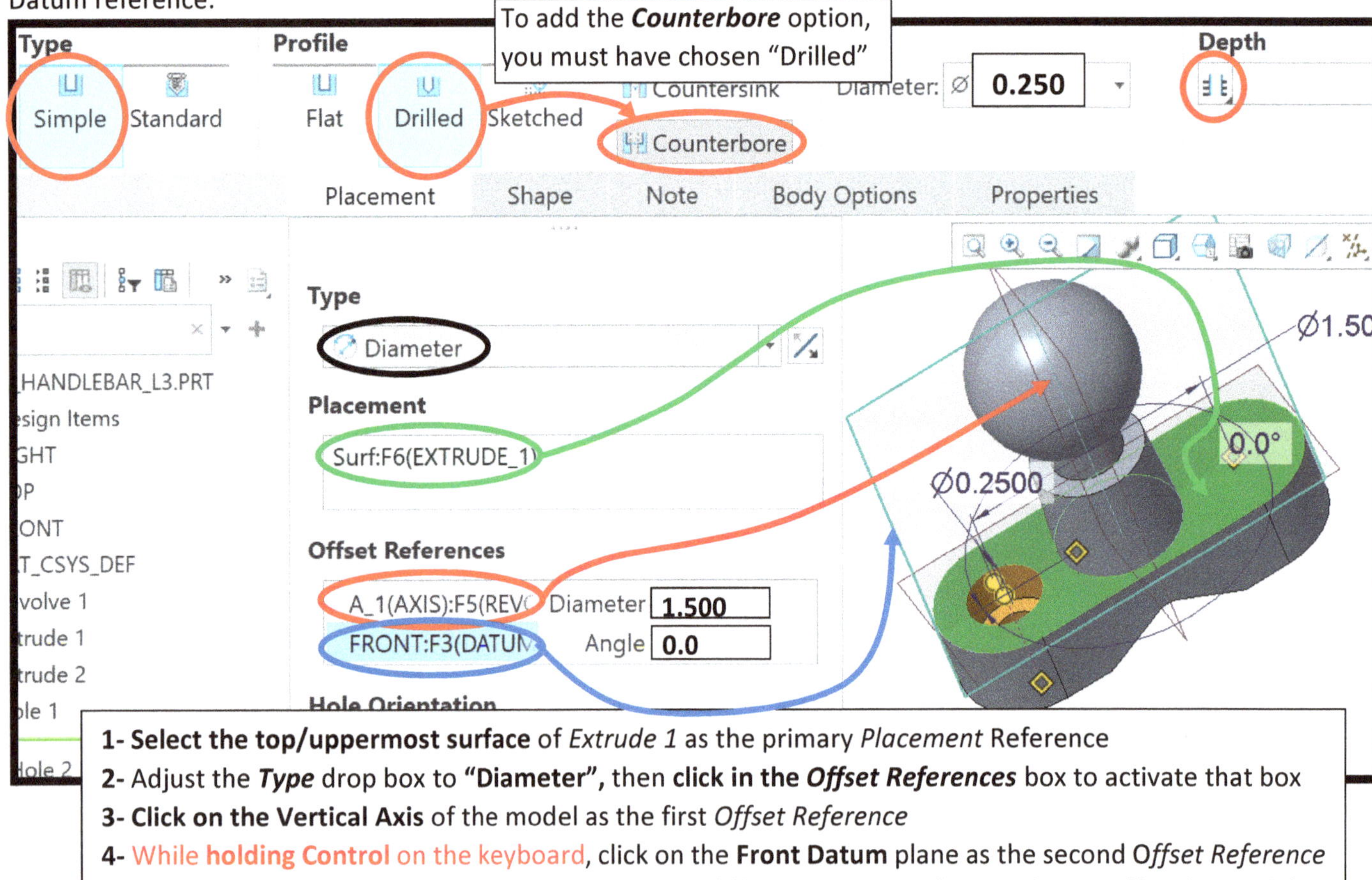

1- **Select the top/uppermost surface** of *Extrude 1* as the primary *Placement* Reference
2- Adjust the *Type* drop box to **"Diameter"**, then **click in the *Offset References* box** to activate that box
3- **Click on the Vertical Axis** of the model as the first *Offset Reference*
4- While **holding Control** on the keyboard, click on the **Front Datum** plane as the second *Offset Reference*
5 - Then adjust the *Offset Reference* values to 1.500" (for the Diameter) & 0.0 degrees (for the Angle)
6- Adjust the **Hole Depth** and **Diameter settings** if not already done.

***Error Notice:** If you get an error when trying to set the Hole Diameter, it is likely because you are trying to make the hole larger than the counterbore, which is not possible. If so, make sure you are typing in the correct value, or refer to the next page to set the Counterbore size before changing the Hole Size.*

Continue to the next page to adjust the Shape tab options of the hole tool.

Step 10 - 2 - Adjust the Shape Tab settings to set the Counterbore size. You must have toggled on the **Counterbore** option in order to see these options in the Shape tab.

- Select the **Shape Tab**: set the **Counterbore depth to 0.250"** and the **Counterbore diameter to 0.4375"** as shown in the image below.

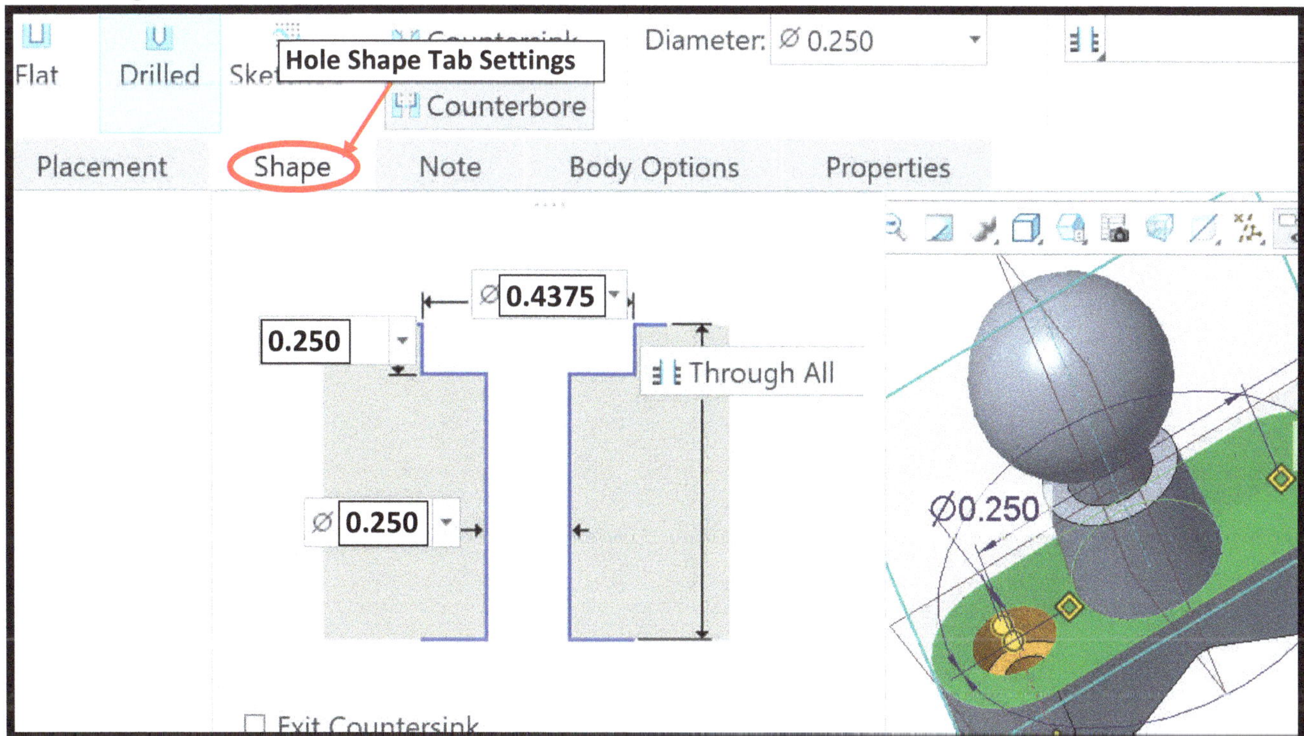

Step 10 - 3 – Press the Checkmark to accept the Hole.   

Step 11 – Use the Mirror Tool to copy **Hole 2** across the **Right datum** to the other side.

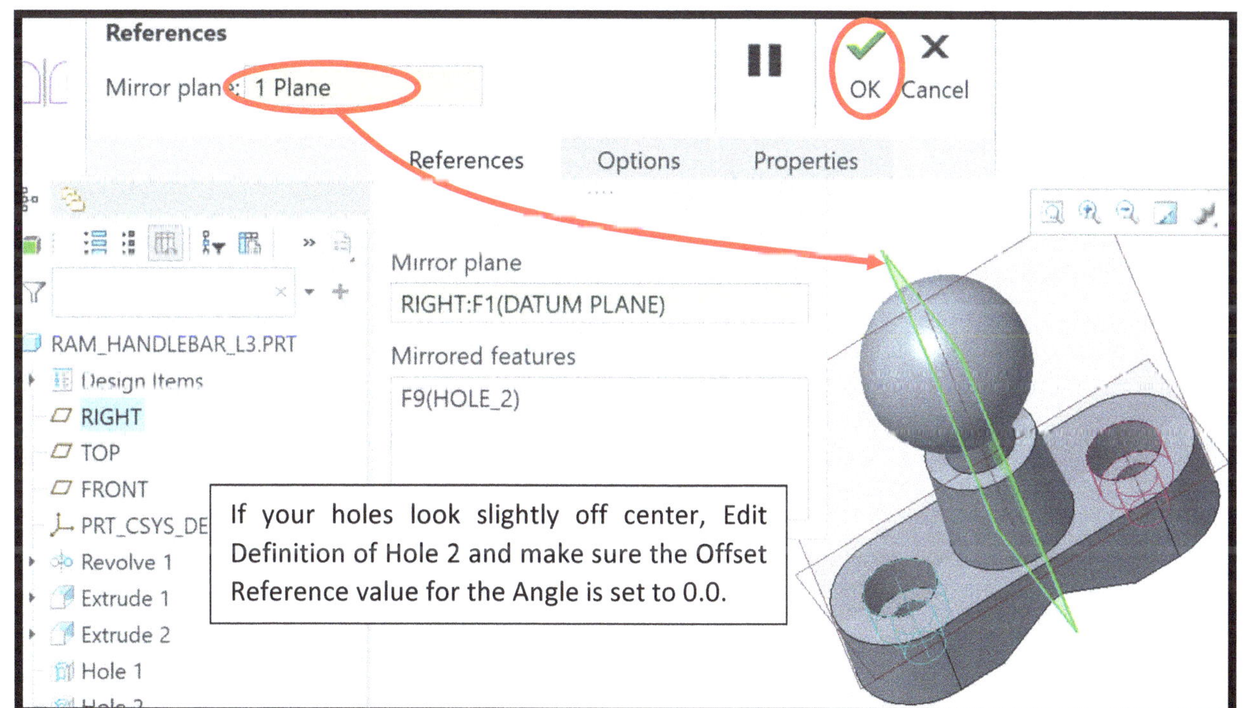

- **LMB** to select *Hole 2* from the model tree – select the **Mirror Tool** – select the **Right datum** – **checkmark** to accept the Mirror Feature.

Step 12 – Set the display properties of <u>each datum</u> so that we may use them as a **Geometric Tolerance** and be able to change them to **ASME style** in our drawings. We must "<u>**Set**</u>" & <u>**Rename**</u> each datum to match the style of this course using letters for names of the Datums.

- **LMB to select a Datum** from the model tree - **hold RMB – Properties** - press SET ("-A-") – select the **name box** – **type in a new name** according to the list below – **OK**.
- **Repeat for each Datum**.

Right Datum: Rename to **B** - Press "Set" - OK
Top Datum: Rename to **C** - Press "Set" - OK
Front Datum: Rename to **A** - Press "Set" – OK

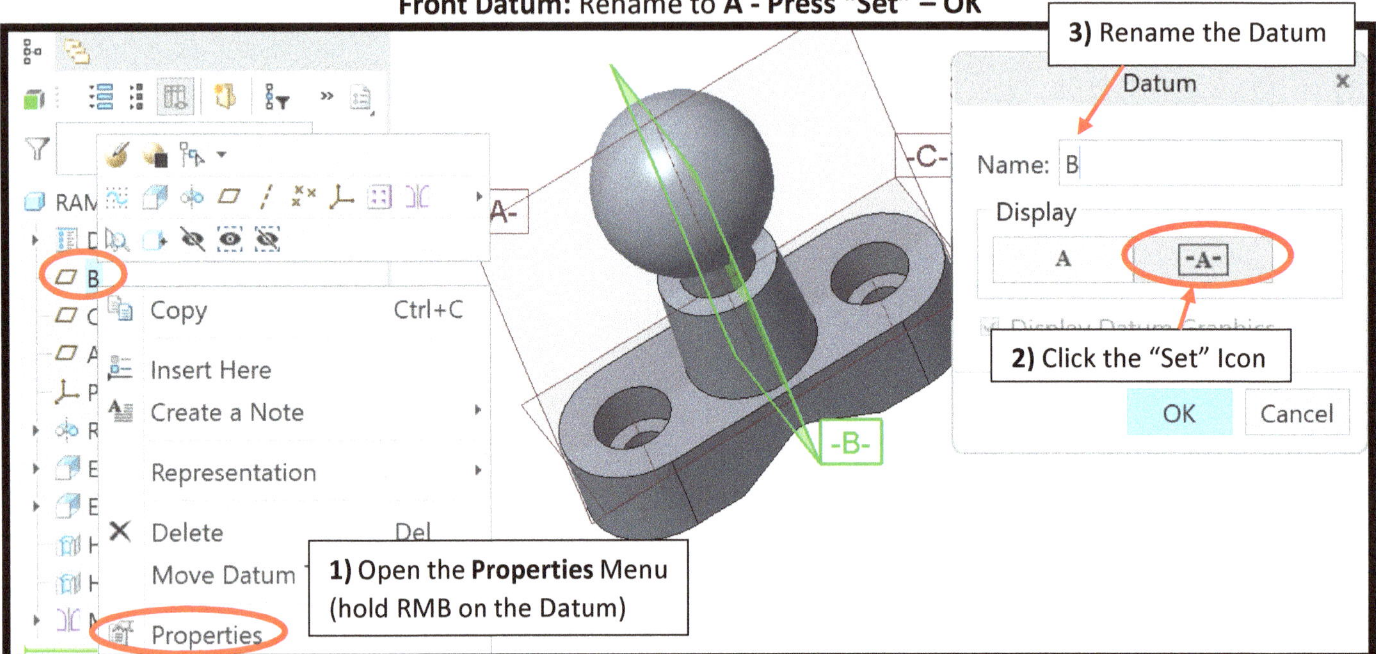

After setting and renaming the datums you should see the datum names or "tags" on the model screen, even if the Display Filters are toggled off. *If you do not see all 3 datum names, you likely did not press "Set" for that datum*.

Step 13 – Add a **Round** to the Revolve 1 edges. Round

- Use the **Round Tool** – set the **radius value to 0.050"** - Open the **Sets tab** in the toolbar – use the **LMB** to select the **3 edges*** of the revolve.

*Select the **edge lines**, not surface lines!*

Note - if you need to remove or edit one of the Rounds you can see a list of all of them under the "Sets" option of the Round toolbar and can then select them to edit the values or delete them if needed.

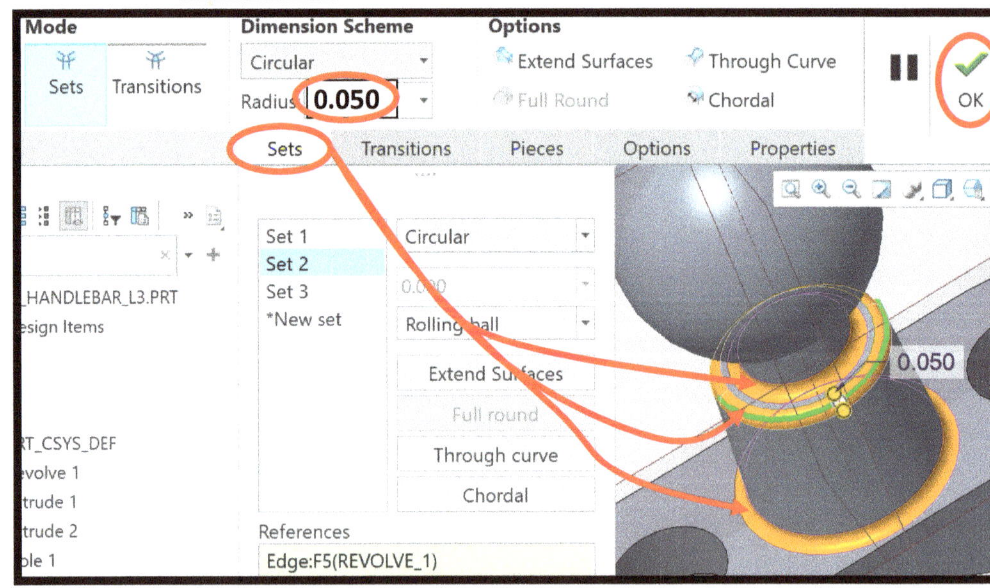

Step 13 - 2 - Press the **green checkmark** to accept the rounds.

Step 14 – Create a Cross-Section of your model to utilize in the drawing.

- **Select the "View Tab"** in the very top toolbar – **Section Tool** – Planar Option - in the **Reference tab** choose the datum to section on (use **Datum A**) – in the **Properties Tab** rename the section as "A" – **press the checkmark** to accept it.

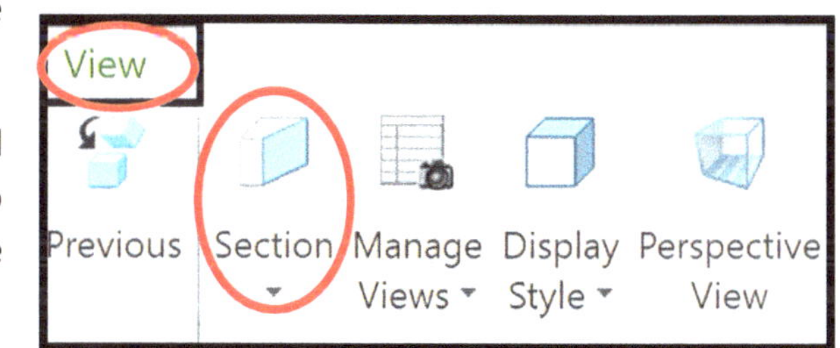

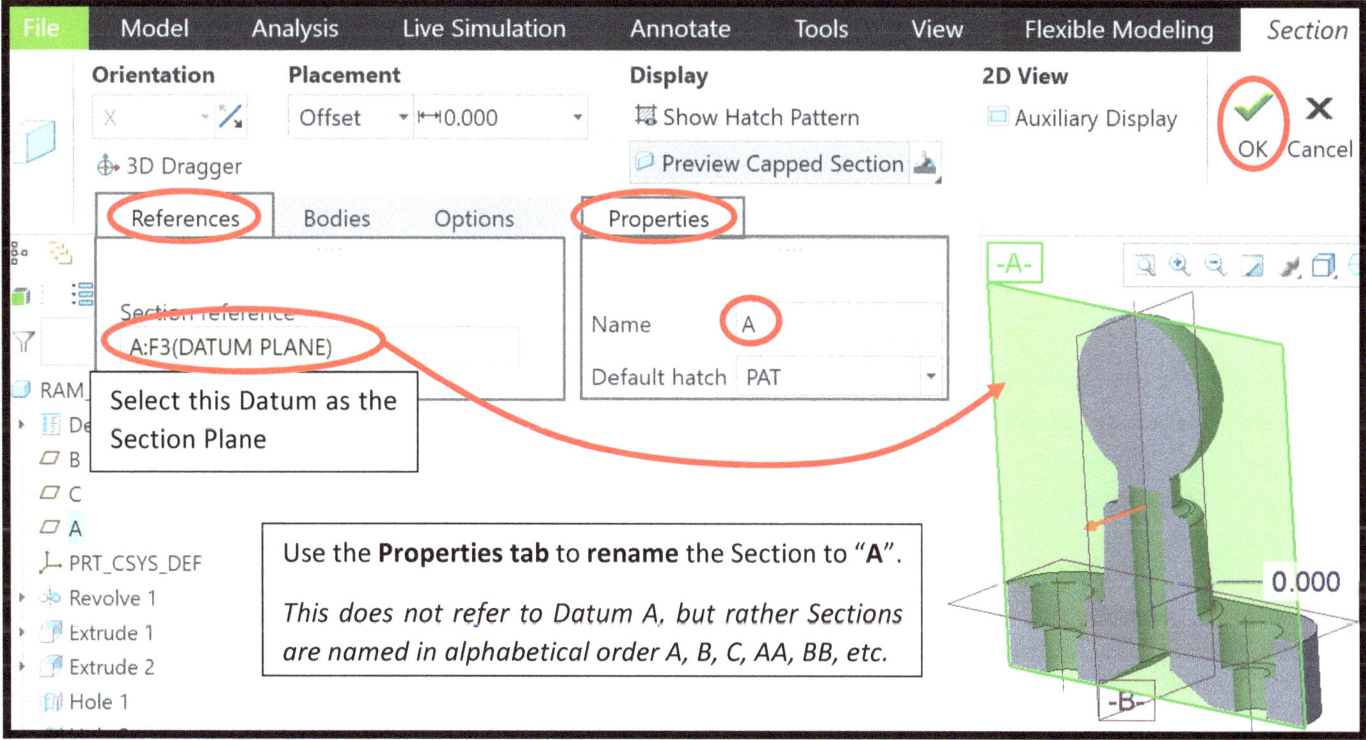

Step 14 – 2 - You can <u>toggle off</u> the Section view so that the full model is visible in the graphics window

- **Select Section A** from the model tree – in the pop-up menu select **Deactivate**.

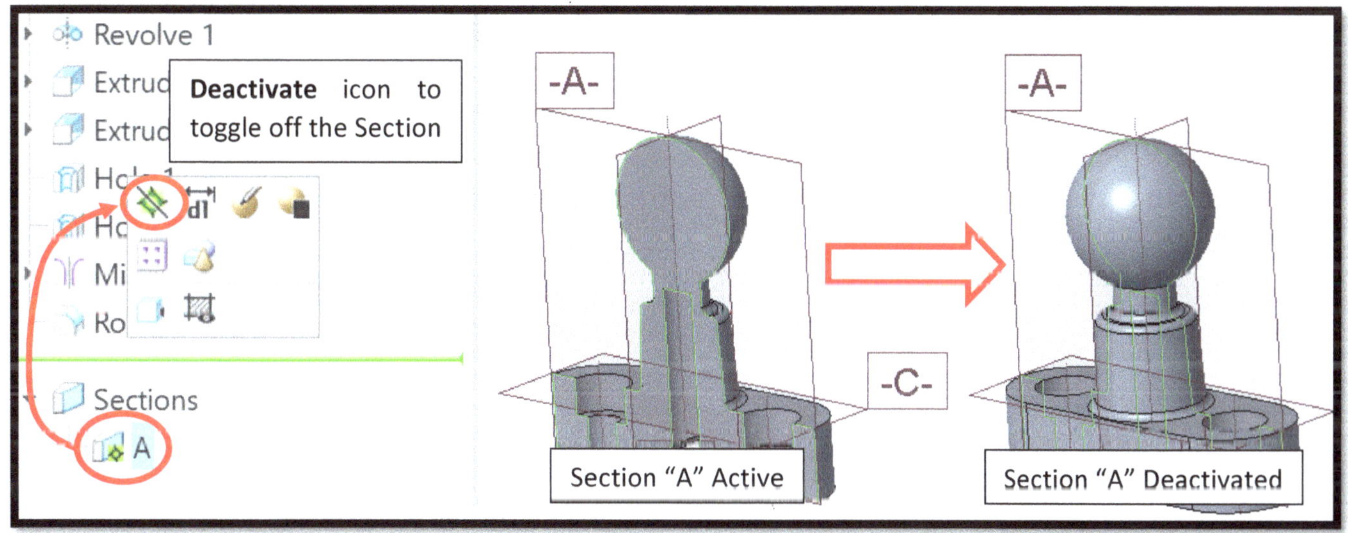

Step 15 - Save your part model, then move on to the next part.

Part #2 - Ram Handle Bottom

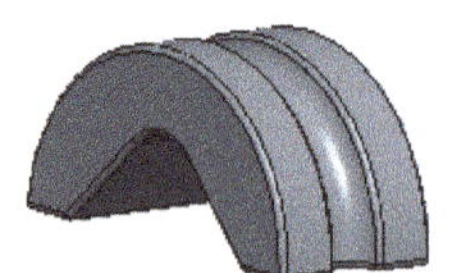

Step 1 – If this is a new session of CREO, set your working directory to your ME101 Directory.

- **File – Manage Session – Set Working Directory – Navigate to your ME101 Directory - OK**

Step 2 – If this is a new Session in CREO, adjust the options so that **3 decimal places** will be shown.

- **File - Options – Sketcher** – set the "**Number of decimal places for dimensions**" to **3 – OK – No.** *Don't bother saving a Configuration File, as you would need to load it each session to be useful.*

Step 3 - Create a new **Part** called "**Ram_Handlebottom_L3**". *You cannot use spaces in a filename!*

- **File - New – select Part as the file type – type in the "*Ram_HandleBottom*" - press OK.** A pop-up menu may appear asking you to choose a "**Template**" – if so, choose "**inlbs_part_solid_abs**" - **press OK**

Step 4 - Change the Material of the part to **PVC** *Look under Legacy Materials*

- **File – Prepare - Model Properties -** in the Material option select **change** – select the Legacy Materials folder - **double click** on "**PVC.mtl**" – press **OK** – press **Close** to close out the Model properties menu.

Step 5 - 1 – Create the first Extrude using the **Front Datum** as the sketch plane. (*You can do this without an image!*)

- Click on the **Extrude Tool** in the top toolbar - select the **Placement** tab – select **Define**.
- Select the **Front datum** as the Sketch Plane by selecting it in the graphics window. Leave the **Reference & Orientation** settings as default (*Right & Right*).

Step 5 - 2 – Press **Sketch** to enter sketch mode for this extrusion. Then select the **Sketch View** icon from the **Quick Toolbar** at the top of the Graphics Window.

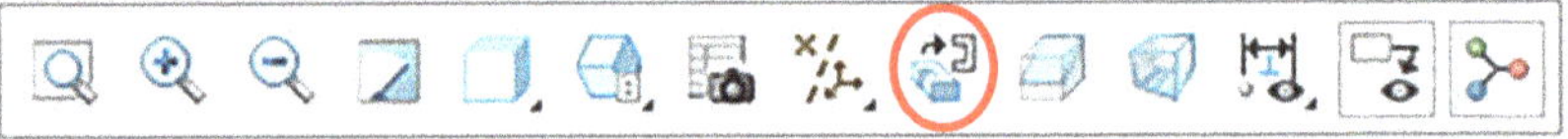

Step 5 - 3 – Sketch the shape shown below.

- Use the **Circle Tool** to create a circle centered on the origin.
- Then use the **Delete Segment tool** to delete out the bottom half of the circle. Then use the **Line Tool** to place a single horizontal line on the bottom of the sketch to complete the closed-loop.
- Set the **Radius Dimension of the half circle to 0.750"**.

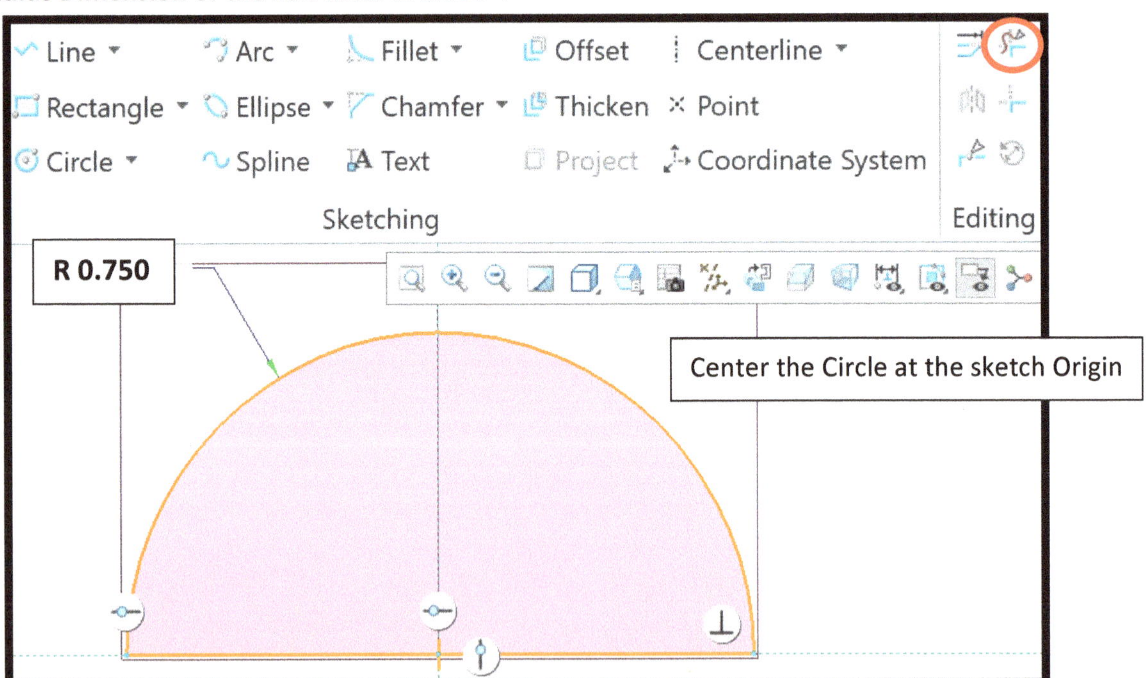

Step 5 – 4 - Checkmark to accept the sketch.

Step 5 – 5 - Set the **Depth, Direction, and Material Removal options** in the Toolbar to complete the Extrude options.

- Select the **Depth option** to be **"Extrude Symmetrically on Both Sides"**
- Set the **Depth Value** to **0.750**.
- **Press the Checkmark** to accept the Extrude

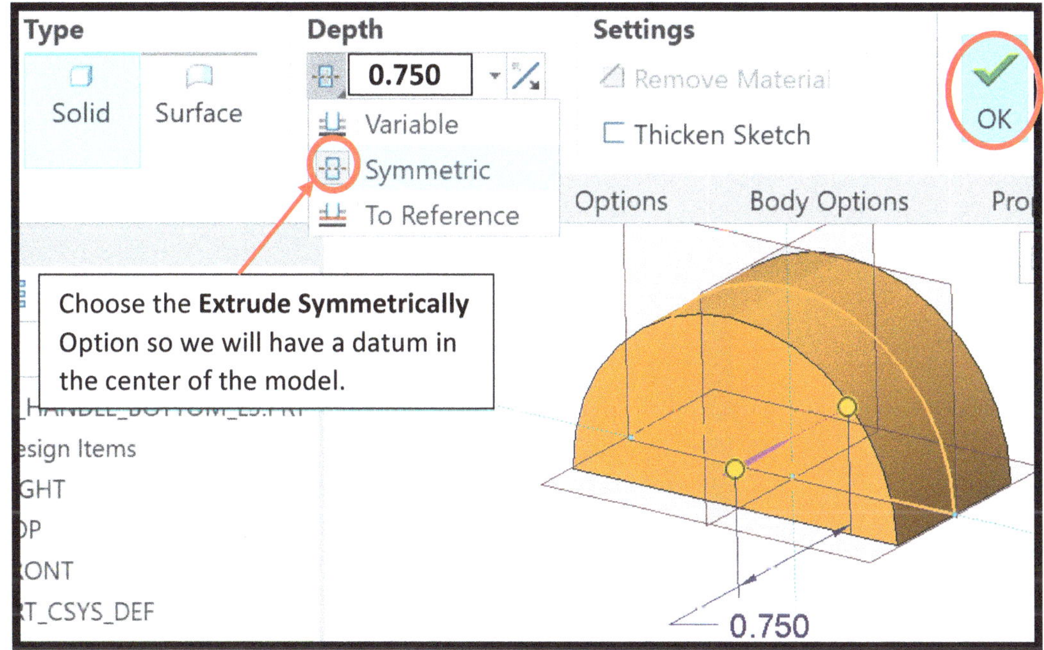

Step 6 - 1 - Create a Revolve that will add a groove to the outside surface of the model.

Revolve - Click on the **Revolve Tool** in the top toolbar - select the **Placement** tab – select **Define** - Select the **Top** datum as the Sketch Plane either by clicking the Top Datum in the Model Tree or in the Graphics Window. Leave the **Reference & Orientation** as default (*Right & Right*).

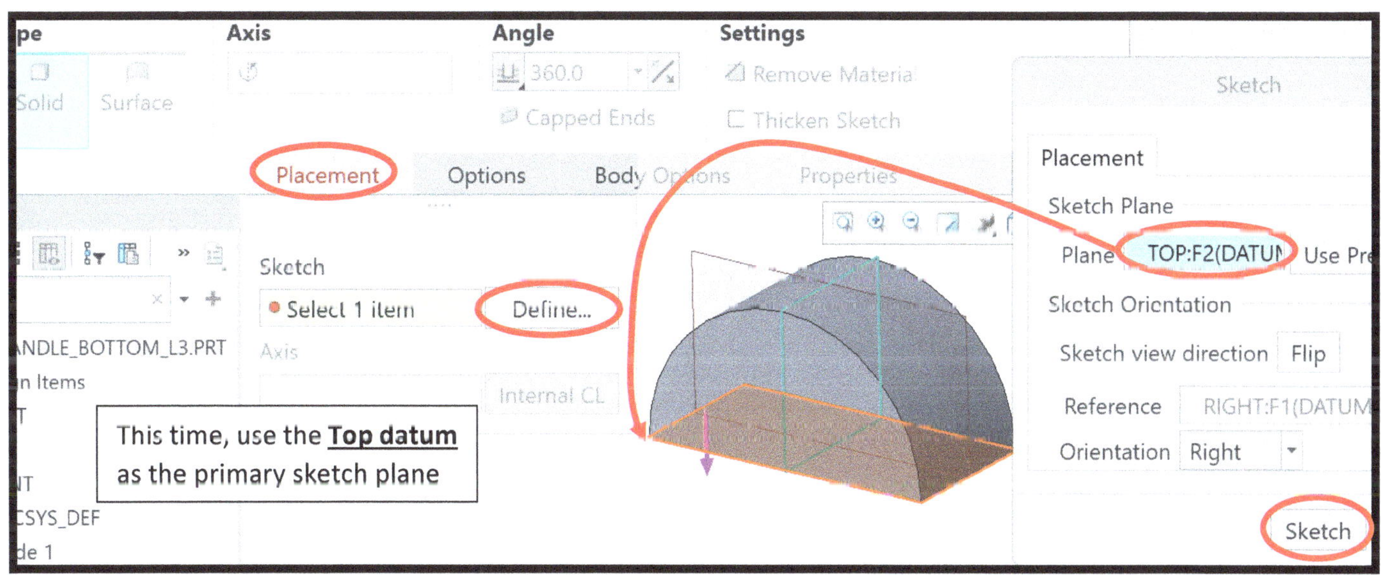

Step 6 - 2 – Press **Sketch** to enter sketch mode for this extrusion. Then select the **Sketch View** icon from the **Quick Toolbar** at the top of the Graphics Window.

Step 6 - 3 – Hold RMB in a blank area of the sketch to bring up the pop-up menu and choose the **Axis of Revolution Icon** - Place a **vertical Axis of Revolution**, then **sketch a Circle** on the outside edge of the component with a **Diameter of 0.220".**

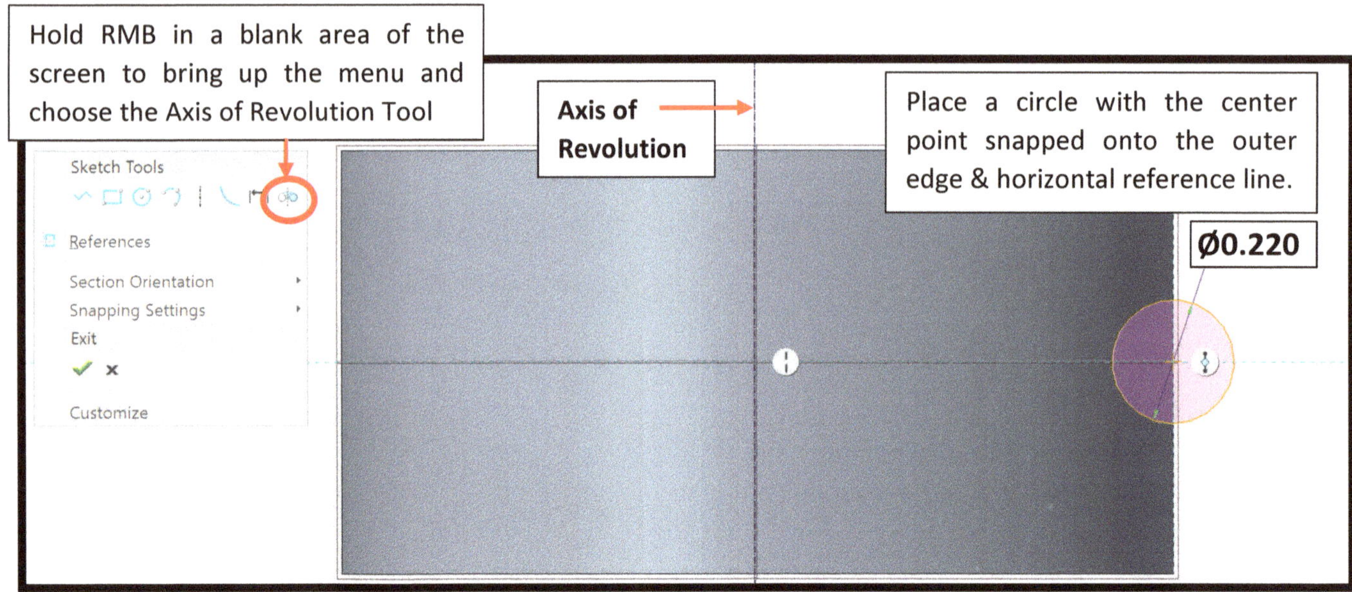

Step 6 - 4 - Press the **Green Checkmark** to accept the sketch. The solid model should default the revolve option to 360 degrees and you should see the Revolved solid material on the screen. *If your Revolve is not visible on the screen it is likely because the sketch is missing an **AXIS of REVOLUTION**. (Use Placement - Edit to return to sketch mode, if needed)*

Step 6 - 5 – Set the Revolve Toolbar options.

- Toggle on **Remove Material** (if it did not do so by default automatically)
- Select the **Revolve option** to be **360.0 degrees**
- **Press the green checkmark** to accept the Revolve.

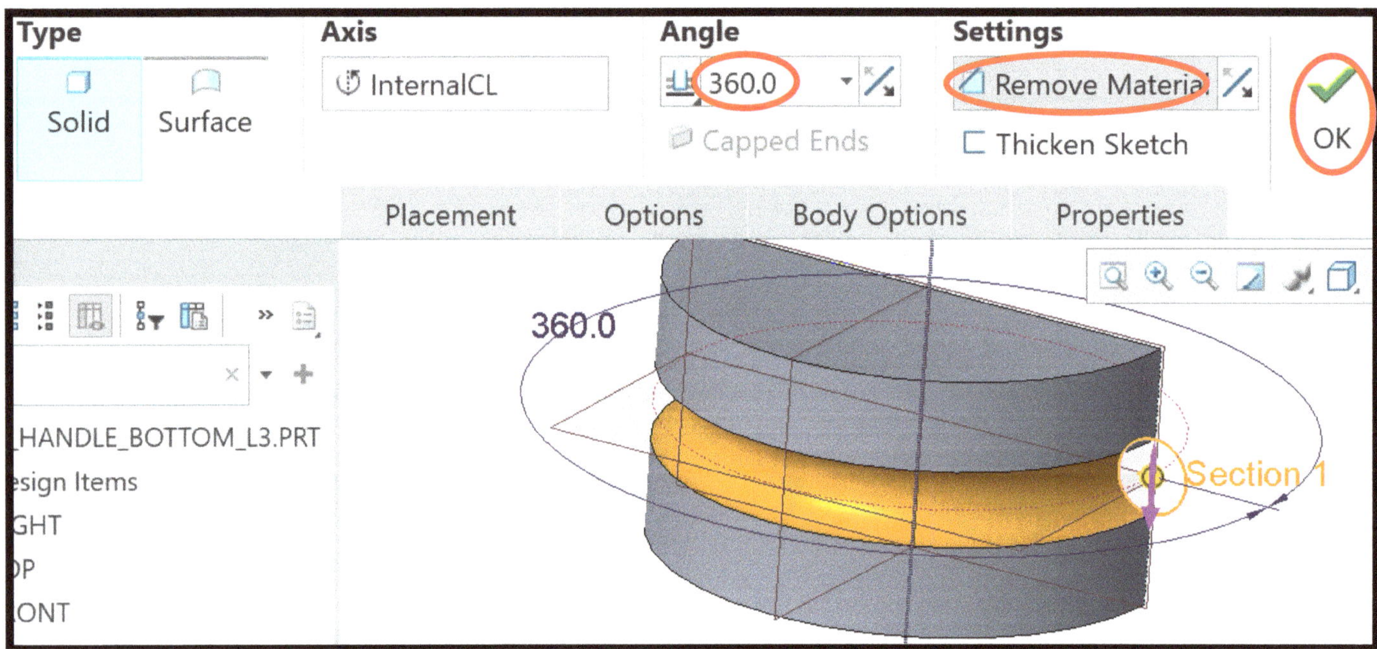

Tip: If you don't see the Revolve on the model, you likely did not place an Axis of Revolution when sketching! Use Placement – Edit to return to sketch mode to fix it.

Step 7 - 1 – Use the **Extrude Tool** to remove material from the part for a handlebar to sit against.

- **Extrude Tool** - Placement tab – **Define** – select the **Front Datum** as the sketch plane, and leave the *Reference & Orientation* as the default values (*Right & Right*).
- **Press Sketch**
- **Press the 2D Sketch View Icon**

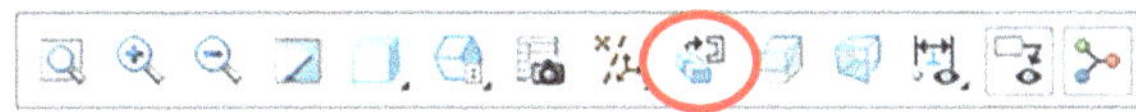

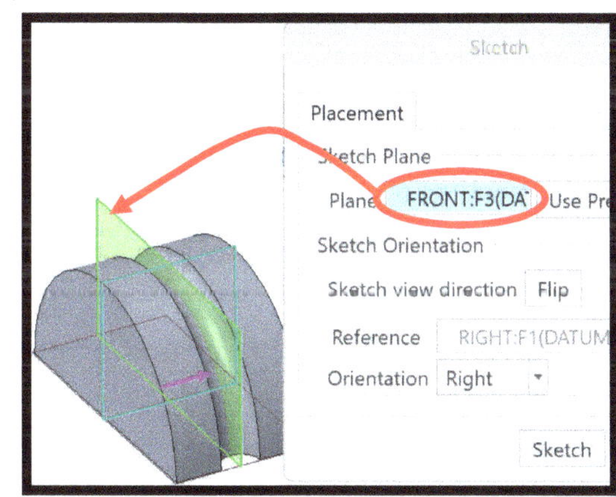

Step 7 - 2 – Create the closed loop sketch shown below, symmetric about the vertical centerline.

1) **Add a Vertical Centerline:** Use the **Centerline tool** to place a vertical centerline that crosses through the origin to use for symmetry.

2) **Sketch the Triangle:** Use the **Line Tool** to sketch the triangle shape so that the bottom line is "snapped" onto the bottom edge of the model

3) **Fillet Tool:** use the **Fillet Tool** – click the intersecting lines to add the fillet to the top of the triangle

4) **Symmetry Constraints***: Use the **Symmetry Constraint tool** to add symmetry to the Fillet "end" points. *If the symmetry constraints were added before the Fillet was created then you may need to remake them as the Fillet may remove the constraint references.*

5) **Dimension the Sketch:** tool to create a height dimension from the **top of the fillet curve** to the bottom line (0.417"), the radius of the **fillet (0.200"),** and the bottom **width of the triangle (1.000").**
   ****You may get a Resolve Sketch notice; if so, try deleting a Perpendicular Constraint if you have one*

6) **Coincident Constraint:** Only needed if the **Fillet center point moves off center**, try setting a **coincident constraint** on the center point of the fillet to the vertical centerline.

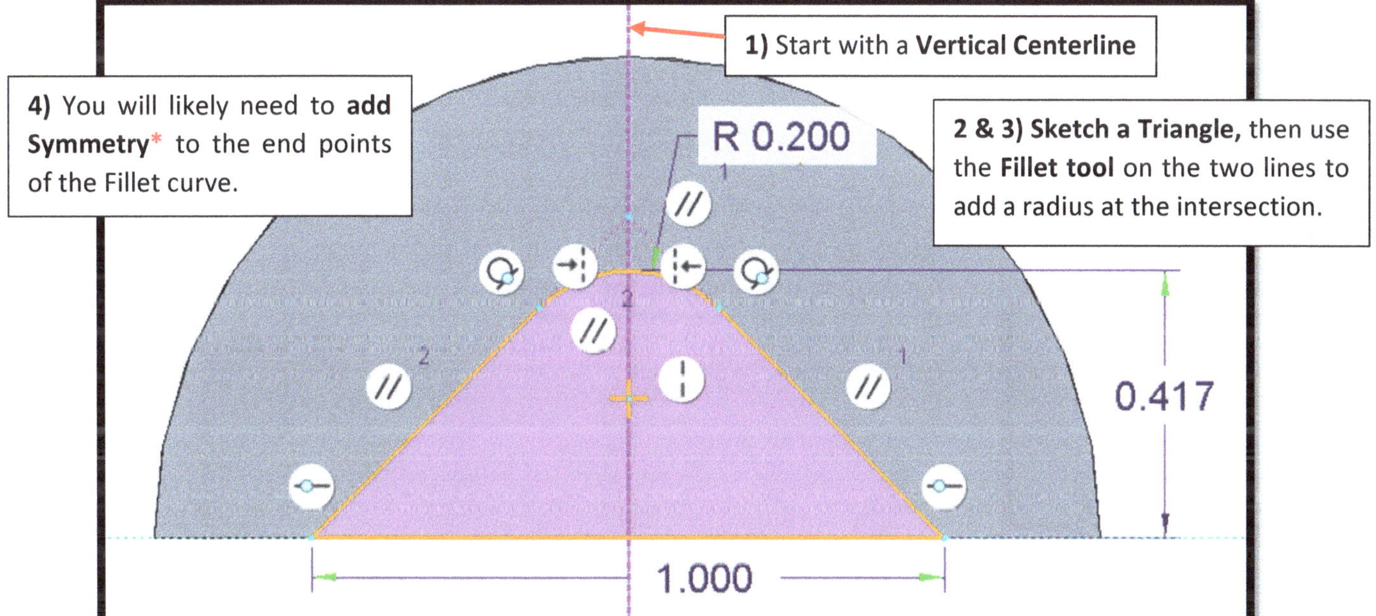

Note: Depending on how you made your sketch you may have different constraints than shown. As long as you only have 3 dimensions, the shape is symmetric, and the center of the Fillet is on the centerline it is OK. If your bottom line is not snapped onto the bottom edge of the model you will need to add a coincident constraint.

Step 7 - 3 - Set the **Depth, Direction, and Material Removal options** in the Toolbar to complete the Extrude options.

- Select the **Depth option** to be "**Extrude Symmetrically on Both Sides**"
- Set the **Depth Value** to **5.000** (*greater than the width of the model*).
- **Press the Checkmark** to Accept the Extrude

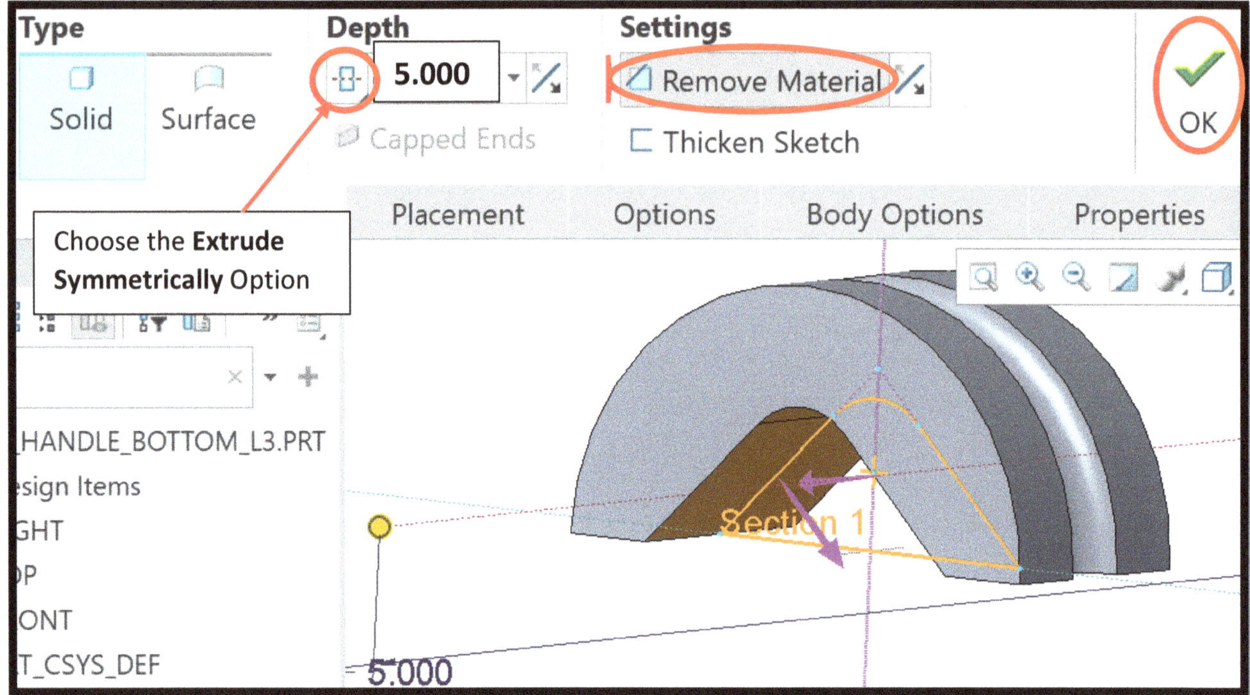

Step 8 – In the top toolbar, use the **Round Tool drop arrow** and choose the **Auto-Round** feature to automatically round all edges. Set the value to be **0.025" and press the Checkmark** to accept the Round Settings. It may take a few seconds to process the auto round.

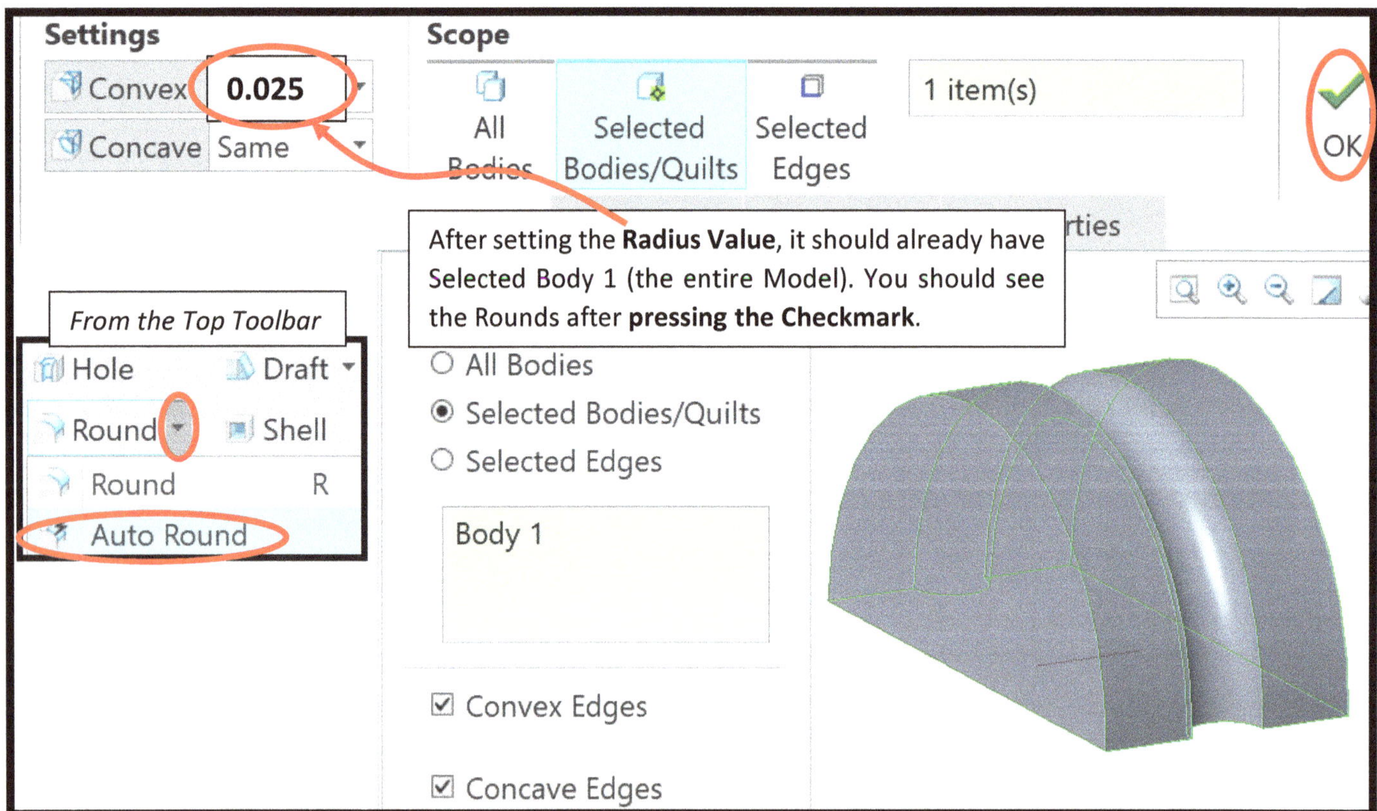

If it fails, a common mistake is using the wrong round value (i.e. 0.25" instead if 0.025" which is too large).

Step 9 – Set the display properties of **<u>each datum</u>** so that we may use them as a **Geometric Tolerance** and be able to change them to **ASME style** in our drawings. We must "**<u>Set</u>**" & Rename each datum to match the style of this course using letters for names of the Datums.

- **LMB to select a Datum** from the model tree - **hold RMB – Properties - press SET ("-A-")** – select the **name box** – **Type in a new name** according to the list below – **OK**.
- <u>**Repeat for each Datum**</u>.

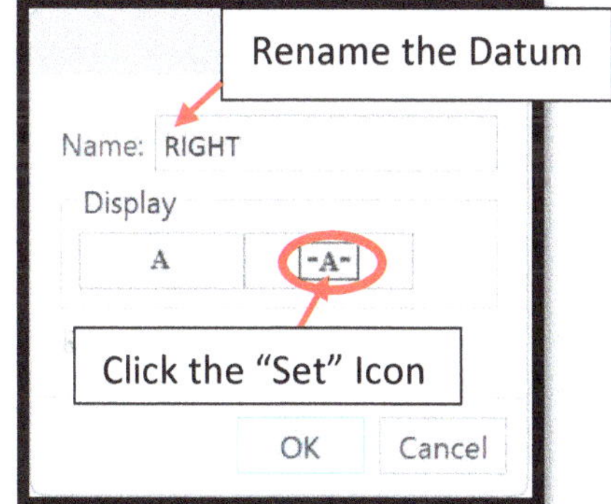

    RIGHT: Rename to "**B**" - Press "**Set**" - OK

    TOP: Rename to "**C**" - Press "**Set**" - OK

    FRONT: Rename to "**A**" - Press "**Set**" – OK

Double Check: After pressing the Set Button on each datum, you should see the Datum "tags" on the screen. If you do not see all 3 datum names, you likely did not press "Set" for that datum.

Step 10 – Save your part. The finished component should look similar to the image below (with View Display of *Shading with Edges* shown in the Quick Toolbar).

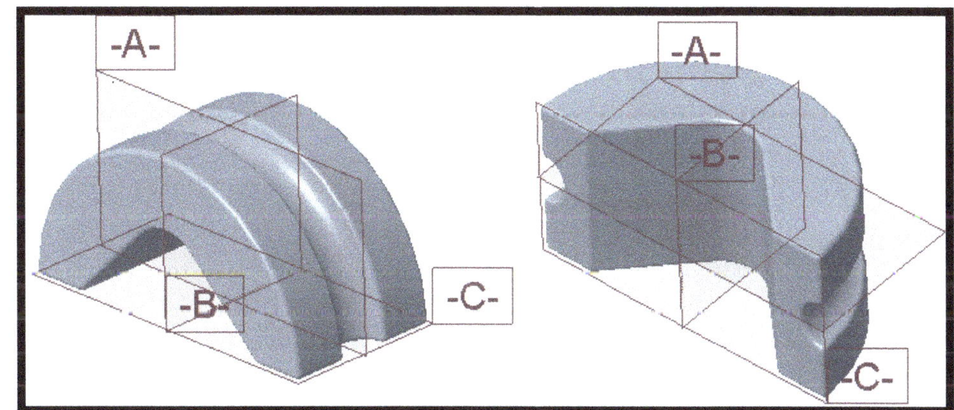

Step 11 – Create Drawings of the two parts that match the lesson keys.

- **Start a new Drawing File: - File – New – Drawing – provide a drawing name of "*L3_Ram_Handlebar*" – OK**.
- In the next menu choose **browse** to specify that the *Ram_Handlbar_L3* will be the drawing's model.
- Leave the Template option as the default *C_drawing* (*do not try and choose the UND format yet!*)

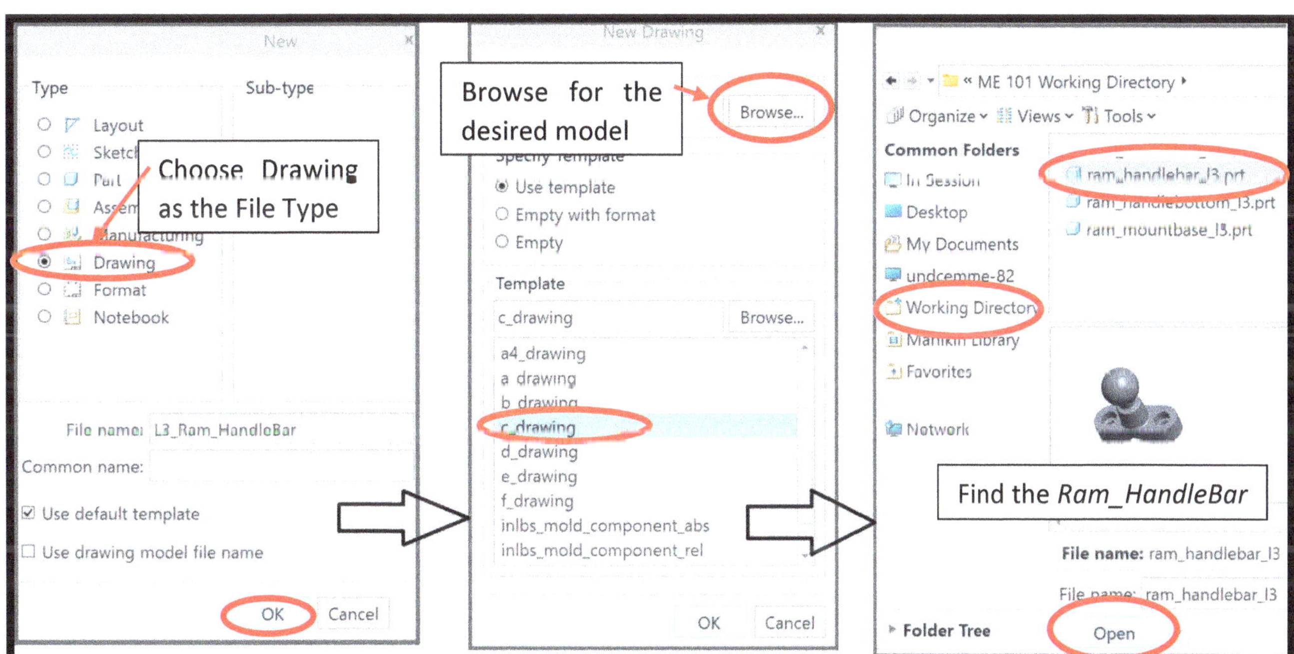

Step 11 – 2 – Follow the steps as needed in the ***Basic Drawing Steps*** (in the Appendix) to create a drawing that matches the Drawing keys posted in Blackboard. Refer to the specific details listed on the drawing keys as you work through the drawing steps.

> For the *Ram_Handlebar*: Skip the *Basic Drawing* **Step 9 (Peel Back Datum Tags)** for this lesson.

> For the *Ram_HandleBottom*: Skip the *Basic Drawing* **Step 8 (Cross-Section), Step 9 (Peel Back Datums Tags), & Step 10 (Show Centerlines & Notes),** as this model does not need centerlines nor a cross-section.

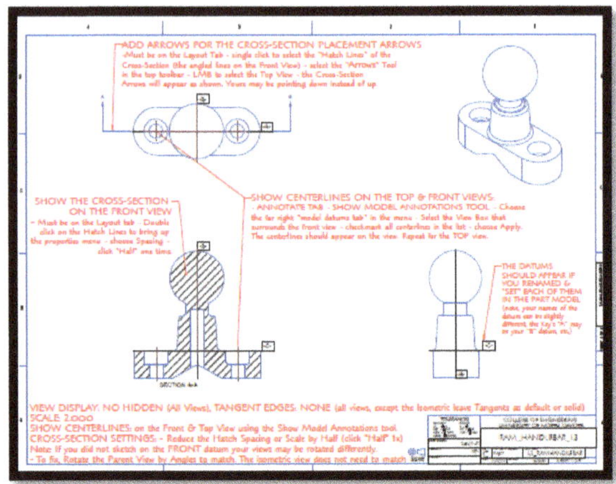
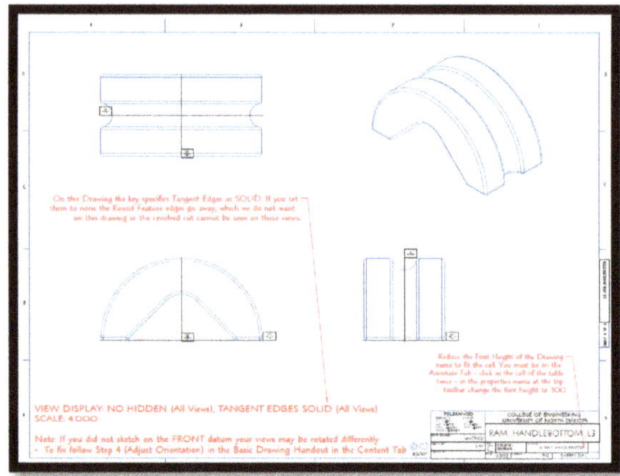

(Check Blackboard for the actual drawing keys so you can read the details!)

Step 13 - Save & Submit your Work:

Save the **Drawing & Part files in CREO**, and then **Save As** a **PDF** of each drawing to submit your drawing PDF for grading.

- **File -** hover your mouse over the **Save As button – select Quick Export (PDF)** – a PDF should open up on the computer - save that PDF to your working directory.

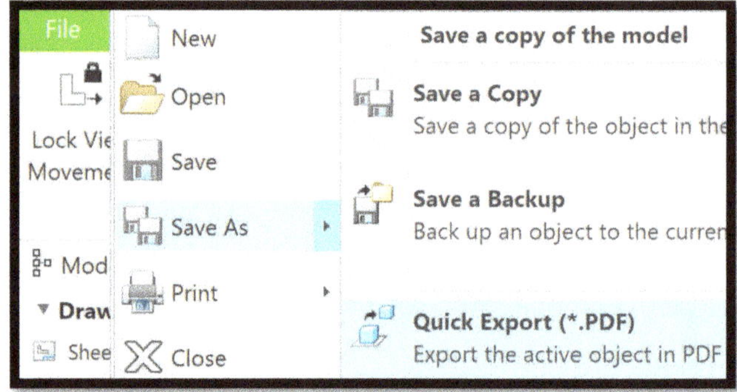

Submit all requested deliverables in the same submission. You must submit a PDF of the drawing for grading, not the .DRW files.

Optional: For further practice, if time allows, complete the ***Ram_MountBase*** part model steps posted in Blackboard. This model is not required and does not need to be submitted, but can be used to practice your modeling skills if you still have time in class.

End of Lesson 3

Lesson 4 – Patterns & Ribs (Gas Can Bracket)

Lesson 4 is to create a mounting bracket that connects the gas can with the frame on a motorcycle. Patterns will allow features to be copied in either axial or linear directions and the Rib tool will be used to easily create a structural "Rib" support instead of doing so with the extrude tool.

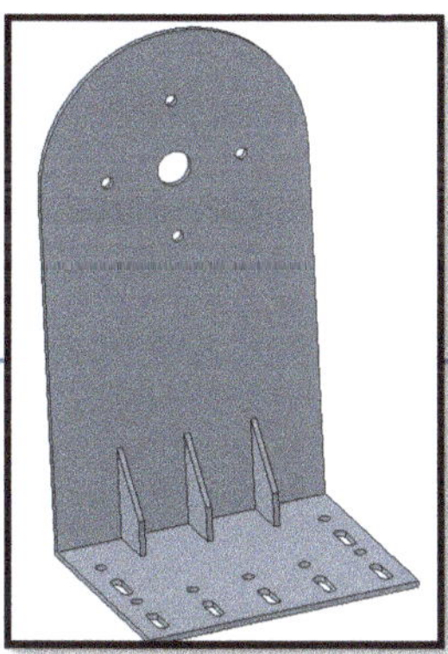

Patterns: Patterns create multiple copies of a feature, typically by specifying the number of total "instances" of the feature to be added and a spacing along a reference to place each subsequent copy. Most patterned features will be exact copies of the original. Changing the size or placement of the original feature will update all copies as well.

If you want to pattern multiple features, such as a slot and a hole together, you must first Group the features together in the Model Tree. Select all desired features to copy and choose Group. The group will act as a single feature.

There are multiple types of Patterns that can be used but the easiest method is to use a **Direction type** (if creating a linear pattern in one or two directions to build a grid of copies) or an **Axis type** (if rotating copies a specific angle about an axis).

You can also toggle off instances in a pattern to skip copies that are not desired.

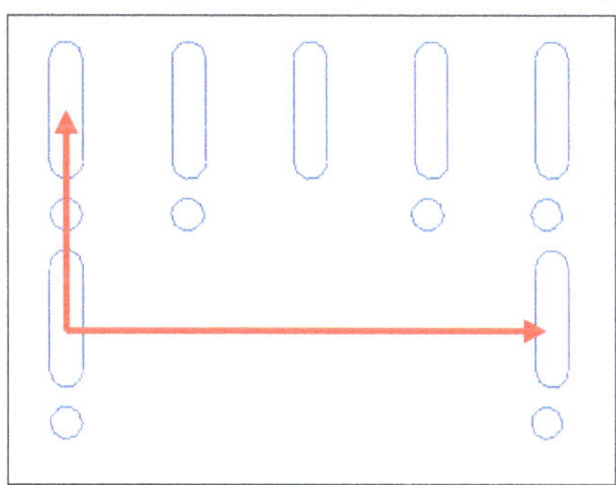

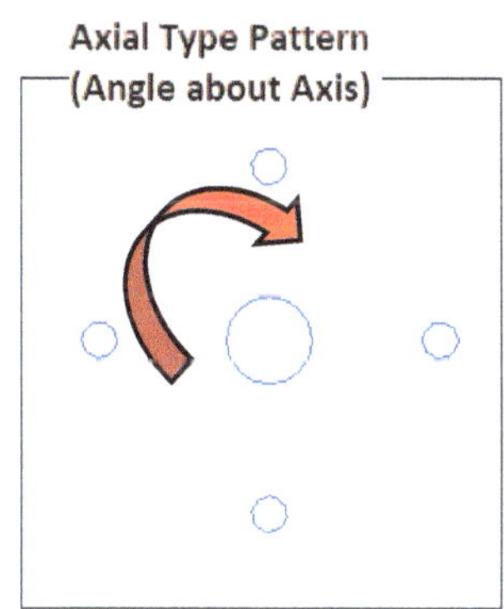

Rib Features:

The Rib tool can create either Profile Ribs or Trajectory Ribs, which provide greater stiffness to a part and reduce stresses on thin-walled parts. The Rib Tool does not utilize a "Closed Loop" sketch, as the sketch for the rib only needs to define the trajectory or profile of the rib, with the rib tool automatically filling in material based on the sketch lines and the setting for a rib thickness.

A **Profile Rib** will fill in material between the sketched profile and any solid surface beneath it, while a **Trajectory Rib** will build ribs along the path of the sketched trajectory down to any solid material below it, such as webbing within a hollow part.

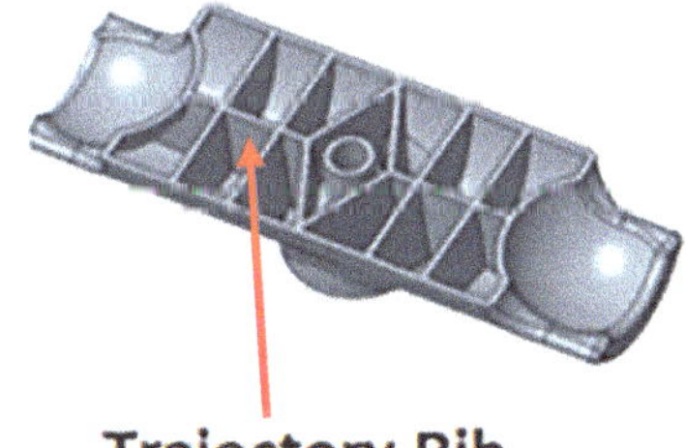

Profile Rib   **Trajectory Rib**

Getting Started: Open the **CREO** *Parametric* software and follow the lesson steps carefully.

Step 1 - Set your working directory to your ME101 Working Directory Folder.

- File – Manage Session – Set Working Directory – Navigate to your ME101 Directory folder – OK

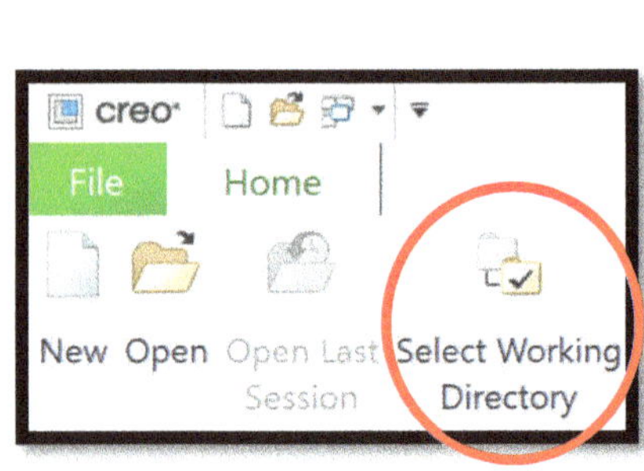

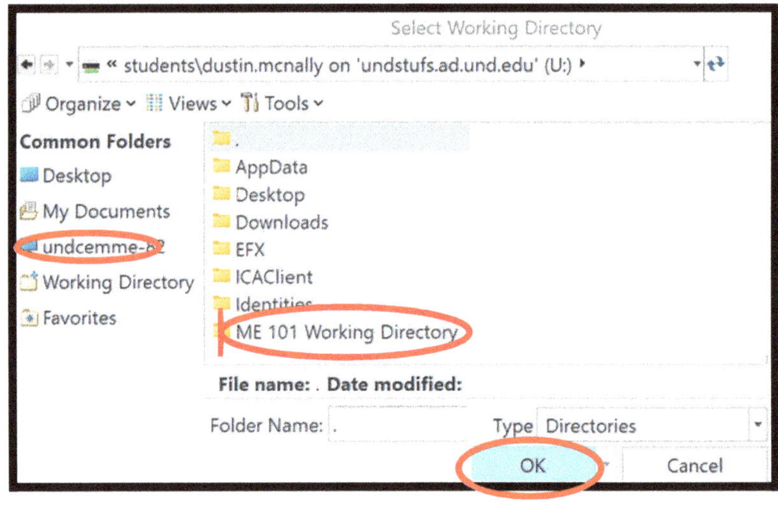

Step 2 - Create a new **Part** called "**GasCan_Bracket_L4**". *You cannot use spaces in a filename!*

- File - New – select Part as the file type – type in the "GasCan_Bracket_L4" - press OK.

A menu may appear asking you to choose a **"Template"** – if so, choose **"inlbs_part_solid_abs"** - **press OK.** *The template menu may not appear in the student home version as it defaults to the desired inlbs_part_solid_abs already without needing to be selected.*

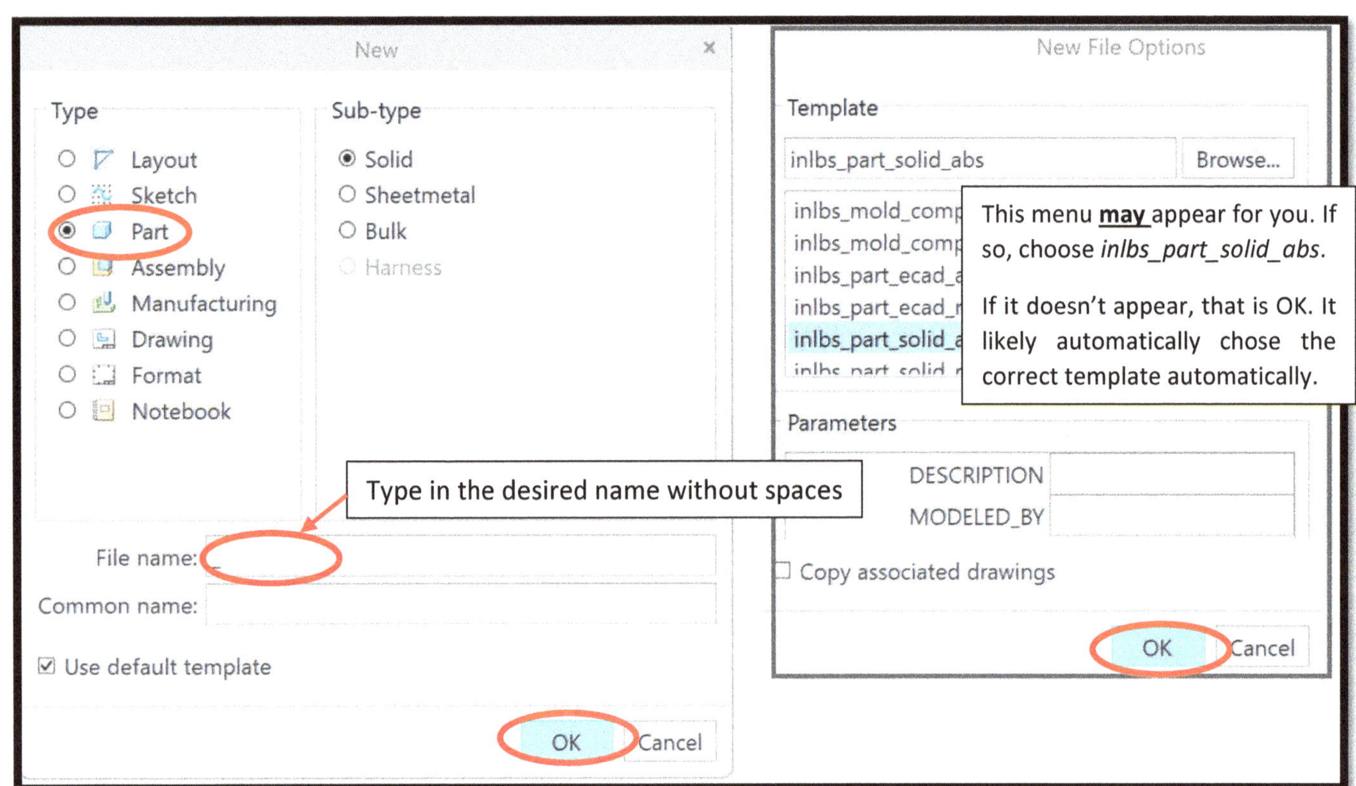

Step 3 - Change the Material of the part to **AL6061** *Look under Legacy Materials*, and be sure your units are in the CREO Default (should be inch-lbm-Second).

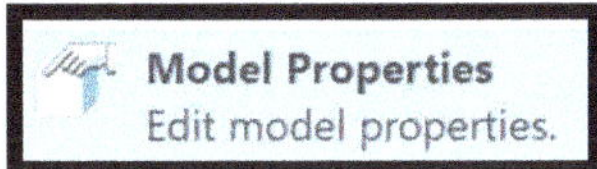

- **File – Prepare - Model Properties -** in the Material option select **change** – select the Legacy Materials folder - **double click** on "**AL6061.mtl**" – press **OK** – press **Close** to close out the Model properties menu.

Step 4 – Adjust the options so that <u>3 decimal places</u> will be shown in sketcher.

- File – Options – Sketcher – set the "<u>**Number of decimal places for dimensions**</u>" to **3** – **OK** - **No***
* *Don't bother saving a Configuration File, as you would need to load it each session to be useful.*

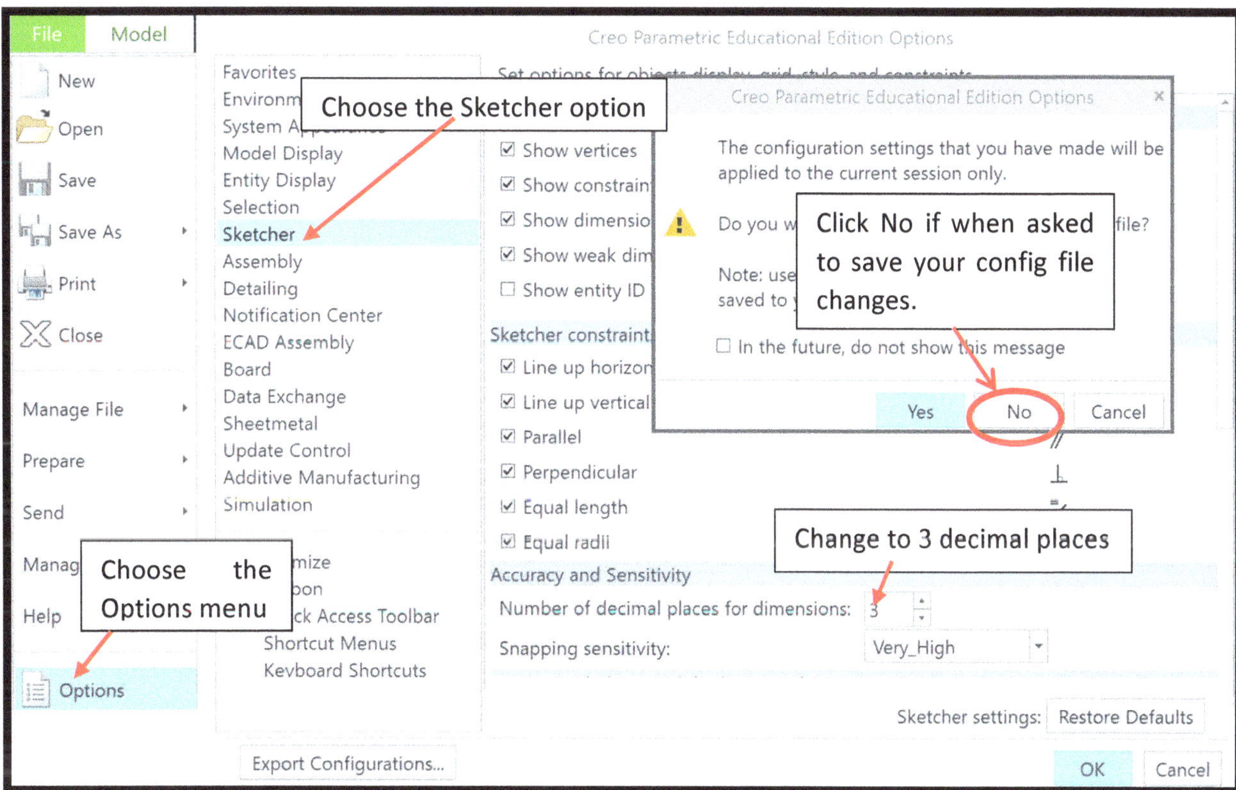

Step 5 - Create an **Extrude** to create the first feature using the **Front Datum** as the sketch plane.

- Click on the **Extrude Tool** in the top toolbar - select the Placement tab – select **Define**- choose the Front Datum and leave the **Reference & Orientation** as default (*Right & Right*).

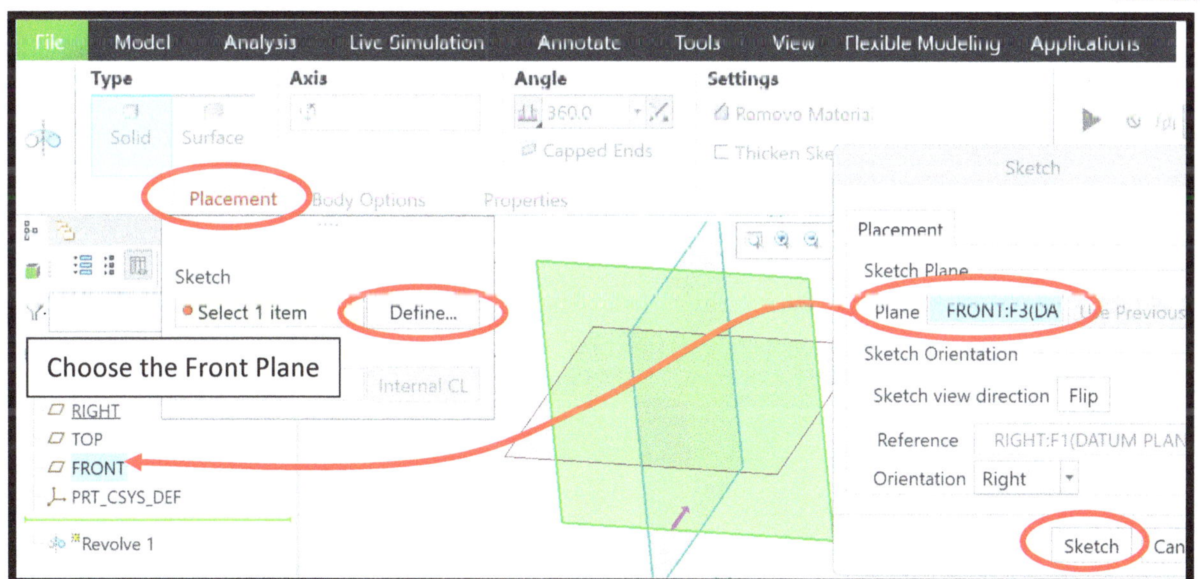

Step 5 – 2 - Press **Sketch** to enter sketch mode for this extrusion. Then select the **Sketch View** icon from the **Quick Toolbar** at the top of the Graphics Window.

Step 5 - 3 – Create the closed loop sketch.

1) **Sketch the "L" closed-loop shape**, with the sketch starting at the origin and coincident with the vertical and horizontal references.
2) **Set an Equal Lengths constraint** on the short line of each "leg" of the L shape, so that the shape is the same thickness.
3) **Dimension the sketch** so that the vertical height is **8.000"**, the horizontal width is **5.000",** and the thickness is **0.250".**

   It should be placed coincident with the origin so there should not be any other dimensions.

4) **Press the checkmark** to accept the sketch.

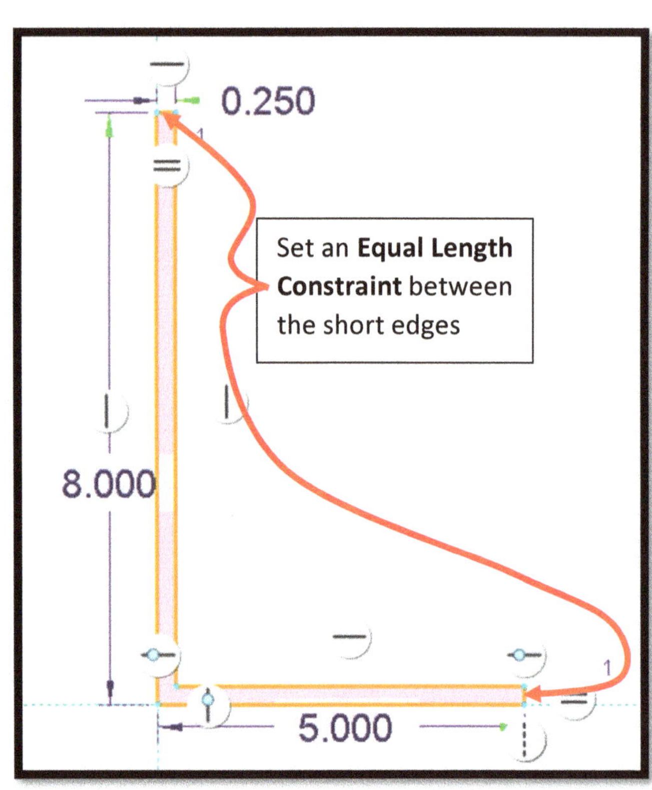

Step 5 – 4 - Set the **Depth, Direction, and Material Removal options** in the Toolbar to complete the Extrude options.

- Select the **Depth option** to be "**Extrude Symmetrically on Both Sides**"

- Set the **Depth Value** to **6.000**.

- **Press the Checkmark** to Accept the Extrude

Check that your model is symmetric or extruding on both sides of the Front datum plane. If not, use Edit Definition to bring back the Extrude 1 toolbar to adjust the direction setting.

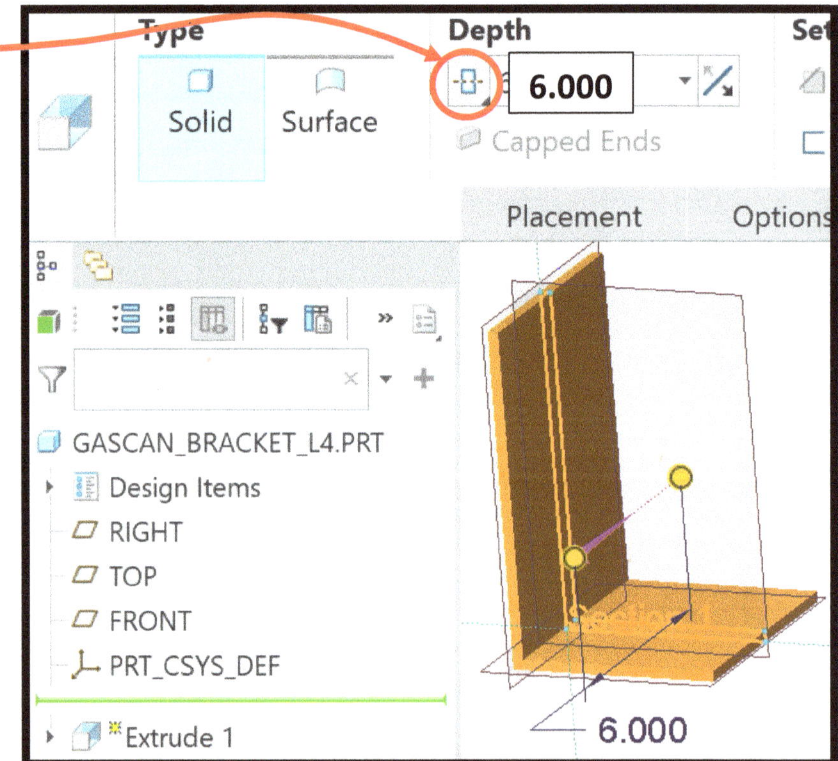

Step 6 – Create an Extrude extend the bracket with more material in a half circle shape, using the **Right Datum** as the sketch plane (or the vertical flat surface of the bracket will also work).

- Click on the **Extrude Tool** in the top toolbar - select the **Placement** tab – select **Define**- choose the **Right Datum** and leave the **Reference & Orientation** as default (*Top & Top*).

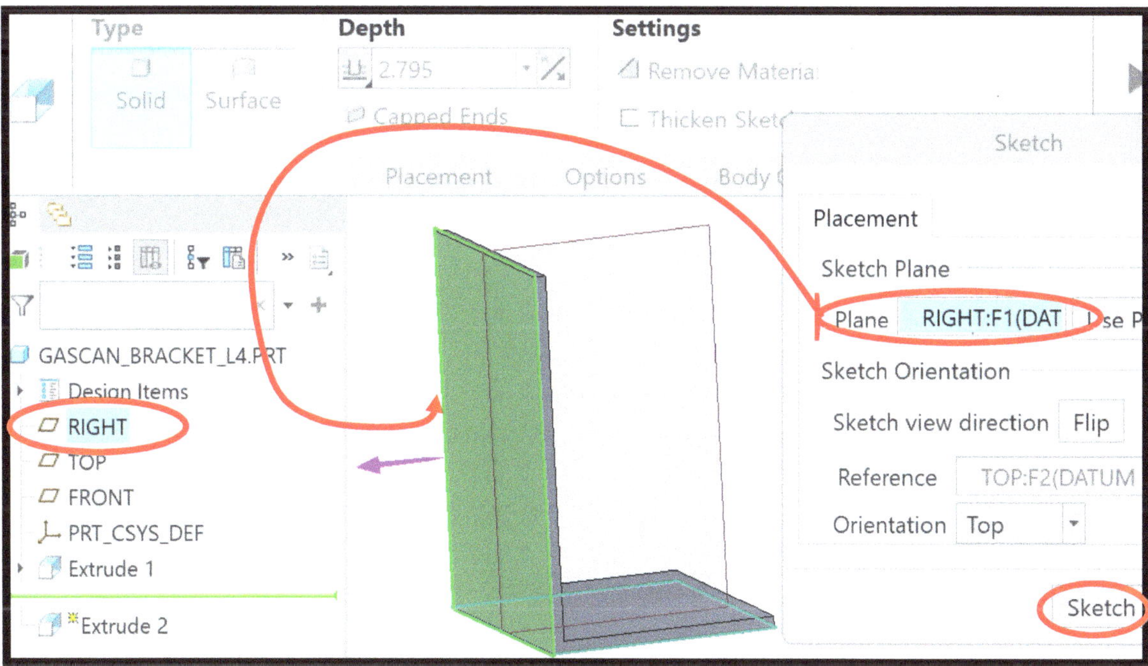

Step 6 – 2 - Press **Sketch** to enter sketch mode for this extrusion. Then select the **Sketch View** Icon from the **Quick Toolbar** at the top of the Graphics Window.

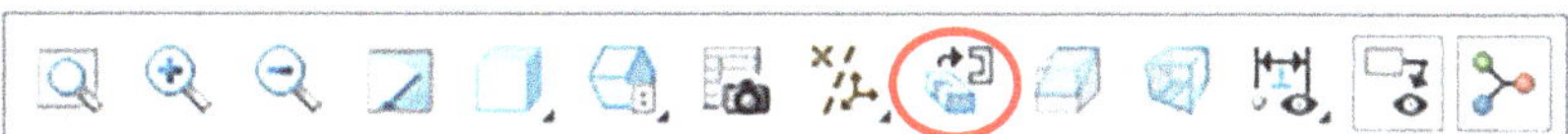

Step 6 – 3 – **Use the Sketch References** tool to add the outer edges of the sketch as snap references.

- **Hold RMB** in a blank area – choose the **References Tool** – **LMB click to add the left, top, and right edges** as reference lines (light blue color). **Press Close** in that menu.

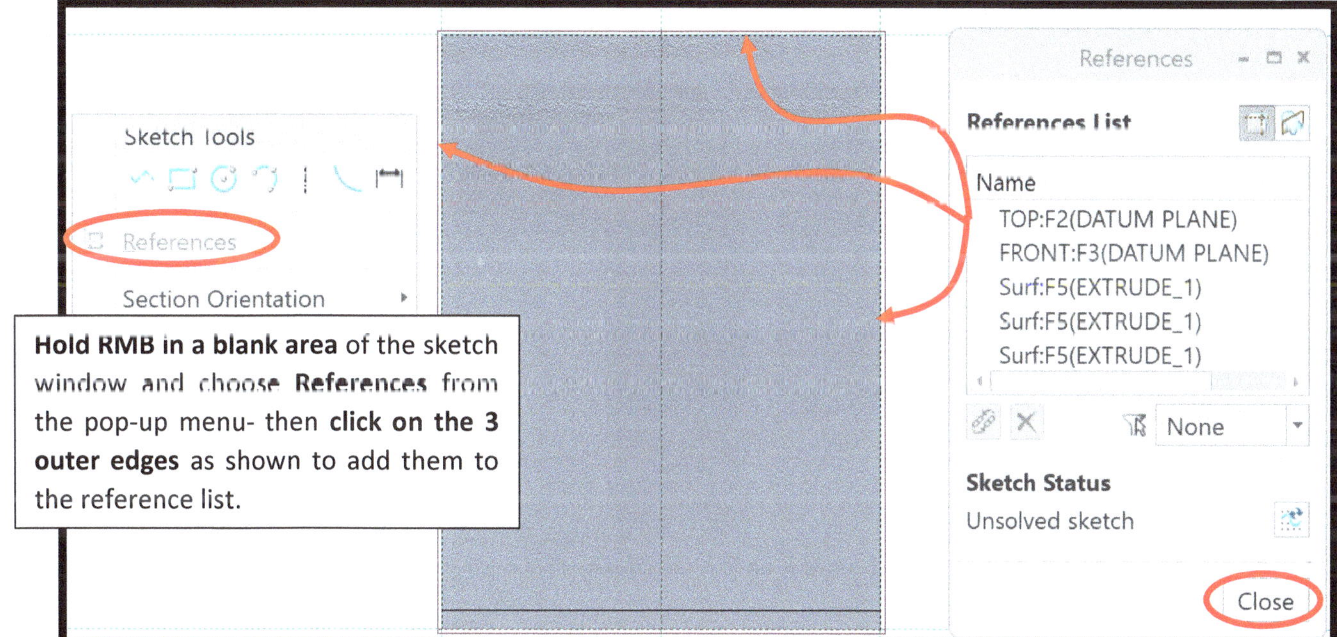

Step 6 – 4 - Sketch the closed loop shape using the **Center & Ends Arc** with a **horizontal Line** on the bottom to make it a closed-loop.

- Place the **Center & Ends arc** on the center of the top edge of the model. Then snap the start and ends points of the arc onto the corners of the model. If they will not snap on the corners, you may need to use a Coincident Constraint to force them together.
- No Dimension are needed for this sketch as the constraints on the corner define the Radius of the arc.
- **Press the checkmark** to accept the sketch.

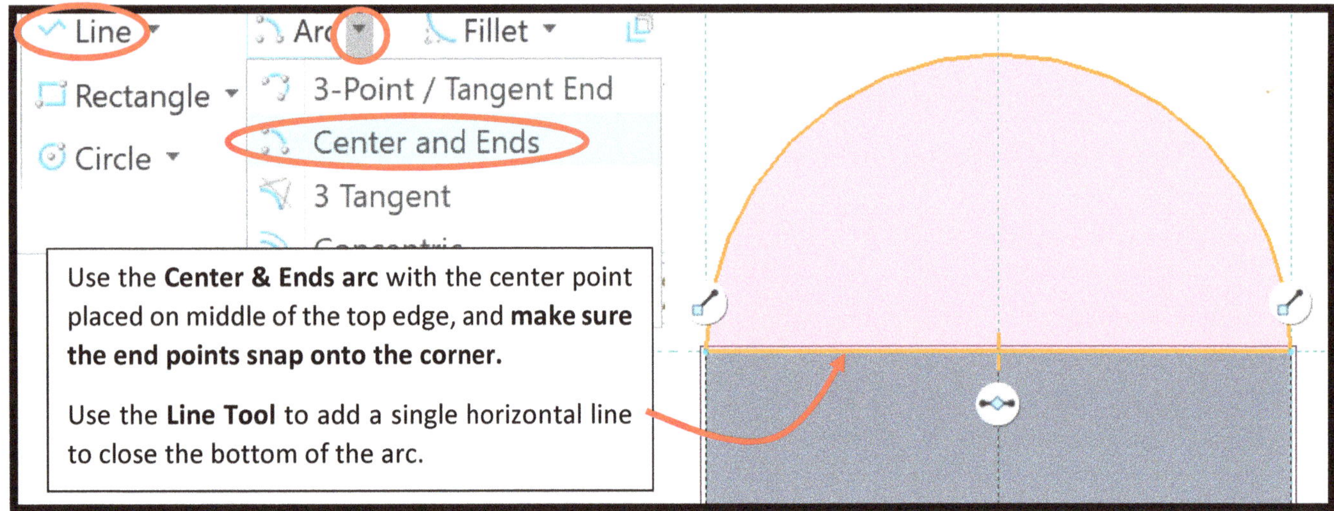

Step 6 – 5 - Set the Extrude toolbar options.

- **Set the Direction option to "*To Reference*"** - select the **<u>front surface of the bracket</u>** so that the depth will continue until it matches the surface of Extrude 1.
    - o   **Tip**: If your sketch started on a different sketch plane than the Right datum, you may need to choose the rear surface instead.
- **Press the Checkmark** to accept the Extrude

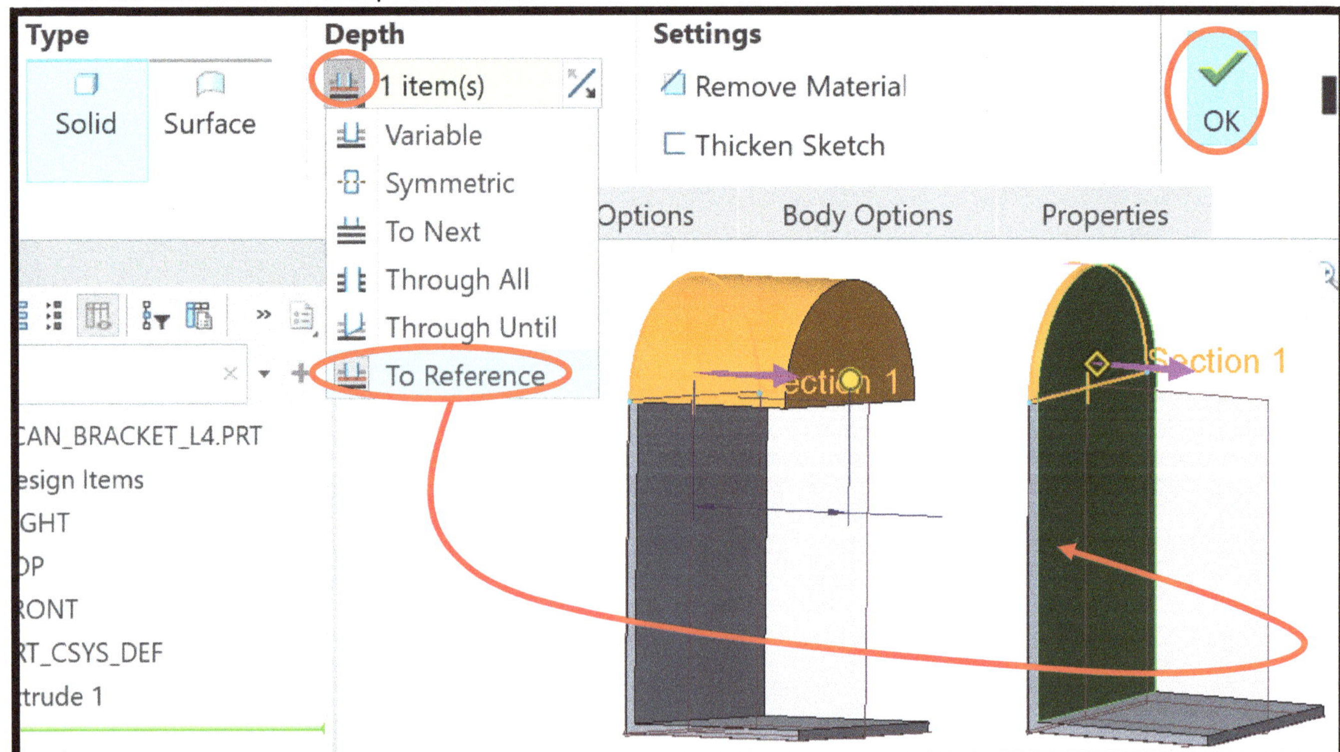

Step 7 - Go to **Blackboard and complete Module 8 (Rib Tool)** to learn about the Rib feature.

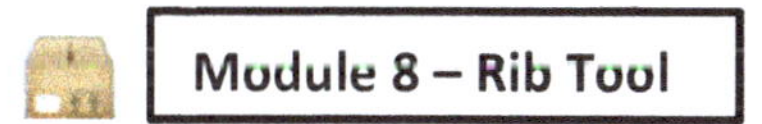

Step 8 – **Create a new Datum Plane** to use as the sketch plane that a Rib will be placed on.

- Select the **Plane Tool** – in the *Datum Plane Menu* click the **Front datum as a reference** – enter an **offset value of 1.500"** – **Press OK**.
    - o Your new datum can be on either side of the Front datum.

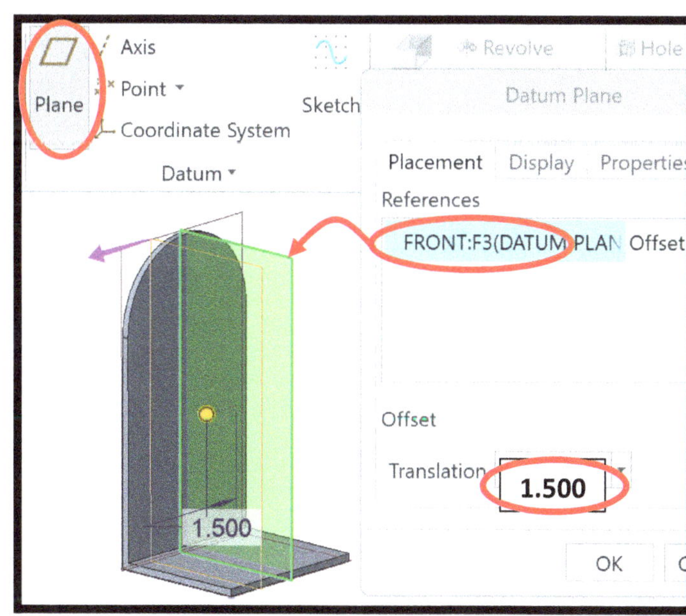

Step 9 - Create a Profile Rib on the new Datum (DTM1). Make sure you choose the "Profile" Rib option.

- **Click in a Blank area** of the screen so that no datum is currently selected.
- Select the drop-down arrow of the Rib icon – select **Profile Rib**.
- Select the **References Tab** and choose **DTM1** as the sketch plane.

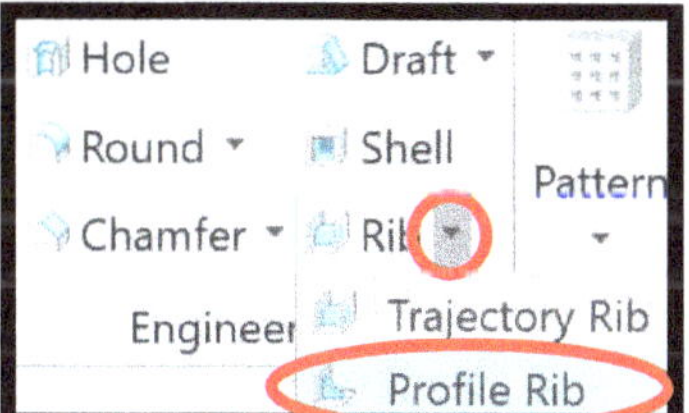

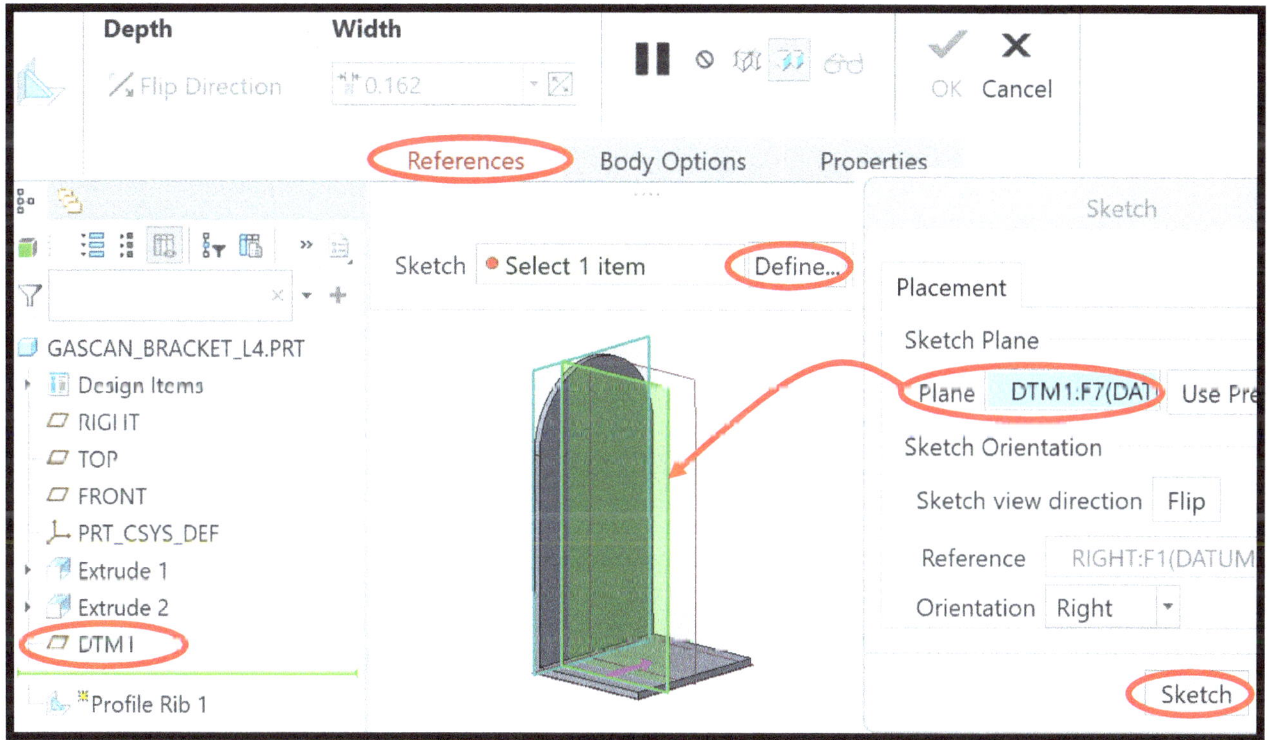

Step 9 – 2 - Press **Sketch** to enter sketch mode for this extrusion. Then select the **Sketch View** icon from the **Quick Toolbar** at the top of the Graphics Window.

Step 9 – 3 - Sketch the rib outline as shown below.

1) Use the **Line tool** to create an **<u>Open-Loop</u>** for the outer edge of the rib. *Do not sketch a closed-loop!*
2) Create the dimensions and set the values as shown below for the **height of the Rib (2.000")**, the **depth of the Rib (1.000")**, and the **height of the front vertical line (0.750")**.
3) **Press the green checkmark** to accept the rib sketch.

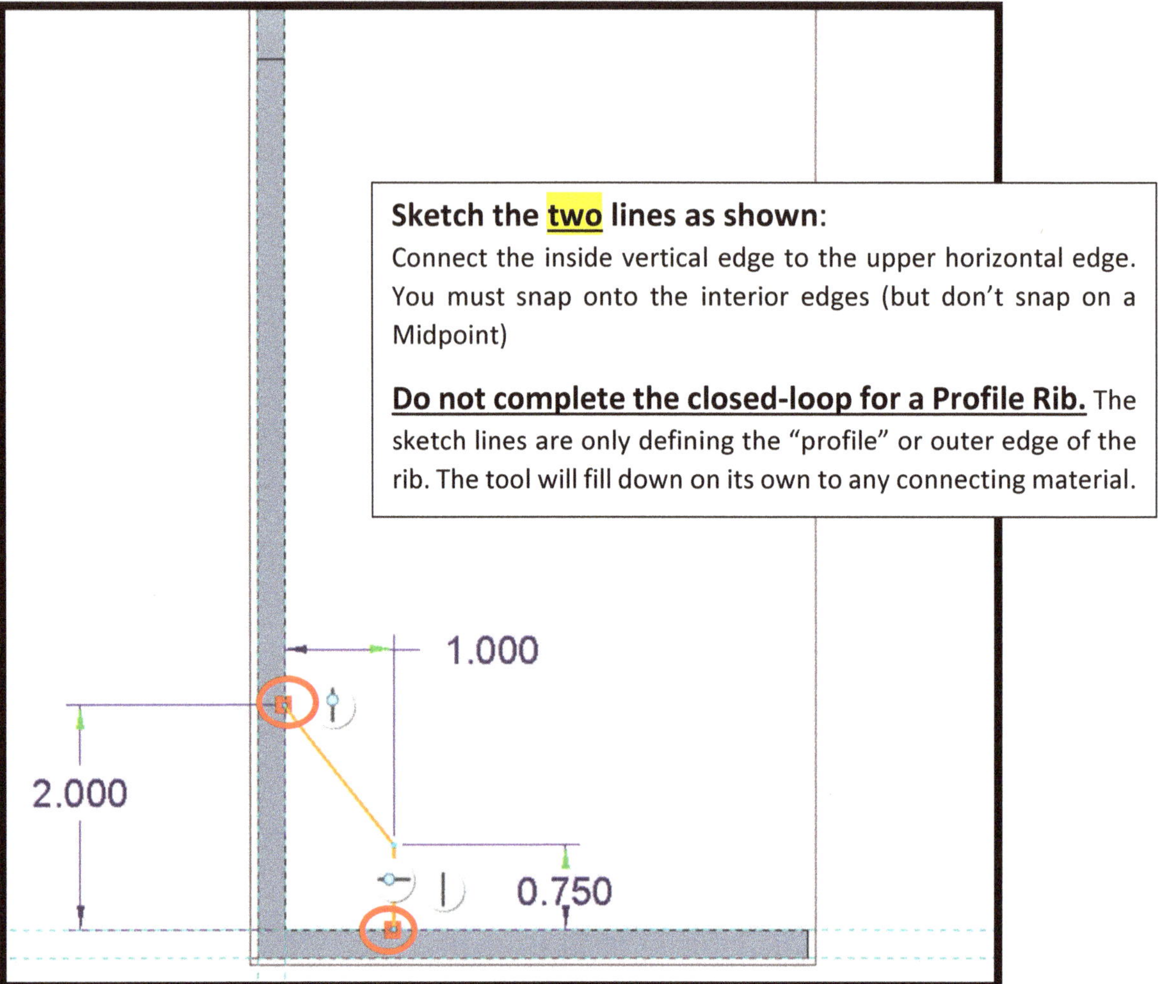

Tip: If your rib does not appear on the screen, you may have sketched the rib incorrectly, chosen the wrong sketch plane, sketched an open loop, or perhaps did not choose the *Profile* Rib tool!

Step 9 - 4 - Adjust the **Rib Thickness** in the top toolbar to **0.250"**.

- **Press the Checkmark** to accept the Rib.

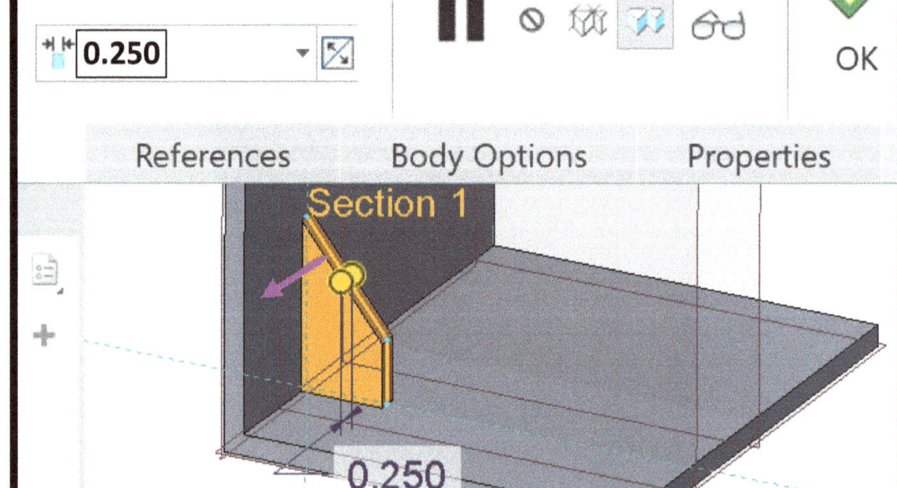

Step 10 - Pattern the Rib using the **Pattern Tool**, using a **Direction type pattern**.

1) **Start the Pattern: Click on** *Profile Rib 1* in the model tree so it is highlighted and select the **Pattern Tool** from the top toolbar.

2) **Choose the Pattern Type: Change the Pattern Type to "Direction",**

3) **Select the Pattern References:** Click on the leading **edge of the model** as the direction reference. Be sure to choose the edge, and not a surface, that runs the direction of the desired pattern. *A surface direction reference will run the pattern perpendicular to that surface!*

4) **Setup the Pattern Settings:** Set the **number of members to 3 and the spacing to 1.500".** This creates 3 total instances, including the original feature, spaced out 1.5" from each other in the direction of the reference edge.

5) **Toggle the Pattern Direction: Toggle the Direction Arrow icon** as needed so the Pattern copies are placed on the model instead of away from it.

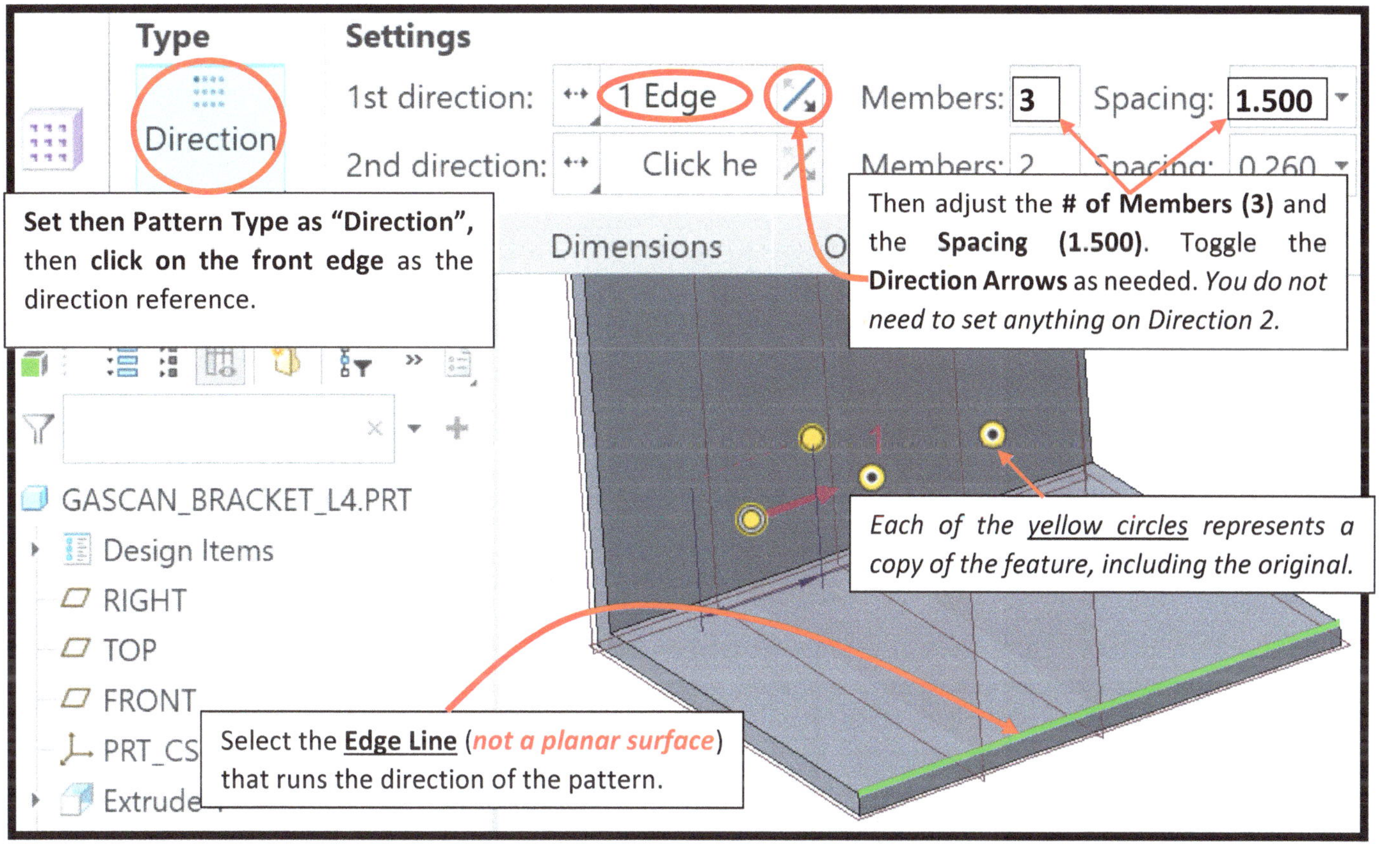

Step 10 – 2 - Checkmark to accept the Pattern. The 3 ribs should appear in the model. If not, Edit Definition of the Pattern and check the direction arrow and reference to be sure the ribs are patterned correctly.

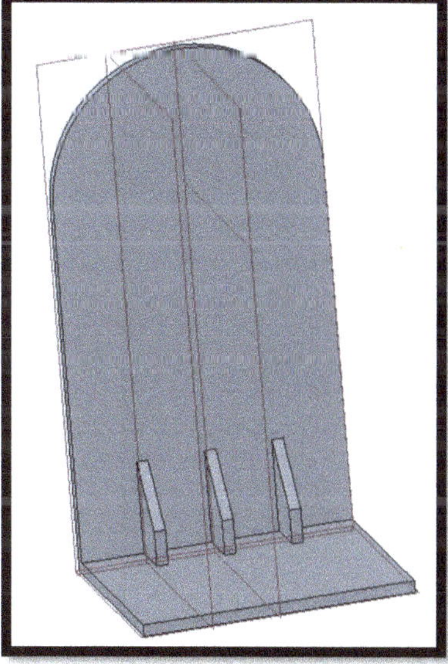

Edit Definition of a Pattern vs. a Feature:

Need to make a change to the Pattern or the underlying feature? Use Edit Definition on the correct item in the Model Tree. You can view the underlying features within a Pattern by using the drop-down arrow next to the pattern item in the model tree and then use Edit Definition as needed to make changes to the Feature. Any changes to the underlying feature should update automatically to all copies as they are by default dependent copies.

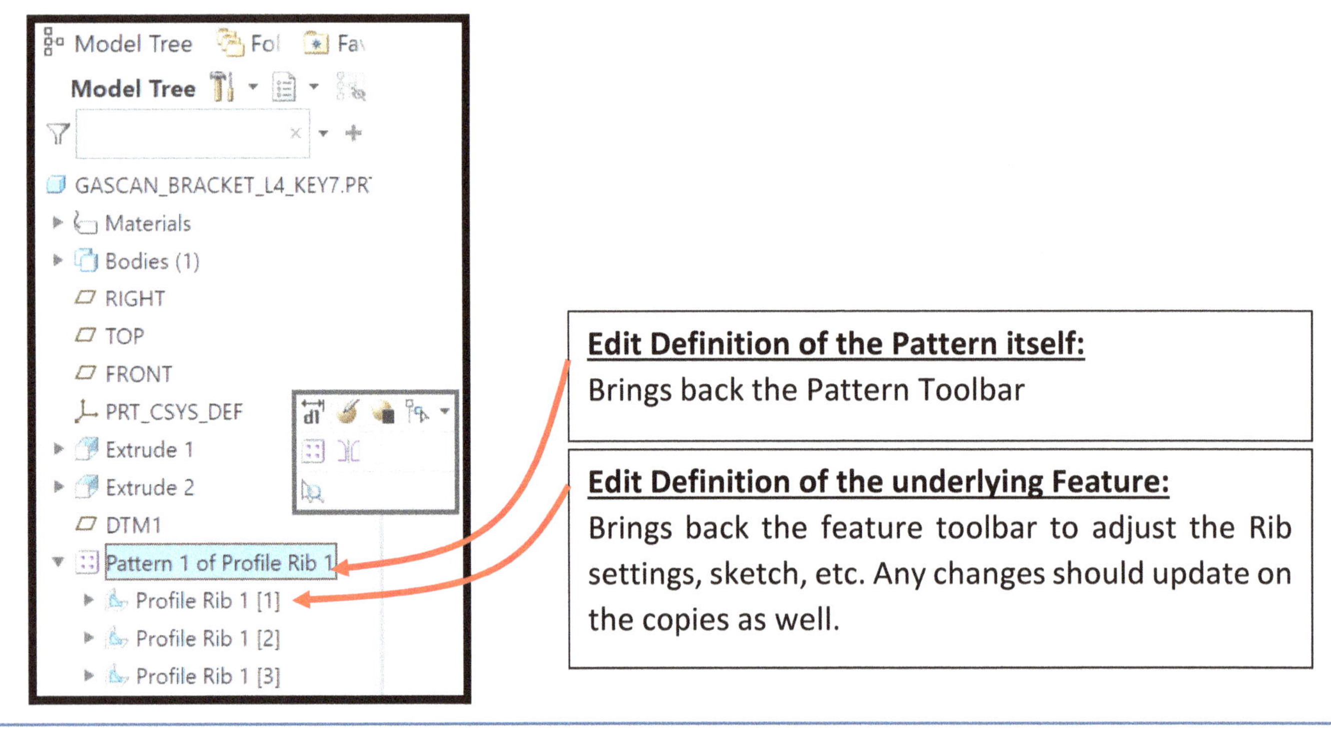

Edit Definition of the Pattern itself:
Brings back the Pattern Toolbar

Edit Definition of the underlying Feature:
Brings back the feature toolbar to adjust the Rib settings, sketch, etc. Any changes should update on the copies as well.

Step 11 – Create a **Datum Axis** at the center point of the half circle on the top of the bracket that will be used for a Coaxial Hole placement reference in the next step.

- **Select the Axis Tool** in the top toolbar – **click on the curved edge of the Extrude 2 arc** in the graphics window – the axis at the center of the arc should appear - **press OK.**

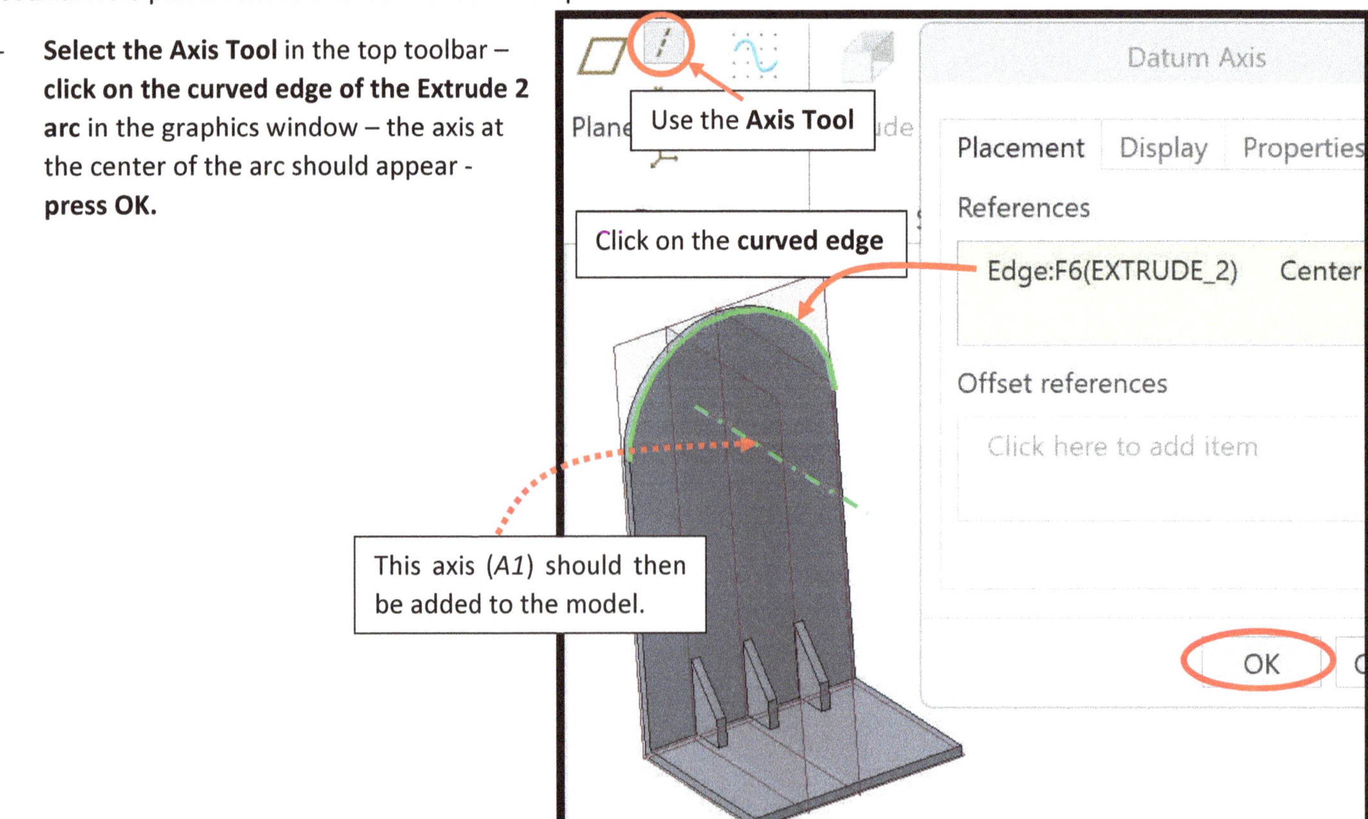

Step 12 – Use the **Hole Tool** to create a Coaxial hole through the center of the half circle 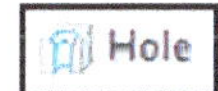

Hole Type: **Simple**

Profile Option: **Flat**

Toolbar Settings: **Hole Diameter = 0.750, Hole Depth = Thru All** (*'Drill to intersect all surfaces'*)

Placement Type: **Coaxial**

Placement References: - Hold CTRL on the keyboard to choose <u>two primary</u> references: Select the **Axis A1** and the **front planar surface** of the model as shown below.

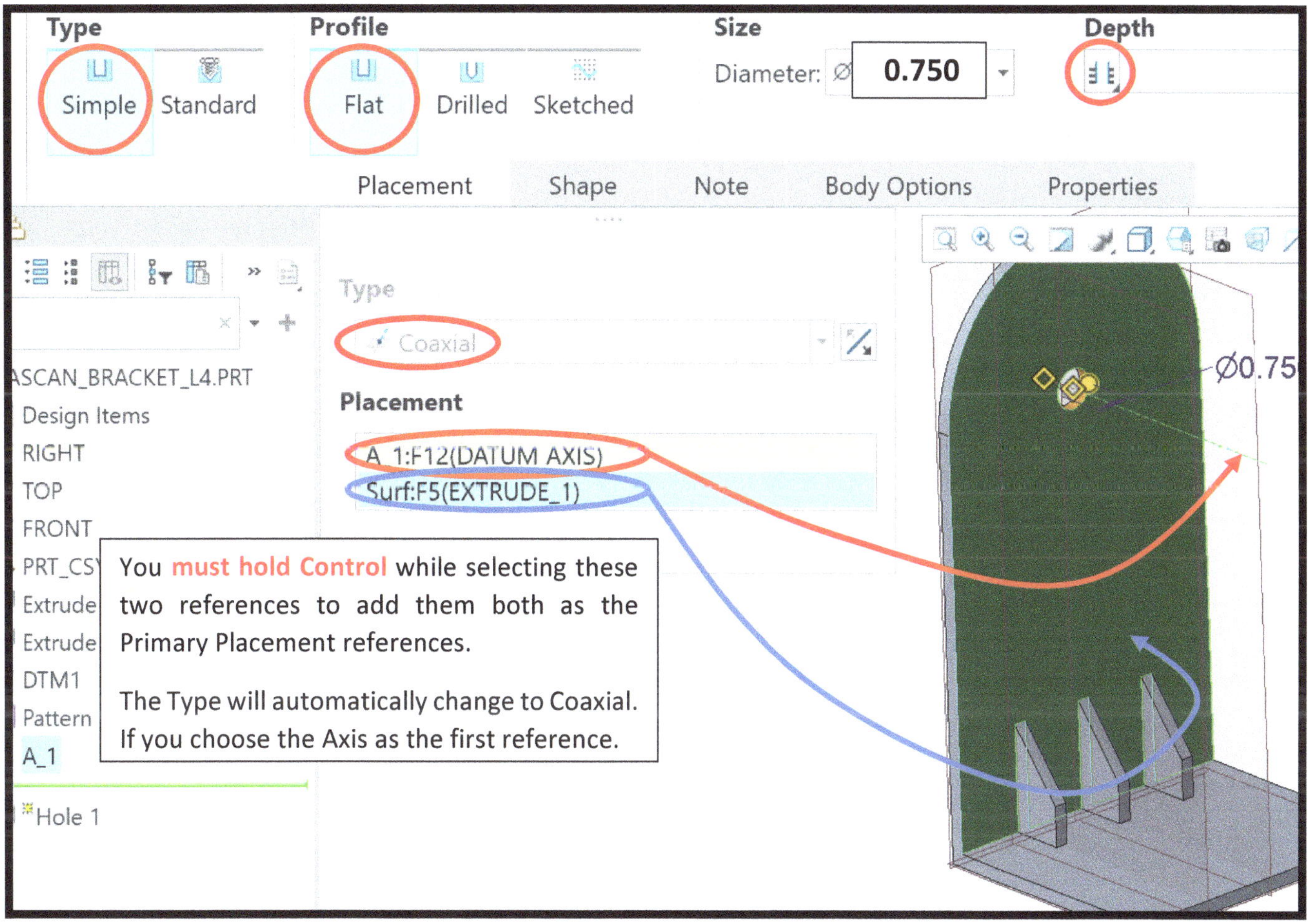

Step 12 -2 – Press the checkmark to accept the hole.

Step 13 – Use the **Hole Tool** to create a threaded hole on a diameter around the previous hole.

Hole Type:　　　　　　**Standard**

Profile Option:　　　　**Tapped**

Shape Tab Options:　　　**"Thru Thread"** setting

Toolbar Settings:　　　　**Hole Size = "3/8-16", Hole Depth = Thru All** (*'Drill to intersect all surfaces'*)

Placement Type:　　　　**Diameter**

Placement References:
- **Primary Reference**: select the **front face surface** of the model
- **Offset References: Adjust the Type to Diameter** instead of Linear, and hold control on the keyboard while selecting the references.
    - o **Select Axis (A1) with a value of 3.000" & Front Datum with a value of 0.0 degrees**

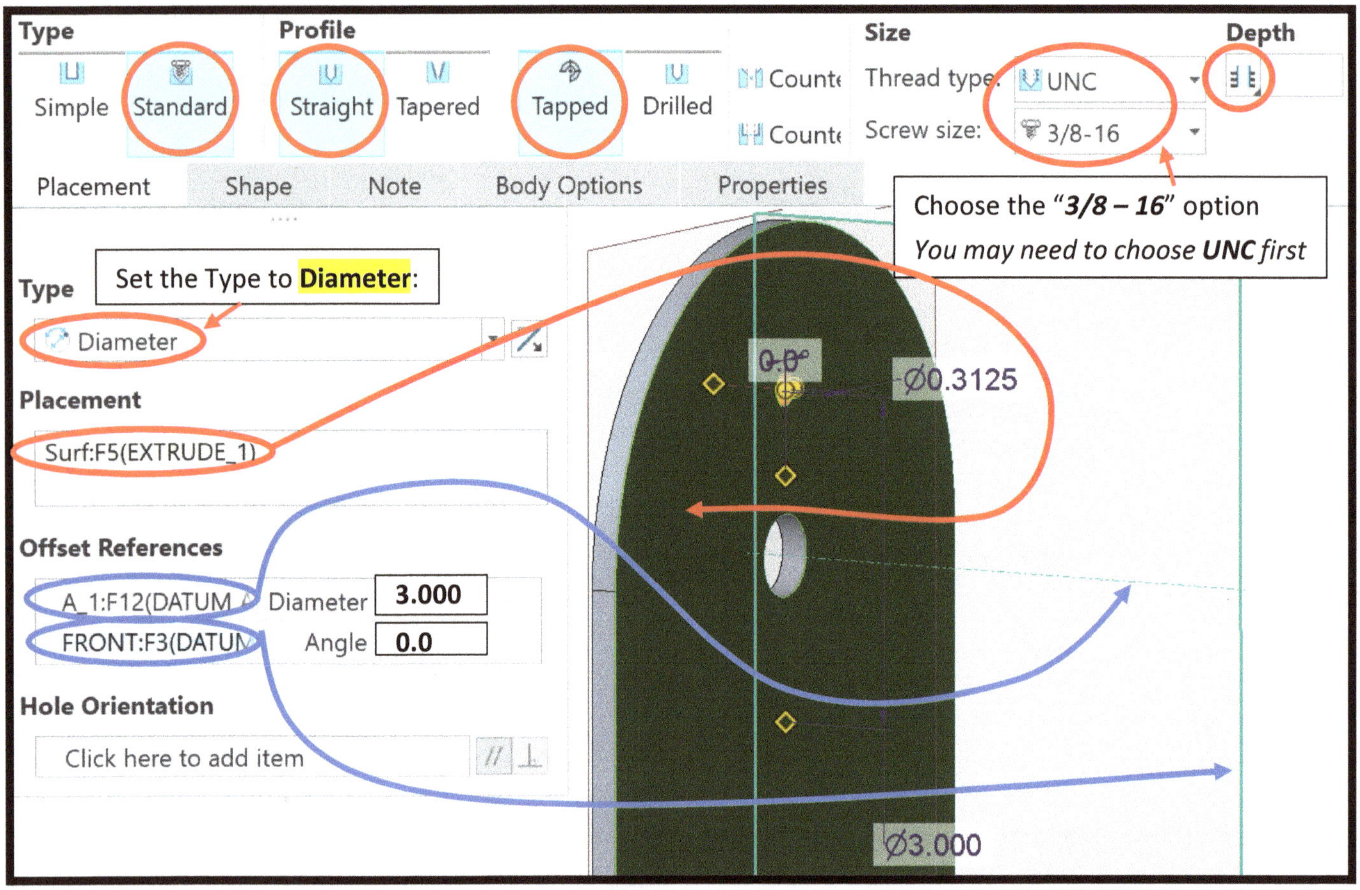

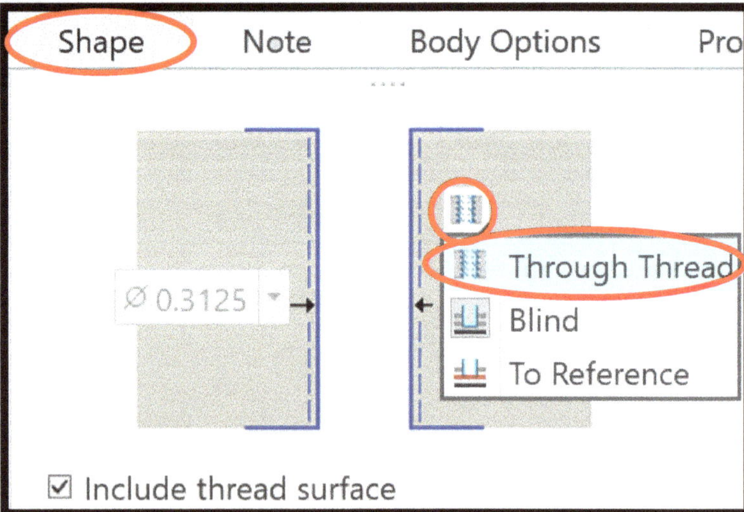

After setting the hole options above, **open the Shape Tab** and change the **Thread Depth Option to 'Through Thread".** This will thread the hole the entire depth of the hole so that a bolt can be threaded through. **This option is only available if you setup the toolbar properly with a Standard Type hole.**

Step 13 – 2 – Press the checkmark to accept the hole. Note that your hole may appear at either the 12 O'clock or 6 O'clock position around the axis, either will be fine.

Note that the model will not visually show the threads of the hole to reduce the graphics processing load on the computer, but the **outline of the threads is visible** when you are actively selected on the *Hole 2* in the model tree.

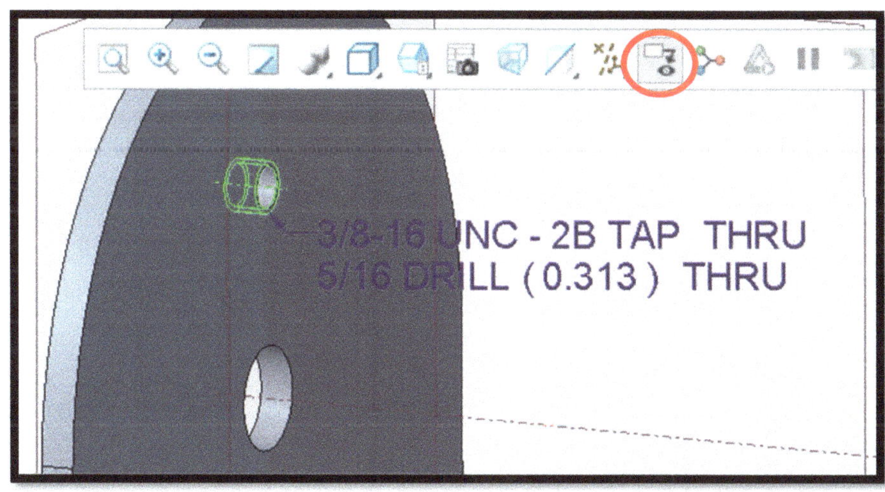

Since this is a Standard Type hole, **a Note detailing the Thread, Tap Size, and Depth is created**. You can toggle this off with the Annotation Display tool.

Step 14 – Pattern the hole using an **Axis type pattern** reference about Axis A1 to create 4 total copies of Hole 2.

Pattern

- **LMB to select Hole 2** from the Model Tree – select the **Pattern Tool** – **change the reference type to Axis** – in the reference box **select Axis A1** from the model tree or graphics window – **adjust the # of instances to 4** – set the **spacing to 90 degrees**.

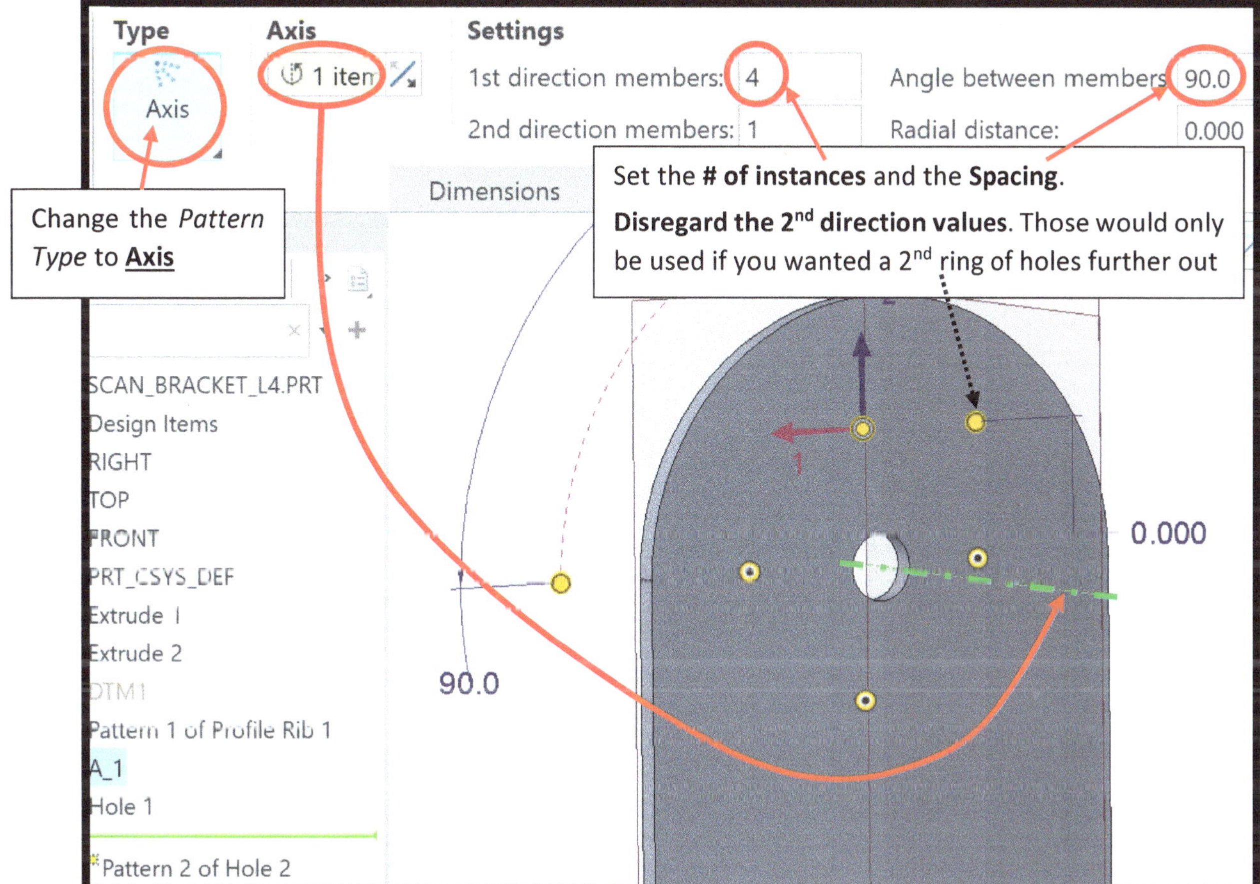

Your 4 instance circles may appear to be floating away from the bracket, which is fine. The reason is that the holes are set to drill "through all" surfaces, so the hole depth is as wide as the entire bracket model, not just the thickness of vertical wall.

Step 14 – 2 – Press checkmark & you will see the holes appear properly on the model.

Step 15 – Use the **Extrude Tool** to create a slot in the bottom part of the model. Use the **planar surface** of the bracket (upper surface of the horizontal portion of the bracket) as the primary sketch plane. Leave the *Sketch Orientation* references as default (*Surface & Right*)

Extrude

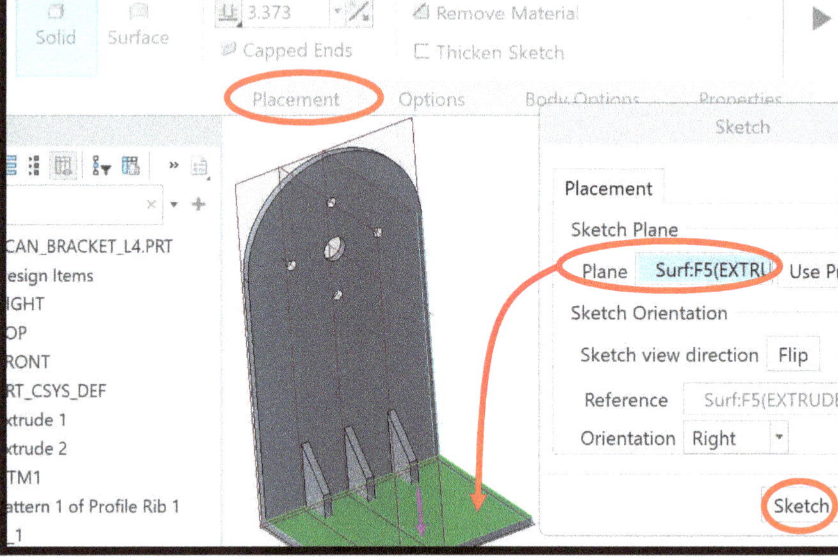

Step 15 - 2 – Sketch the **closed-loop shape** in the bottom left corner of the bracket as shown in the zoomed-in image below.

1) **Add a Centerline:** Add a **vertical centerline dimensioned 0.500"** from the left edge. Make sure it is not placed snapping onto any other entities otherwise you may get a resolve sketch menu and need to delete a constraint.

2) **Sketch the Shape**: Sketch the shape using the **Rectangle tool,** placed so it is **Symmetric** about the centerline**.**

   Tip: You may need to zoom in so it doesn't snap on other lines.

3) Use the **Center & Ends Arc tool** to add the arcs on each end. Make sure that the start & end points snap onto the corner.

4) **Delete Segment**: Use the **Delete Segment tool** to remove the top and bottom lines of the rectangle.

5) **Dimension the Sketch: Create the Dimensions** and set the values as shown, with an overall **height of 0.800"**, **width of 0.250"**, and **distance from the bottom edge of the model as 0.400"**.

   If you have excess Dimensions leftover, you may need to set **Symmetry.**

6) **Checkmark to accept** the sketch.

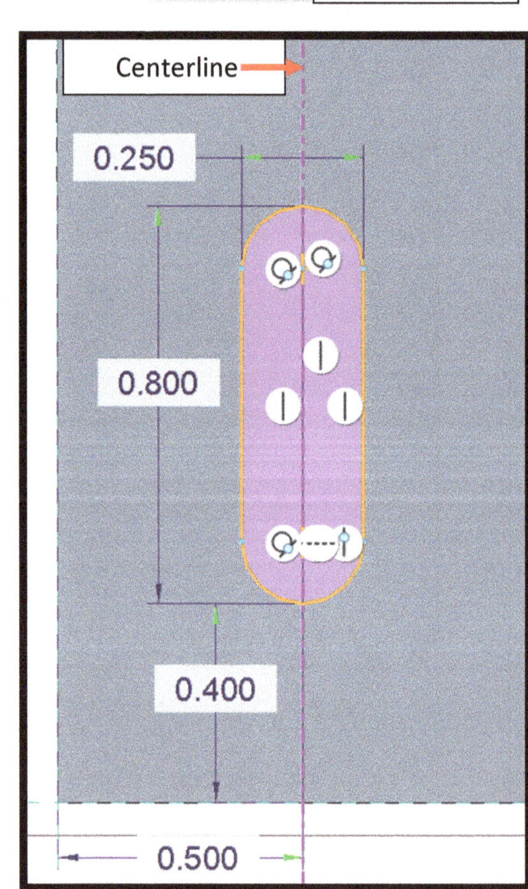

Step 15 – 3 - Set the **Depth, Direction, and Material Removal options** in the Extrude Toolbar.

- Toggle the extrude **Direction Arrow icon** to flip the extrude direction to go into the Model material.
- Toggle on **Remove Material** (if it did not do so by default automatically)
- Set the **Depth option** to the **Through-All Option** (i.e. *"Extrude to intersect with all surfaces"*)

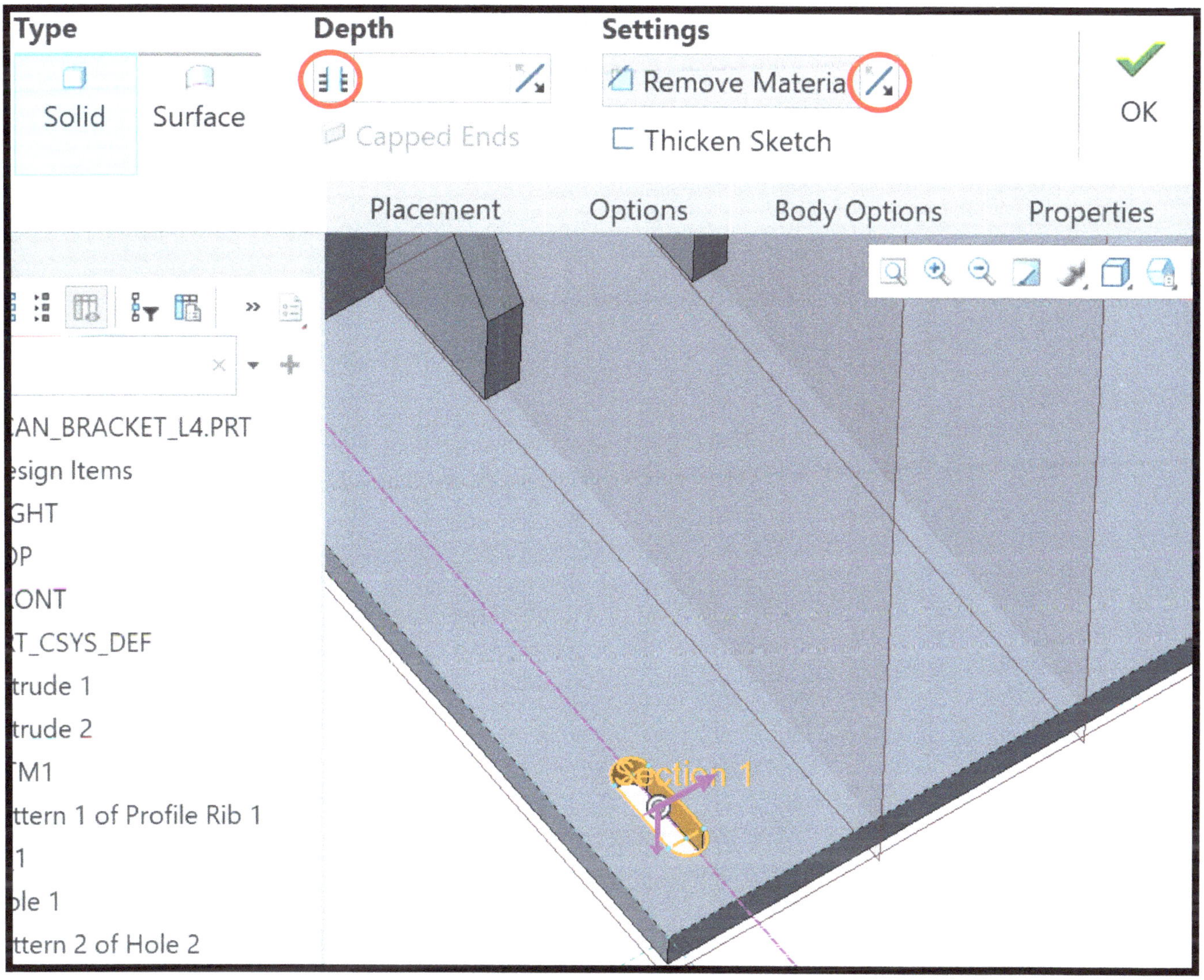

Step 15 – 4 - Press the Checkmark to accept the Extrude.

Step 16 - Use the **Hole Tool** to create a simple, non-threaded hole next to the slot.

Hole Type: Simple

Profile Option: Flat

Toolbar Settings: Hole Depth = Thru All (*'Drill to intersect all surfaces'*)

Placement Type: Linear

Placement References:
- **Primary Reference**: select the upper **planar surface of the bracket base**
- **Offset References:** Hold control on the key board when selecting the two offset references.
  - o   Select the **left edge of the bracket with a value of 0.500"** & the **front edge of the bracket with a value of 1.875"**

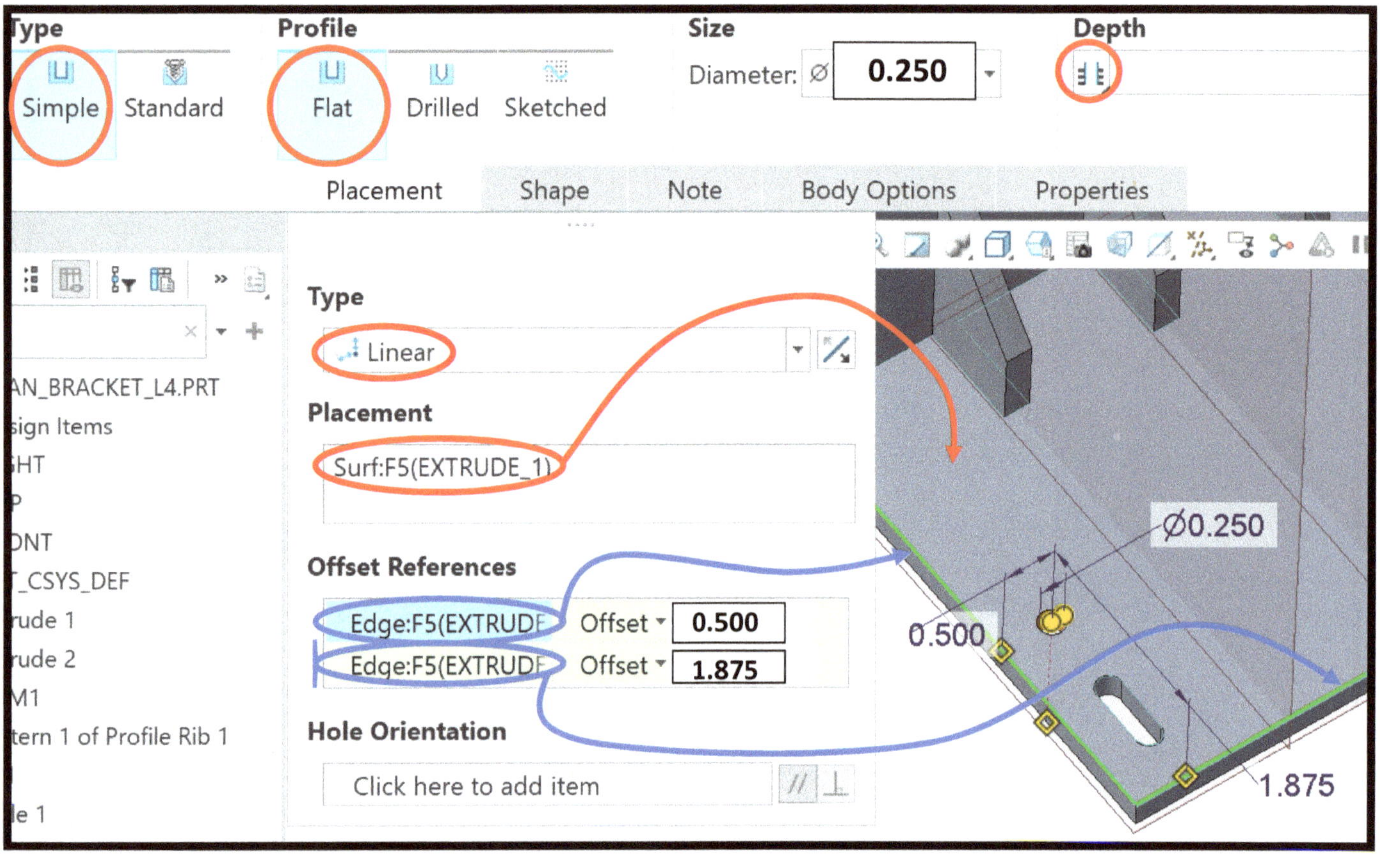

Step 16 - 2 – Press the Checkmark to accept the Hole.

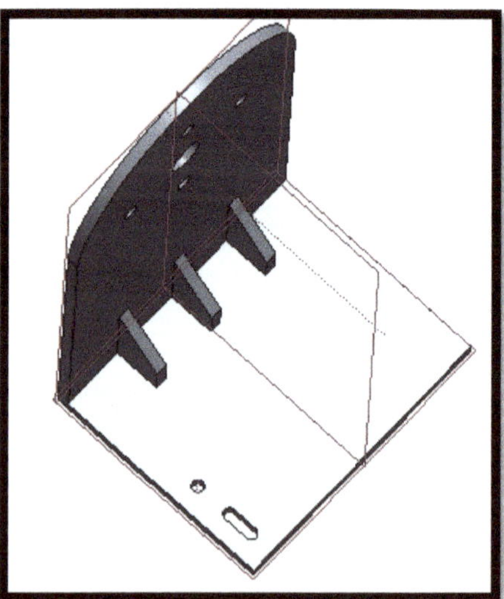

Step 17 - Group the slot and hole features together so they can be patterned together as a unit.

- While holding Control on the keyboard, **LMB click to select the slot (Extrude 3) and previous hole (Hole 3)** – in the pop-up menu **select Group.** Make sure that only those two features are grouped!
- **Knowledge Tip: To cancel a group,** select the group name in the Model Tree – choose Ungroup.

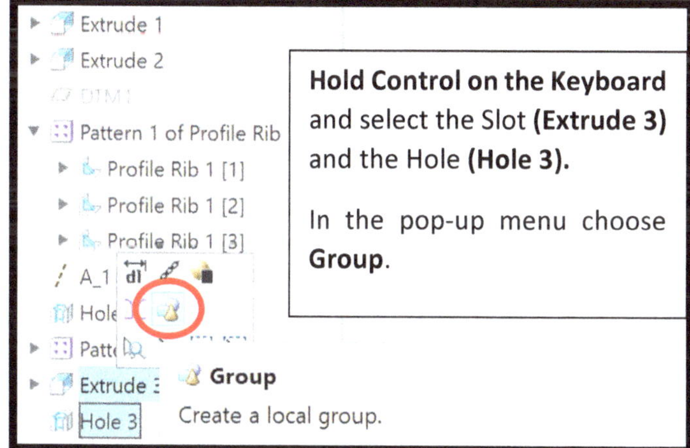

Hold Control on the Keyboard and select the Slot **(Extrude 3)** and the Hole **(Hole 3).**

In the pop-up menu choose **Group**.

Step 18 - Create a Pattern of the Group created in the previous step to make 2 rows of copies.

1) **Select the Group** *(with Extrude 3 & Hole 3 inside it)* from the Model Tree and select the **Pattern tool**
2) **Choose the Pattern Type:** Set the type as "**Direction**".
3) **Set the 1st Direction Reference: Select the short <u>edge</u> of the bracket** (choose an <u>edge</u> that runs the desired direction, not a planar surface) – set the **# of members to 2** – set the **spacing value to 1.875"** – toggle the **Arrow Direction** as needed to pattern the correct direction towards the model.
4) **Set the 2nd Direction Reference: Click in the second reference box - select the long <u>edge</u> of the bracket** - set the **# of members to 5** – set the spacing value to **1.250"** – **toggle the Arrow Direction** as needed.
5) **Toggle off un-wanted Copies**: Click on the yellow circle for the 3 instances that are close to the ribs.
6) **Press the Checkmark** to accept the Pattern.

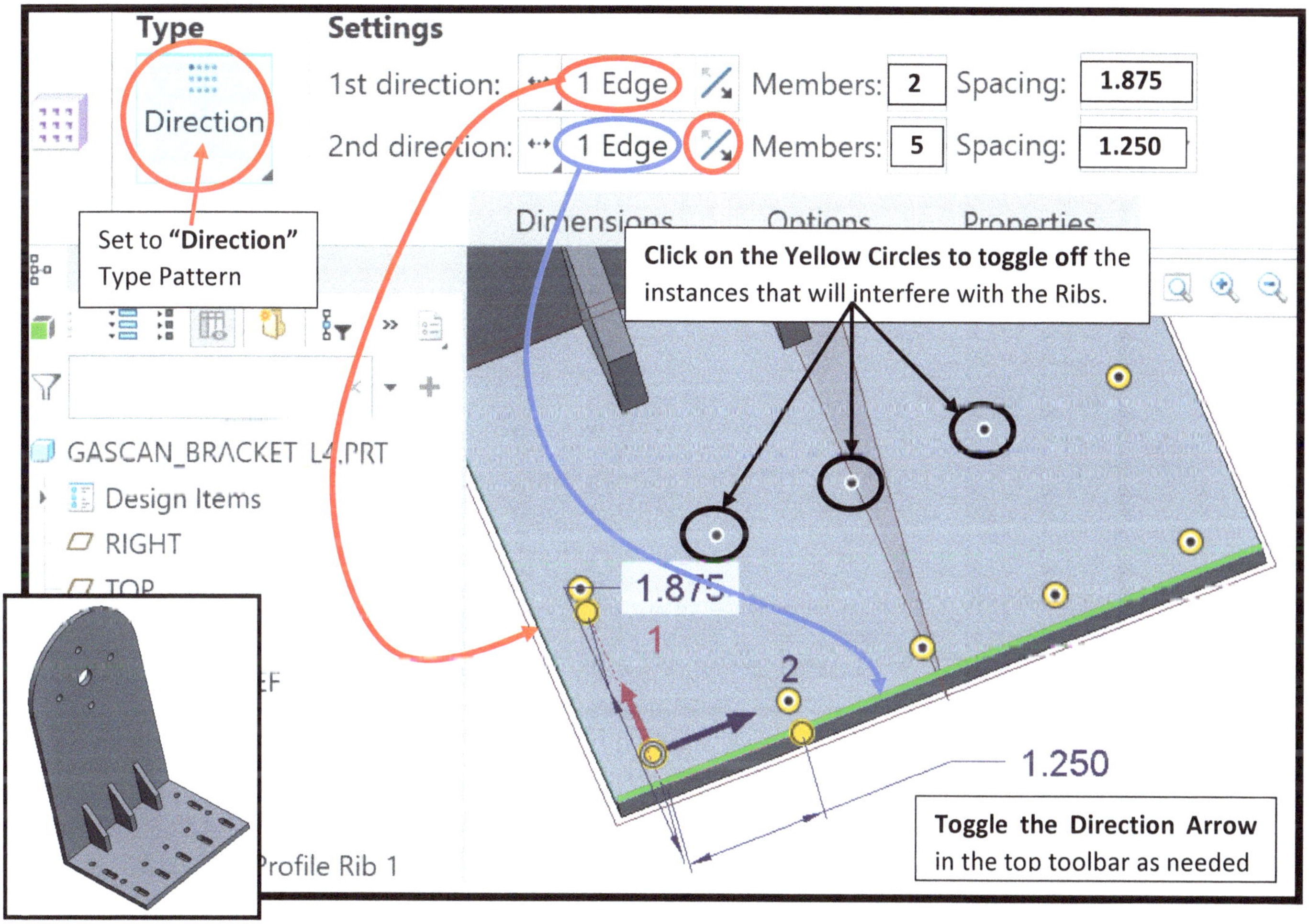

Step 19 –Save your model before continuing to the next step.

Design Revision:

Follow the steps on the following pages to make a few design revisions.

The bracket and rib thickness are thicker (and therefore costlier) than they need to be for the design criteria of this bracket. An analysis shows that a thickness of 0.125" will be appropriate for this part, as well as using ribs that are only 0.125" thick.

Also, the four mounting holes around the half circle should be changed to a smaller size to allow a ¼-20 sized bolt to fit through *without* having to be threaded in (a nut will be used on this bolt in the future)

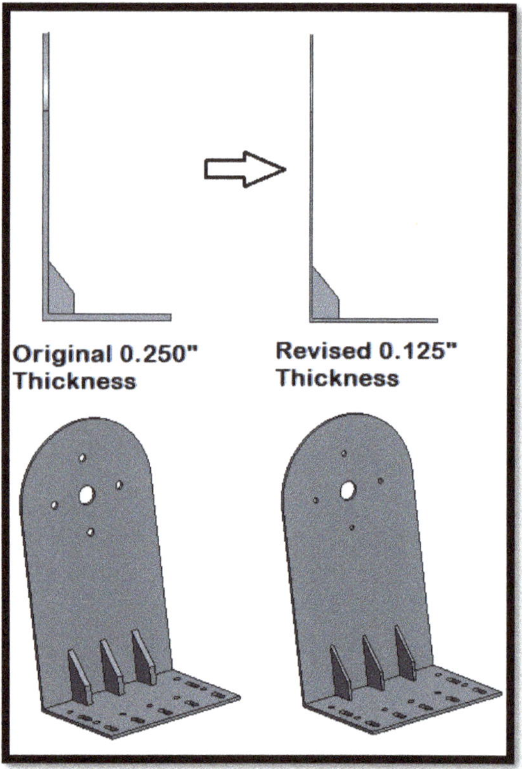

Step 20 – Adjust the Bracket Thickness. When Extrude #1 changes thickness then Extrude #2 should change appropriately if the Extrude #2 depth was set to "up to referenced surface" as instructed in the steps.

- **Select Extrude 1** from the Model Tree and choose "**Edit Dimension**" ("d1" icon in the pop-up menu) – this brings up the sketch dimensions on the model – find the thickness dimension and **change it to 0.125"**. You need to press **Control + G on the keyboard** to force an update on the model.
- *Alternatively, you could use Edit Definition - Placement –Edit to get back to the sketch to make the change.*

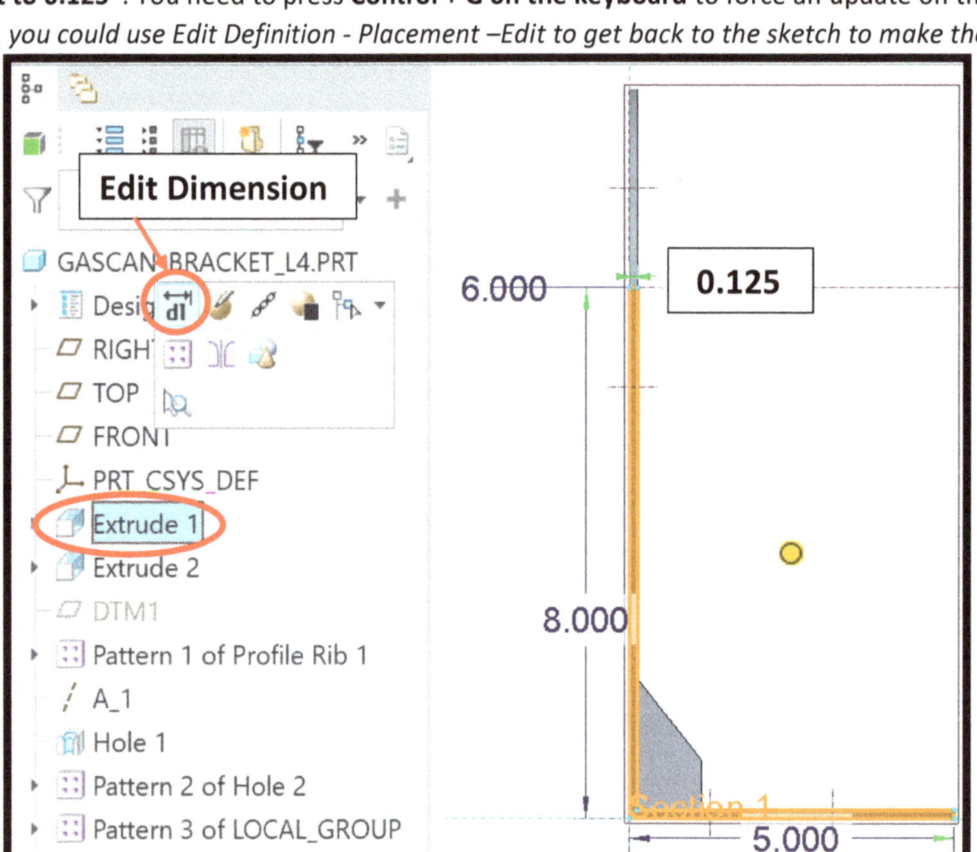

Note that Extrude 2 should also change thickness as it had a Depth of "To Reference Surface". If your Extrude 2 did not change thickness you should Edit Definition of it and adjust the depth setting to the correct "To Reference" setting.

If your model has any failed features (colored red in the Model Tree), use Edit Definition of each feature and look for errors. Most likely is that Extrude 1 or Extrude 2 may have an error on the sketched shape or extrude depth setting.

Step 21 – Adjust the Rib Thickness:

- Select the **drop-arrow of Pattern 1** in the model tree – select **Profile Rib 1** – choose **Edit Definition** in the pop-up menu- adjust the thickness setting to **0.125"** - **Checkmark** to accept the rib changes.

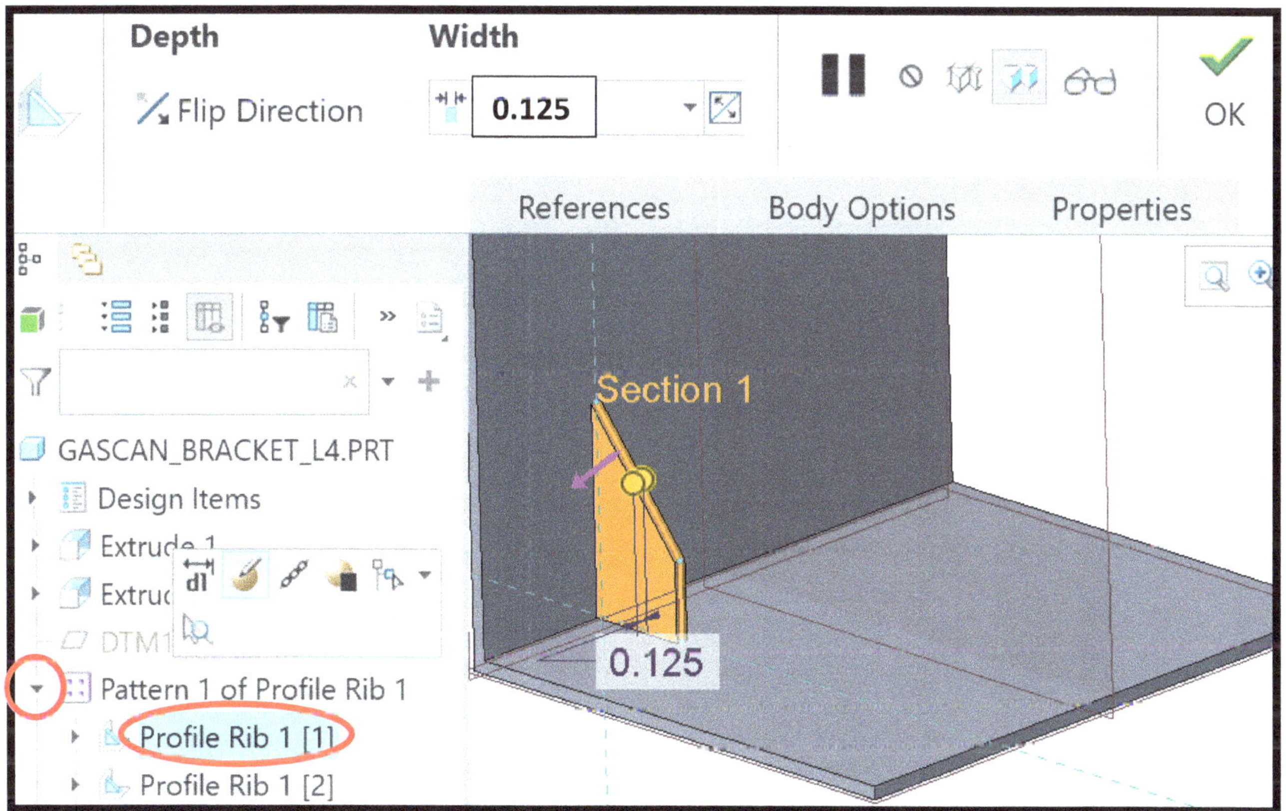

All 3 ribs should update automatically since they are dependent copies of the original Profile Rib 1.

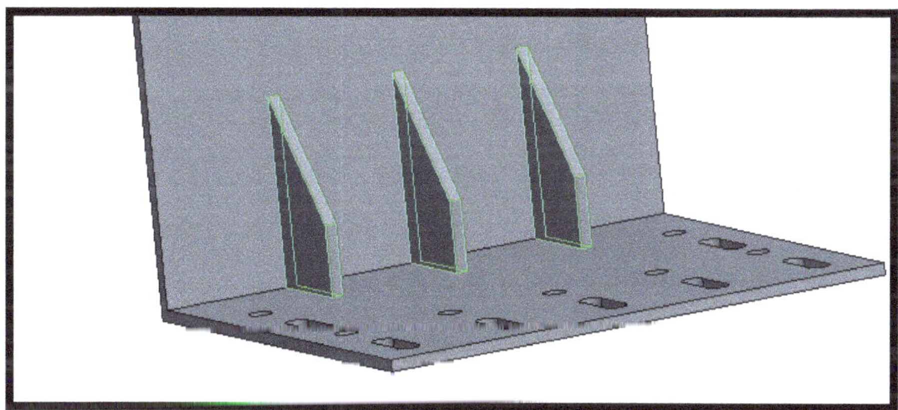

Step 22 - Adjust the Hole 2 settings to reduce the size and remove the tapping or threads of the hole.

- **Select the drop-down arrow on Pattern 2** in the Model Tree - **select Hole 2** and choose the **Edit Definition icon** in the pop-up menu – **Toggle Off the "Add Tapping" option**– **Toggle On the "Clearance"** option – Adjust the Screw Size to **"¼ - 20"** in the screw size option list – under the **Shape tab** check that the option for **Close Fit** is selected by default
- **Press the checkmark** to accept the hole changes.

The holes should now be slightly smaller, without threads. This will allow a ¼-20 sized bolt to fit through with a close fit so it can be tightened against a nut or some other hardware during assembly.

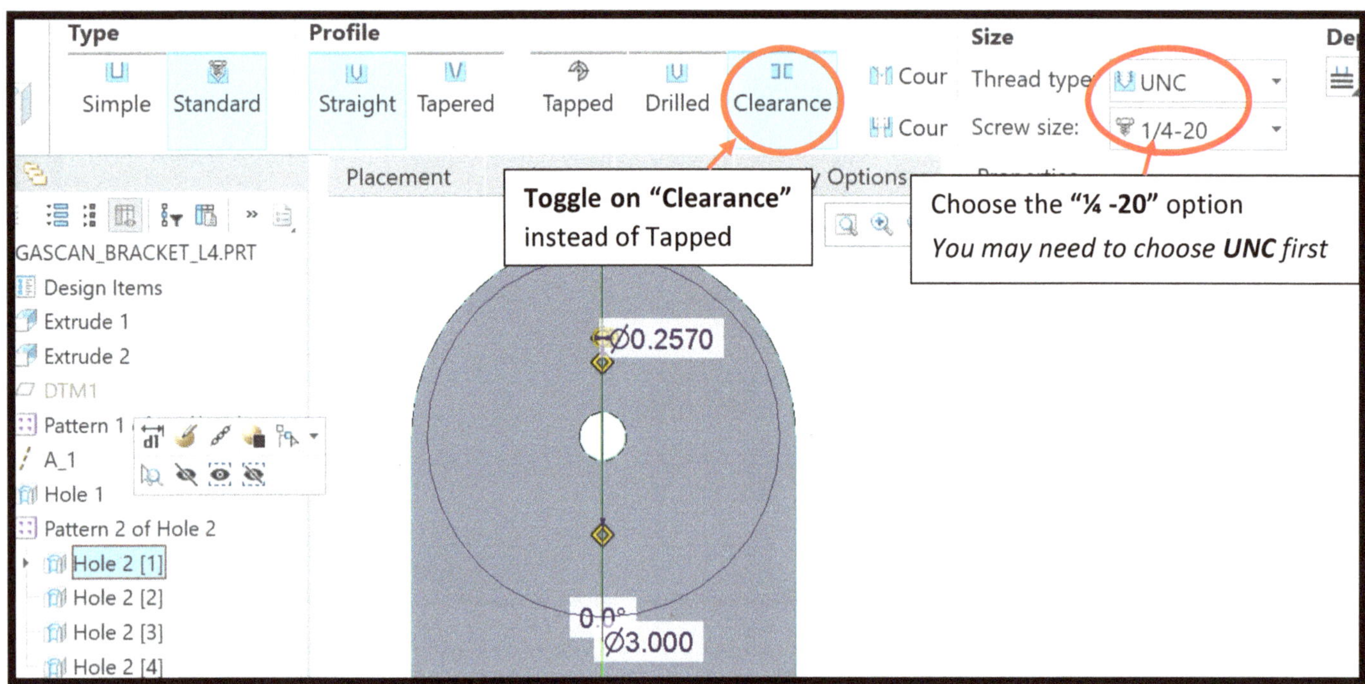

Step 23 - Set the display properties of <u>**each datum**</u> so that we may use them as a **Geometric Tolerance** and be able to change them to **ASME style** in our drawings. We must "**Set**" & Rename each datum to match the style of this course using letters for names of the Datums. You do not need to Set or Rename DTM1.

- **LMB to select a Datum** from the model tree - **hold RMB – Properties - press SET ("-A-")** – select the **name box – Type in a new name** according to the list below – **OK**.
- **Repeat for each Datum**.

RIGHT: Rename to **B - Press "Set" - OK**

TOP: Rename to **C - Press "Set" - OK**

FRONT: Rename to **A - Press "Set" – OK**

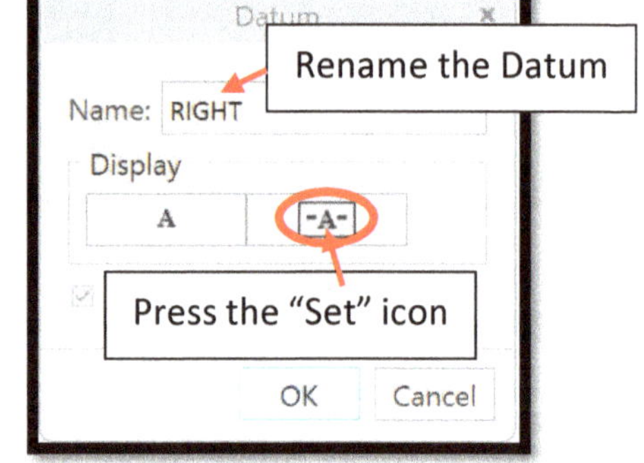

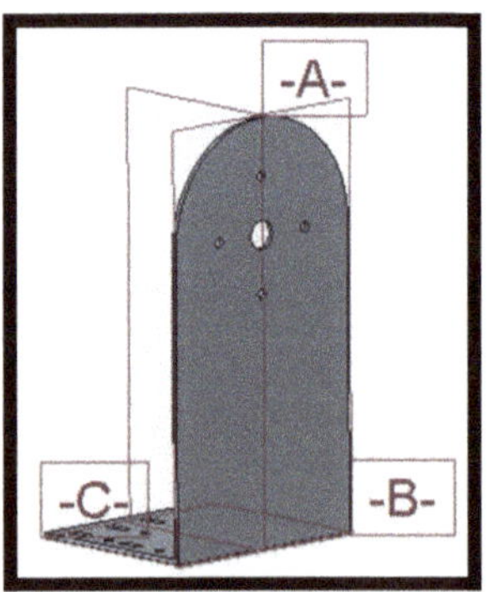

After setting and renaming the datums you should see the datum names or "tags" on the model screen. If you do not see all 3 datum names, you likely did not press "Set" for that datum.

Step 24 – Save **your completed part model.**

Step 25 - Create a Drawing of the part to match the Key:
- **File – New – Drawing – use "*L4_GasCan_Bracket*" for the name – OK.**
- Leave the Template as default (C-size) – you will choose the UND format in the Sheet Setup drawing step later on, don't do it here – **press OK** to start the Drawing.

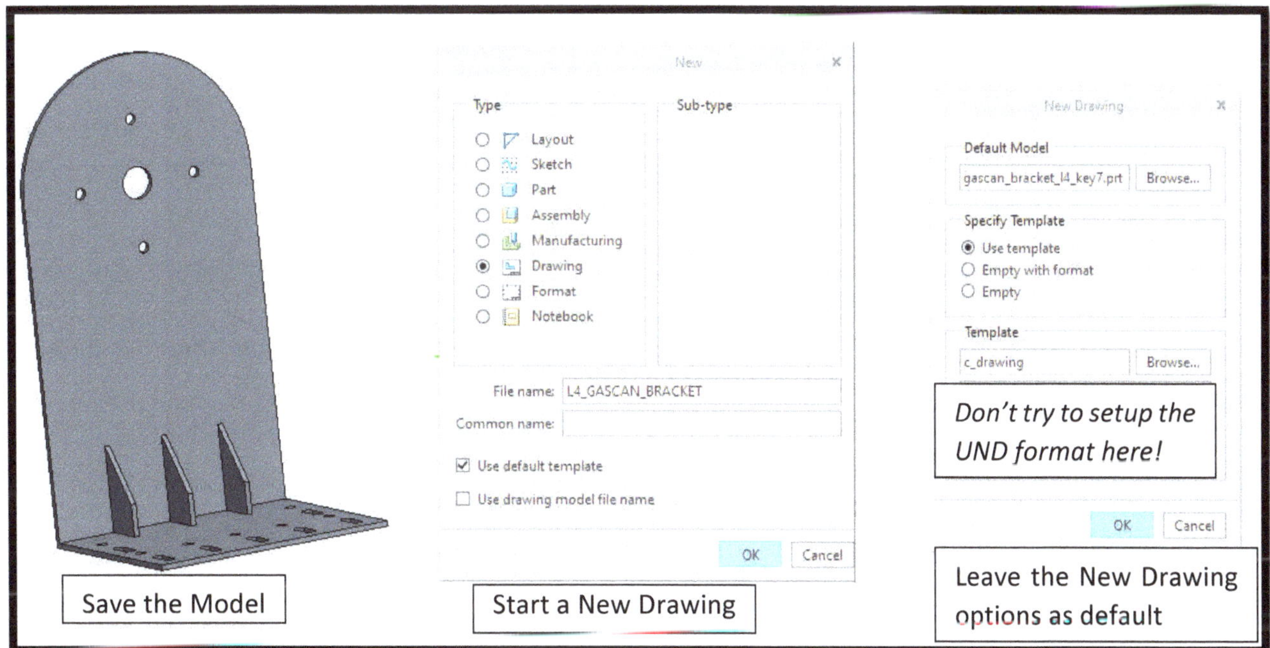

Step 26 – Follow the Steps in the ***Basic Drawing Steps*** (in the Appendix), <u>**excluding step 8**</u> (Cross-Section) as there is not any "interior" geometry to show a cross-section for. New for this lesson you will be expected to clean up the datum tags (**Step 9** of the *Basic Drawing Steps*) and add a Standard Hole note (**Step 7** of the *Detail Drawing Steps*).

Step 27 – Save your Drawing, check that it matches the lesson key posted to Blackboard, and create a PDF **to submit for grading.**

End of Lesson 4

Lesson 5 – Detail Drawings (Ram Extension)

Lesson 5 is to create an extension component for the Ram Mount system that allows a GPS mount or similar to pivot and swing into new positions. This lesson will also focus on creating Detail drawings which include dimensions and other details.

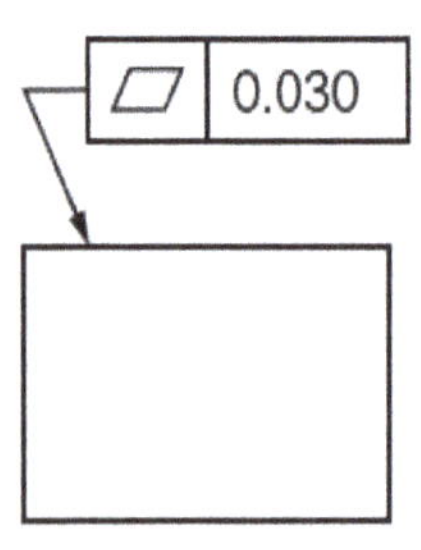

Why create 2D Part Drawings?

Many companies rely on 2D paper drawings of parts and assemblies of components even though the models are created in 3D. The older engineers/drafters may still design components completely in 2D. Some of them may even take your expertly crafted 3D models and remodel them in 2D using their outdated software because they are unwilling to train or use new software.

Drawing Details & Standards:

Drawings will often follow a company, industry, or national standard that specifies drawing details and layout styles. For this class you should use a "Clean & Organized" style; use your judgement to show enough details that the design can be easily understood and created in a machine shop but not so many dimensions that the drawing becomes cluttered, especially on 3D printed parts.

Example: This image below has properly positioned dimensions, with shorter dimensions nested inside of the larger ones. Dimension lines, when possible, are not crossing over each other, and multiple notes and dimensions are used to detail the size or placement key features. Datums, if shown, are ASME Style.

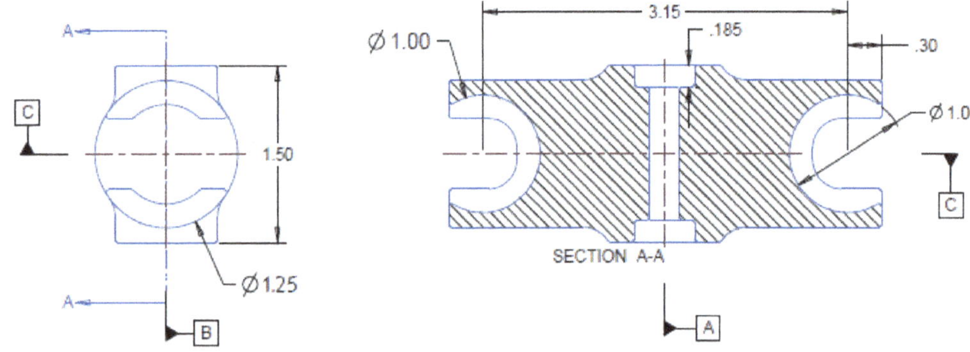

Geometric Tolerances: Many detail drawings have a variety of symbols and notes that specify *Geometric Tolerances* to the machinist that must be met for the part to be created correctly (surface roughness, concentricity, weldments, etc.) Geometric Tolerances, as well as dimension Decimal Places, communicate the level of precision and accuracy these features must be manufactured to be acceptable, with a **tighter or more precise tolerance requiring more time, skill, and cost to manufacture**. In CREO you can specify various types of geometric tolerances but we will not cover them in this class as they are mainly used in a professional or manufacturing environment and are too complex for a beginner.

TOLERANCES

DECIMALS	ANGLES
X.X $= \pm 0.1$	X $= \pm 5$
X.XX $= \pm 0.01$	X.X $= \pm 0.5$
X.XXX $= \pm 0.001$	X.XX $= \pm 0.50$
X.XXXX $= \pm 0.0001$	

Decimal Places & Tolerance: The UND template specifies tolerance of dimensions by their number of decimal places. As more decimal places are shown, the manufacturer will need more time, skill, and money.

For this class the decimal places should match the key, but in your own projects you may find that using a standard of 2 or 3 decimal places is fine, and can change the setting on any fractional dimension to more decimal places or use Fractional display (e.g. 1/8th thick sheet is 0.125 so would show 3 decimal places of 0.125", instead of only 2 decimals of 0.13").

Getting Started: Open the **CREO _Parametric_** software and follow the lesson steps carefully.

Step 1 - Set your working directory to your ME101 Working Directory Folder.

- Flle – Manage Session – Set Working Directory – Navigate to your ME101 Directory folder – OK

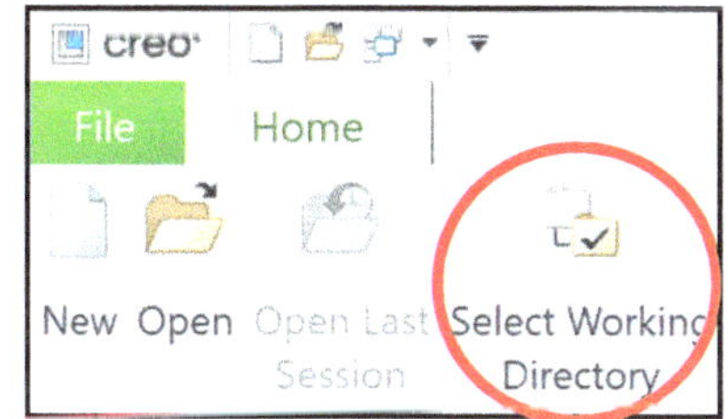

Step 2 - Create a new **Part** called **"Ram_Extension_L5"**. *You cannot use spaces in a filename!*

- File - New – select Part as the file type – type in the "Ram_Extension_L5" - press OK.

A menu may appear asking you to choose a **"Template"** – if so, choose **"inlbs_part_solid_abs"** - **press OK.**

Step 3 - Change the Material of the part to **AL6061** *Look under Legacy Materials*, and be sure your units are in the CREO Default (inch-lbm-Second).

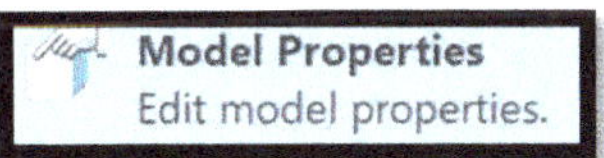

- File – Prepare - Model Properties - in the Material option select **change** – select the Legacy Materials folder - **double click** on "AL6061.mtl" – press **OK** – press **Close** to close out the Model properties menu.

Step 4 – Adjust the options so that <u>3 decimal places</u> will be shown in sketcher.

- File – Options – Sketcher – set the "**Number of decimal places for dimensions**" to **3 – OK - No***
** Don't bother saving a Configuration File, as you would need to load it each session to be useful.*

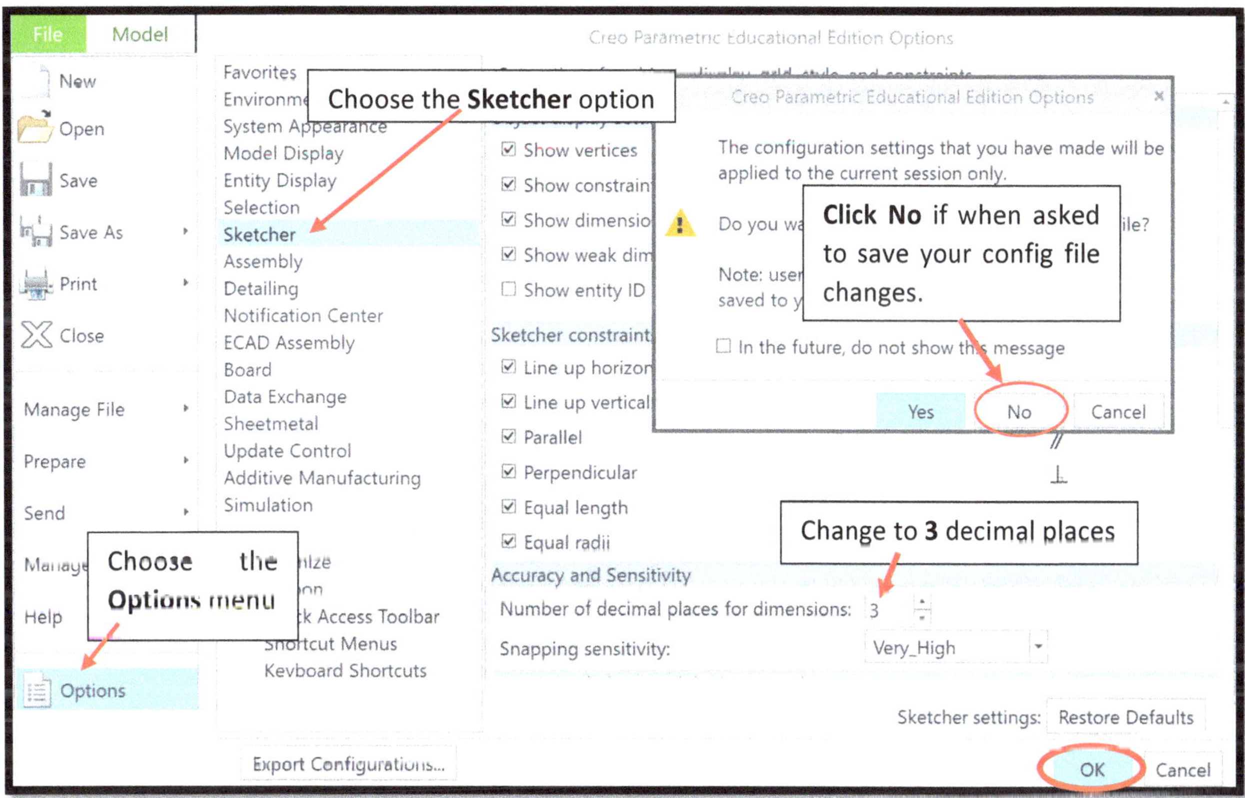

Step 5 - Create the first Extrude sketched on the **<u>Front</u> Datum**. (*image not shown*)

Sketch Shape: A circle centered on the origin with **a circle diameter of 1.250"**.

Extrude options: "Extrude Symmetrically on Both sides" option, with a **Depth of 3.750"**

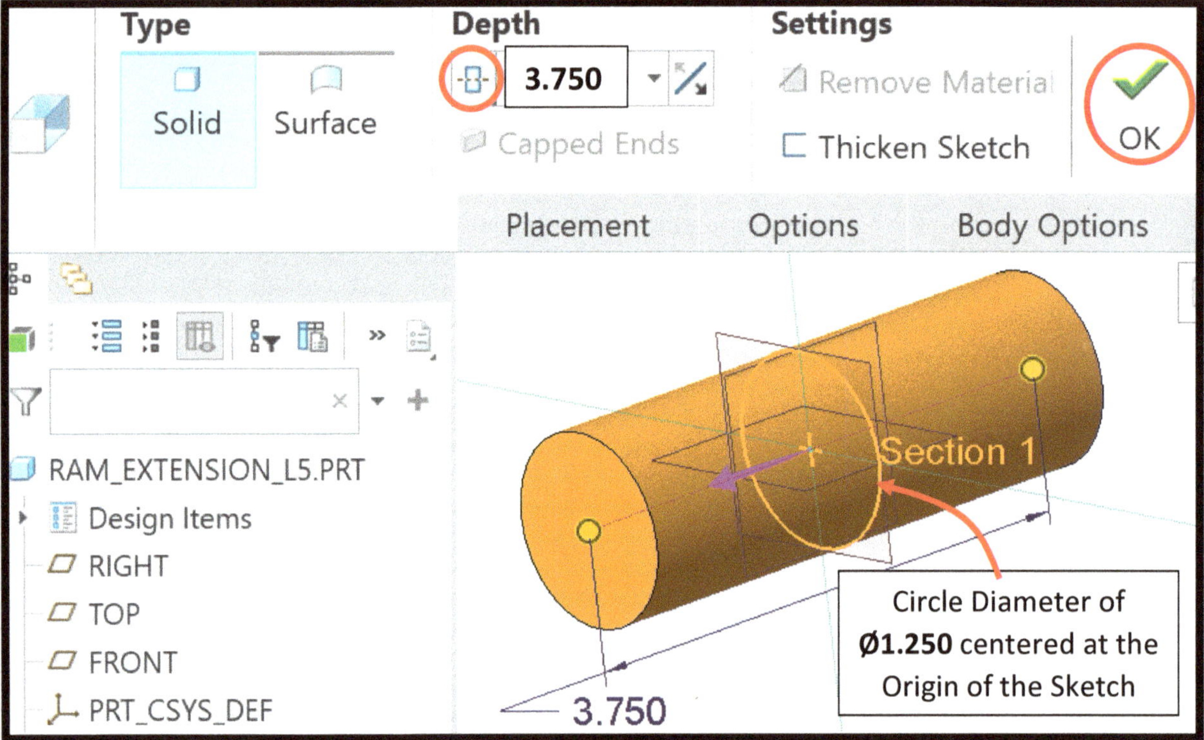

Step 6 - Create a second Extrude sketched on the **Top Datum**.

Sketch Shape: A circle centered on the origin with a **circle Diameter of 0.900"**

Extrude Toolbar Options: "**Extrude Symmetrically**" option, with a **Depth of 1.500"**, and **Toggle Off Remove Material**

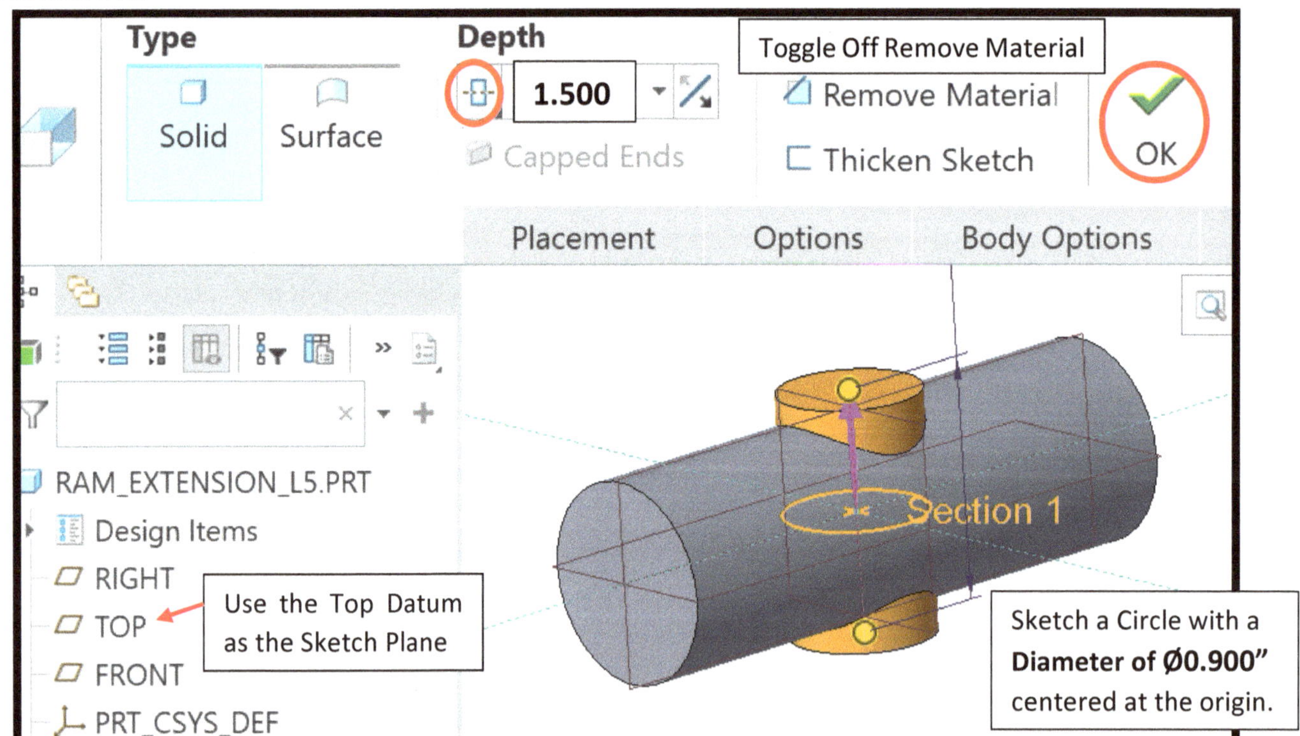

Step 7 – Use the Round Tool to add a Radius of 0.250" to the edges between the two extrudes. Select the two edges as shown. The rounds should automatically continue 360 degrees around the cylindrical surface.

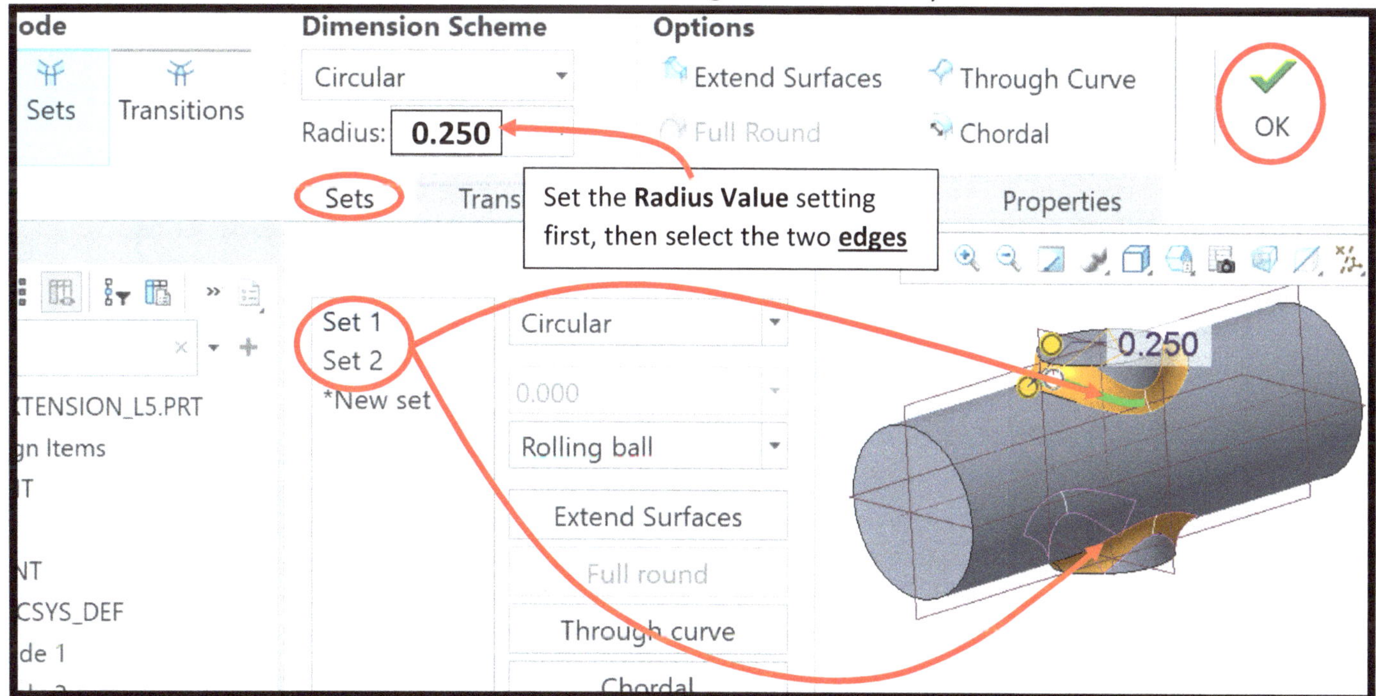

Step 8 – Create a **Revolve** to remove a spherical volume from one end of the cylinder, sketched on the **Right Datum. (– Placement – Define – select the Right Datum)**

1) <u>Axis of Revolution:</u> **Add an Axis of Revolution** along the horizontal reference line.

   - **Hold RMB in a blank area – Axis of Revolution Tool – LMB to place the start and end points**

2) <u>Sketch the Shape:</u> Sketch a **Center & Ends Arc** with the center point on the Axis of Revolution. The arc curve should start on the left reference edge* and stop on the Axis of Revolution. *Note – do not snap the arc point on to the Midpoint of the left edge, as that constraint will need to be removed if it does get set.*

3) <u>Complete the Closed-Loop</u>: Use the **Line Tool** to add a left vertical line and a bottom horizontal line.

4) <u>Set Sketch Dimensions</u>: **Create dimensions for the** distance from the **left edge to Center of the Arc (0.300")** and the **Arc Diameter (1.000").** Be sure to set a **Diameter** using the Dimension tool: **double-click** on the arc curve then press MMB to set the value position.

5) **Checkmark to accept the sketch.**

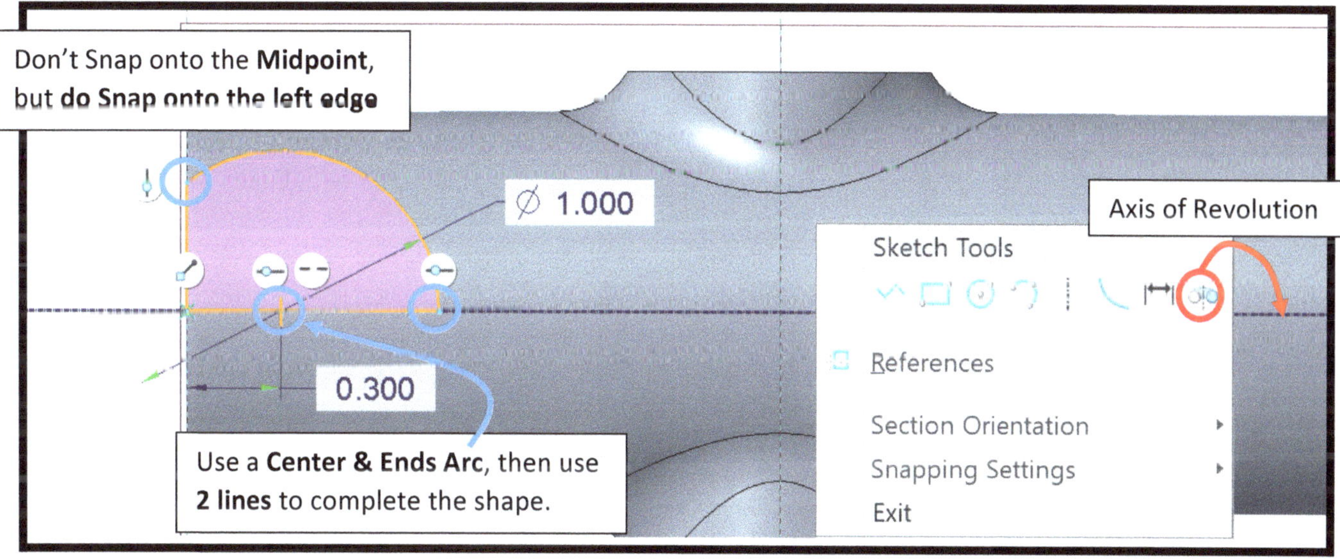

Tip: If you are unable to place an Axis of Revolution you likely did not choose the Revolve Tool!

Step 8 – 2 - Set the Revolve Toolbar Options.

- **Set the revolve rotation to 360 degrees.**
- **Toggle on Remove Material,** if not already done so by default.
- **Checkmark to accept the Revolve.**

If your Revolve is not visible on the graphics window, you likely did not add an Axis of Revolution to the Sketch, or did not have the sketch snapped on to the outer edge of the model.

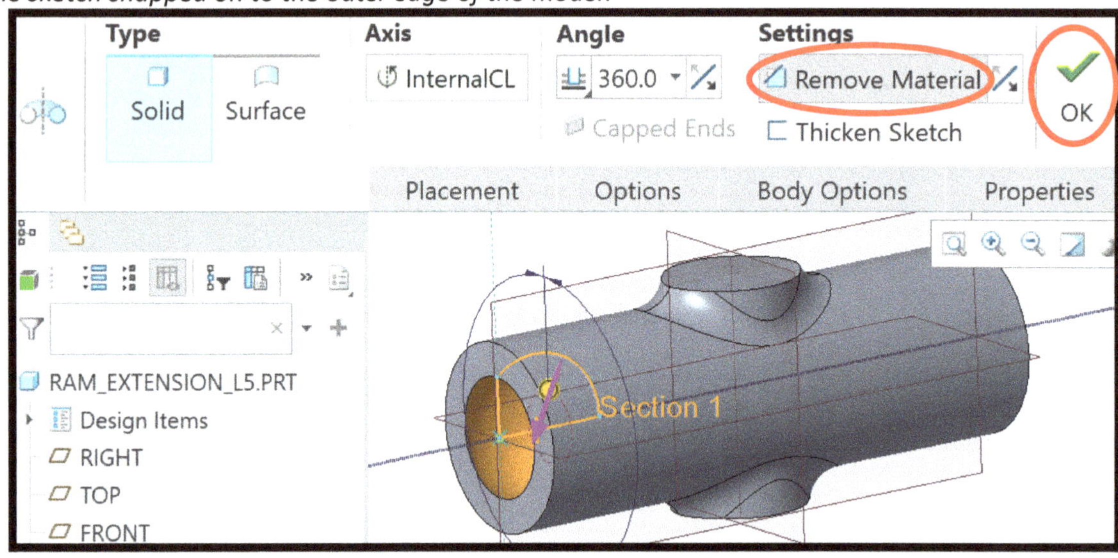

Step 9 – Create an **Extrude** to cut out the sidewall of the cylinder end, sketched on the **Right Datum**.

1) **Add a Horizontal Centerline** & **Toggle on Hidden Line Display** in the quick toolbar.
2) **Sketch the Shape:** Create a **Rectangle** and use the **Fillet Tool** to add a radius to the upper and lower right corners. *The rectangle start point should snap on the left edge but do not let it "snap" onto any hidden geometry or Mid-Points, you may need to zoom in to do so.*
3) **Set Constraints:** Set an **Equal Lengths** constraint to both Fillet radius curves. If you have more than 3 dimensions than you may need to add symmetry. The image below did not need to as it had a Midpoint of the rectangle snapped on the centerline instead.
4) **Sketch Dimensions: Overall Width (0.600"), Height (0.600"),** and **Fillet Radius (0.250").**
5) **Checkmark** to accept the sketch.

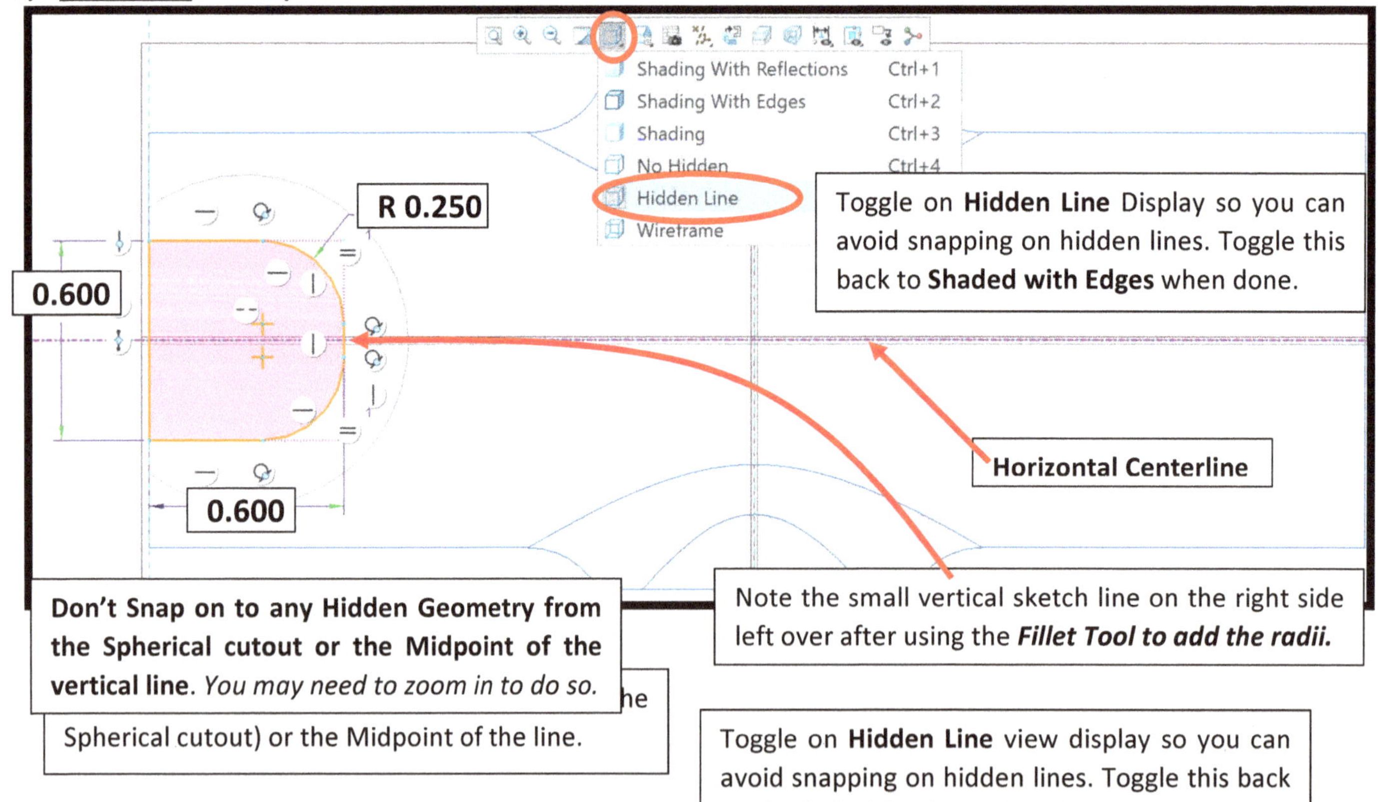

Step 9 – 2 – Set the **Depth, Direction, and Material Removal options** in the Toolbar to complete the Extrude options.

- Toggle the View Display back to **Shading with Edges** in the Quick Toolbar

- Select the **Depth option** to be "**Extrude Symmetrically on Both Sides**"

- Set the **Depth Value** to **2.000**.

- **Toggle on Remove Material**, if not done so by default.

- **Press the Checkmark** to Accept the Extrude

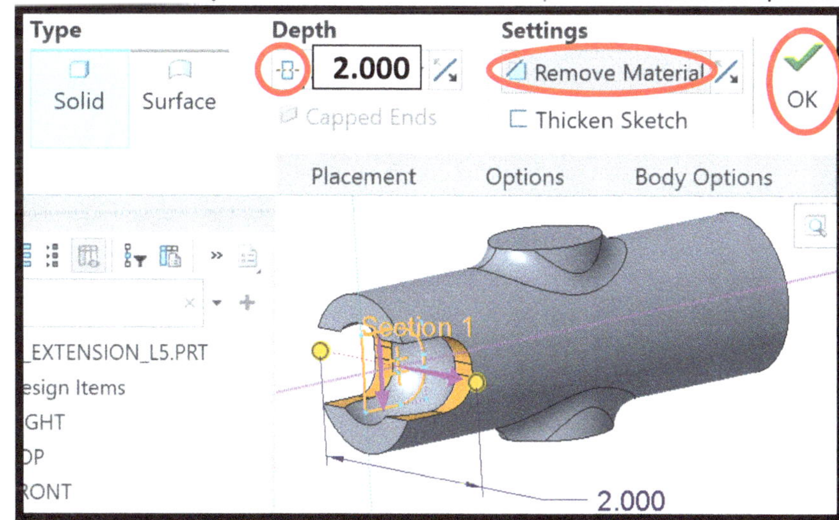

Step 10 – Mirror the Revolve 1 & Extrude 3 Cutouts to the opposite end of the cylinder, by adding them as a Group and using the Mirror Tool

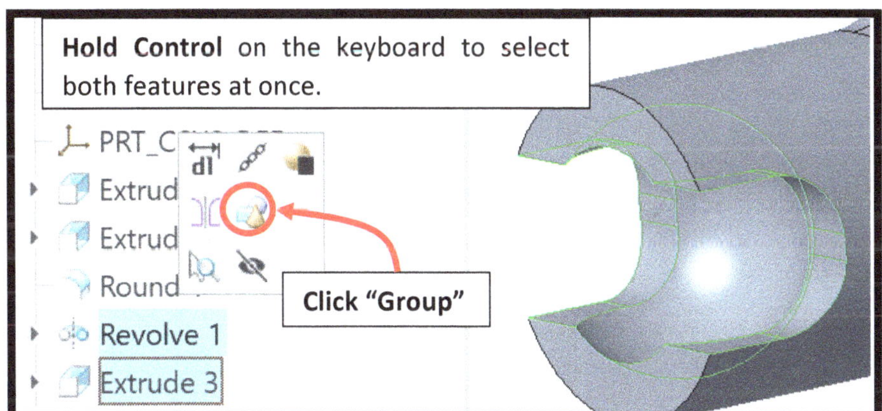

Step 10 – 1 - Group the Features:

- **Hold Control** on the keyboard and Select both **Extrude 3 & Revolve 1** in the Model Tree – select the **Group icon** in the pop-up menu.

Make sure no other features were added to the Group.

Step 10 – 2 - Mirror the Group:

Click on the Group in the model tree - Mirror Tool - click on the **Front Datum** to mirror the Group to the opposite side of the part.

If the copies do not Mirror properly, check that Extrude 1 was setup to extrude symmetrically so that the Front Datum will be in the center of the model.

Another issue could be if the Revolve 1 or Extrude 3 are not snapped on the left edge of the model.

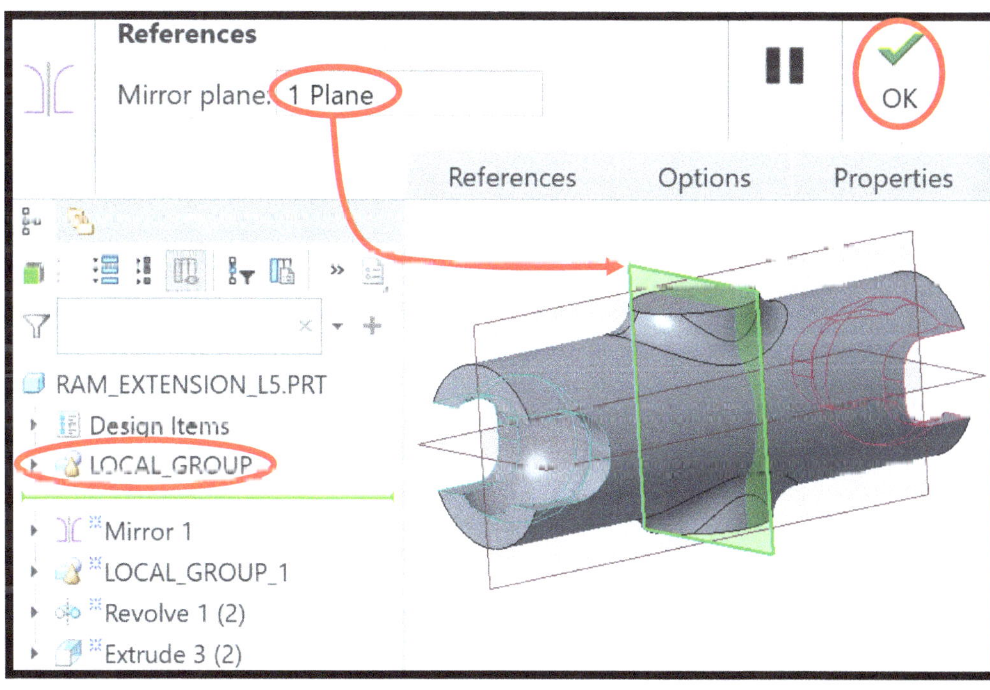

Step 11 – Use the **Hole Tool** to create a **Standard, Coaxial hole** through the Extrude 2 cylinder. This hole will allow a threaded bolt to slide through without threading.

Hole Type: **Standard**

Profile Options: **Toggle on the Clearance Option** (this removes the Tapping/Threads from the hole)

Toolbar Settings: **Screw Size** set to **"¼ -20"**
                                        Hole Depth = Thru All (*Drill to intersect all surfaces*)

Placement Type: **Coaxial**

Placement References: Hold CTRL on the keyboard to choose two primary references: select the **Axis A1** and the **planar surface** of Extrude 2 as shown below. *Your Axis Display filters must be toggled on in the Quick toolbar.*

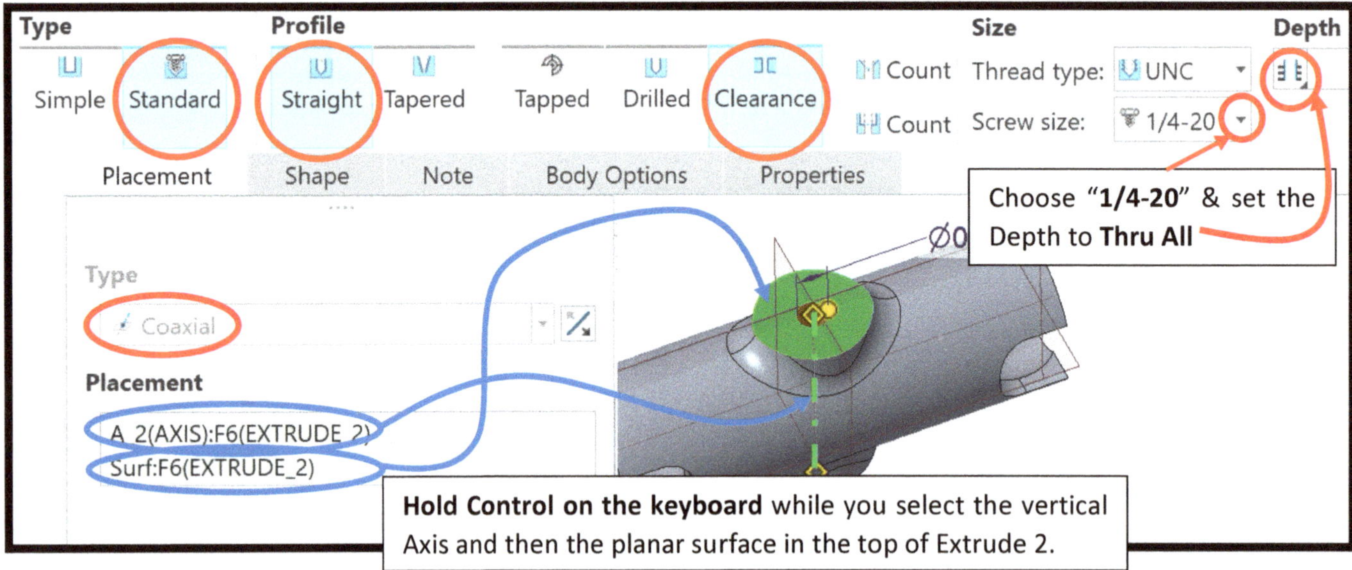

Tip: Select the Axis first when creating a Coaxial Hole. Also make sure to choose the "Standard" type first so you can setup the rest of the hole size and tapping options properly.

Step 11 – 2 –Press the Checkmark to accept the Hole.

Palette Tool:

The next step will make use of a pre-built "Palette" of polygon sketch shapes. This makes it much faster to sketch out common shapes, such as hexagons, as these pre-built shapes come in with the proper constraints in place. The user only needs to drag the desired shape onto the screen, adjust its position, set the scale, and then finalize any dimensions. This is also very useful if you are using your own sketch repetitively, as the Palette tool will find any saved sketches (.sec files) from your working directory.

Note: By default, the imported shape does not snap onto entities when you drag it onto the screen. Complete your final placement to snap onto references as needed **after initially dragging** it onto the screen.

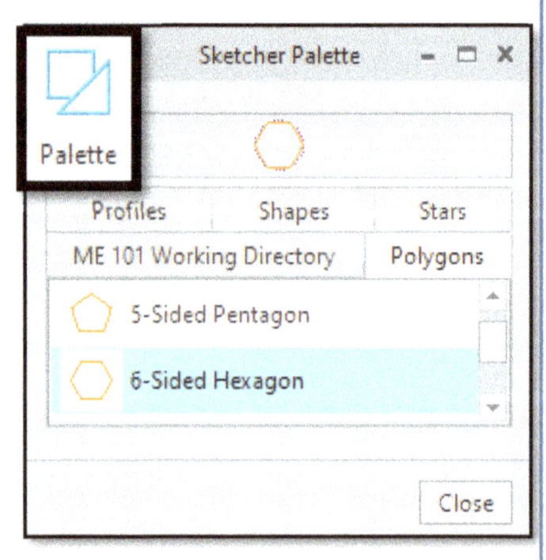

Step 12 – Create an **Extrude** sketched on the <u>upper flat surface</u> of the Extrude 2 cylinder.

1) <u>**Use the Palette Tool:**</u> When you get to sketch mode click on the **Palette Tool** in the sketch toolbar. Click on the **Polygon Category tab** and find the **"6-sided Hexagon"** – click and hold on the Hexagon to drag and drop it on to the screen in a blank area. *It will not snap references while dragging it on the screen!*

2) <u>**Position the Hexagon:**</u> **Click & Hold on the center circle or "Drag handle"** of the Hexagon, and drag it until it snaps on to the Origin. By default, the **"flats" of the hexagon should be parallel** with the length of the model. If your shape is different, use the **rotation symbol** on the hex shape as needed.

3) <u>**Close the Import Section toolbar:**</u> You do not need to adjust the scale setting. **Checkmark to accept the Import.**

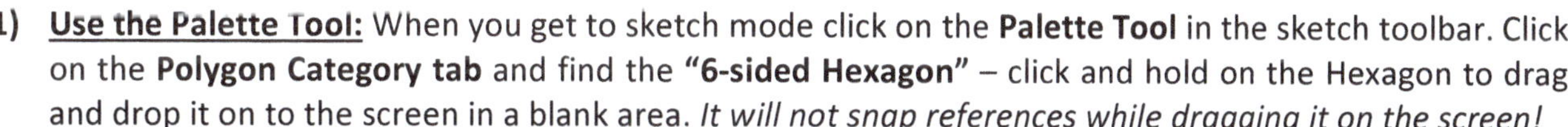

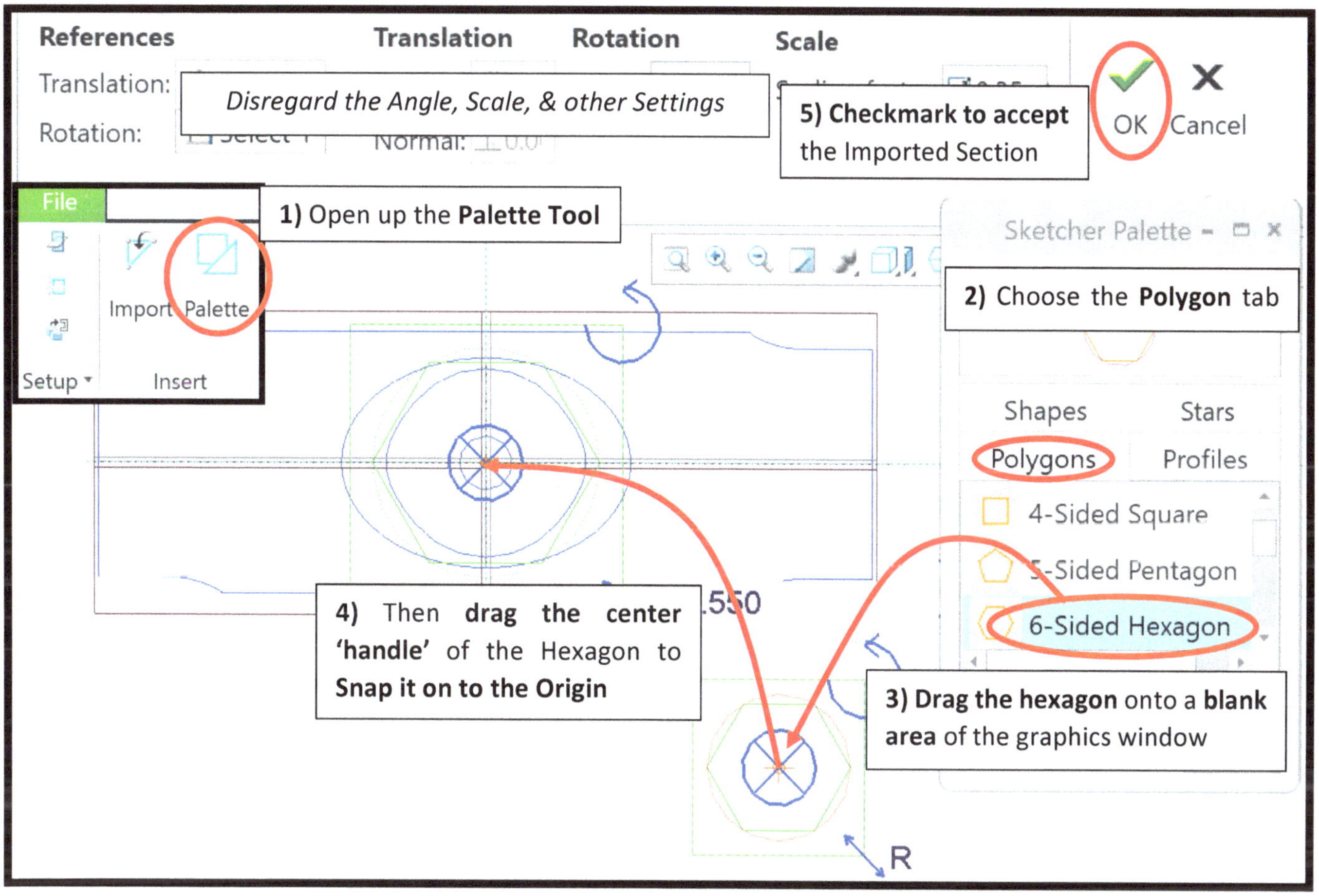

4) <u>**Set Coincident Constraints:**</u> If your shape did not snap onto the origin properly (if you notice any weak dimensions), use the Coincident Constraint tool to snap the center point of the Hex onto the vertical or horizontal centerlines.

5) <u>**Dimension the Sketch:**</u> Use the **Dimension Tool** to set the dimension for the size of the Hex Shape from the top to bottom edges **(0.455")**. Do not set a dimension from "point" to "point", but from "flat" to "flat" of the Hex.

You will likely get a Resolve Sketch error, and will need to delete out the dimension for the line length.

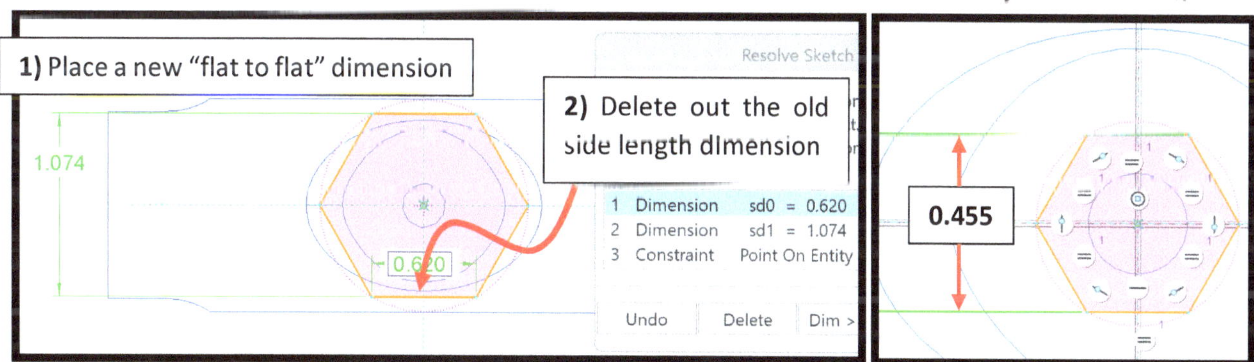

If you have any weak dimensions, you likely need to set a **Coincident Constraint** from the center point of the hex to the vertical or horizontal reference lines.

Step 12 – 2 - Set the **Depth, Direction, and Material Removal options** in the Toolbar to complete the Extrude options.

- **Toggle on Remove Material,** if not already done by default.
- **Set the Extrude Depth to 0.185"**
- **Toggle the Direction Arrow,** if needed
- Leave the **Extrude Depth** option as the default *"Extrude by specified value"*

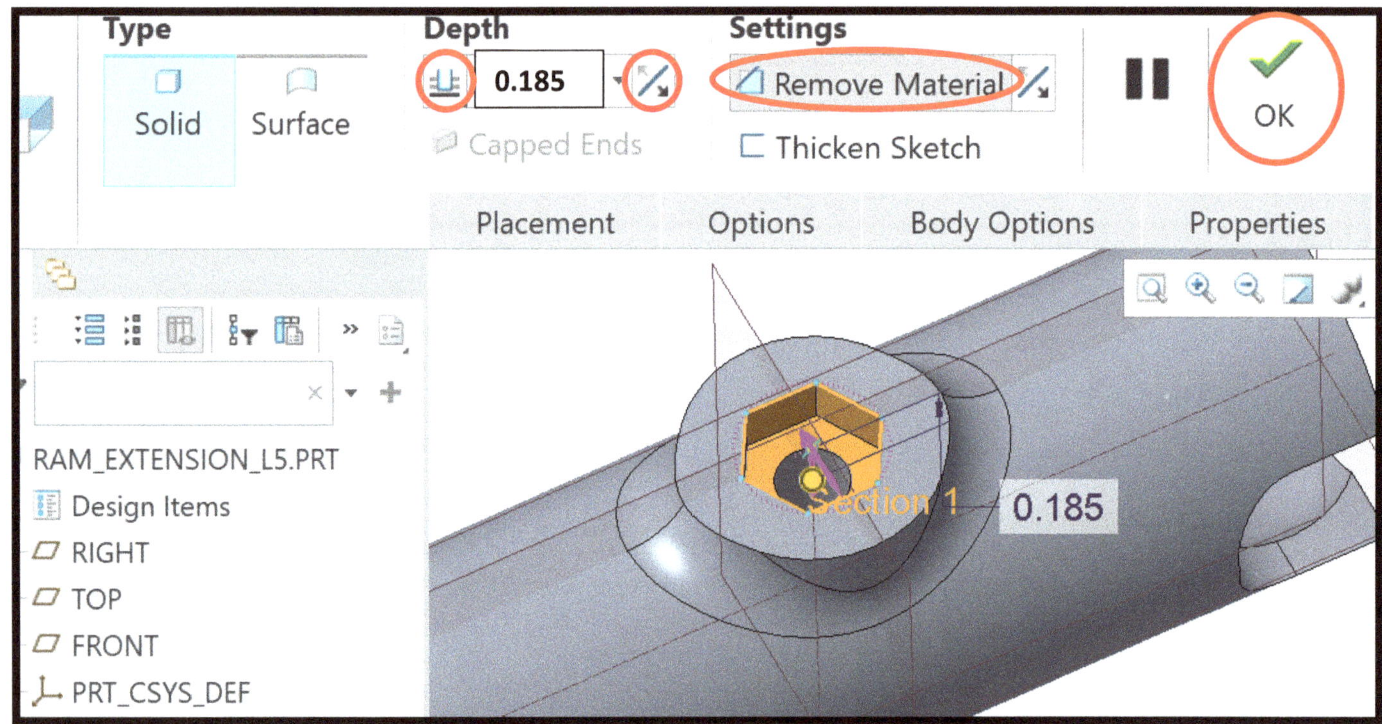

Step 13 - **Mirror the Extrude #4** to the opposite side of the cylinder, using the **Top Datum** as the Mirror Plane. *If your Extrude will not mirror properly, check that Extrude 1 has the sketched Circle centered on the origin.*

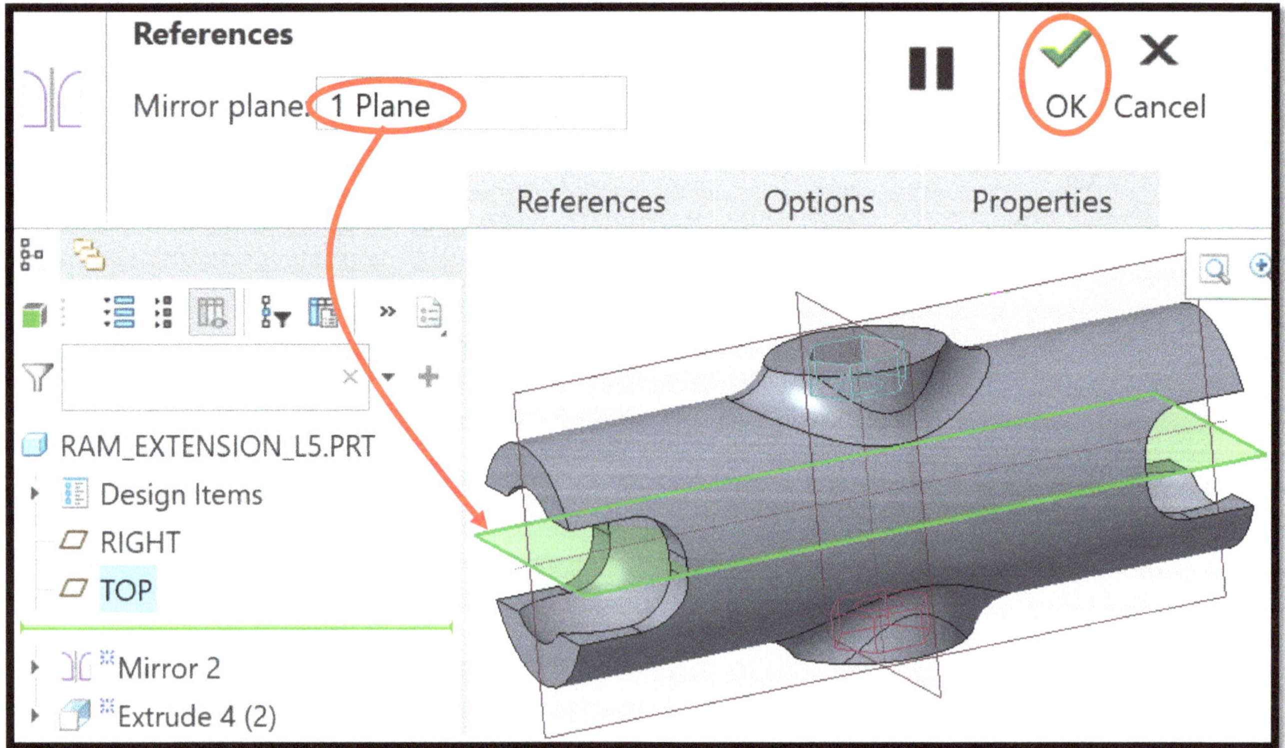

Step 14 - Set the display properties of <u>**each datum**</u> so that we may use them as a **Geometric Tolerance** and be able to change them to **ASME style** in our drawings. We must "<u>**Set**</u>" & **Rename** each datum to match the style of this course using letters for names of the Datums.

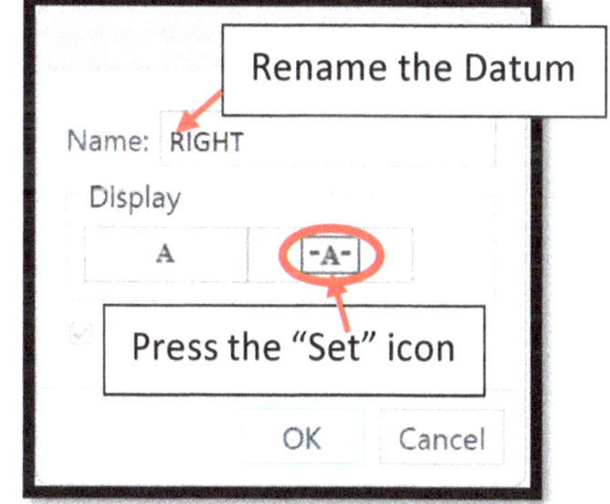

- **LMB to select a Datum** from the model tree - **hold RMB – Properties - press SET ("-A-")** – select the **name box – Type in a new name** according to the list below – **OK**.
- **Repeat for each Datum**.

    RIGHT: Rename to **B - Press "Set" - OK**

    TOP: Rename to **C - Press "Set" - OK**

    FRONT: Rename to **A - Press "Set" – OK**

After setting and renaming the datums you should see the datum names or "tags" on the model screen. If you do not see all 3 datum name tags, you likely did not press "Set" for that datum.

Step 15 – Add an Auto-Round to the part model with Radius of 0.025":
- Round Tool Drop Box Arrow – choose **Auto Round** – set the **radius to 0.025"** – press the **checkmark**.

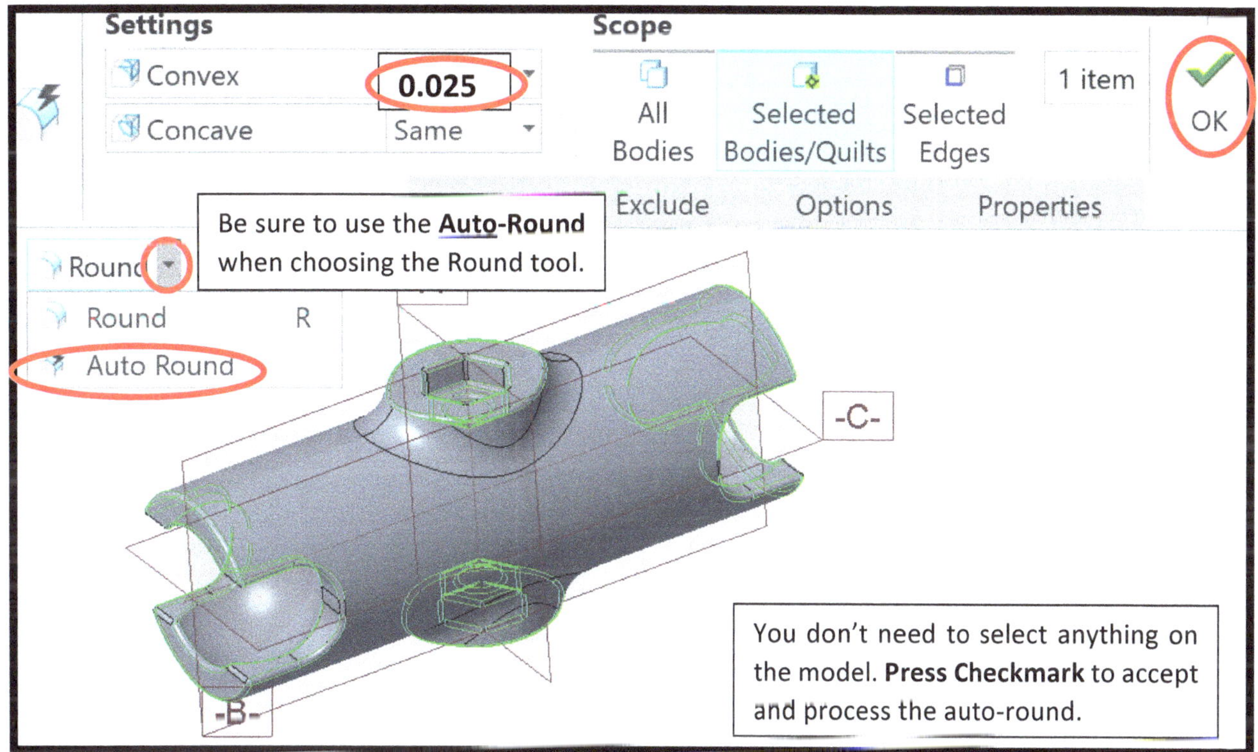

Step 16 - Save your part model to your Working Directory.

Step 17 – Create a Drawing of the model to match the key.
    - **File - New - Drawing – "*L5_Ram_Extension*" – OK**

Step 18 – Complete the following ***Basic Drawing Steps*** (appendix) **Steps #2 through #7**. Don't worry about the other views yet, they will be covered on the next page.

☐ Turn off Display Filters   ☐ UND Template (Sheet Setup)   ☐ Scale   ☐ Set the View Displays
☐ Rotate Front (Middle) View to match the key (if needed)   ☐ Adjust the positions of the Front/Top/Right views

    After completing the *Basic Drawing Steps* **#2 through #7** continue onto the next page to add in the new views.

Step 19 – Add the Left View using the **Projection View tool.**

- **Click in a blank** area to deselect from any view.
- Select the **Projection View tool** (*Layout tab*) – LMB click on the **Front View** as the reference - LMB an area to the left side to place the projected view.
- **Turn off Lock View Movement** and drag the views to their approximate positions on the screen.

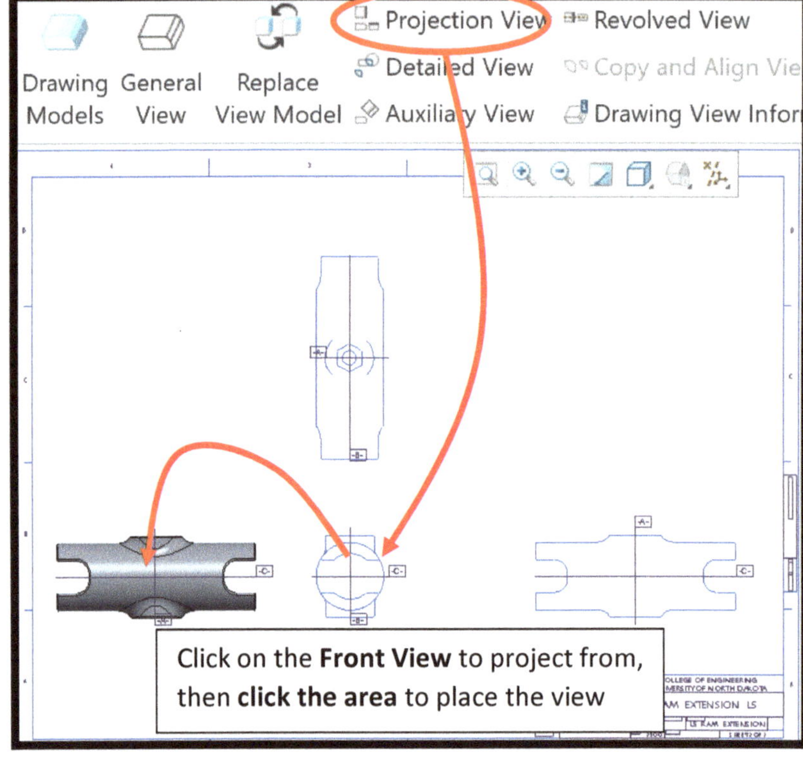

Step 20 – Add the Detail View.

- **Select the Detailed View tool** (Layout Tab)
- **LMB on the arc edge** of the side cutout to act as the Center point of the Detail View.
- **LMB click to add spline points** to define the border of the new detail view. Each click adds a point of a spline curve
- When the points are close to completing a circle **press the MMB** to accept the boundary (*your spline does not need to be a complete loop, it can be slightly open at the end*).
- **LMB click in the upper left area** of the drawing to place the new view in position.

- *Tip: Read the command prompt* *instructions in the bottom left corner of CREO as you go through this step so you know what CREO is asking for*. If you are stuck, you may need to press MMB or "Control + A" to exit the tool.

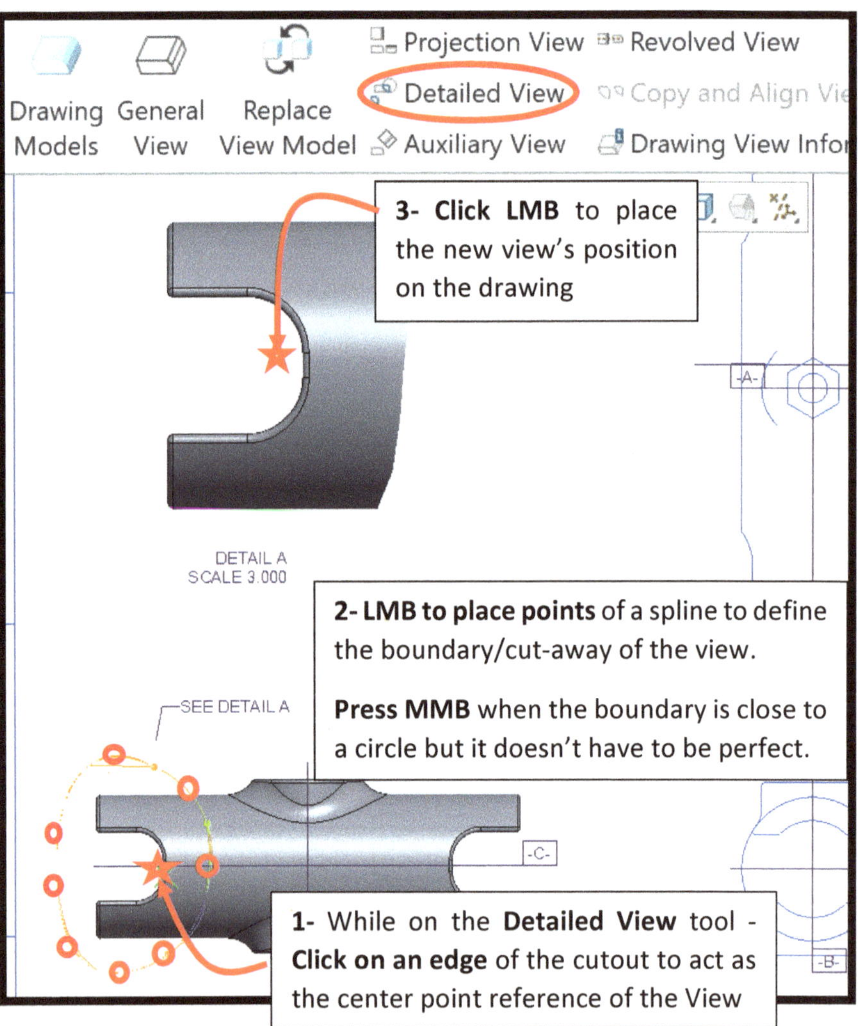

Step 21 – Set the Datum display option to be "ASME Style":

- **File - Prepare - Drawing Options** – **"Detail Options" change** – search for **"gtol_datums"** – click a few times on the Value drop-down box to see the full list & then select **"std_asme"** – Add/Change - OK.

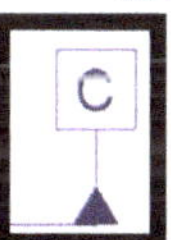

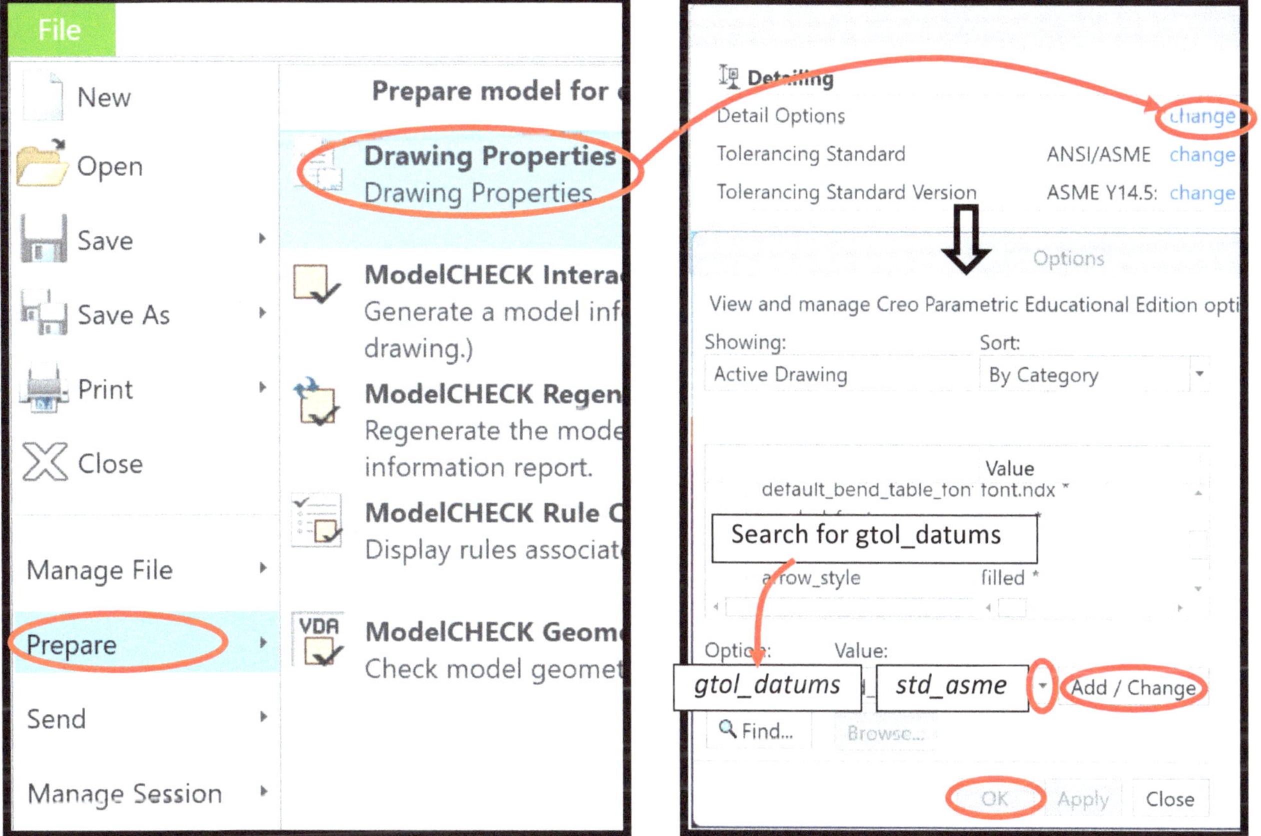

Step 21b – Peel Back the Datum tag lines on all views. You must be on the Annotate Tab to select the Datums, then use the drag handles (small boxes) to move them off of the views. If your datums do not appear on the drawing go back to step 14 and "Set" them. *Tip: The "C" datum of the Detail view is likely so far over it will be across the Top view. Make sure to peel that back to the Detail View.*

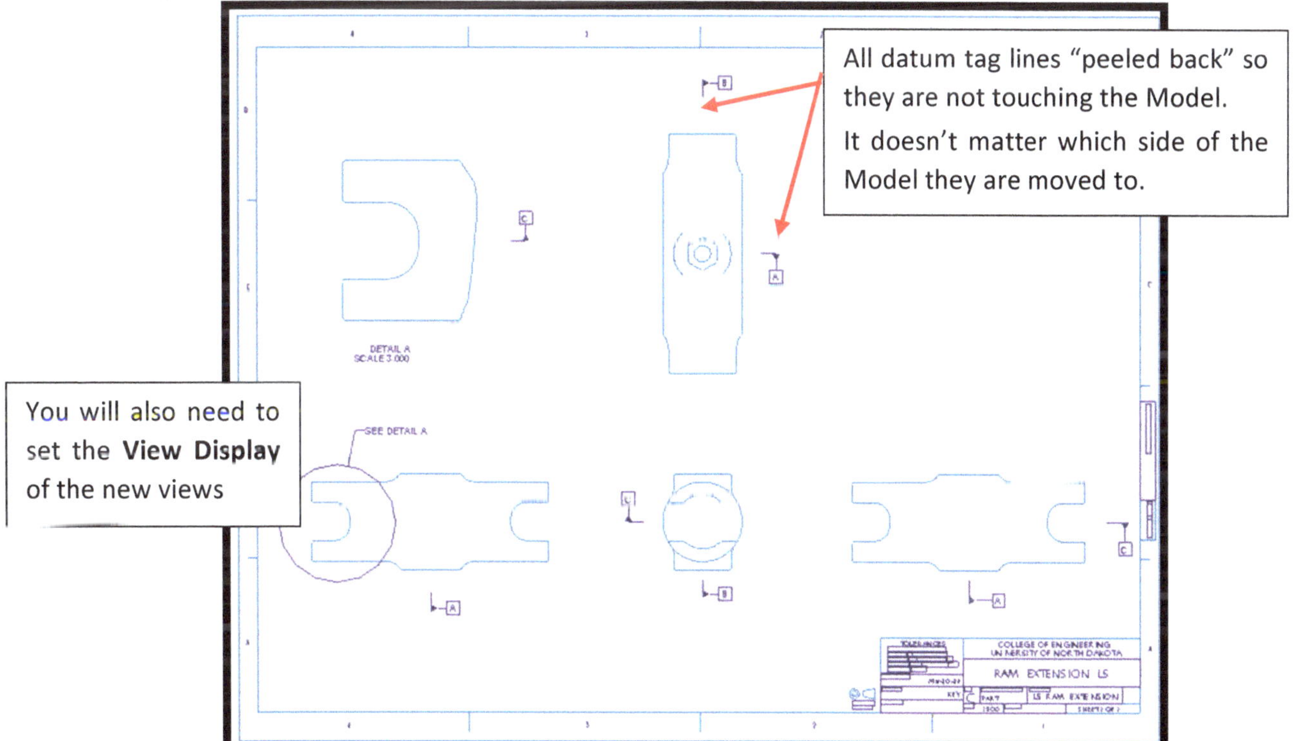

Step 22 – Insert a second sheet to show the isometric view on by itself.

- **Select the (+) sign** at the bottom left of the screen – a prompt will appear asking for the name and material for the UND format on the new sheet – add the General View in place – adjust the View Display and Scale of this new sheet.

- To switch back to the Sheet 1, use the Sheet tabs at the bottom of the drawing.

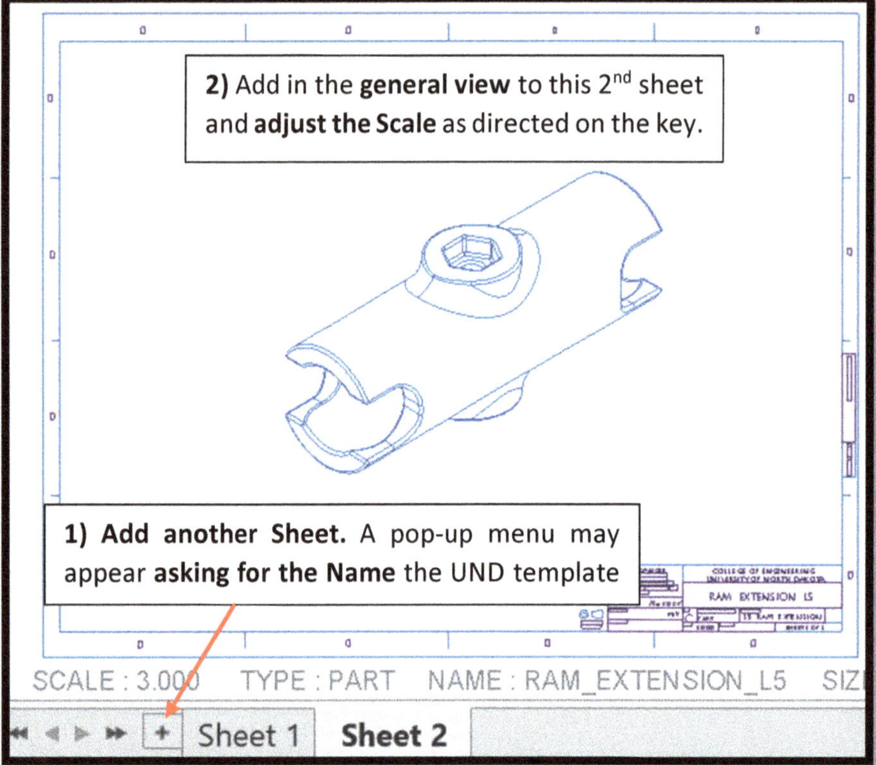

Step 23 – Complete the **remaining Steps #6, #7, & #8, in the *Basic Drawing Steps*** (in the Appendix) before moving on to the next page.

☐Step 6 - **Reposition the views** as needed to match the key so views have enough room to place dimensions and do not touch the drawing edges.

☐Step 7 - **Setup the View Display** on all views to match the key.

☐Step 8 – Create the **Cross-Section** on the **Right** view. Then **set the Scale & Add Arrows** for the section line. The Key will tell you what Scale to set the hatching to.

Continued on next page:

Step 24 – Show the Axis/Centerlines on most views (except the Detail and Isometric Views):

- **"Show Model Annotations Tool"** (*Annotate tab*) – **Model Datum** tab (*far right one in that menu*) – select the **view "box"** that surrounds the entire view (*not the model itself*) – **Checkmark All** – apply.
- **Repeat for the Front/Top/Right/Left views** (centerlines should not be added to the Isometric view!). The Detail View does not need a centerline.

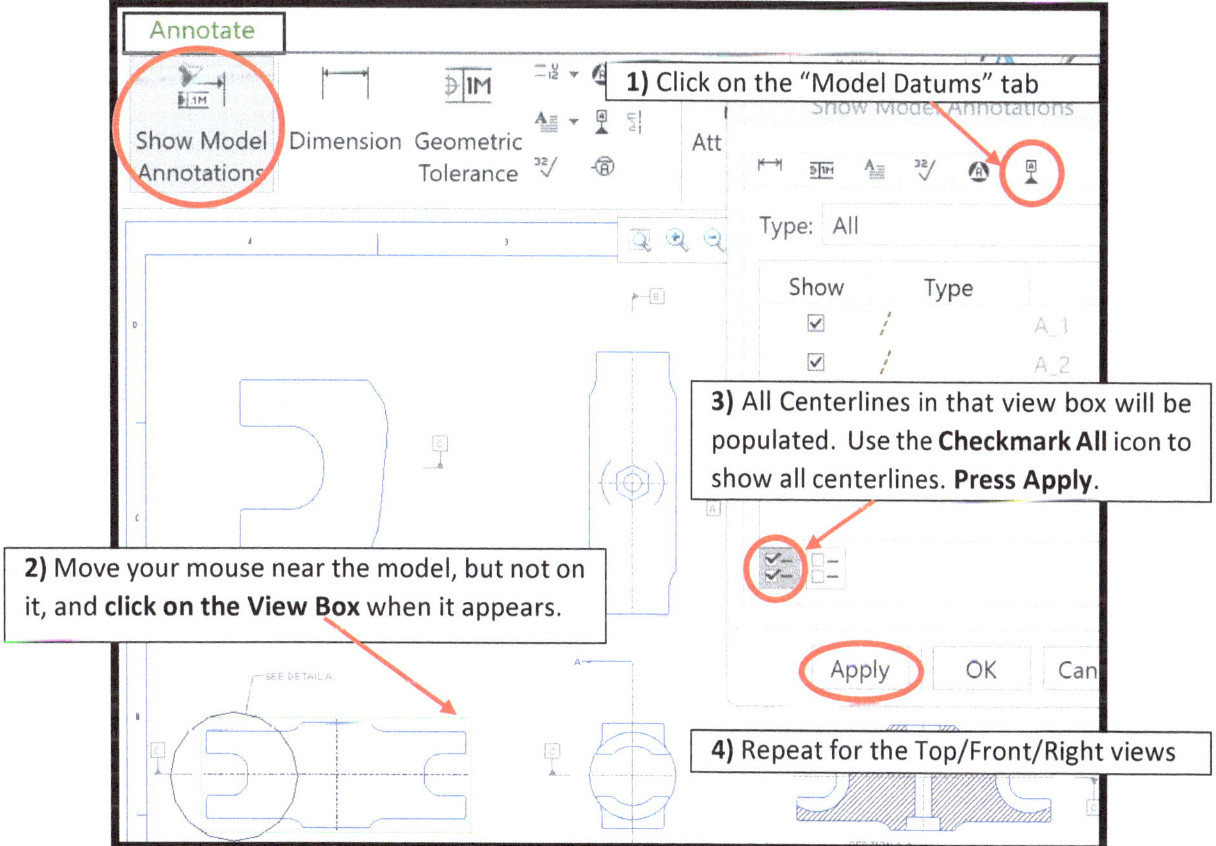

Step 25 – Insert the General Note at the top of the drawing: - **Annotate Tab** – open the **drop box on the Note Tool** – select **Unattached Note** – LMB to place where it will be on the drawing – copy & paste or type in the text:

NOTE 1: ALL EDGES ARE ROUNDED TO 0.025" UNLESS NOTED OTHERWISE

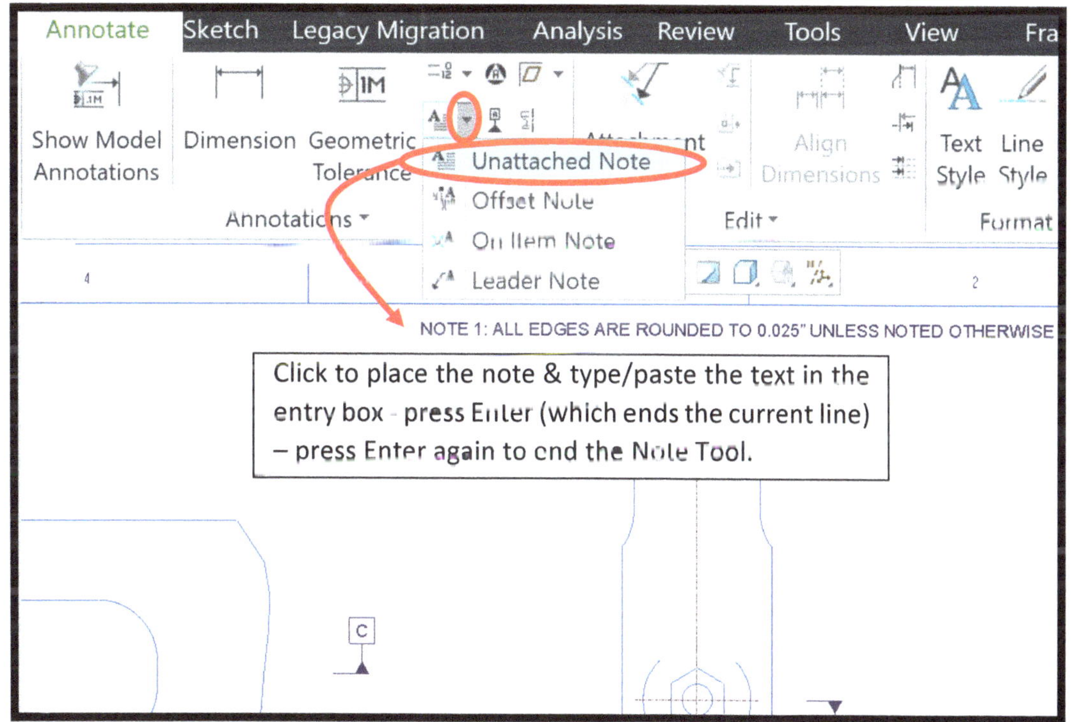

Step 26 – Show the Standard Hole Note on the Top View using the **Show Model Annotations** tool – click on the **Note category tab** – select the View Box around the Top View – **checkmark to accept the note** – OK.

Tip: If the hole note does not appear, you likely did not click on the View Box (don't click on the model!). Or your Hole may not be set to the "Standard" type when creating it.

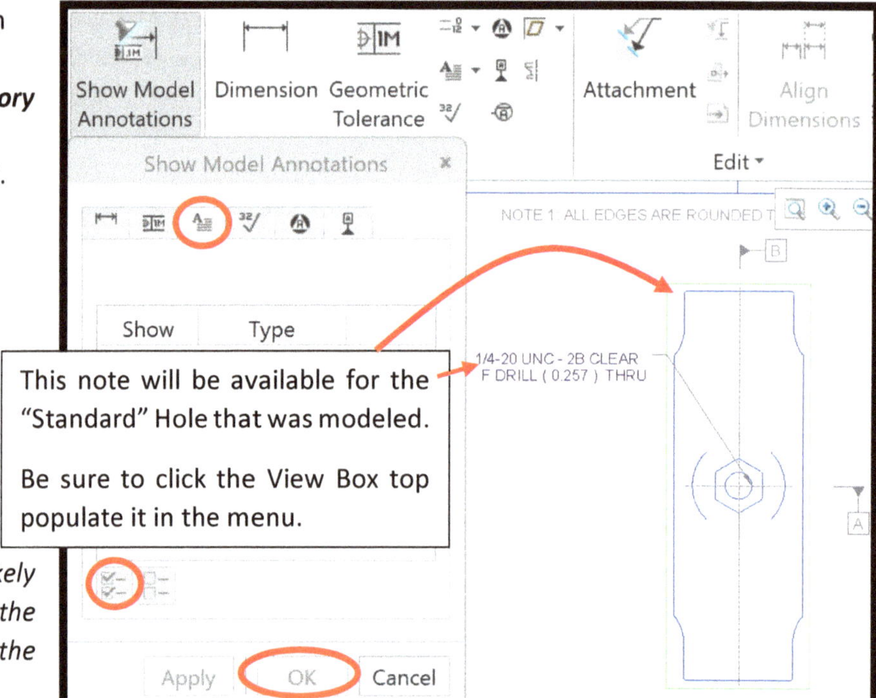

Step 26b – Edit the attachment of the Note Arrow so it does not cross over the center of the hole

- **Annotate Tab** – click in a blank area – **select the Attachment tool** in the top toolbar – **select the Note** – choose **Change** Reference in the pop-up menu (you may need to accept a warning message) – **click a new position** of the hole edge to have the note point to.

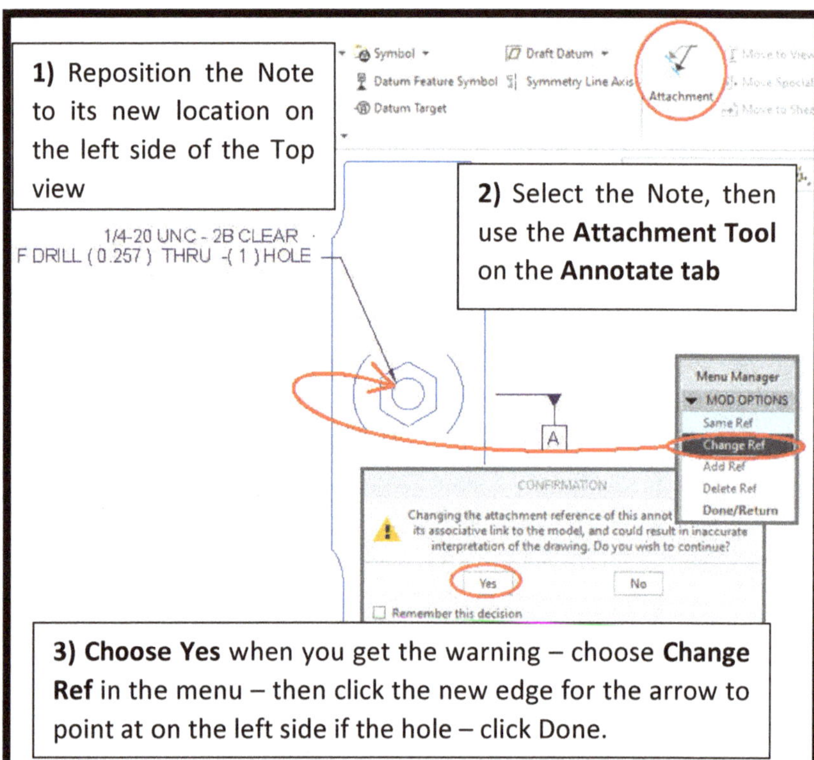

Step 27 – Refer to the ***Detail Drawings Steps*** (appendix) to complete the rest of the detail drawing steps to match the key. Most of the steps have already been completed in the steps above except the following:

☐ **Complete Step 3** to "**Create**" or "**Show**" **dimensions** to match the key. On this lesson, the key has color coded the dimensions that must be "Created" in an Orange color. The rest of the dimension should be "Shown".

☐ **Complete Step 4** to **Flip Diameter** dimensions to a single arrow style

☐ **Complete Step 5** to adjust the **Decimal Places** of all dimensions to match the key

☐ **Check your drawing against the Key** (without instructional notes). Then save and submit your deliverable.

End of Lesson 5

Lesson 6 - Blends & Shells (Nozzle)

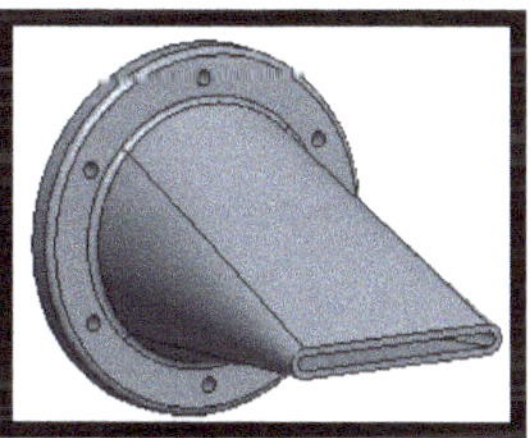

Lesson 6 is to create a nozzle which is inspired from fittings for an extrusion machine for pushing out material into certain shapes (pasta, jerky, etc.). This model will utilize the **Blend** and **Shell** tools.

Blend Tool: The Blend tool creates a shape that has a changing cross-sectional area by connecting various sections together by their sketch "vertices". All Blend sections must have the same number of vertices as well as have a Start Point that aligns appropriately to keep the sections from twisting as they blend.

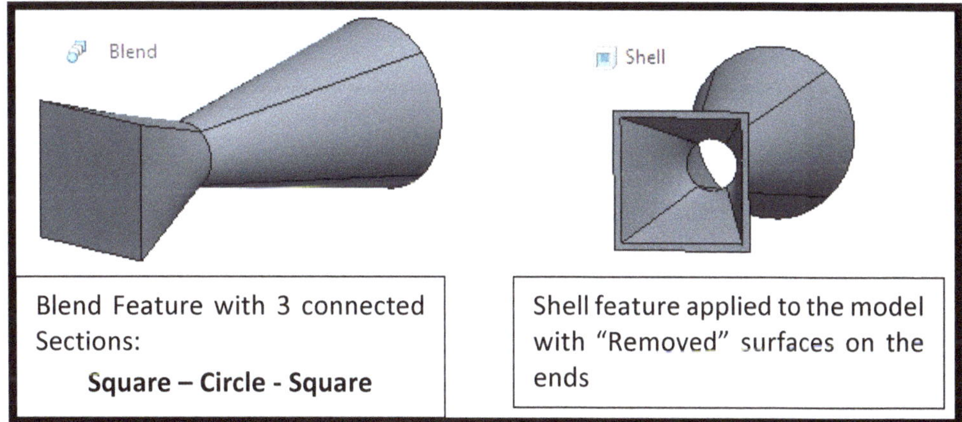

Blend Feature with 3 connected Sections: **Square – Circle - Square**	Shell feature applied to the model with "Removed" surfaces on the ends

Shell Tool: The Shell tool hollows out the entire model to a specific wall thickness. The Shell Tool also allows specific surfaces to be removed or to have a different wall thickness than the rest of the Shell.

Shell Feature Order of Operations: Re-ordering features in the model can affect the outcome of the Shell Tool. If the Shell feature is re-ordered in the Model Tree to come after a hole, a "hole boss" will remain as the shell is following the hole and leaving a wall thickness around it as it shells out the model. Similarly, if a Shell comes after a round, the inner shell surface will follow the round curvature.

Features can only be "re-ordered" in the Model Tree as long as they are not placed earlier in the sequence than other features it is using as a placement or dimensions reference (i.e. A hole cannot be moved before an extrude that the hole is placed on).

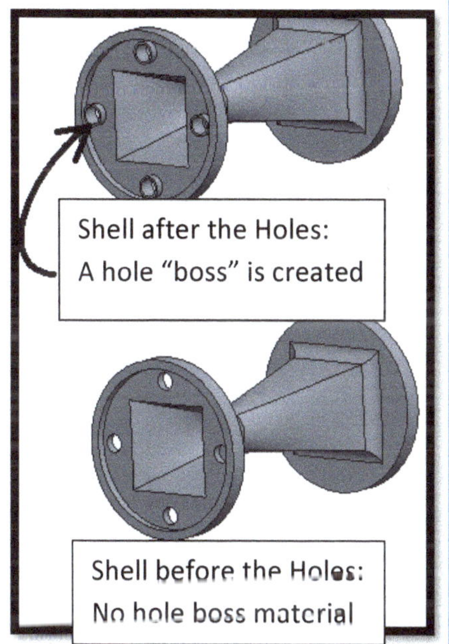

Shell after the Holes:
A hole "boss" is created

Shell before the Holes:
No hole boss material

Getting Started: Open the **CREO *Parametric*** software and follow the lesson steps.

Step 0 – Complete the **Blend Module and Quiz** in Blackboard if you have not already done so.

Step 1 - Set your working directory to your ME101 Working Directory Folder.

- **Flle – Manage Session – Set Working Directory – Navigate to your ME101 Directory folder – OK**

Step 2 - Create a new **Part** called "**Blend_Nozzle_L6**". *You cannot use spaces in a filename!*

- **File - New – select Part as the file type – type in the "Blend_Nozzle_L6" - press OK.**

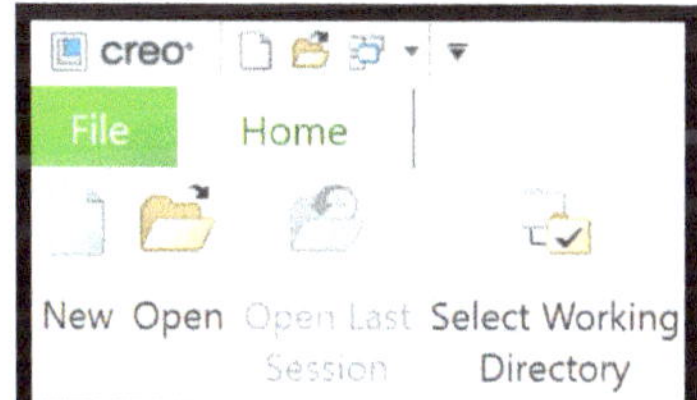

Step 3 - Change the Material of the part to **PVC** *Look under Legacy Materials*, and be sure your units are in the CREO Default (inch-lbm-Second).

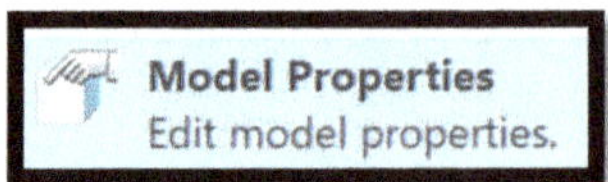

- **File – Prepare - Model Properties -** in the Material option select **change –** select the Legacy Materials folder - **double click** on **"PVC.mtl"** – press **OK** – press **Close** to close out the Model properties menu.

Step 4 - Create an **Extrude** to create the base "disc" feature. Use the **Front** **Datum** as the sketch plane.

Sketch Shape: Place a circle centered on the origin with **a Diameter of 6.00"**
Extrude Toolbar options: Depth Option of single Direction (**"Extrude from Sketch Plane by Value"**), with a **0.50" Depth**
Press OK to finish then Extrude

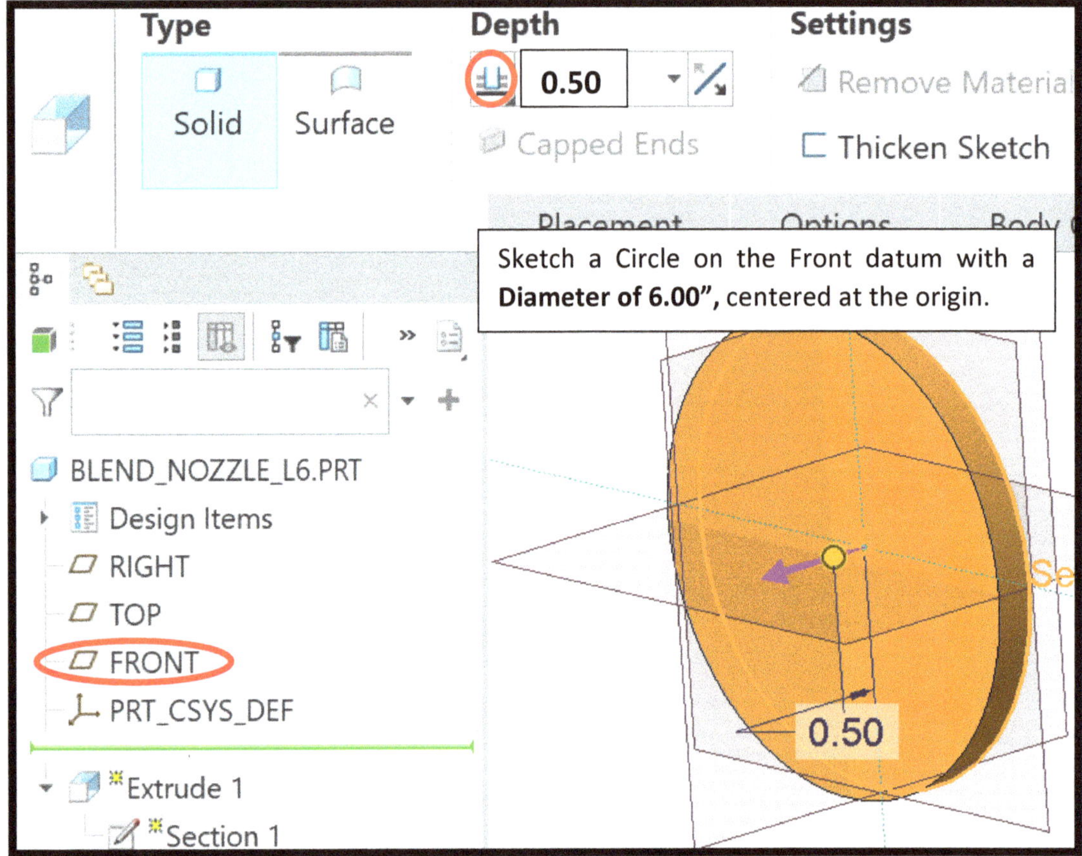

Step 5 – Set the display properties of <u>each datum</u> so that we may use them as a **Geometric Tolerance** and be able to change them to **ASME style** in our drawings.

- **LMB to select a Datum** from the model tree - **hold RMB – Properties -** **press SET ("-A-")** – select the **name box – Type in a new name** according to the list below – **OK**.
- **Repeat for each Datum**.

  RIGHT: Rename to **B - Press "Set" - OK**

  TOP: Rename to **C - Press "Set" - OK**

  FRONT: Rename to **A - Press "Set" – OK**

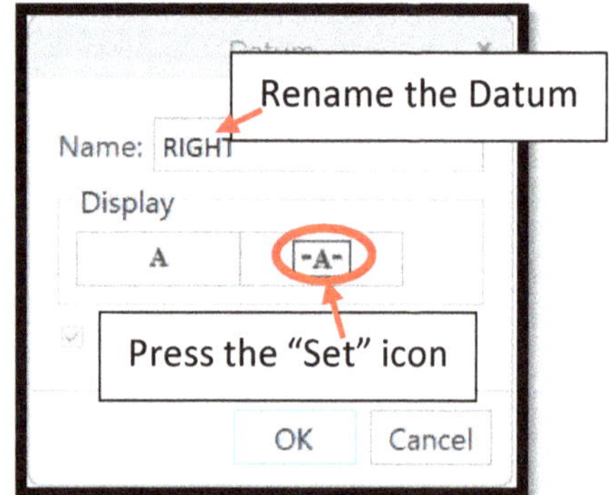

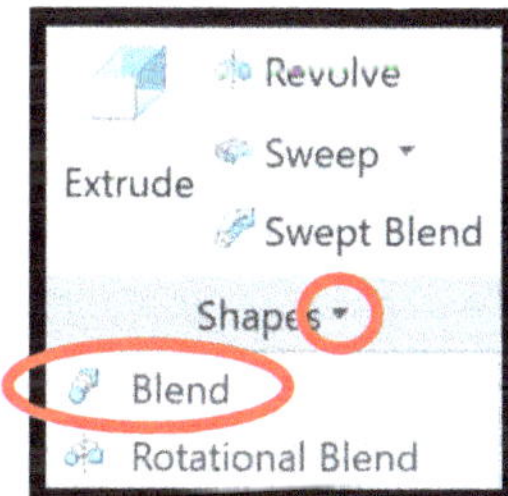

Step 6 – 1 - Click in a blank area of the screen so you are not selected on any datum or surface. Then start the **Blend tool** (*under the **Shapes** drop-down box in the top toolbar.*)

Step 6 – 2 - Open the **Sections** tab and **Define** the first section to be placed on the **planar face surface** of the disc. **Click Sketch** to accept the sketch plane for the first cross-sectional shape of the Blend.

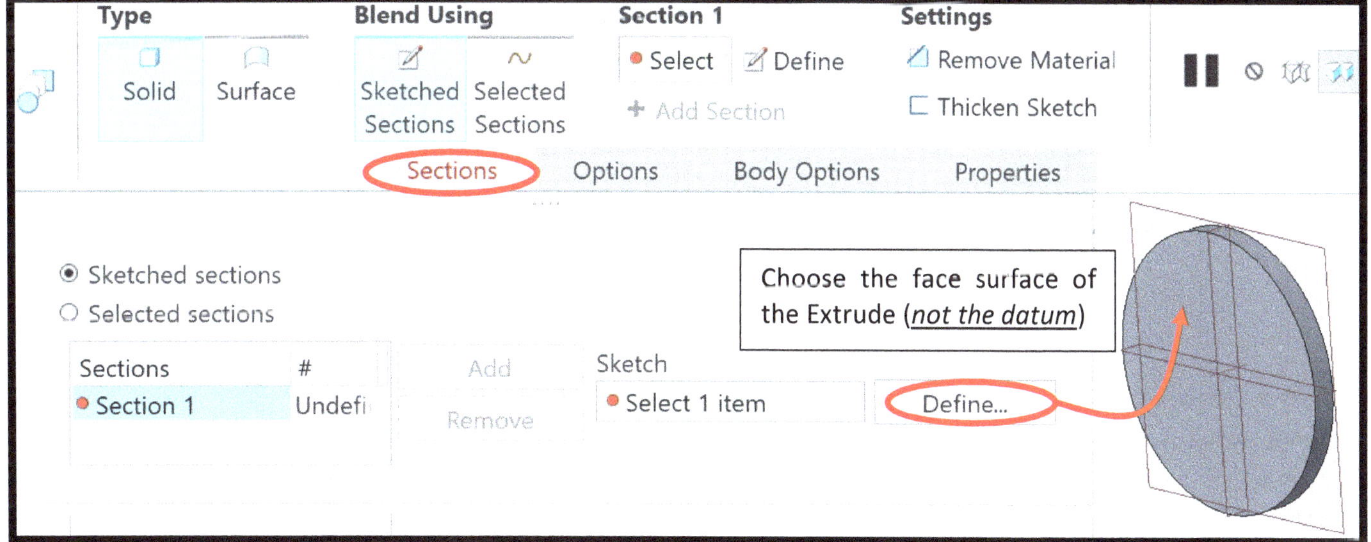

Step 6 – 3 - Create the Sketch shown below using Arcs to approximate a circle, as this Blend will require 4 Blend Vertices. *Do not use the Circle tool for this closed-loop, you must use the **Arc tool** to create the circle In sections.*

1) **Centerlines: Place two centerlines at 45° angles** through the origin that will be use to position the arcs.

2) **Center & Ends Arcs**: Use 4 **Center and Ends Arcs** to create the circle. Place the center of the arcs at the origin, and place the start and end points of the arcs on the centerlines covering **90°** in each arc. ***Do not snap*** these arcs onto *the edge of the circle from the previous step, leave them short of the edge.*

3) **Dimensions**: Create a **Diameter** Dimension of **4.25"** (**Dimension Tool** – **double click** on the arc curve – MMB to place the dimension). *Do not use a Radius dimension, as your drawing key needs to show the Diameter.*

4) **Blend Start Point: Set your Start Point** position (*the point with the Yellow Arrow*): **Select Tool - Click on the upper left arc point** - hold RMB to bring up the menu – choose the "**Start Point**" item in that menu. It does not matter which direction the arrow points to, as long as it is on the correct point.

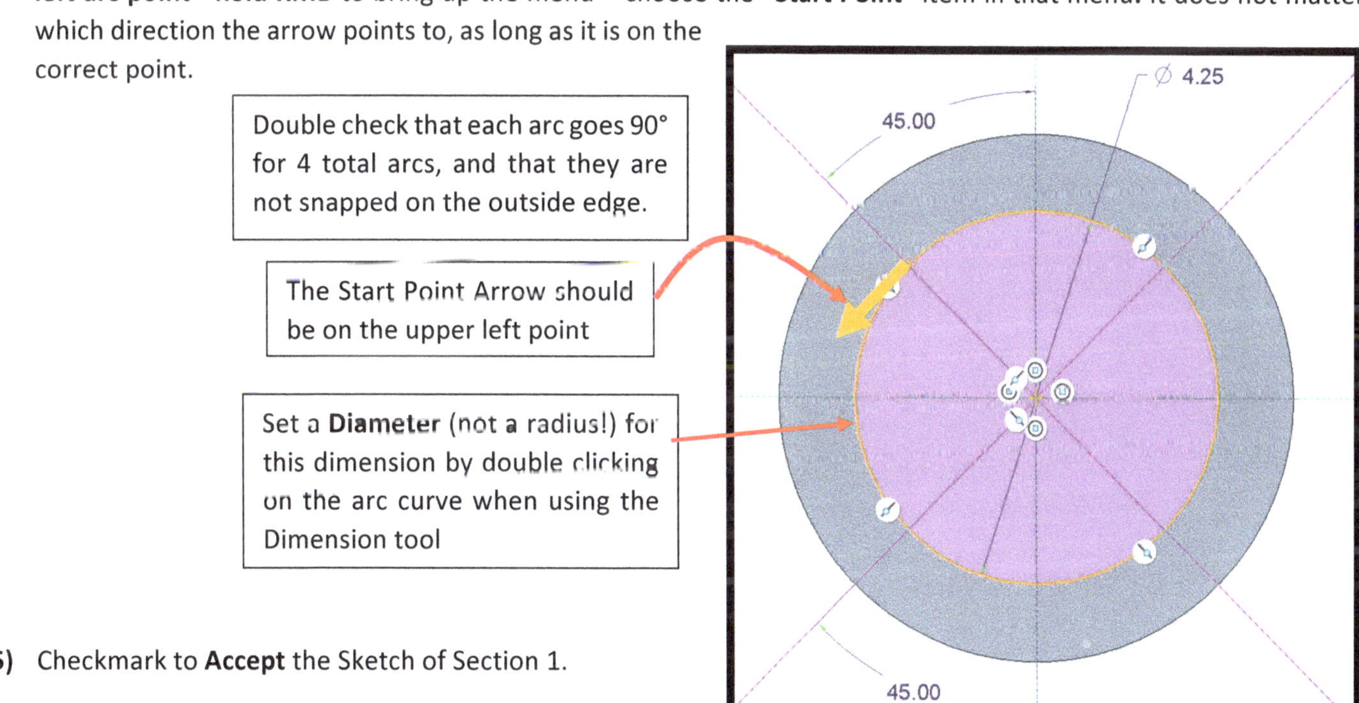

5) Checkmark to **Accept** the Sketch of Section 1.

Step 6 – 4 - Open the Sections tab in the Blend Toolbar. Check that your Section 1 has 4 blend points listed in the menu. If not, edit *Section 1* to correct the sketch to only have 4 points. Otherwise, under the *Section 2* options set the **offset distance to 6.00"** and **click Sketch**.

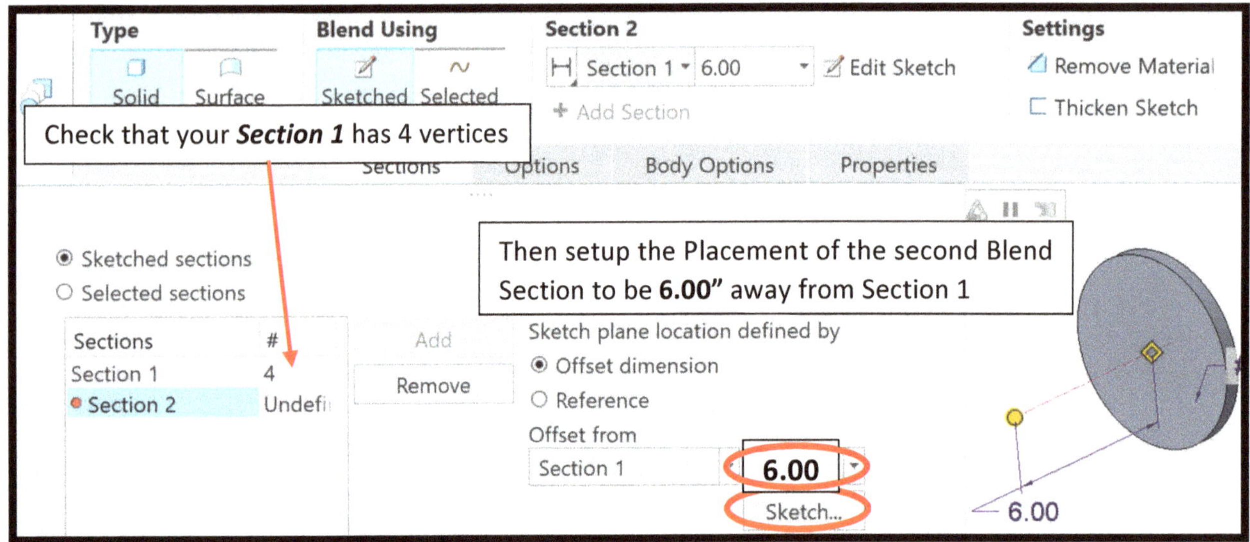

Step 6 – 5 - Create the sketch for Section 2. This sketch must also have **4 vertices** to match the number in Section 1.

1) Add Centerlines: Place a vertical centerline, then add a horizontal centerline **1.00" below** the horizontal reference.

2) Sketch the Rectangle: Place a **Rectangle** about the offset horizontal centerline. *Do **not** to snap the rectangle onto the edges of the previous sketches.*

3) Symmetry Constraints: Set a **Symmetry Constraint** about both the vertical & horizontal centerlines, if needed.

4) Center and Ends Arcs: Add **Center & End Arcs** to each end of the rectangle.

5) Delete Segment Tool: Delete the left and right vertical sketch lines of the rectangle, leaving a closed-loop.

6) Dimensions: Create a **Diameter** Dimension of **0.50"** (Dimension Tool – double click on the arc curve – MMB to place the dimension). Do not use a Radius dimension! **Set the width (corner to corner) of the rectangle to 3.50"**

7) Start Arrow: Set the Start Arrow to match Section 1: **Hold RMB on the point -** *Feature Tools - Start Point.*

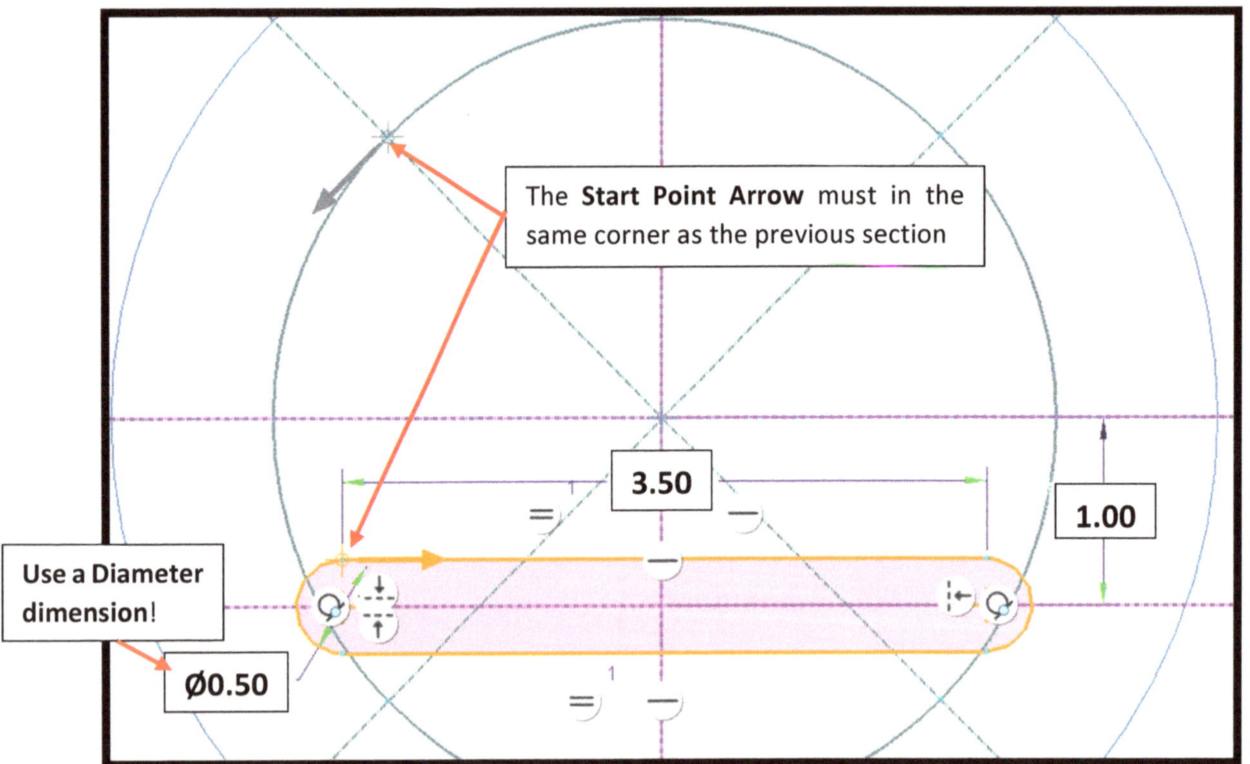

8) Press the Checkmark to accept Section #2.

If you notice an error or the blend does not appear on the screen, use Edit Definition and open the Section tab to Section #1 & Section #2 for the correct # vertices, similar start point locations, dimensions, etc.

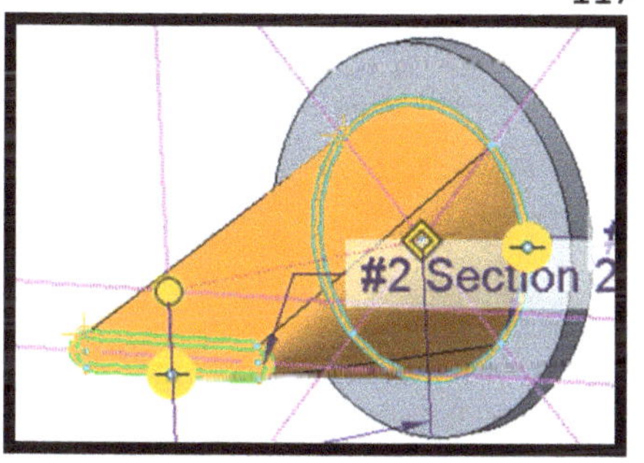

Step 6 - 6 – Checkmark to Accept the Blend.

Step 7 – Use the **Shell Tool in the top toolbar** to shell the part to a **thickness of 0.125"** while also removing the front & rear surfaces. 

- Shell Tool - References Tab – click in the "Removed Surfaces box" - hold control on the keyboard & LMB to select the two face surfaces to remove (*The rear of the disc and the face of the nozzle*) - **Adjust the Shell Thickness to 0.125** - Press the **Checkmark** to accept the Shell feature.

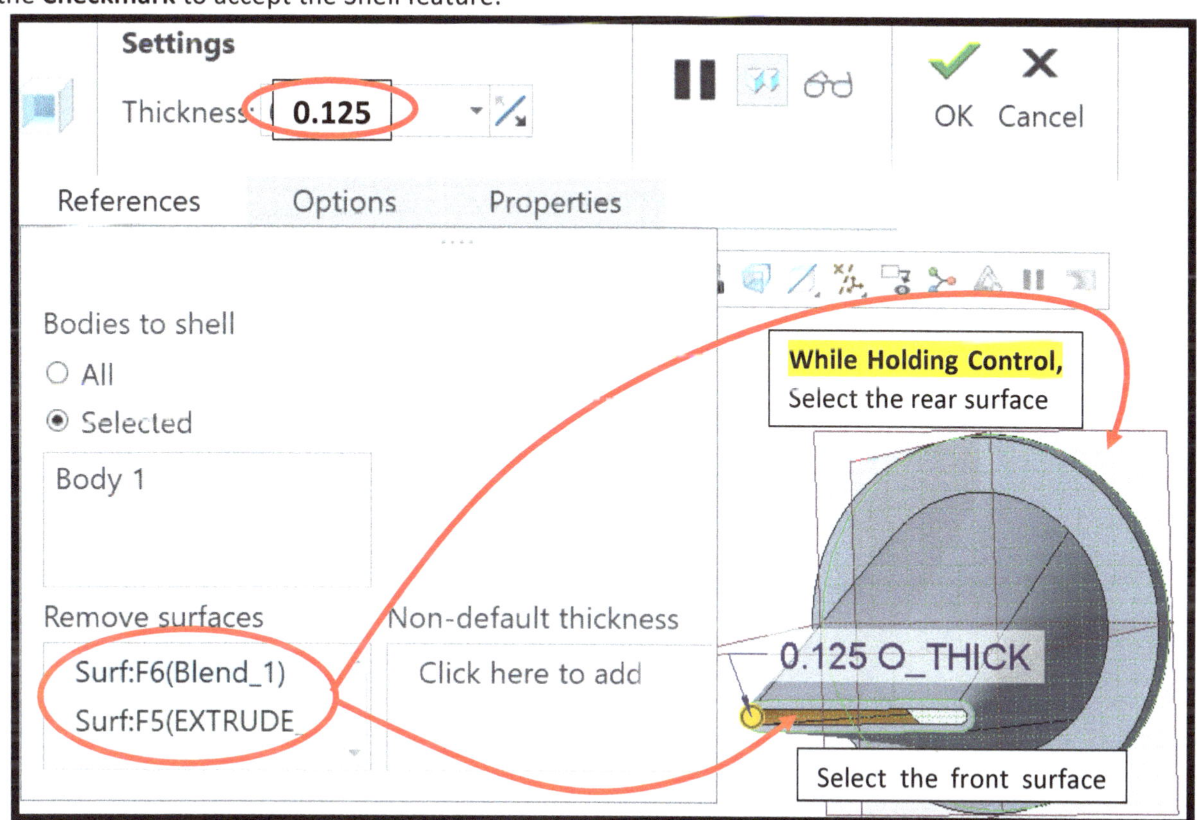

Step 8 – Use the Round Tool to round two edges to **0.125"** as shown at the base of the Blend and the outer front edge of the disc.

- Double check that both Rounds under the "Sets" tab have the correct radius.
- Do not put a round on the interior inside edge of the shelled surface as it will interfere with the Shell re-ordering step later on.

Checkmark to accept the Round.

Click in a blank area to deselect the pointer

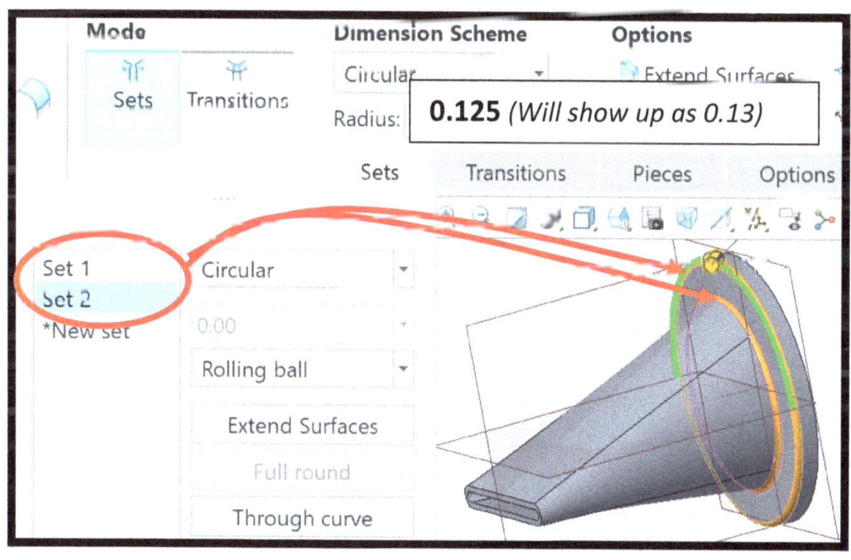

Step 9 – Create a Hole on the front of the Disc. This will be re-ordered in the Model Tree later, so do not use the rear "shelled" surface as a placement reference.

Hole Type: Simple

Profile Options: Flat

Toolbar Settings: Hole Diameter = **0.250"**

Hole Depth = **Thru All** (*Drill to intersect all surfaces*)

Placement Type: Change to **Diameter** type

Placement References:

- *Primary Reference:* **front face of the Disc/Extrude #1** (Do not use the shelled rear surface!)
- *Offset References:* **Adjust the Type to Diameter** instead of Linear, and hold control on the keyboard while selecting the references.
  - o **Select Axis (A1)** with a value of **5.00"** & **Right Datum** with a value of **0.0°**
  - *Tip: the axis is short as the disc of Extrude 1 is thin, but you can zoom in and rotate the view to see it*

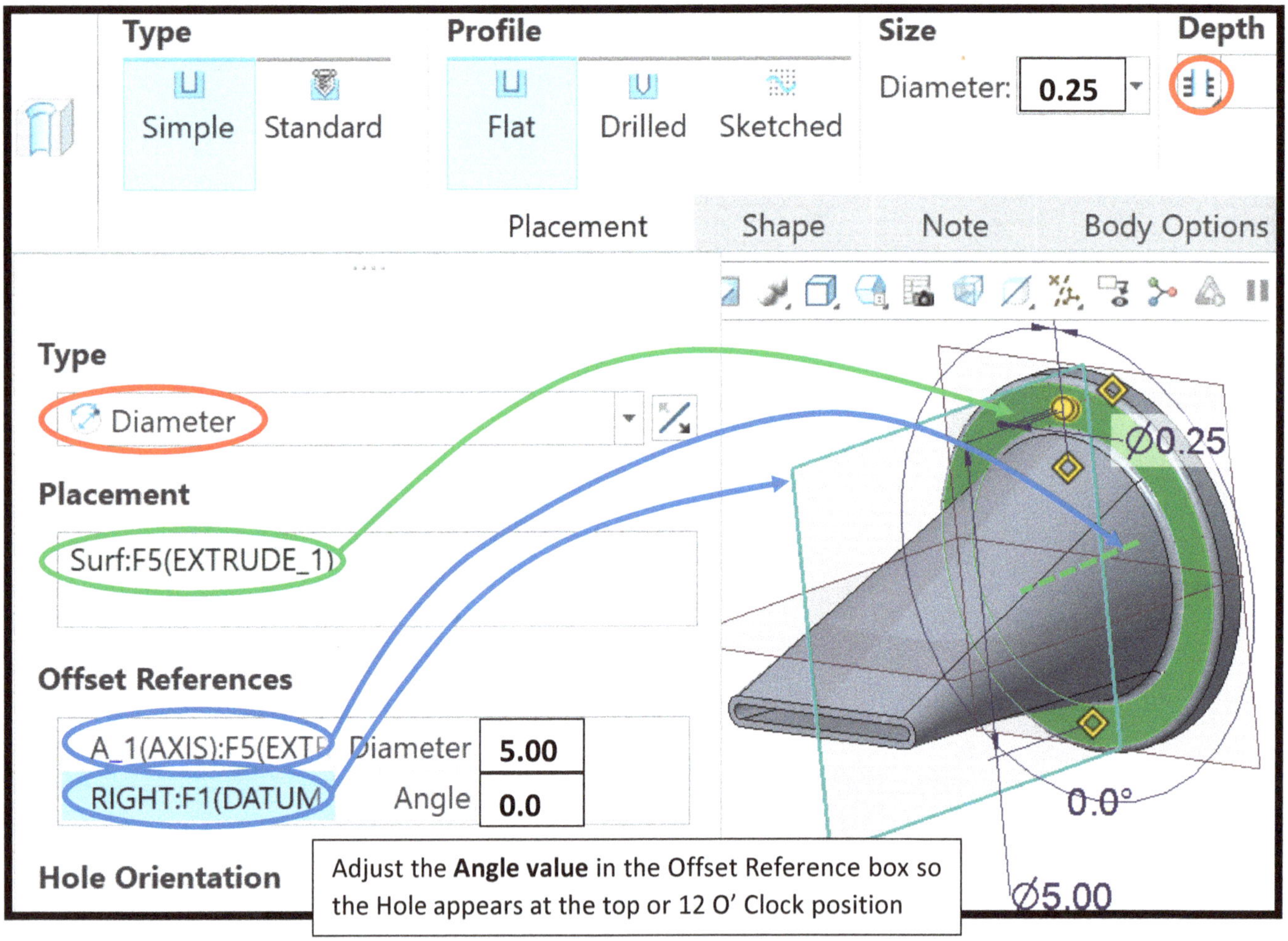

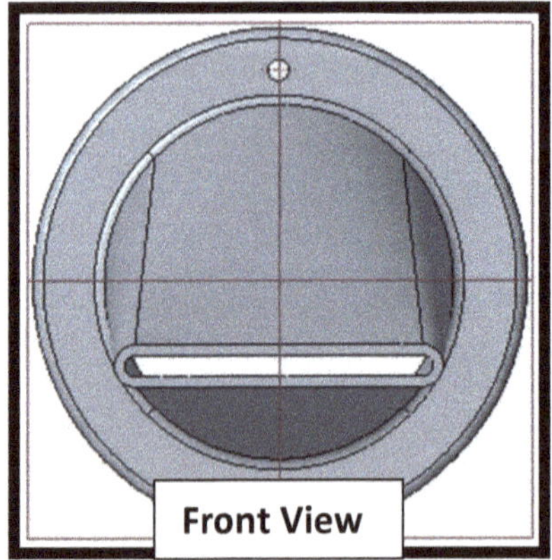

9 – 2 - Checkmark to accept the Hole.

Step 10 – Use the Pattern Tool to make copies of Hole 1. (Select Hole 1 in the Model Tree – Pattern Tool)

Pattern Type: Change to **Axis Type** – select the center Axis of Extrude 1 as the axis reference (zoom in to find the axis)

Pattern Options: 6 Members with an Angle Spacing of 60°

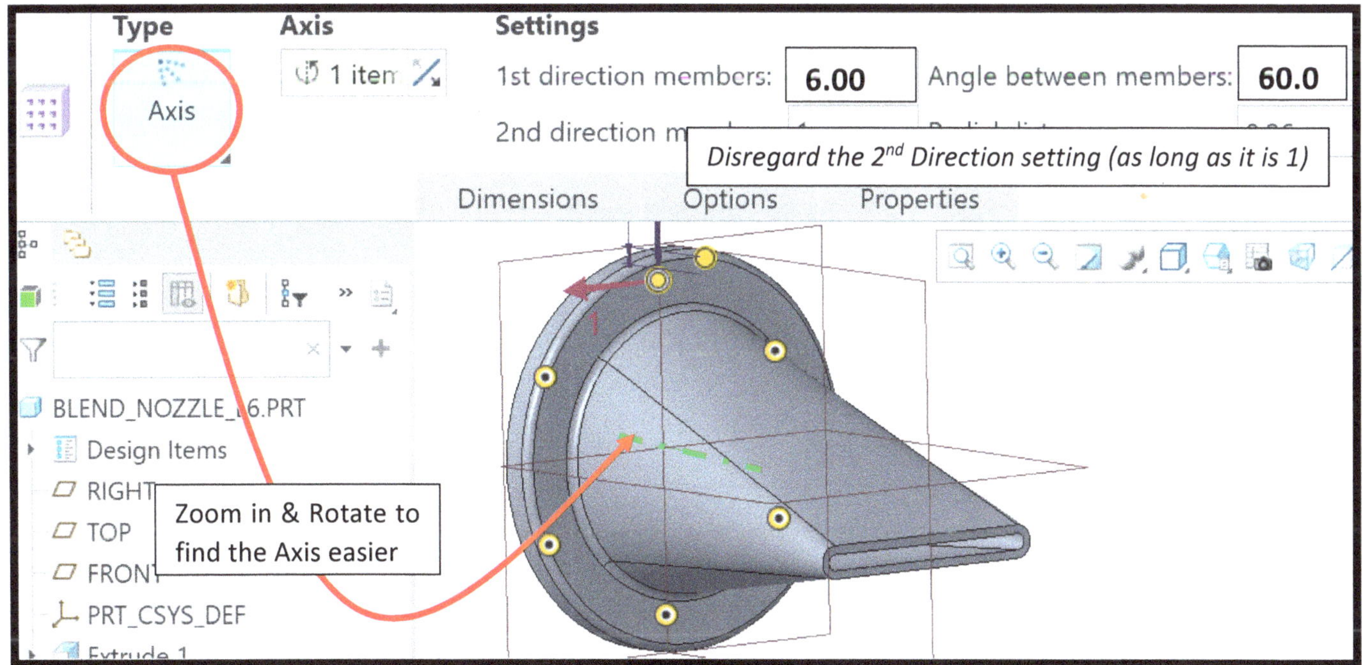

Step 11- Reorder the Shell in the Model Tree so that the Shell comes after the Holes and Rounds *.

- Select the **Shell 1 feature** in the Model Tree - hold LMB and drag it to a new position in the tree order.

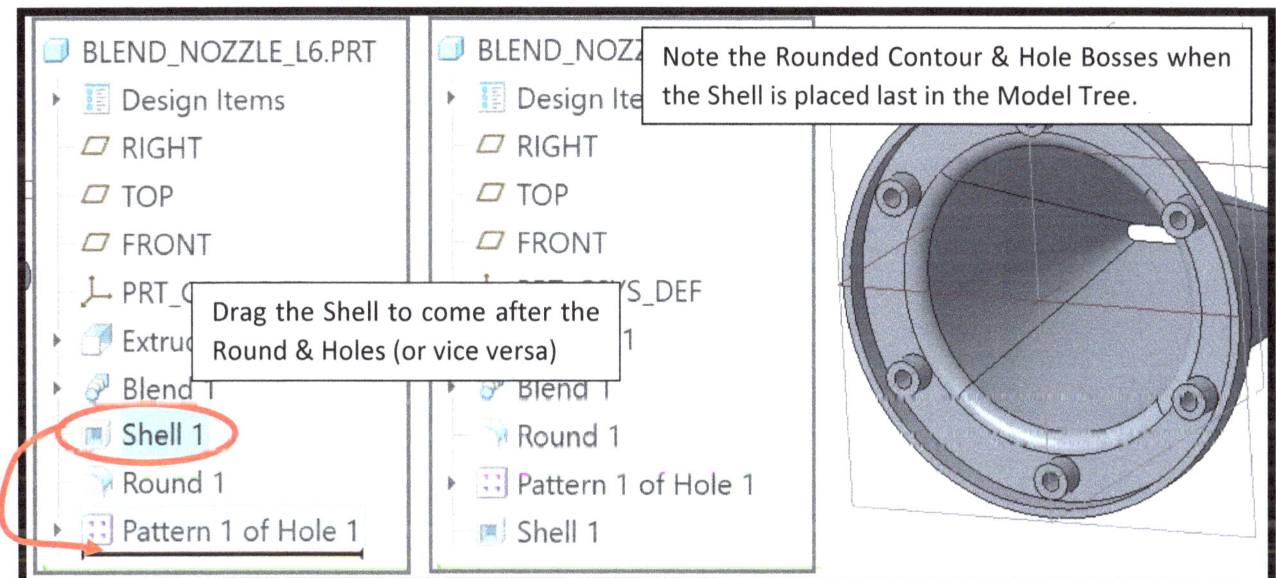

Tip: When dragging the Shell let go of it when the last Feature highlights (*Pattern 1*).

*If the model fails to regenerate when re-ordering or will not let you order the features:

- There could be interference with the Hole Boss and the edges. Check that your Round dimensions, Blend Section #1 Diameter value, Extrude #1 Diameter value, and Hole 1 offset references are setup correctly.
- Try moving the Round or Pattern before the Shell – if one will not move, that indicates an issue.
- Check that you have not used the "shelled out" rear surface of Extrude 1 as a Reference for the Round or Hole.
- As a last resort you can Delete the Shell, then remake it so it is placed at the end of the model tree.

Step 12 – **Save your Part File** & create a **New Drawing:** – File – New - Drawing – *"L6_BlendNozzle"*

Step 13 – **Setup the Drawing to match the Key** by following the ***Basic Drawing Steps*** and the ***Detail Drawing Steps***.

When detailing the dimensions on the part you will want to use a combination of **showing driving dimensions using the Show Model Annotations** tool and **creating your own Dimension measurements** using the Dimension tool.

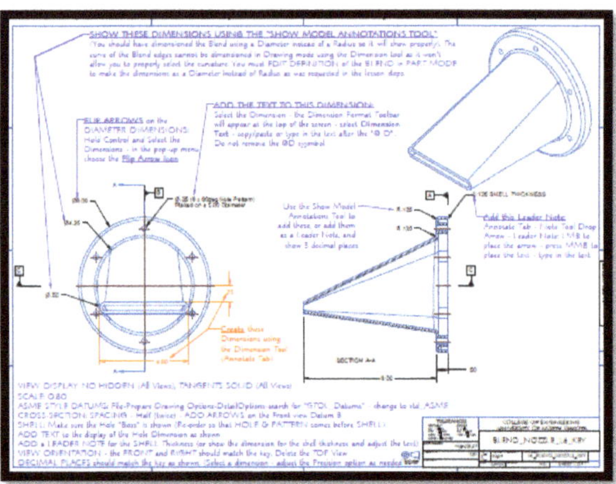

Step 14 - Match the details of the key, save your drawing, and create a PDF for submission.

Other Drawing Notes & Tips:

- **Dimension & Note locations different than the Key?** Your dimensions & notes may appear in different locations than the key, which is fine (i.e. your Shell note or Round Radius dimensions may show up on the bottom instead of the top of the Right view).

- **Cannot dimension Blend Edges?** Note that some dimensions from the Blend must be shown with the *Show Model Annotations* tool as they cannot be "created" on the drawing due to the geometry of the blend edges.

- **Centerlines Style Different than the Key?** Rarely, some students may find that centerlines show up as a circular **Radial style** for the Holes instead of the vertical & horizontal dotted lines. This is OK if this happens to you. If desired, you can change the option that causes this to occur on some student's software:

    - File - Prepare - Drawing Options - *"radial_pattern_axis_display"* set to No.

End of Lesson 6

Lesson 7 – Sweeps & Text (Steering Wheel)

Lesson 7 creates a steering wheel model which will utilize the Sweep feature for the supporting arms that connect the wheel to the center hub. Now that you are getting into more advanced topics in CREO so less detailed step-by-step instructions will be provided for the tools you have been using often. You can review earlier lessons as needed.

Sweeps:

The **Sweep tool** is used to model a shape that has a varying trajectory, such as a bent pipe. The cross-section will "sweep" along the specified trajectory to define the volume. Where the sweep meets other surfaces, the options of Free or Merged ends can be chosen to fill in the gaps where it connects.

Circular Text:

The **Text tool** can be using in sketch mode of an Extrude to create text, either straight or it also can be placed along a curve. When using the Text Tool in sketch mode it will require you to draw a single line – this line will set the orientation and height of the letters. Also, the text tool will have the option to *"place along a curve"*.

Getting Started: Open the **CREO** *Parametric* software and follow the lesson steps carefully.

Step 1 – Complete the **Sweep Module on Blackboard** if you have not yet done so.

Step 2 – Set your **Working Directory** and create a **New Part** called *"Sweep_Wheel_L7"*.

Step 3 – Change the **Material** of the part to **AL6061**.

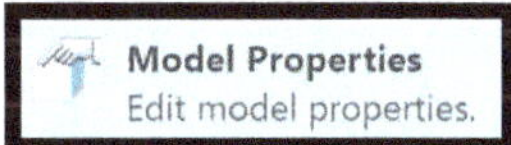

File- Prepare- Model Properties- in the Material option select **change** – double click on **Legacy Materials** - <u>double click</u> on "Al6061.mtl" to set it as the material – press **OK** – press **Close**

Step 4 – Adjust the Options so that <u>3 decimal places</u> will be shown in sketcher.

      File – Options – Sketcher – set the "<u>**Number of decimal places for dimensions**</u>" to **3 – OK** – choose No when asked to save the config file for future use.

Step 5 – 1 - Use the Revolve Tool to create the base of the component, **sketched on the <u>Front</u> Datum.**

Tip: Don't forget the vertical **Axis of Revolution**. (**Hold RMB in a blank area – Axis of Revolution icon**).

<u>**Revolved Dimension Steps**</u>: If needed, use the Dimension Tool – LMB on the Axis – LMB on the Sketch point LMB again on the Axis – MMB to place the revolved dimension.

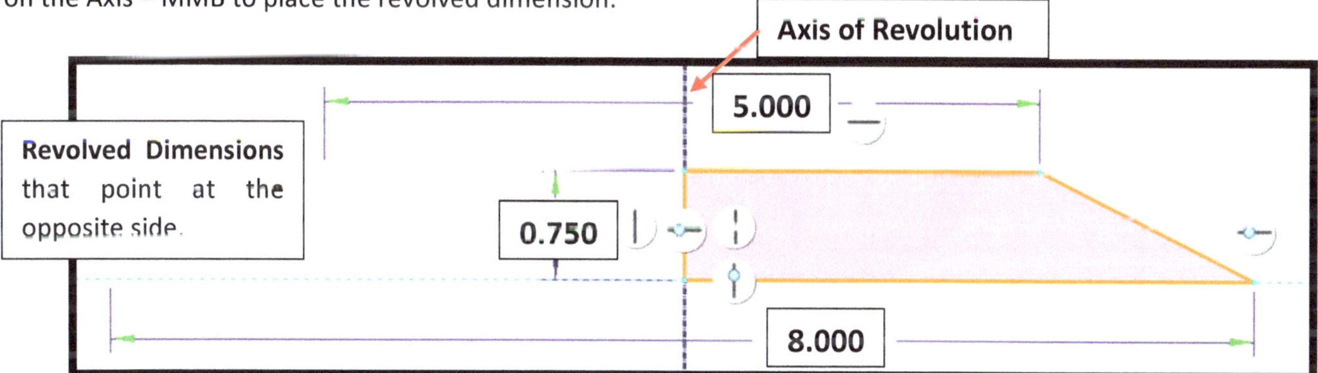

Step 5 – 2 – Set the Revolve Toolbar options to **360°** (it should be by default). If you don't see the Revolve on the model, you likely did not set an Axis of Revolution!

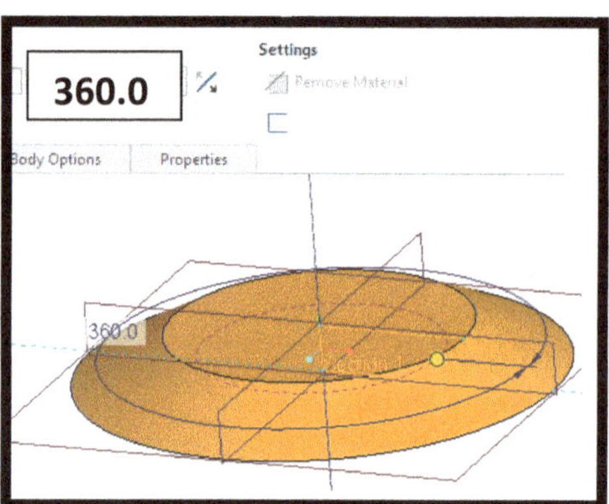

Checkmark to accept the Revolve.

Step 6 – 1 – Use the Revolve Tool again to create the circular handle, **sketched on the Front Datum.**

Tip: Don't forget the vertical **Axis of Revolution**. (**Hold RMB – Axis of Revolution icon**)

The Circle should be **dimensioned from the top edge of the hub**, not from the bottom reference line.

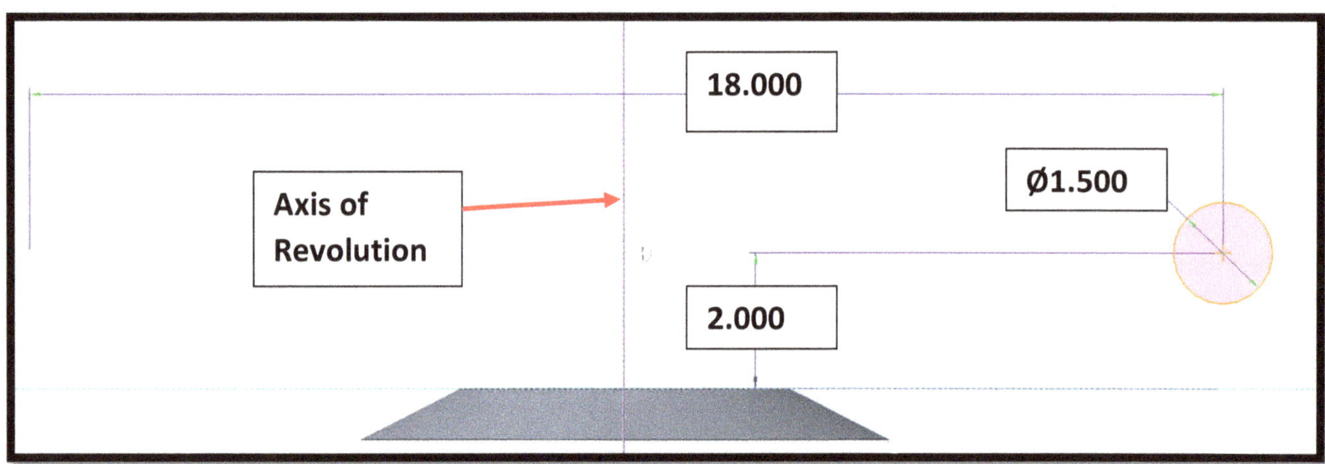

Step 6 – 2 – Set the Revolve Toolbar options to **360 degrees** (it should be by default).

Checkmark to accept the Revolve.

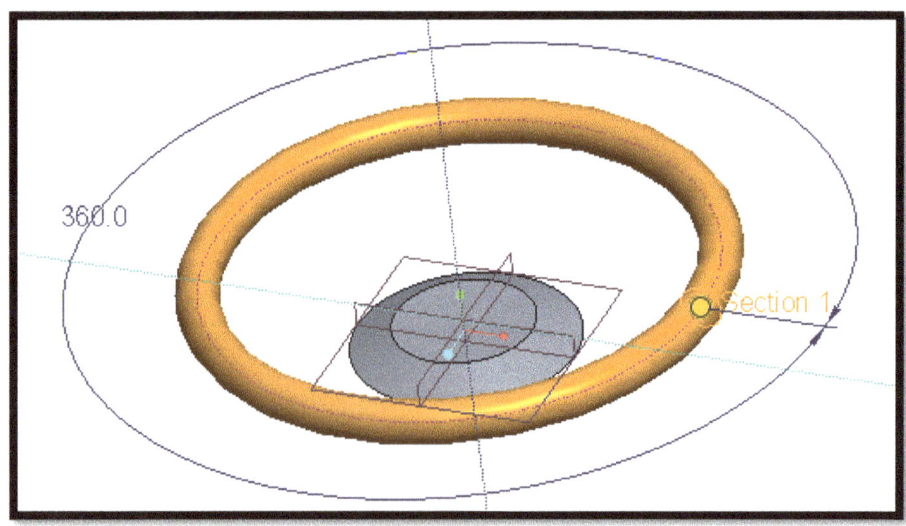

Step 7 – Use the **Sketch** Tool from the top toolbar to sketch a trajectory shown below, placed on the **FRONT** datum. Then press the **2D Sketch View icon** in the quick toolbar to view the sketch plane.

<u>Add Sketch References:</u> Add the **Sketch References** of the *Slanted Edge* and the *Outer Curve* of the Handle Circle as shown below.

- **Hold RMB** in a blank area – **References Tool** – click to add the desired edges as references – OK

<u>Sketch the Trajectory</u>: Use the **Spline tool** in the top toolbar to create the trajectory by setting the spline points as listed below.

- LMB on the Midpoint of the Slanted Line (point 1 below)
- LMB on the Left Horizontal Edge of the Circle of the Handle (point 2 below)
- LMB on the Center point of the Circle of the Handle (point 3 below)
- **Press MMB** to end the Spline tool. No dimensions are required.

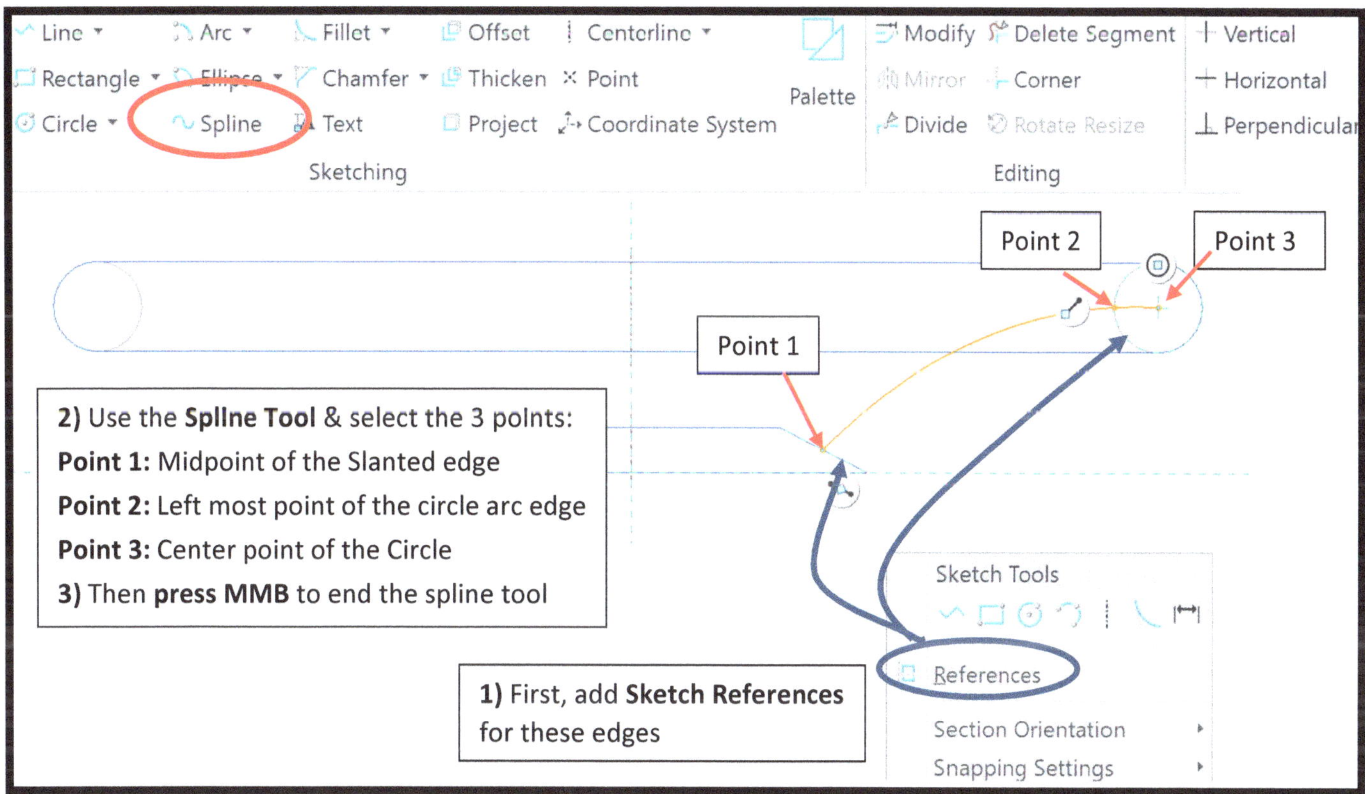

Step 7 – 2 - Checkmark to accept the Sketch. You will see the **sketch line** in the graphics window as a light blue line.

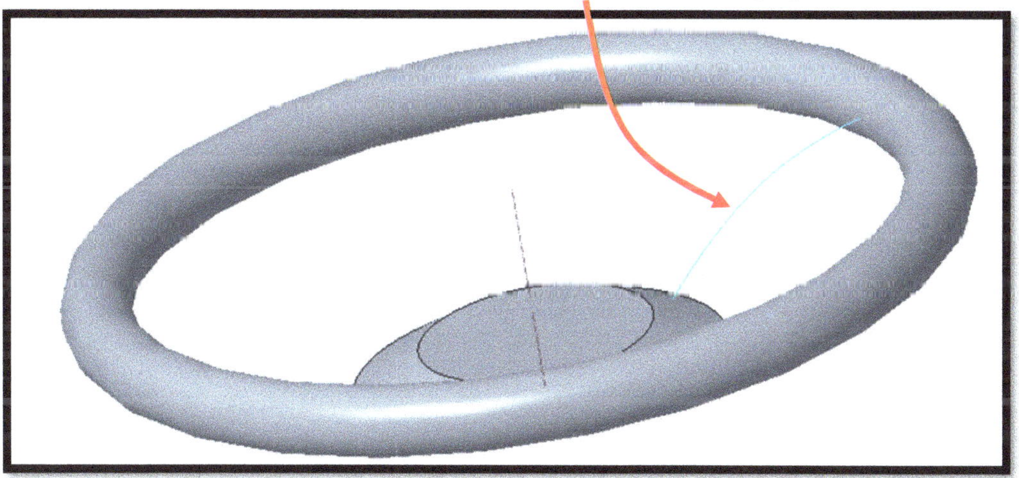

Step 8 - Use the **Sweep** tool to turn the previous sketch into a solid volume.

Choose the Trajectory: Choose the **Sweep tool**, then click the **References tab** and select **Sketch 1** as the trajectory (*if it wasn't already selected*).

Start the "Sweep Section" Sketch Mode: Select the **Sweep Section icon** in the toolbar to enter **sketch mode** for the Cross-Section sketch.

View the Trajectory Plane: Press the **Sketch View** icon in the quick toolbar to view the trajectory origin.

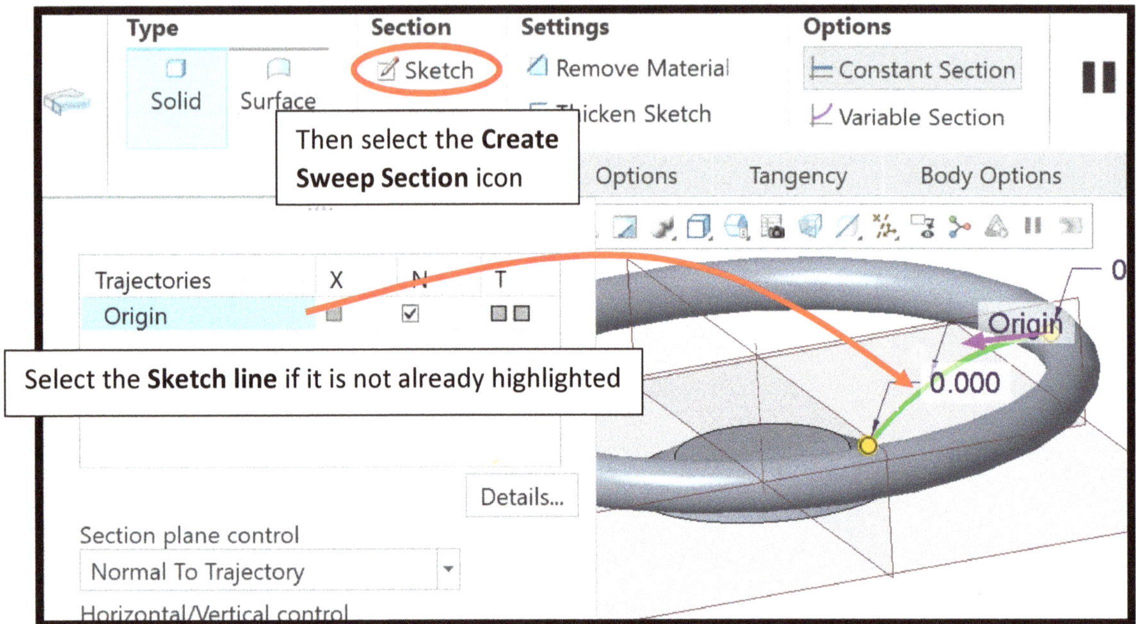

Sketch the Shape: Use the **Ellipse** tool and place an ellipse symmetric around the **"Trajectory Origin"** (the magenta dotted lines). Your origin may be on the handle (as shown) or at the Revolve #1 base. Either is OK.

- o **Place Point 1** above the origin on the vertical reference.
- o **Place Point 2** on the opposite side of the origin on the vertical reference. Snap in a Symmetric (or possibly Midpoint) Constraint as you place this point.
- o **Set Point 3** for the width to finish the ellipse tool.

Dimensions: Dimension the sketch as shown, so the ellipse length of 1.500" is the width of the completed sweep arm. The thickness will be 0.750".

Checkmark the sketch to accept the sweep cross-section.

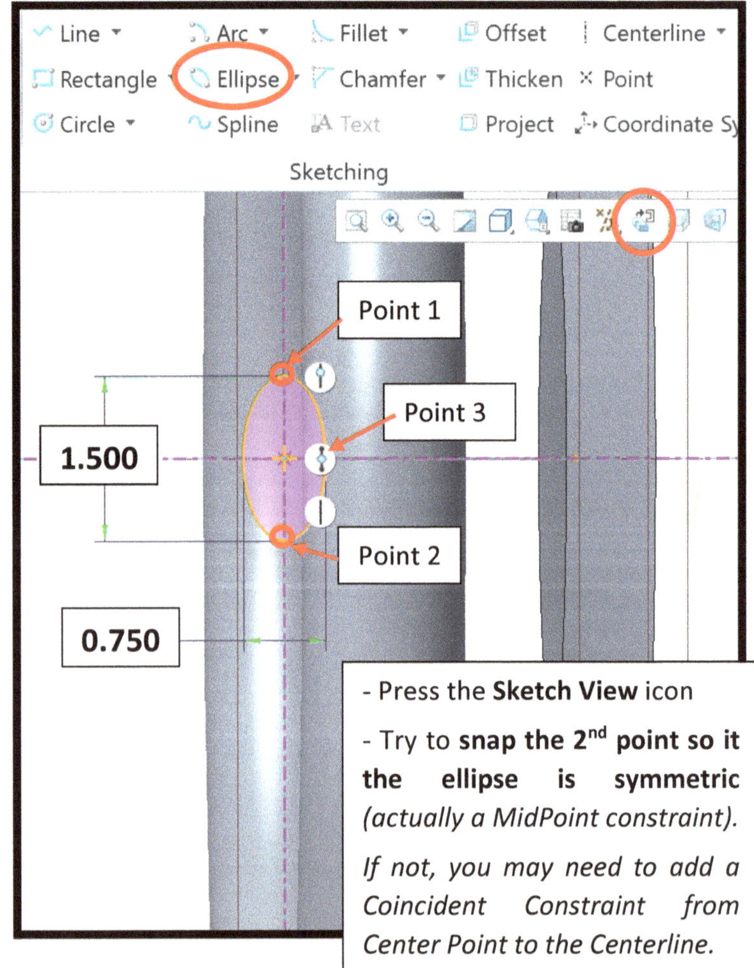

Step 8 – 2 - Adjust the **Options** of the Sweep to be **Merged Ends** to fill in the gap at the Sweep ends.

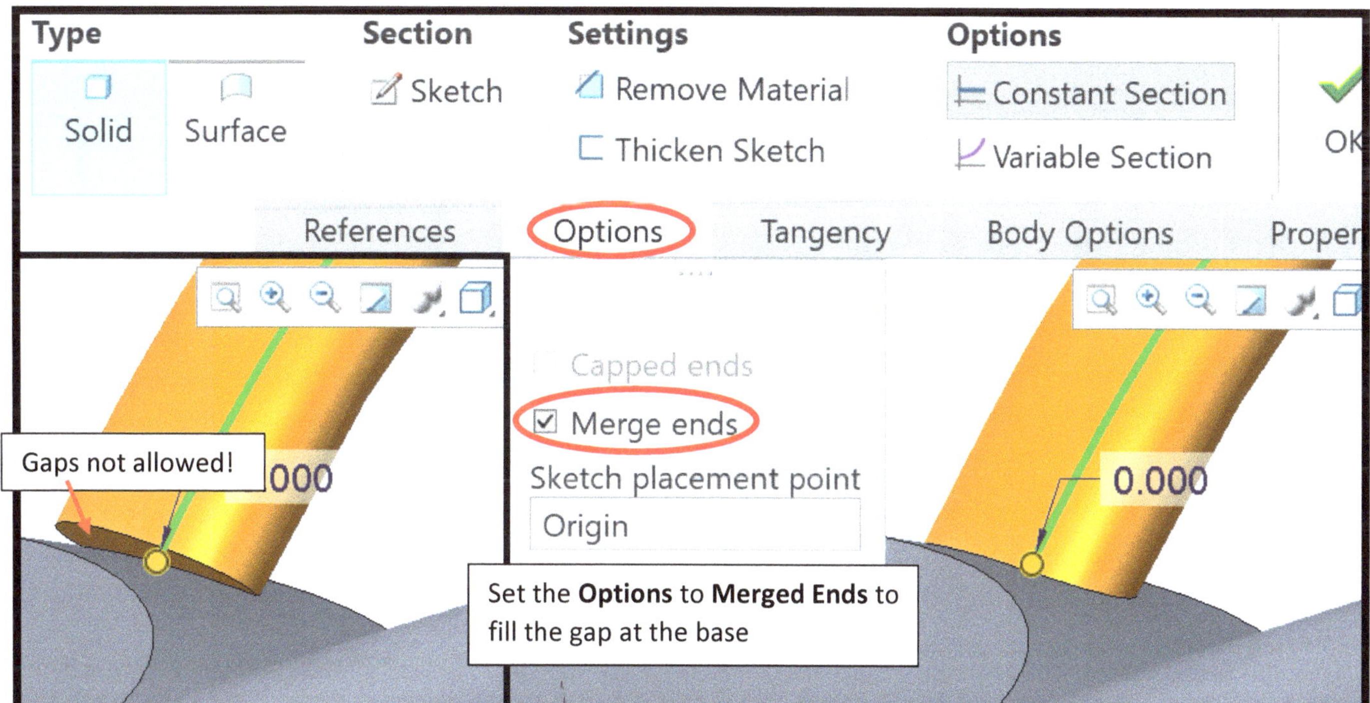

Tip: If your Sweep fails to show on the screen or does not merge the end it is likely an issue with the Trajectory Sketch, or the Ellipse Sketch not being aligned on the Origin. Sometimes the Sweep tool will fail when you select Merged Ends as the geometry if too difficult or too wide of a gap to merge, but that should not be an issue on this model.

- Checkmark & go back and Edit Definition of Sketch 1 to make sure it is connected on the 3 points as shown.
- Edit Definition of the Sweep to get back to the Cross Section (use the **Create or Edit Section icon)** and check that the ellipse is centered around the Trajectory Origin (pinkish lines) without any weak dimensions.

Step 8 – 3 - Checkmark to accept the Sweep Feature.

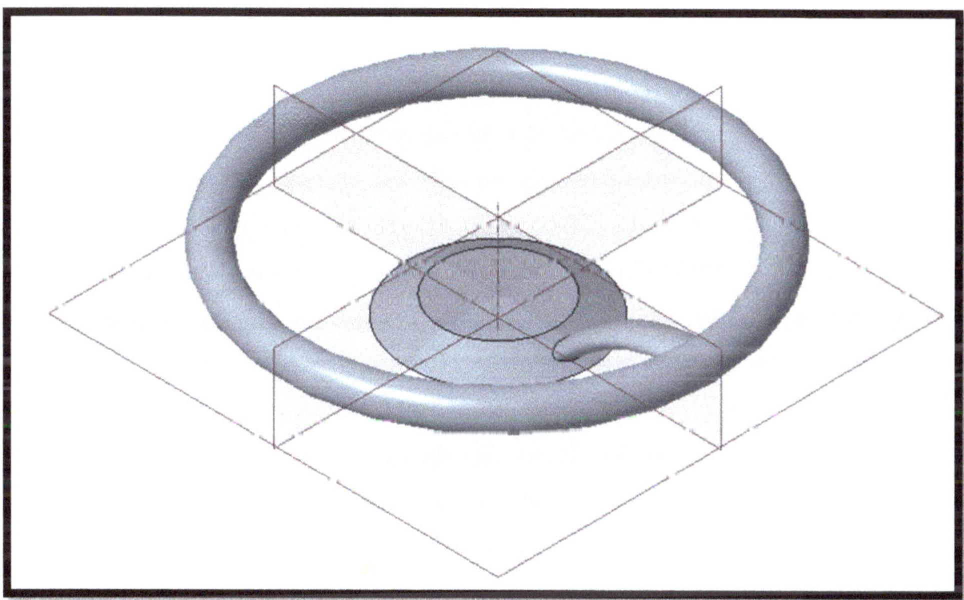

Step 9 - Set the display properties of <u>each datum</u> so that we may use them as a **Geometric Tolerance** and be able to change them to **ASME style** in our drawings.

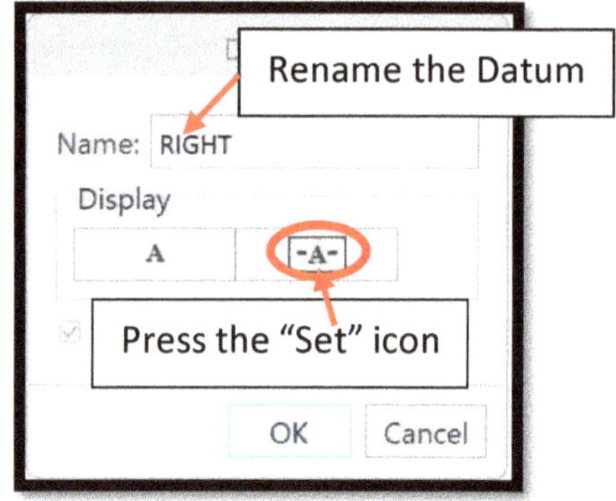

- **LMB to select a Datum** from the model tree - **hold RMB – Properties - press SET ("-A-")** – select the **name box – Type in a new name** according to the list below – **OK**.
- **Repeat for each Datum.**

   RIGHT: Rename to **B - Press "Set" - OK**

   TOP: Rename to **C - Press "Set" - OK**

   FRONT: Rename to **A - Press "Set" – OK**

Step 10 – Pattern the **Sweep** feature using the **Pattern Tool** with an **Axis type setting**. Set the pattern options for **4 instances** with a **Spacing of 90 degrees**.

- **Toggle Off** one instance to only show 3 of the 4 copies. (Toggle off the 12 O'clock position copy when the original is at 3 O'clock.)
- Use the *Saved Orientation* tool in the quick toolbar to view the Top view to check that the correct copy is off.

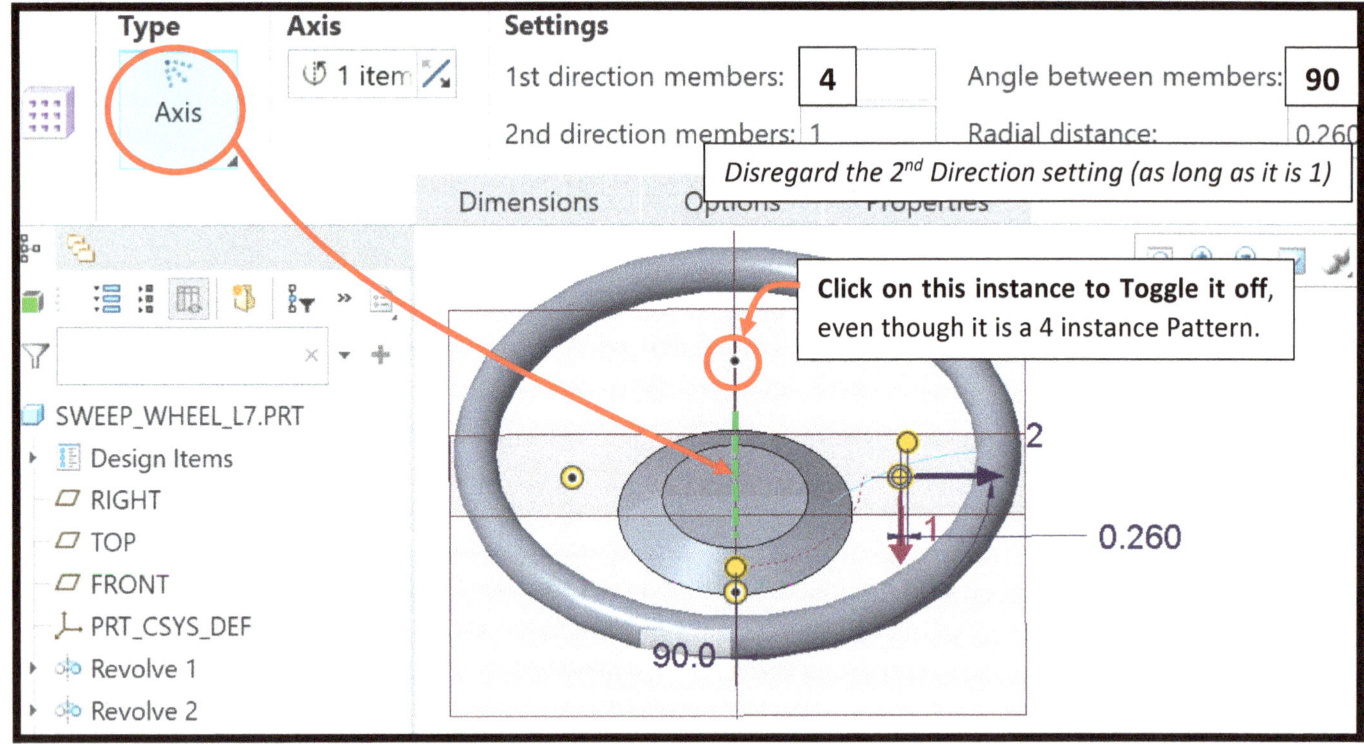

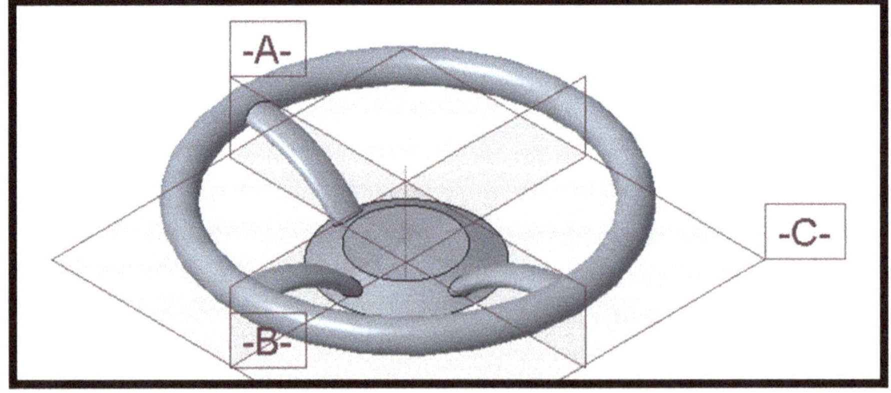

Step 11 – Create the Hole on the outer edge of the Hub. Note that this hole is not drilling into the slanted edge at an angle, but rather straight down from the top of the hub.

Hole Type: Simple

Profile Options: Toggle on the **Drilled Profile** and toggle on a **Counterbore.**

Shape Tab Options: Adjust the **Counterbore Diameter** (0.900") & **Counterbore Depth** (0.550")

Toolbar Settings: **Hole Diameter = 0.500", Hole Depth = Thru All** (*'Drill to intersect all surfaces'*)

Placement Type: Diameter

Placement References:
- **Primary Reference**: select the flat top **planar surface*** of the center hub (*not the slanted surface!*)
- **Offset References**: Adjust the Type to **Diameter** & hold control on the key board when selecting the references.
    - Select the **Axis** of the Hub with a **6.500"** Diameter value, and **Datum A** with a **30°** value (*30° toward the "toggled off" sweep instance*)

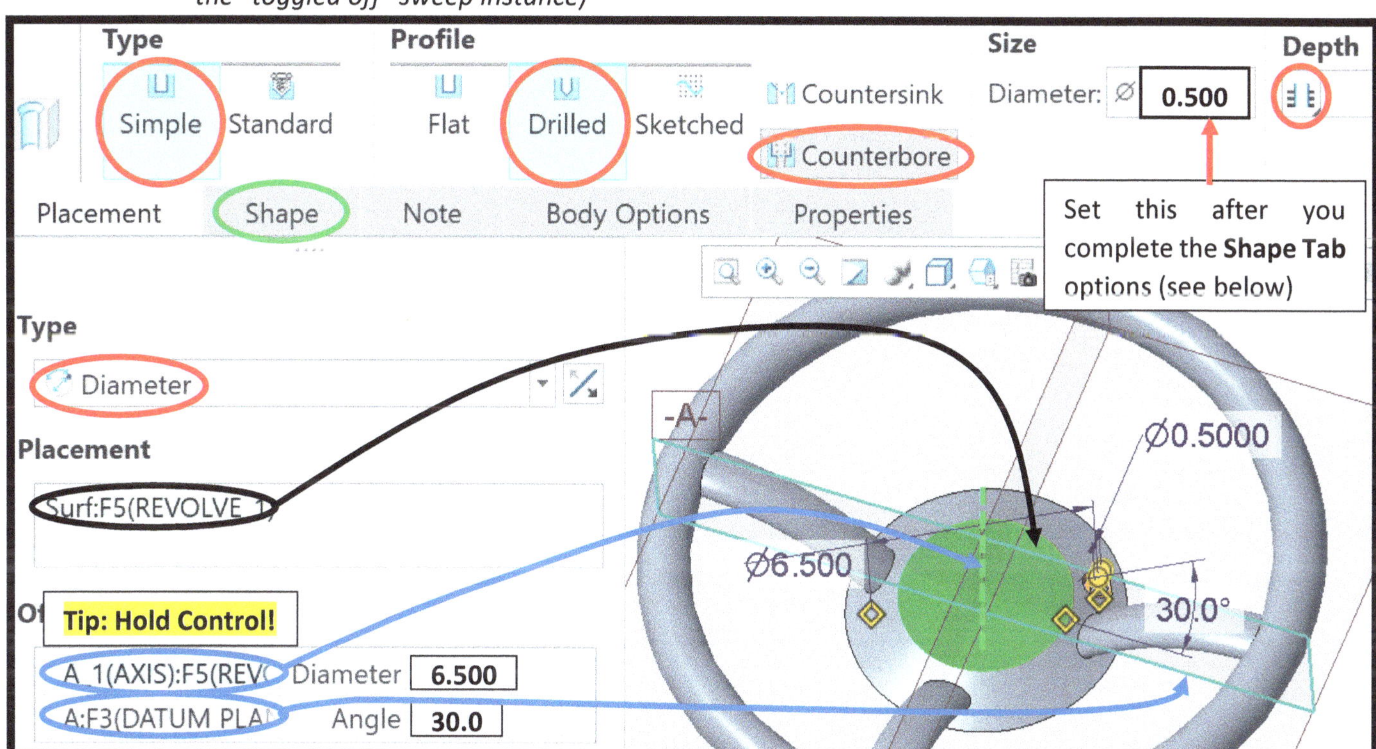

Adjust the Shape Tab:

Toggle on the Counterbore option, open the **Shape Tab**, then adjust the **Counterbore dimensions to be 0.550" deep and 0.900" diameter** of the counterbore.

Tip: CREO will not let you set the counterbore diameter smaller than the hole diameter, so make sure you have the hole and conterbore values entered correctly.

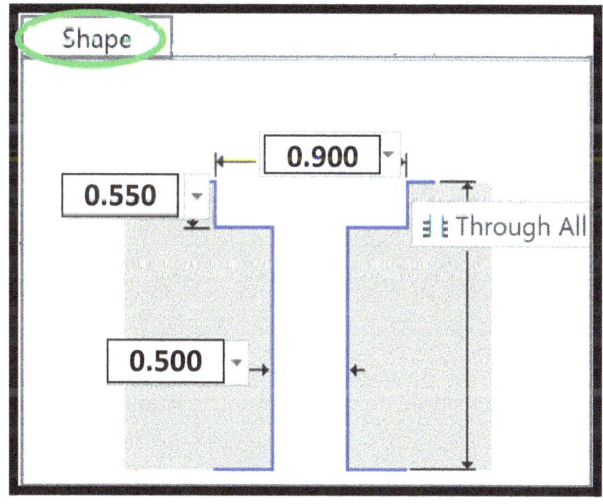

Step 11 – 2 - Checkmark to accept the completed Hole.

Step 12 – Pattern the **Hole** using an **Axis Pattern**. Set the options to **6 instances** with a **Spacing of 60°**.

- **Toggle Off one instance** (the copy that will interfere with the middle Sweep arm) to show only 5 of the 6 copies in the pattern.
- **Tip:** You may need to Edit Definition of your Hole #1 to adjust the Angle *Offset References* value to match the drawing key so that you will have a hole at the 12 O'clock position where the missing Sweep is.

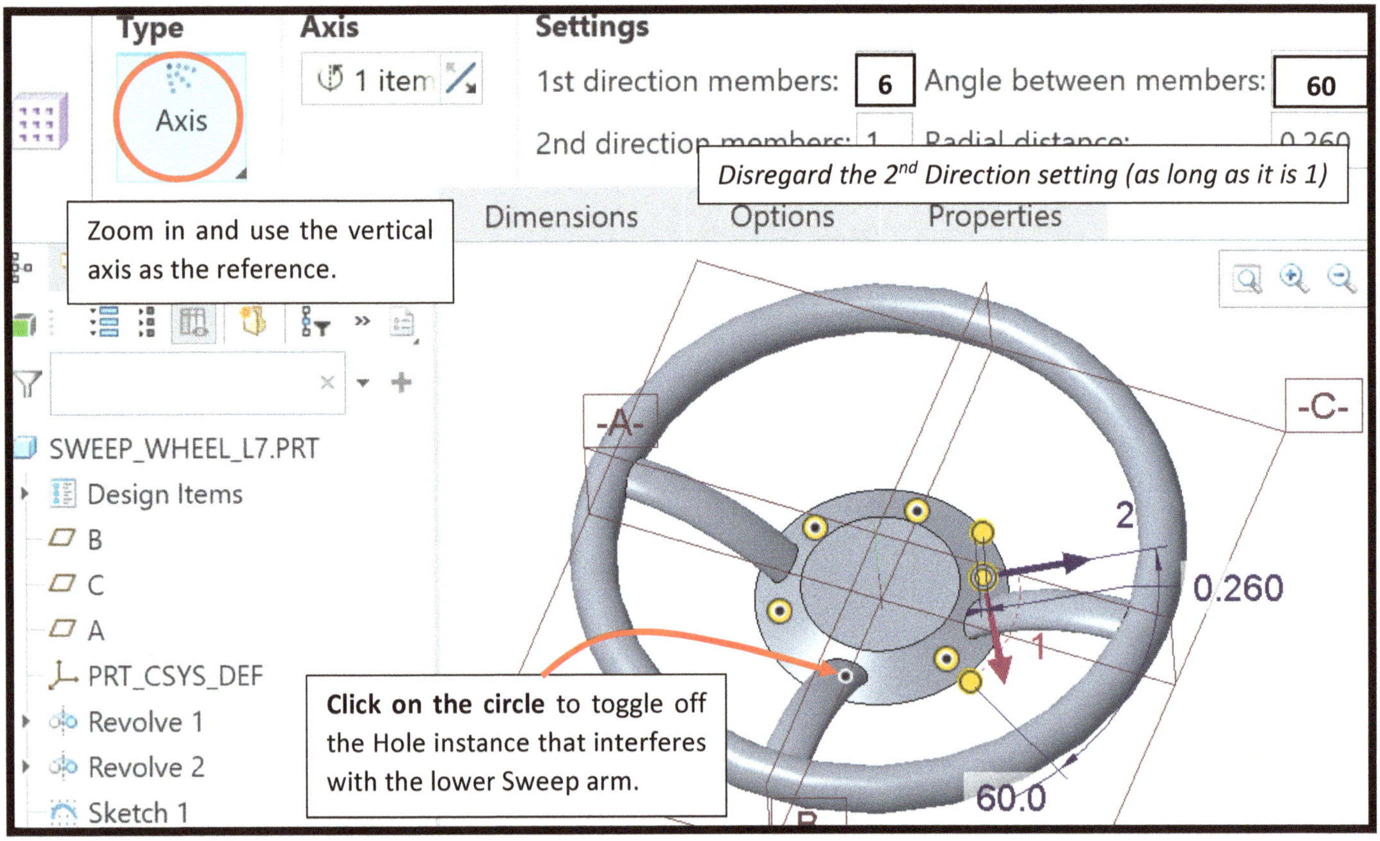

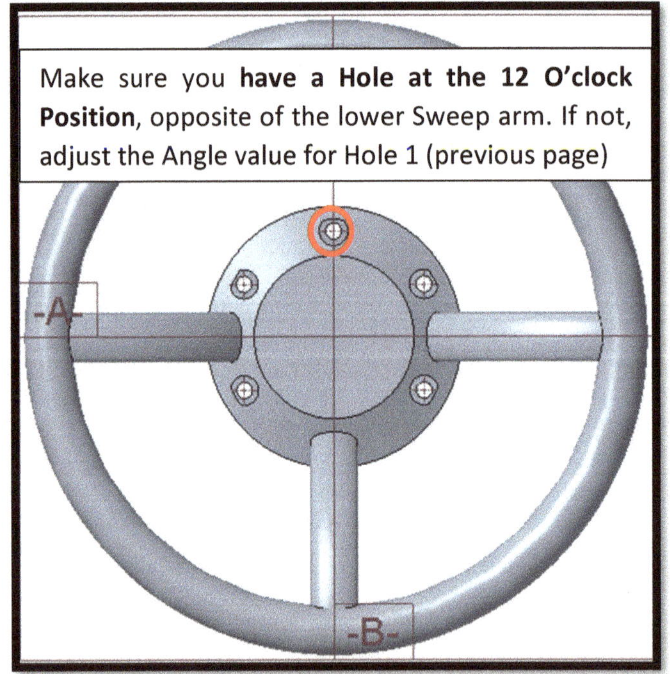

- **Checkmark to Accept the Pattern**

Step 13 – Create the Hole in the center of Revolve 1.

Hole Type: Simple

Profile Options: Toggle on the **Drilled Profile** and toggle on a **Counterbore.**

Shape Tab Options: Adjust the **Counterbore Diameter** (1.900") & **Counterbore Depth** (0.250**)**

Toolbar Settings: **Hole Diameter = 0.900", Hole Depth = Thru All** (*'Drill to intersect all surfaces'*)

Placement Type: **Coaxial**

Placement References:
-    **Primary References**: Hold Control on the key board when selecting the two primary references.
        o   **Axis of the Hub** (*choose this first*)
        o   **Flat top surface of Revolve 1**

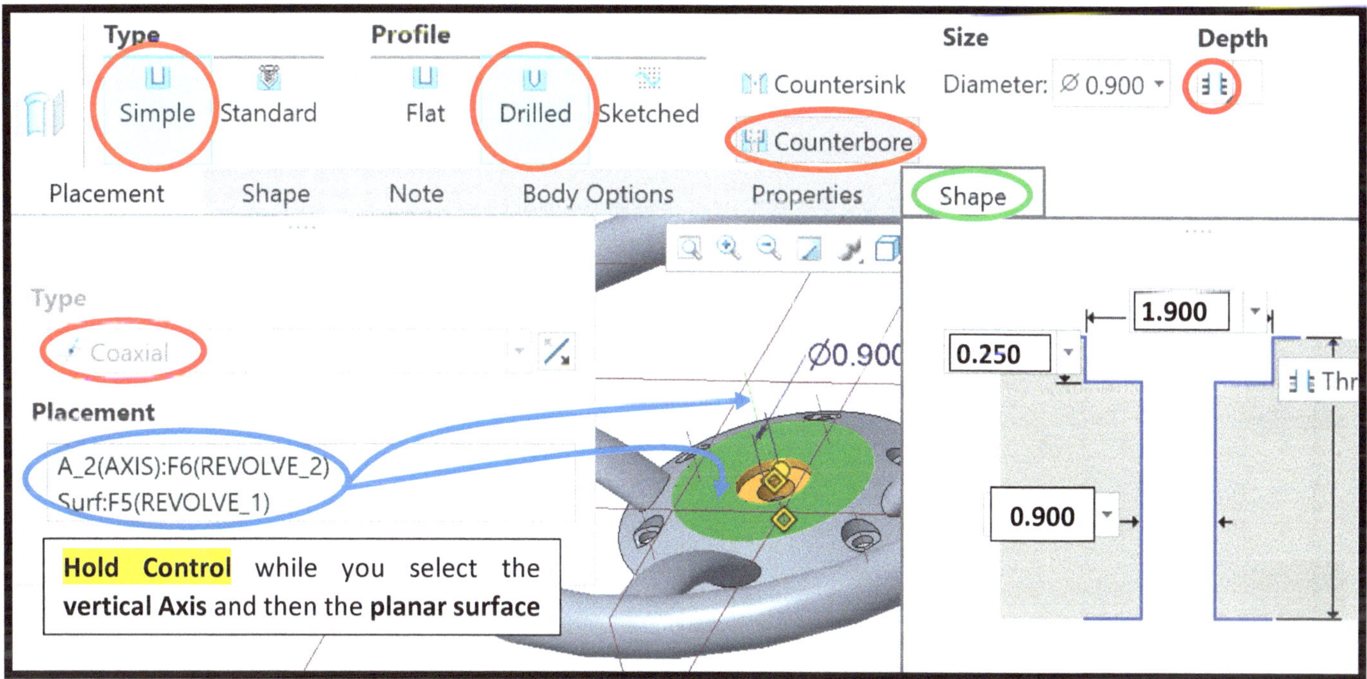

Tip: If you get errors when setting the hole size, you will need to set the counterbore larger than the hole first.

Checkmark to accept the Hole. Your holes should look similar to the image below.

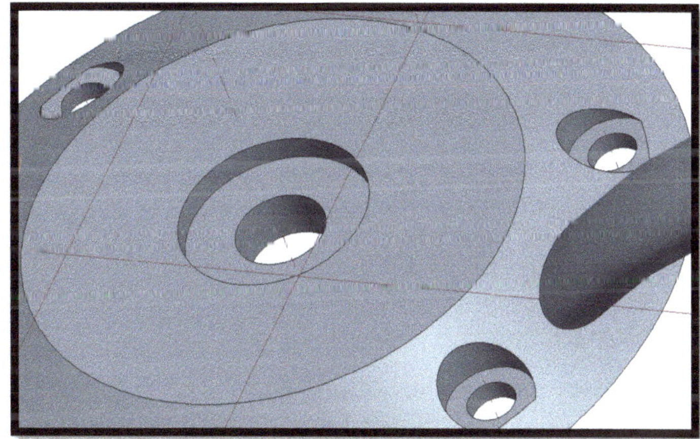

Step 14 – Use the **Extrude tool** to add text to the model. Use the top **surface of the base hub** (*Revolve 1*) as the primary sketch plane.

1) <u>**Construction Mode:**</u> **Toggle on Construction Mode in the sketch toolbar**. Any entities created while Construction Mode is on will be dotted reference lines which are not part of the closed-loop.

2) <u>**Sketch the Reference Circle:**</u> Use the **Circle tool** and place a circle at the origin with **Radius = 1.500".**
   The Circle must be a dotted "Construction Mode" Circle so it is not part of the closed loop!

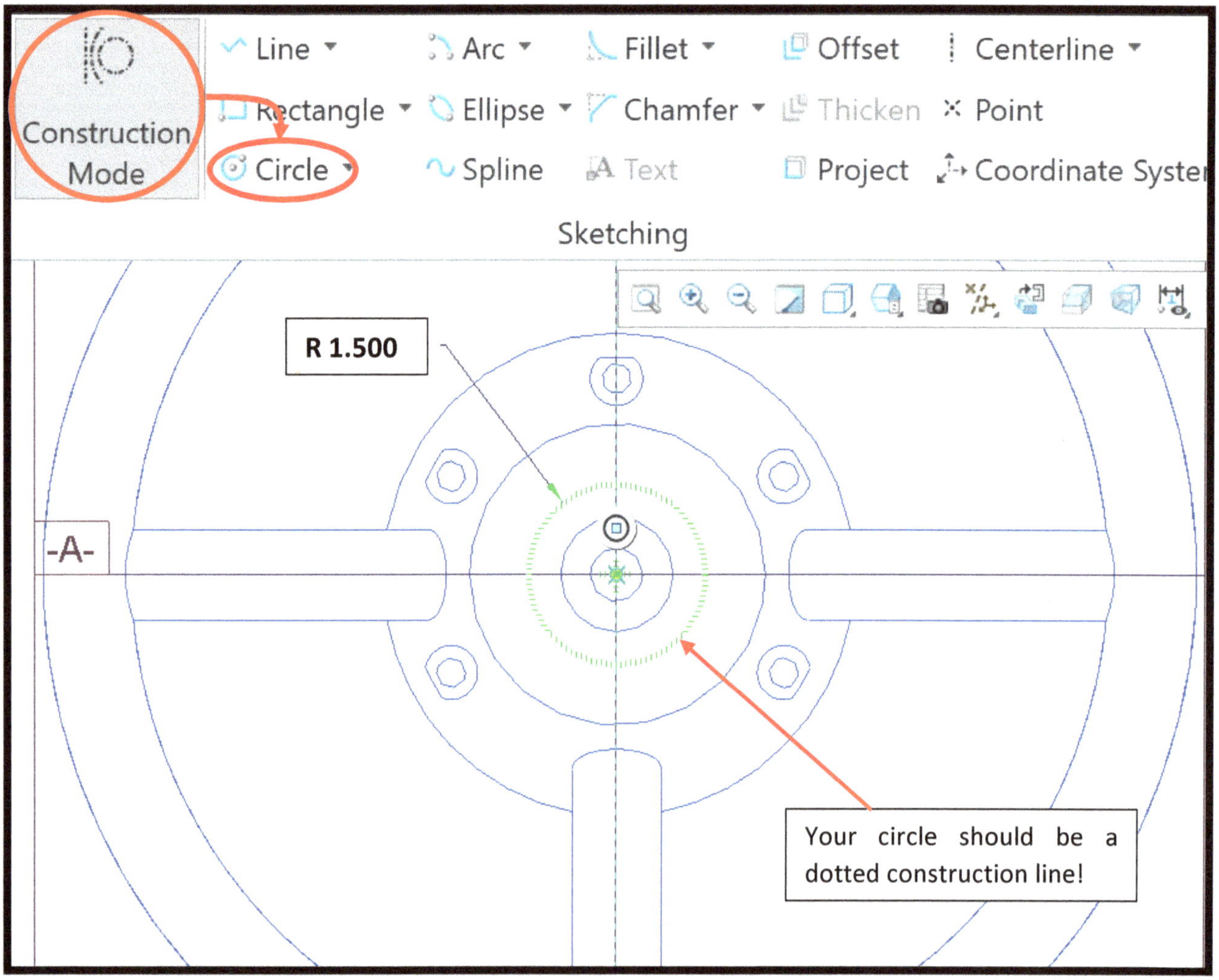

3) <u>**Toggle Off Construction mode**</u>: Press the Construction Mode icon to **toggle it off** so that the next step will have sketch entities that are part of the closed-loop.

Text Tool FYI 	A Text

The text tool is a useful way to add letters to a <u>Sketch</u>. To use the **Text Tool,** you only need to draw a single line. The direction you sketch the line controls the text direction. Read the examples below:

A line starting at the bottom: The text direction is to the right of the line

A line starting at the top: The text direction is to the left of the line.

A horizontal line starting at the left: The text is written down from the line.

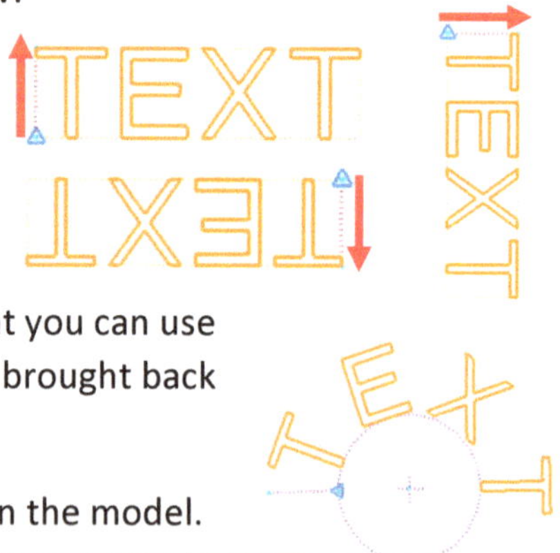

After placing the Text there will be a **small point** at the bottom left of the text that you can use to drag the text or use for a dimension reference. The **Text Menu Options** can be brought back by double clicking on the text, if needed.

You can also use **Place Along a Curve** in the text Menu to have it follow a curve on the model.

Step 14 – 2 - Use the Text Tool to add your desired text to the sketch.

1) <u>Text Tool</u>: Use the **Text tool to sketch a small vertical line**, not snapped onto any geometry. The length of the line made while using the Text Tool will be the initial size of the text letters.

2) <u>Text Menu</u>: **Type in your desired text** (whatever words you want is OK) - then **checkmark the "Place along a Curve"** option – the tool will be waiting for you to **click on the construction circle made previously** as the curve to follow. You may need to **toggle the Direction Arrow** to move the text to the correct side of the circle.

3) <u>Adjust the Dimension</u>: you can adjust the dimension for the **Height of the Text to 0.400"**, then **double click on the text** to bring back the menu to adjust the *Aspect Ratio* as desired so the text slant looks good.

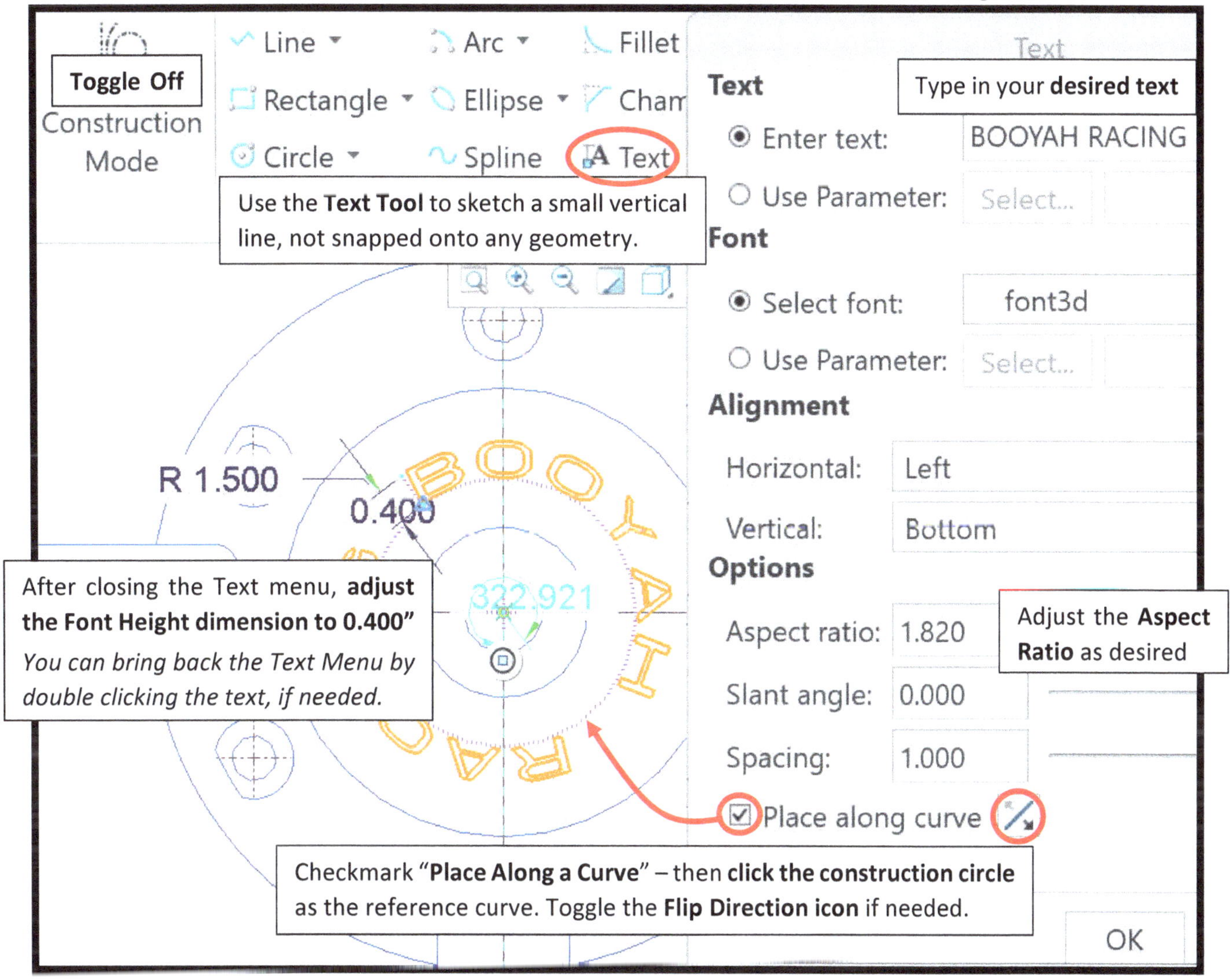

Tip: If your sketch fails, make sure your Circle is a construction circle (dotted lines). Also try adjusting the font, size, and aspect ratio, as some combinations of letters can produce a sketch that have letters that cross over each other.

4) **Accept the sketch** and set the options of the Extrude:
   - **Flip the Direction Arrow** so the text cuts into the hub.
   - Toggle on **Remove Material**, if not done already
   - Leave the Depth option as single direction with **a Depth of 0.250"**
   - **Checkmark** to accept the Extrude.

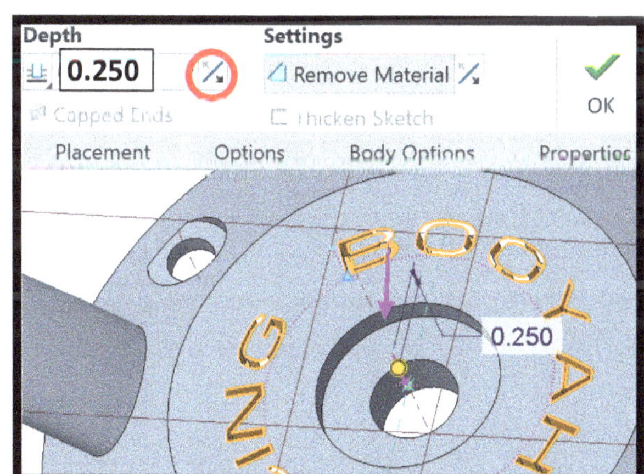

Step 15 – Create a **custom view state** to approximate an Isometric view that approximately matches the key.

Creating a Custom Isometric View:

1) **Rotate the model** on the screen to <u>approximate</u> the **Isometric View** on the drawing key.

2) In the Top Toolbar go to the **View Tab** – select **Manage Views** – in the View Manager change to the **Orient** tab.

3) **In the Orient Tab** select **New** – type in the name for the new view **"MYISO"** – press Enter.

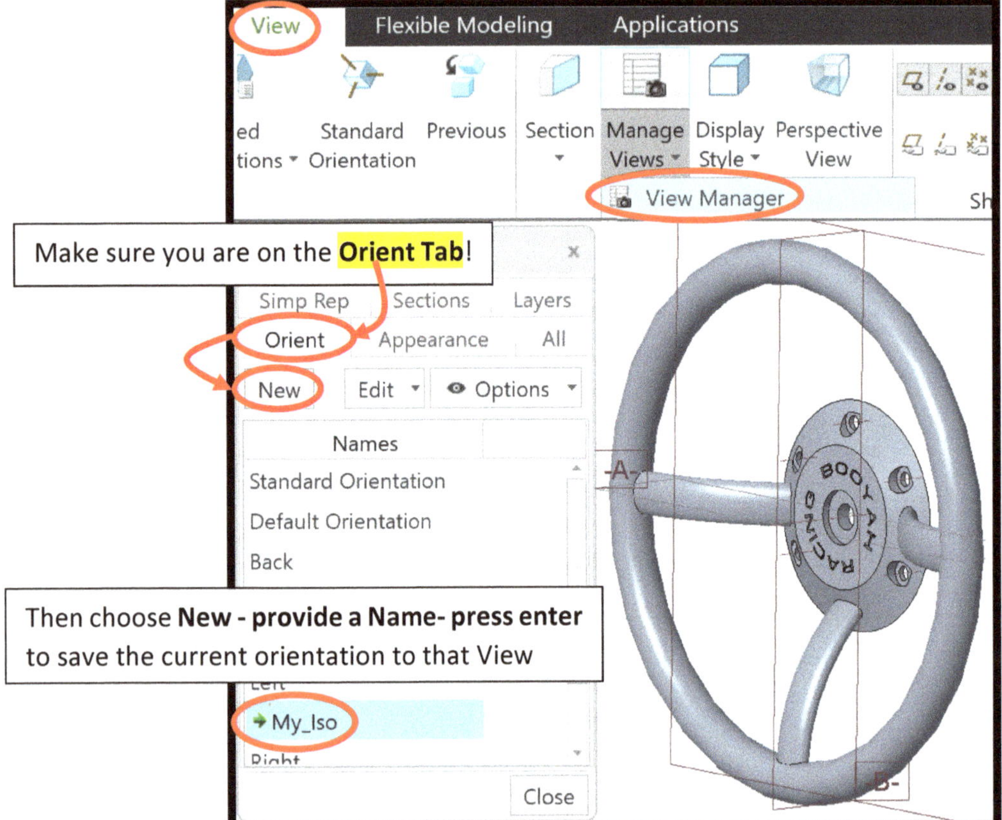

4) <u>**Test out the new view**</u>: Rotate the model in the graphics window – in the quick toolbar select the **View Orientation** tool – select the new ***MY_ISO*** view. The view should rotate to your custom Isometric view.

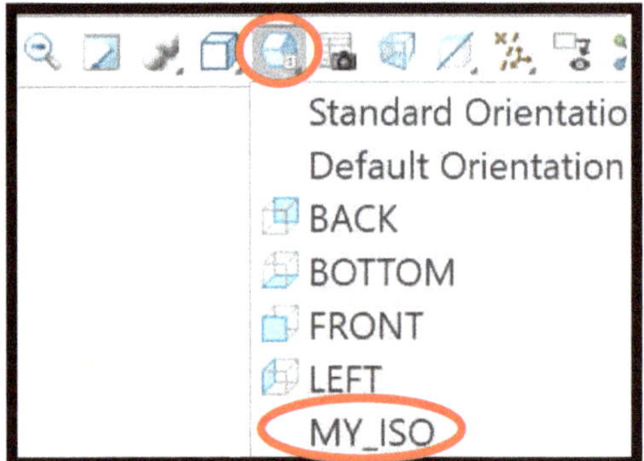

Note – If the view does not show your Isometric view, you likely did not make the view while on the **Orient Tab**! You will need to fix your error by changing the tab you did make the view on (usually the Simple Rep tab) back to its default setting (i.e. back to Master rep setting) and then create the view under the correct Orient Tab as described above.

Step 16 – **Save your Part File** and **create a new Drawing**: – File – New - Drawing – *"L7_SweepWheel"*

Step 17 – **Setup the Drawing to match the key** by following the steps in the *Basic Drawing Step & Detail Drawing Steps* in the appendix as needed to match the key.

Note – When inserting the **General View** – choose the view type as the **MY_ISO** custom view state you created above. If you do not see the *My_Iso* as a view state option, you likely made the custom view state while on the "Simple Rep" tab instead of the "Orient tab" (see note above). Also, sometimes students have saved a backup of the model and then have their drawing using a version of their model that they did not setup custom View state on.

End of Lesson 7

Lesson 8 – Helical Sweeps & Cosmetic Threads (Spring, Bolt, & U-Bolt)

Lesson 8 is to create three different part models utilizing **Helical Sweeps** and **Cosmetic Threads.** This lesson will show how to create a coil spring, threads on a bolt, and how to properly setup Cosmetic threads on a bolt (which is the ideal way to model threads in order to save on processing power for larger assemblies.)

Helical Sweeps:

The **Helical Sweep tool** creates a sweep feature that has a trajectory that rotates about an **Axis of Revolution**. The spacing of each revolution of rotation is provided by a **Pitch** value, which defines the distance between each revolution of the helical. This tool can be used to model threads, coil springs, etc.

Notice on the example image shown on the right, the Trajectory is an **Open-Loop**, and cannot be Perpendicular to the Axis of Revolution. As the Cross-Section follows the trajectory, it will also rotate so that it makes 1 revolution per length of Pitch.

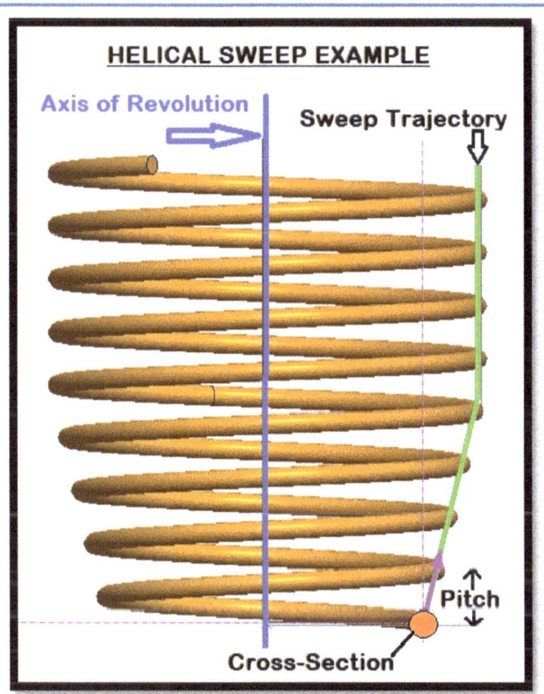

Cosmetic Threads:

The **Cosmetic Thread** feature creates a threaded feature on the model that includes all the details about the thread but is <u>not</u> visualized on the model. This is preferable over modeling a threaded surface using a helical sweep as the Cosmetic Thread creates the details of the threads without actually showing it in the model graphics window. This simplifies the model and lowers computer processing requirements for larger assemblies that may have many threaded surfaces. The details of a cosmetic thread can be quickly shown on a Drawing as all the important details about the threads are available to show as a Note using the Show Model Annotations tool.

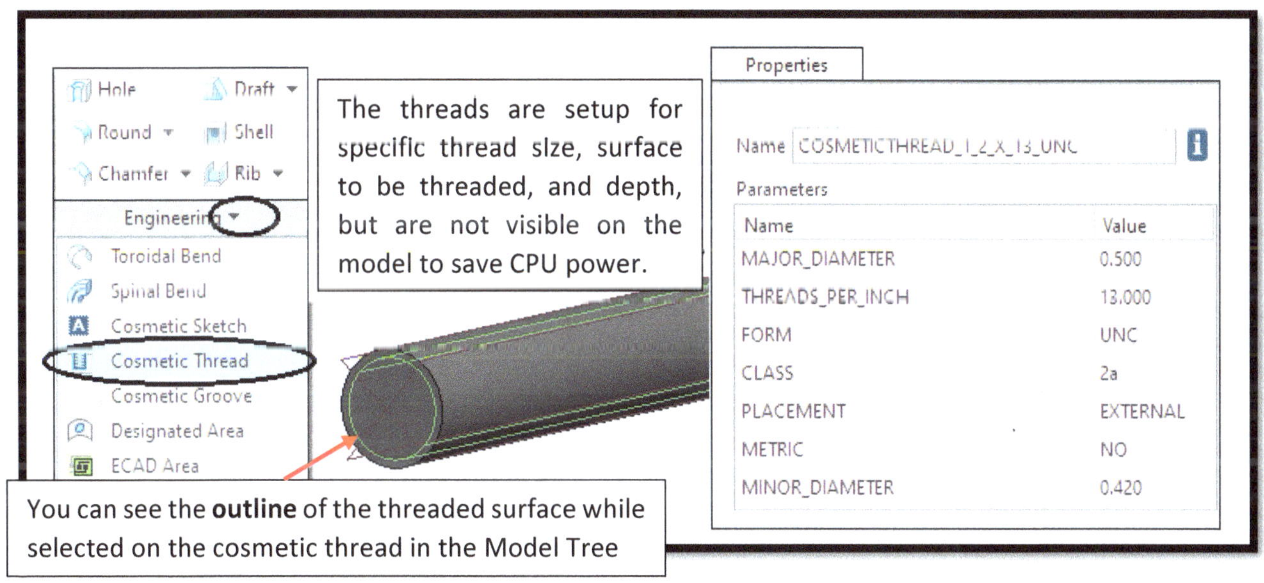

You can see the **outline** of the threaded surface while selected on the cosmetic thread in the Model Tree

Getting Started: Open the CREO *Parametric* software and follow the lesson steps carefully.

Step 1 - Set your working directory to your ME101 Working Directory Folder and start a new part model named *"Helical_Spring_L8"*.

Step 2 – Change the material of the part to **Steel**.

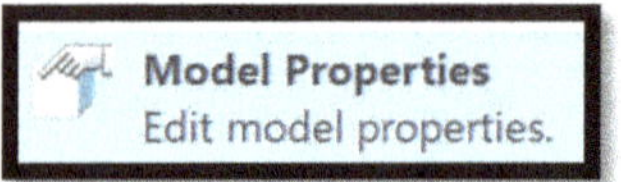

File- Prepare- Model Properties- in the Material option select **change** – double click on **Legacy Materials** - __double click__ on *"Steel.mtl"* to set it as the material – press **OK** – press **Close**

Step 3 – Adjust the options so that 3 decimal places will be shown in sketcher.

- **File – Options – Sketcher** – set the "__Number of decimal places for dimensions__" to **3** – **OK** - **No**

Step 4 - 1 – Use the **Helical Sweep tool** (expand the Sweep tool drop-down menu to select the Helical Sweep feature).

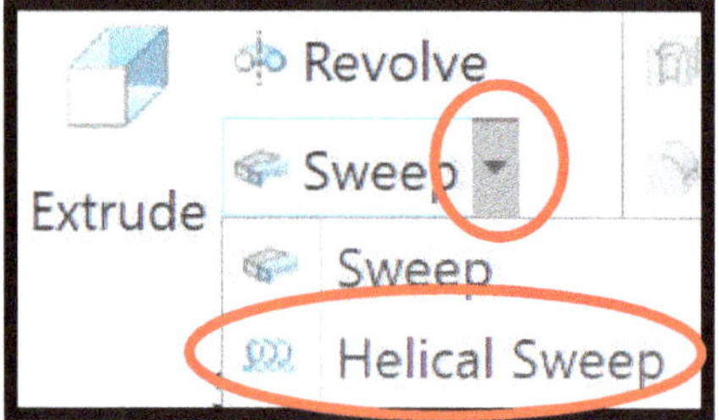

Step 4 – 2 - Sketch a Trajectory ("Helix Profile") for the helical sweep placed on the **Front Datum.** This sketch requires an **Axis of Revolution** and must be an **open-loop** sketch.

1) **Define the Helical Profile:** Click **Define** on the References tab and choose the **Front datum.**

2) **Axis of Revolution:** Add a vertical **Axis of Revolution** along the vertical origin reference. (- **Hold RMB** in a blank area – select the **Axis of Revolution icon**).
If the icon does not appear, check that you are using the __Helical__ Sweep tool, and not the Sweep tool.

3) **Trajectory Sketch: Sketch a single line** starting on the horizontal reference as shown to act as the outer profile of the Spring shape.
The sketch __must not__ be a closed-loop nor be Perpendicular to the Axis of Revolution or it will fail.

4) **Dimensions: Dimension the Line** with a vertical height of 3.000", lower revolved Diameter of 2.000", and the upper revolved Diameter of 3.000".
If your dimensions are not "revolved", you are likely missing the Axis of Revolution.

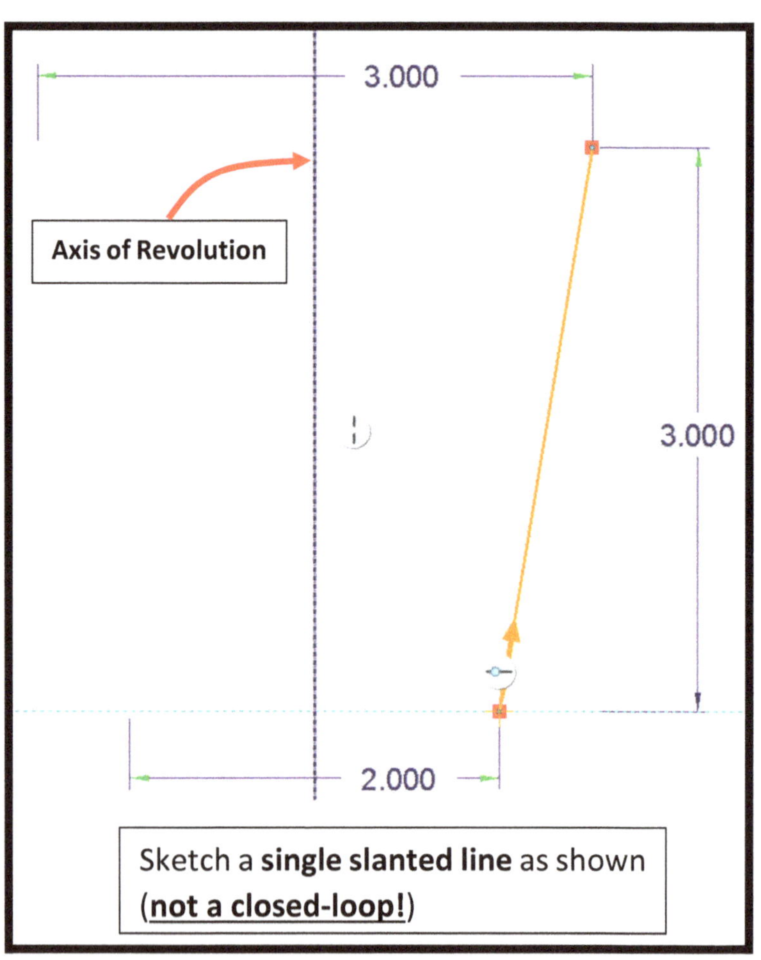

5) **Checkmark** to accept the Sketch

Step 4 – 3 - Press the "**Create or Edit Sweep Section**" **icon** in the top toolbar to enter Sketch mode for the helical sweep Cross-section.

- Press the **2D Sketch View button** in the quick toolbar to view the Trajectory sketch plane properly.

- **Sketch a Circle** at the Trajectory Origin with a **Diameter of 0.100"**.

- **Press the Checkmark** to accept the Sketch.

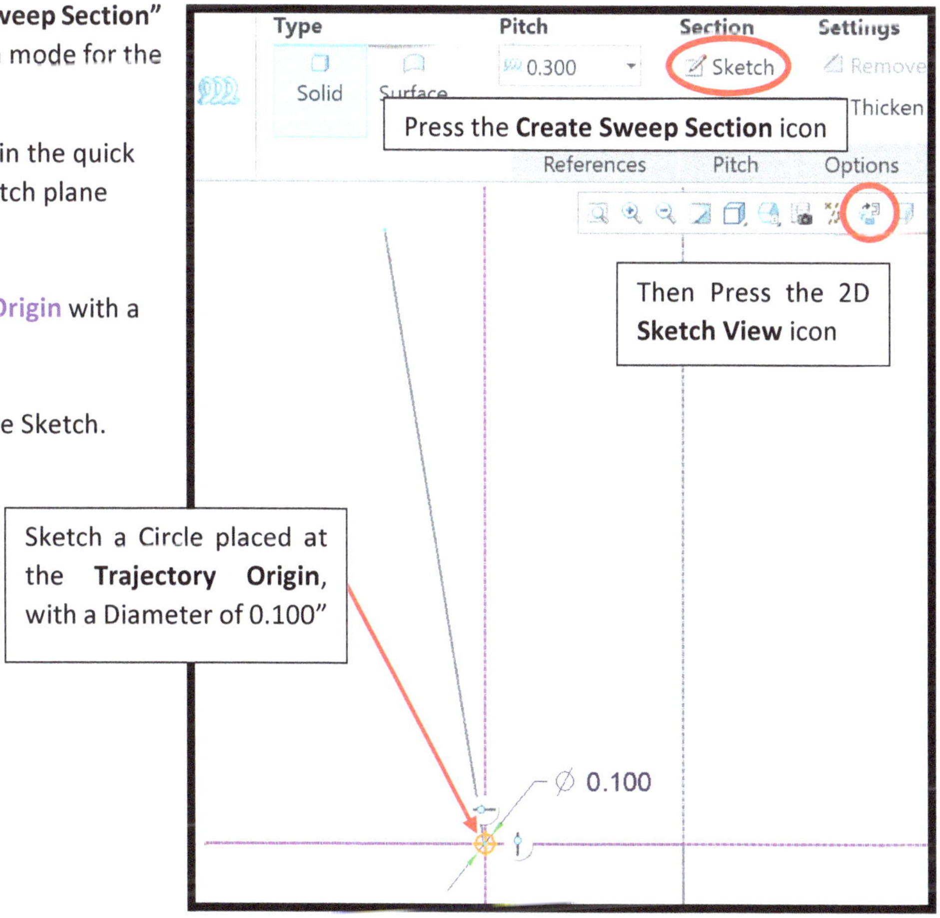

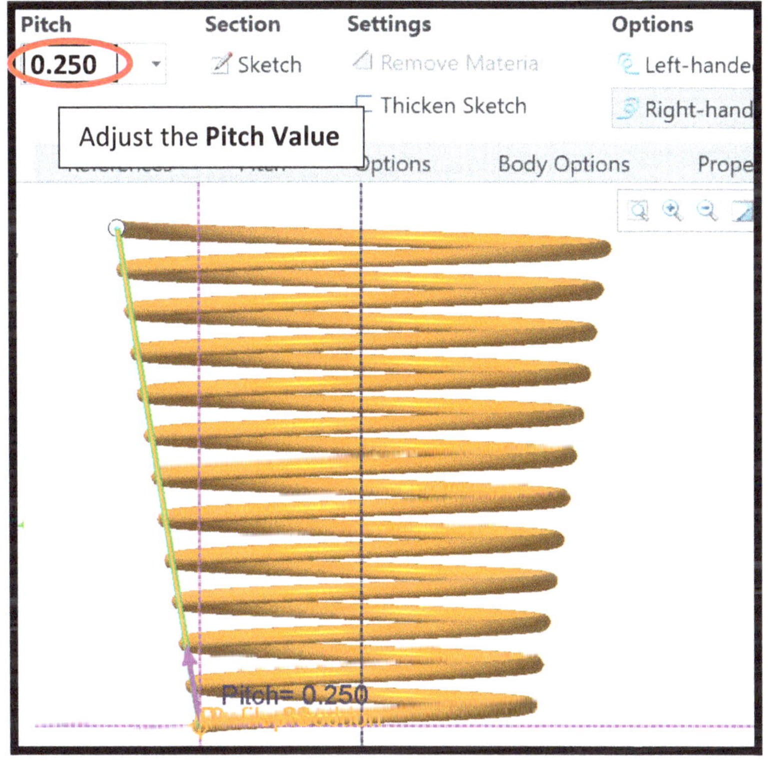

Step 4 – 4 - Adjust the **Pitch to 0.250"** in the top toolbar, which sets the spacing between each revolution.

If your Helical Sweep does not appear on the screen, you may have an incorrect Trajectory sketch. Make sure it is a single slanted line, not a closed-loop. You can open the References Tab – Edit to bring back that sketch and make changes if needed.

Another common issue is students put the Cross Section Sketch in the Trajectory, or vice versa. *Trajectory & Section sketches are never combined!*

Step 4 – 5 - **Checkmark** to accept the Helical Sweep.

Step 5 – 1 - Add a Datum that will be used to "trim" the ends of the spring to create flat surfaces, which is a common option on coil springs.

- Select the **Plane Tool** in the top toolbar – select the **Top Datum** as a reference (which is located at the base of the spring) – set an **offset value of 3.000" inches**. This should place the new datum through the center of the spring cross-section - **OK**. If it goes the wrong direction, set the offset as -3.000".

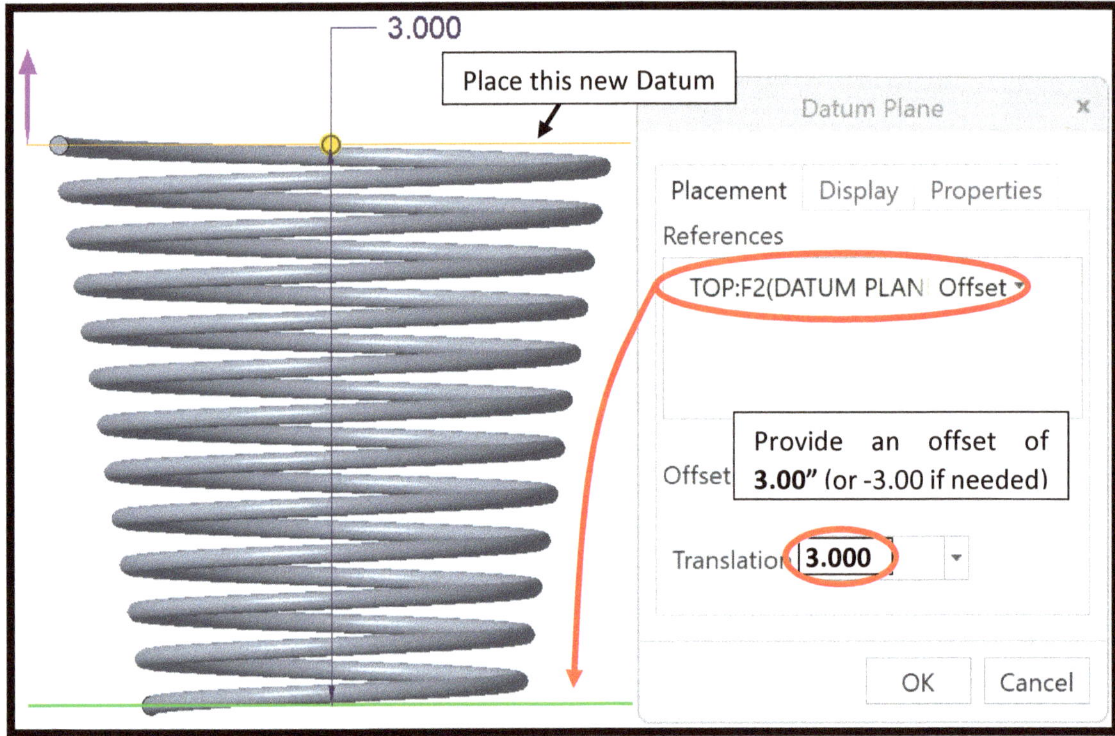

Step 5 – 2 - Use the **Extrude Tool** sketched on the new datum (DTM1). This will cut the top surface of the spring to have a flat surface.

- Sketch the **Extrude on the new datum, DTM1,** and sketch a **circle larger than the spring** (5" or so)
- Set the options to **Remove Material and flip the Direction** so it is removing all material above DTM1

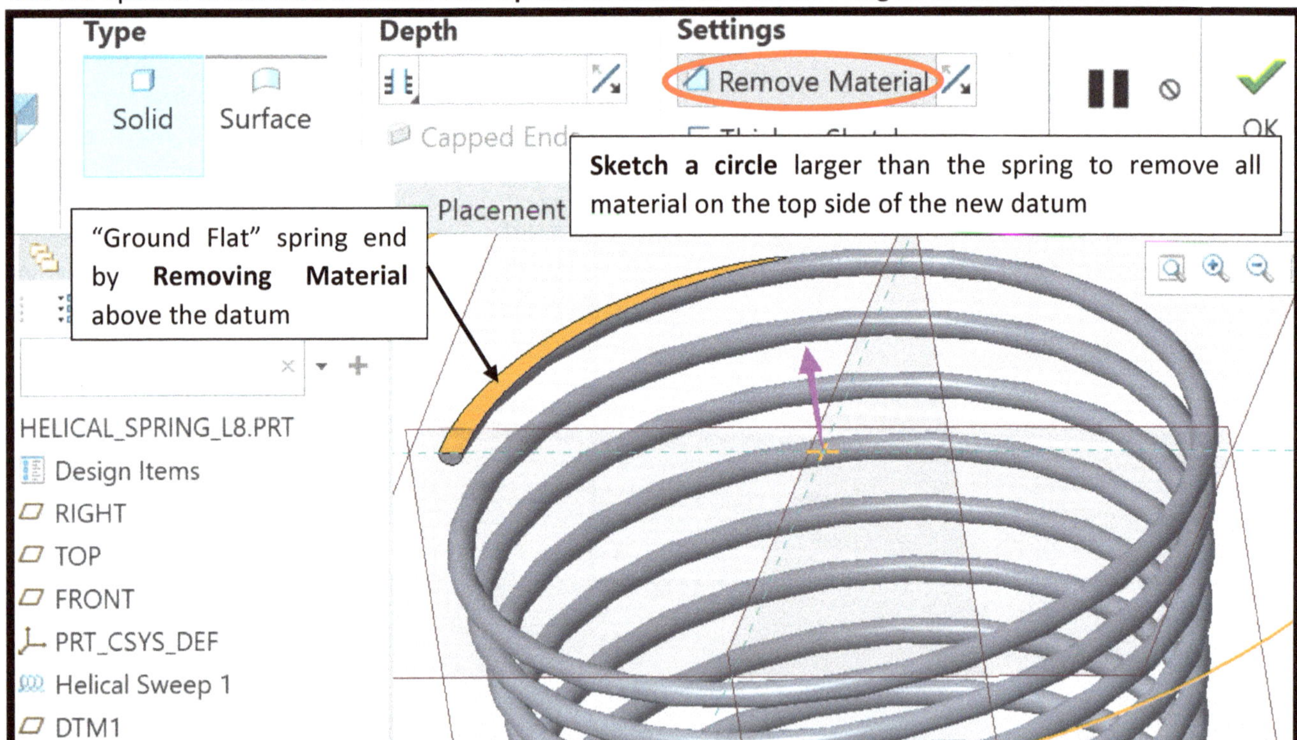

Step 5 – 3 - Create a second Extrude to Remove Material on the bottom of the spring, using the Top Datum as the sketch plane.

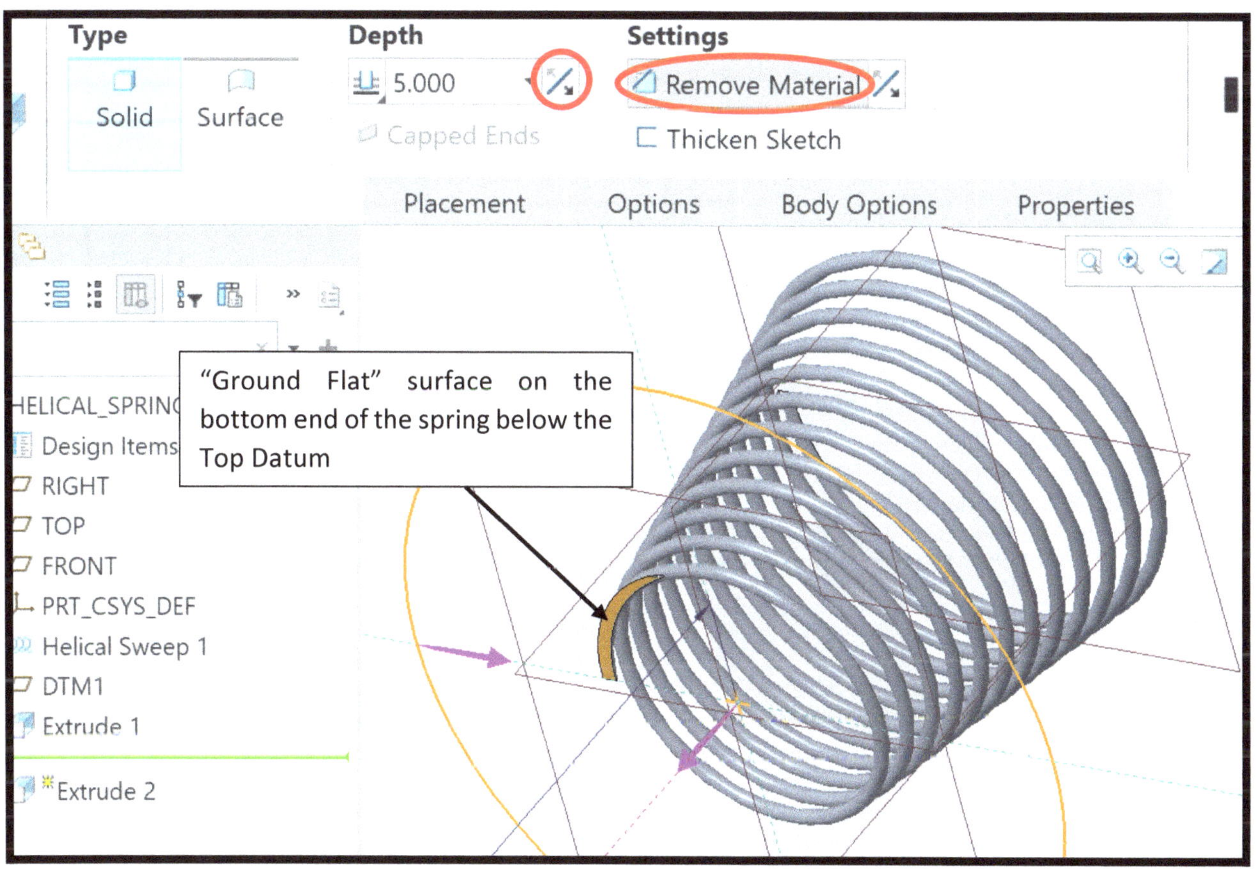

Step 6 – Save the part and move on to the next component. We will not Set & Rename datums on this lesson.

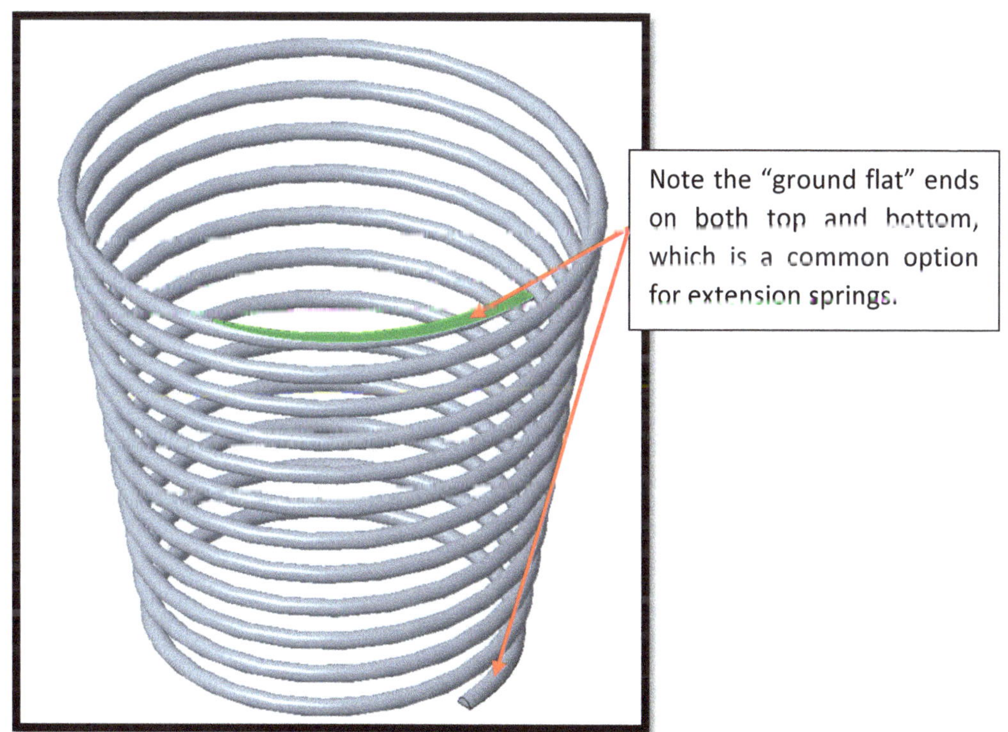

Part #2: Ram Bolt

Step 1 – Set your **Working Directory** (if not done earlier) and create a **New Part** called *"Ram_Bolt_L8"*

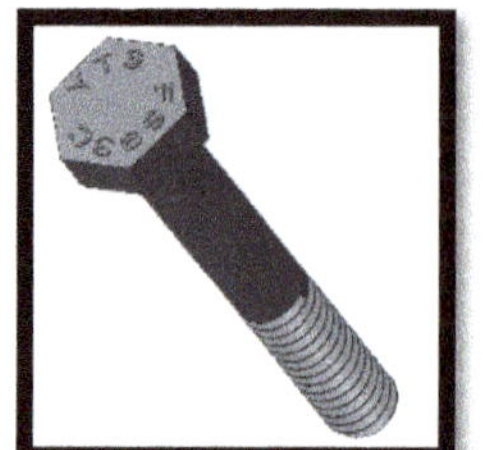

Step 2 – Set the material as **Steel. File - Prepare- Model Properties - change** – double click on **Legacy Materials** - **double click** on "Steel.mtl" to set it as the material – press **OK** – press **Close**

Step 3 – Adjust the options so that 3 decimal places will be shown in sketcher, if you did not already do so in your current session of CREO. - **File – Options – Sketcher** – set the "**Number of decimal places for dimensions**" to **3** – **OK**

Step 4 – Use the Extrude Tool to create the Hexagon extrusion, placed on the **Top** datum plane. 

1) <u>**Import Section:**</u> While in Sketch mode use the **Palette Tool** in the top toolbar - select the **Polygon category tab** in the Sketcher **Palette Tool** – scroll down to find the **6-sided hexagon** – click & hold LMB to drag Hex shape from that list onto the sketch screen – then click and drag the center handle of the Hex to snap it onto the vertical & horizontal centerlines – **checkmark** the *Import Section* at the top toolbar.

2) <u>**Set Coincident Constraints**</u>: If needed, set Coincident Constraints between the horizontal & vertical reference lines to force the Hex shape to be centered on the cylinder. This will remove any extra weak dimensions.

3) <u>**Dimension Tool:**</u> set the dimension for the height of the Hex Shape from "flat-to-flat" to **0.430** – delete out the line length dimension that conflicts with this new dimension in the Resolve Sketch menu.

4) **Checkmark to accept the Sketch**.

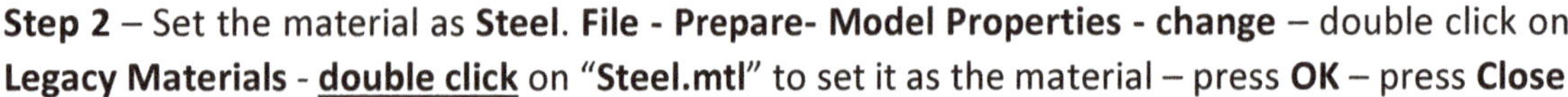

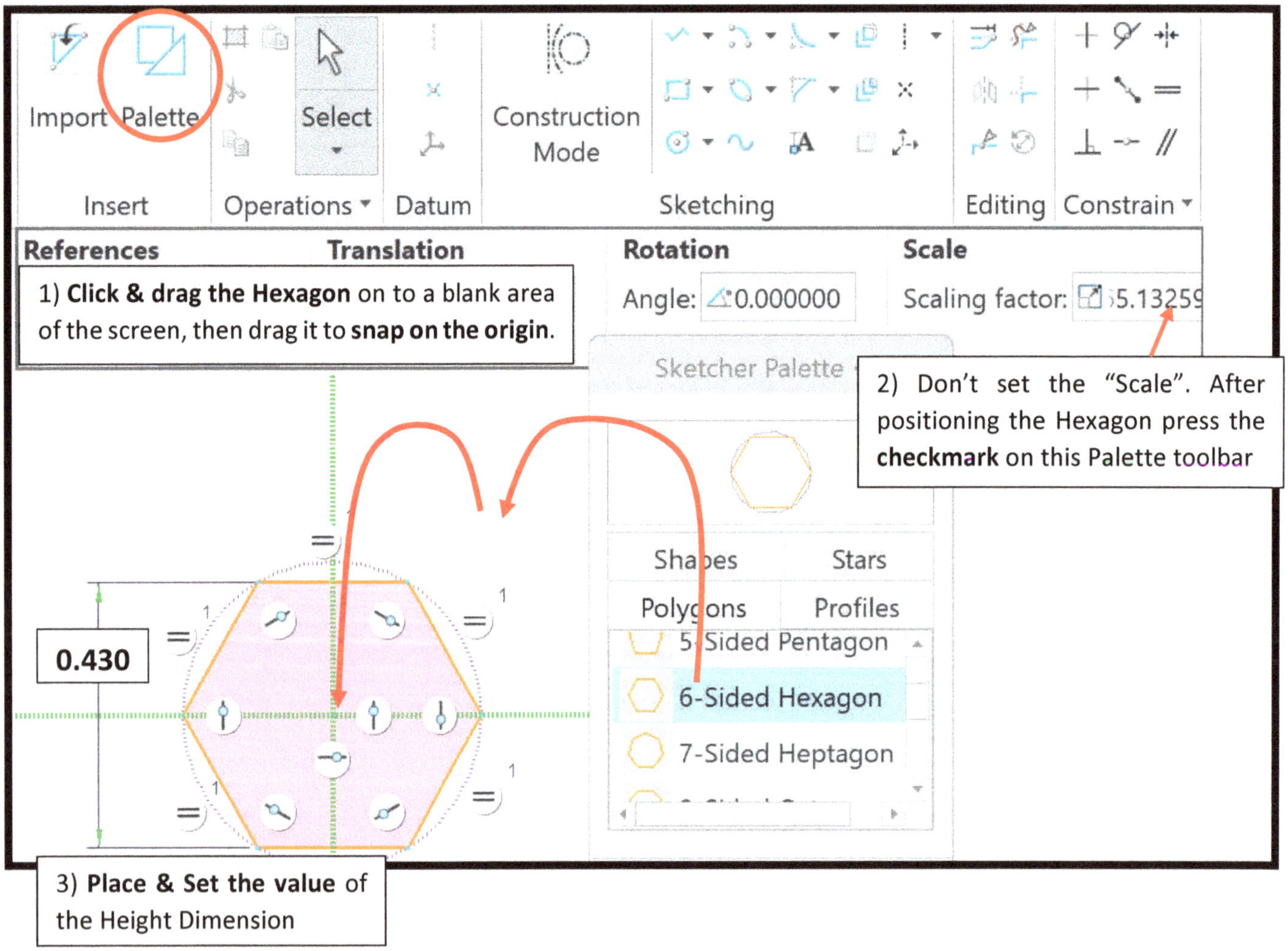

Step 4 – 2 - Set the Extrude Toolbar Options with a Depth of **0.170"**, and Direction should be going **above** the Top Datum.

Checkmark to accept the Extrude.

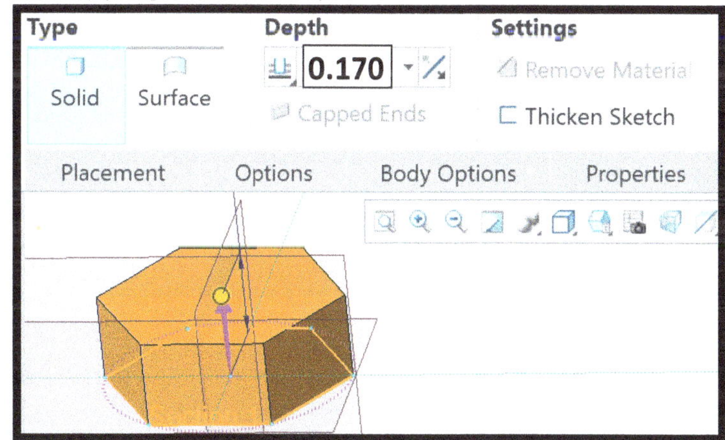

Step 5 – Use the Extrude Tool to create the cylindrical shaft of the bolt below the Hex Head, sketched on the **Top** Datum.

- **Sketch a circle** centered on the origin with **Diameter of 0.245"**
- Set the **Extrude Depth to 1.750"** and **Flip the Direction Arrow** to extrude below the Top Datum (or away from the Hex Head so it is adding 1.750" of length below the Hex head).
- **Checkmark** to accept the Extrude.

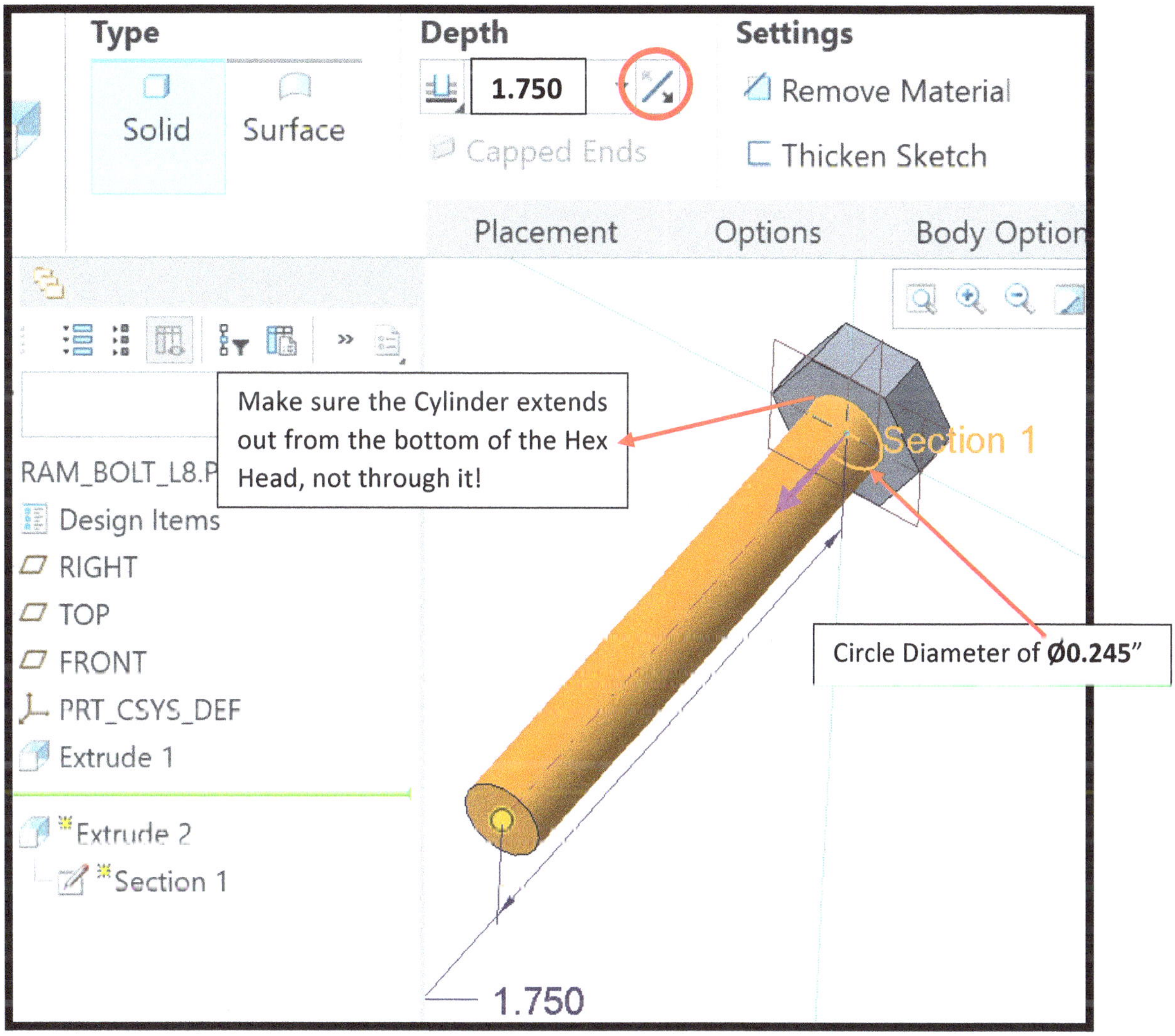

Step 6 – 1 - Use the <u>Helical</u> Sweep Tool to create the thread cutouts as shown below. Use the drop box next the Sweep tool to find the **Helical Sweep tool**.

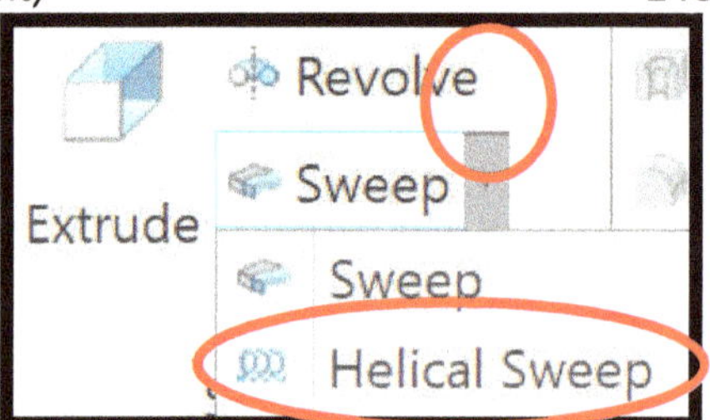

1) **<u>Start the Trajectory</u>: References Tab – Define – select the Front Datum - OK.**
This will bring you to sketch mode for the Trajectory or outer profile of the threads (*which is not a closed-loop*)

2) **<u>Axis of Revolution</u>**: Add a vertical **Axis of Revolution** in the center of the cylinder

3) **<u>Trajectory Sketch Line</u>:** Create a **single Vertical line** along the outer edge reference of the cylinder. **Start the line** about 1/3rd of the way up the cylinder and then going <u>past</u> the bottom edge of the shaft*.

   Note that the **Start Point Arrow** is at the top of the line. If you need to change it, LMB on the top point – hold RMB – Feature Tools - **Start Point** option

   Tip: If your line doesn't snap on the outer edge of the cylinder, use RMB- Sketch References to add it as a reference first.

   * Running the line past the bottom edge ensures that the Sweep cuts out past the end of the cylinder, preventing a "filled thread" at the bottom edge of the part when the sweep stops rotating about its Axis.

4) **Dimensions:** Dimension the **Line** to be **0.750"** from the bottom edge, and a total line length of about **1.000"**

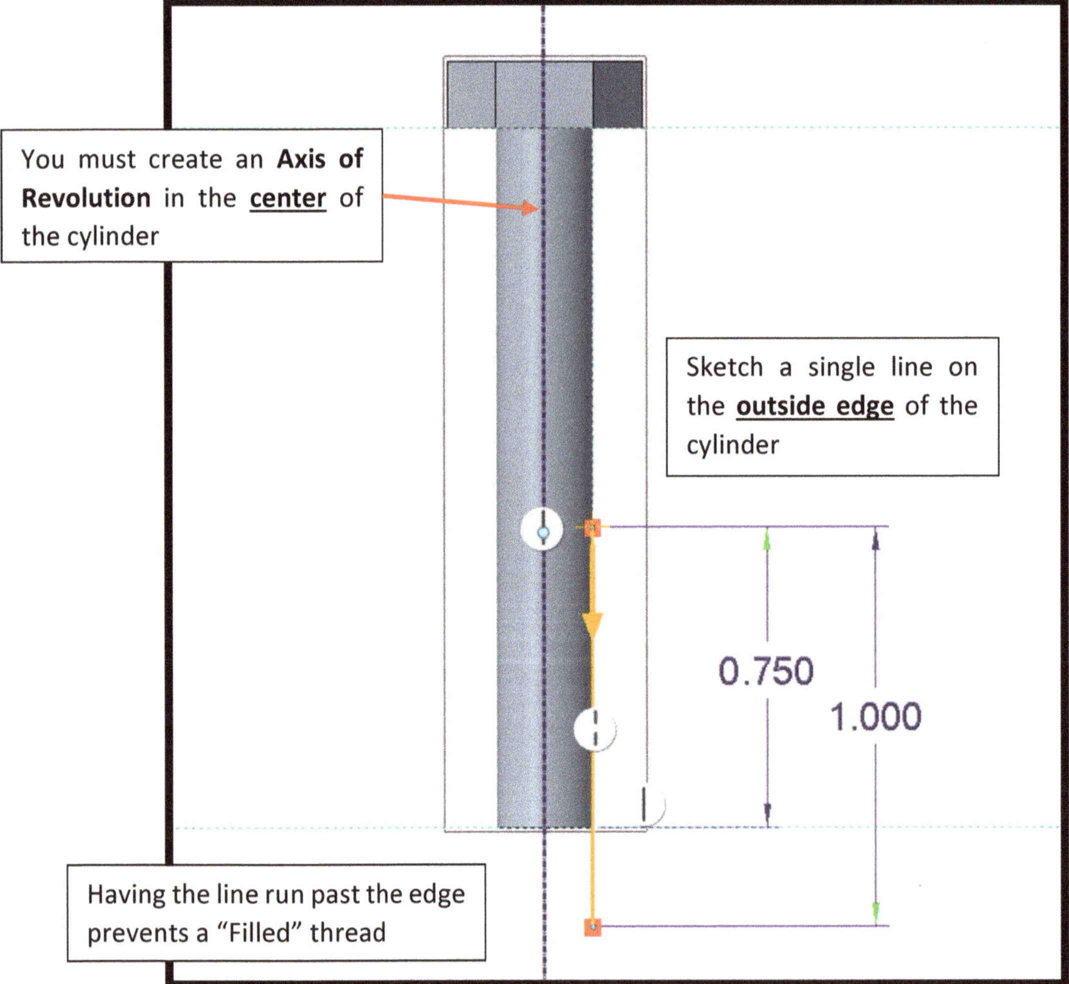

Step 6 – 2 - Checkmark to accept the Trajectory Sketch and return to the Sweep toolbar.

Step 6 – 3 - Use the **Create Sweep Section icon** in the toolbar to enter sketch mode for the Section.

1) <u>Sketch View</u>: Press **the 2D Sketch View** to view the sketch plane of the "Trajectory Origin"
2) <u>Sketch the Section</u>: Use the **Line Tool** to sketch the **four lines of the Closed Loop shape** shown below, starting at the trajectory origin and "cutting" into the cylinder material from the outside edge.
3) <u>Constraints</u>: Set **Equal Length constraints** on the two angled lines.
4) <u>Dimensions:</u> Set the Dimensions as shown with **heights of 0.04375" & 0.0125" and a width of 0.027"**.
   a. *These are specified from a machinist handbook for a ¼-20 UNC thread shape.*
   b. *The values will truncate to 3 decimal places, but the actual dimension will use the entered value*

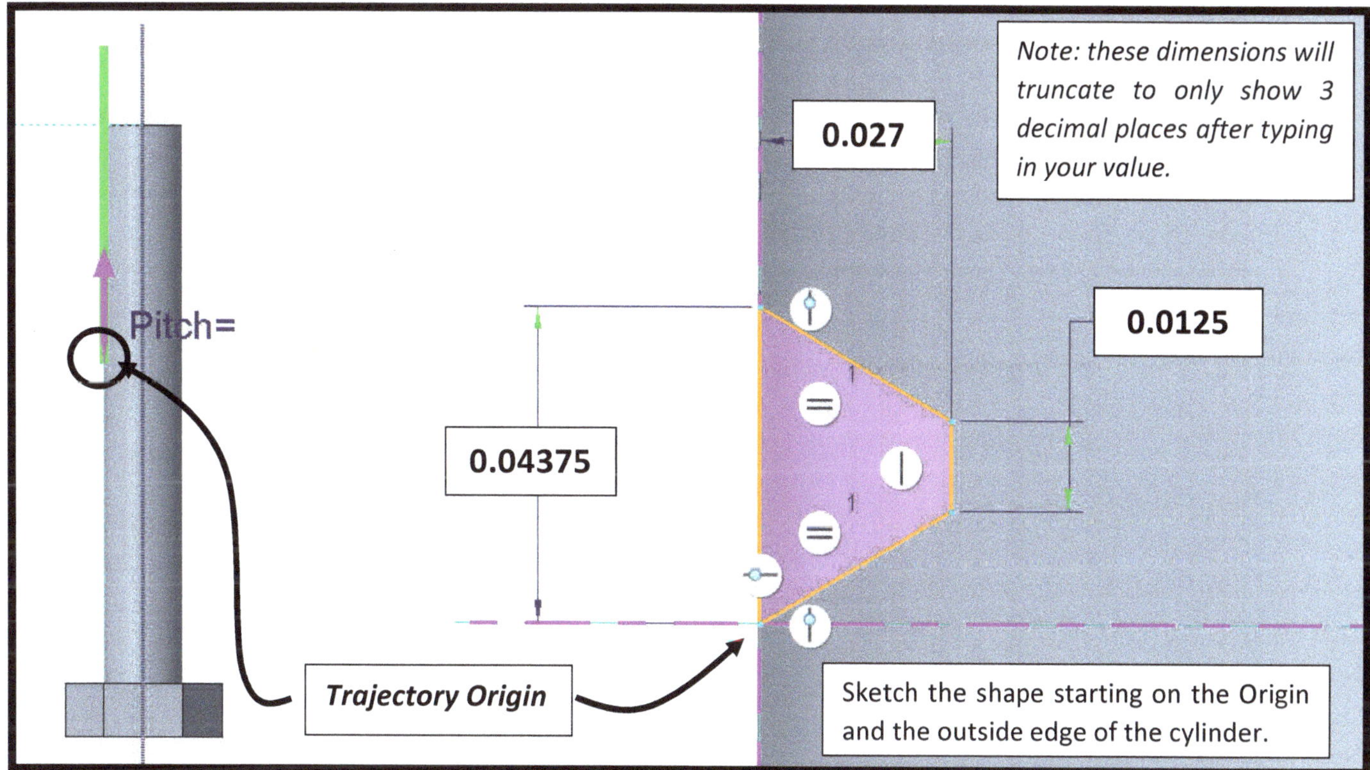

5) **Checkmark** the sketch to accept your sweep section.

Step 6 – 4 - Set the **Pitch Value to 0.050" and toggle on Remove Material** in the top toolbar.
If you need to adjust the Trajectory use the References Tab. If you need to adjust the Section use the Sweep Section icon.

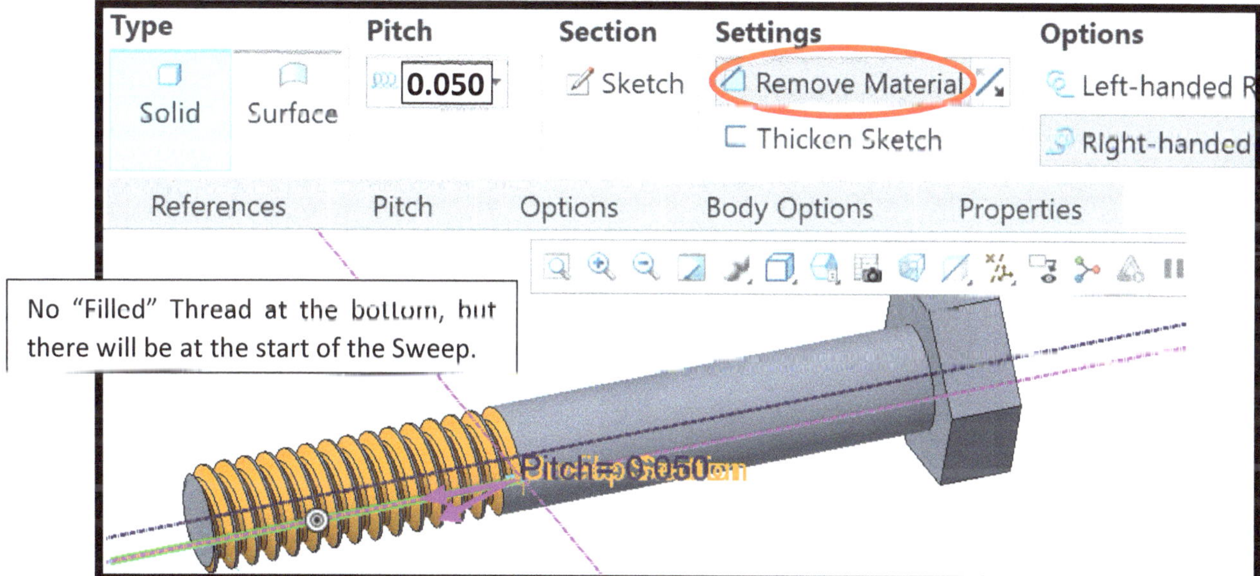

Step 6 – 5 - Checkmark to accept the helical sweep feature.

Step 7 – 1 - **Extrude** the **Circular Text** to the top surface of the Hex Head using the **Extrude Tool**, using the top surface of the bolt head as the sketch plane.

1) **Construction Mode: Toggle On** "Construction Mode" in the sketch toolbar
2) **Sketch the Circle Reference**: create **a Circle with 0.125" Radius**, which should be a dotted line. Do not snap it onto the hidden lines of the shaft or threads.
3) **Toggle Off Construction Mode**: Press the Construction Mode icon to toggle it off so that the next step will have sketch entities that are part of the closed-loop. If you do not toggle this back off, your sketch will not work!
4) **Text Tool:** use the **Text Tool** to sketch a single vertical line on the screen (*not snapped on to any geometry*).
   - Type in the text "**YTS F593C**"
   - Select the "**Place Along a Curve**" option in the Text Menu. Immediately after pressing that option, CREO will be waiting for you to select the dotted construction circle you made earlier.
   - **Flip the Arrow** to place the text on the outside of the curve.
   - **Press OK in the Text Menu** – adjust the **Text height dimension** to **0.060"**
   - **Double click LMB on the Text** to bring back the **Text menu** - Adjust the **Aspect Ratio** to space out the text similar to as shown. **It does not need to be exact, but 1.35 was used**.

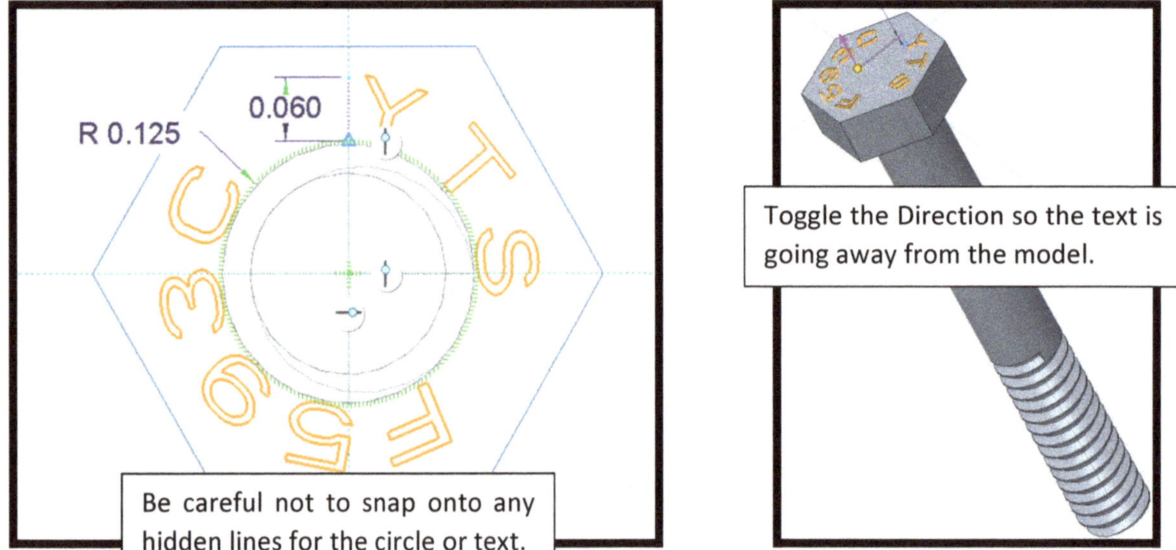

Step 7 – 2 - **Checkmark the Sketch** and set the toolbar options with a **Depth of 0.005"** and toggle the **Direction Arrow** to extrude the text **above** the bolt head. **Checkmark to accept** the Extrude.

Step 8 – Use the Round Tool to round the top outer edges (all 6 Hex edges) of the Hex Head (**R 0.025"**). Accept it and then repeat to create a different **Round** on the bottom edge where the Cylinder and Hex Head intersect (**R 0.010"**). Make sure each round has the correct value by opening up the 'Sets' tab while using the Round tool.

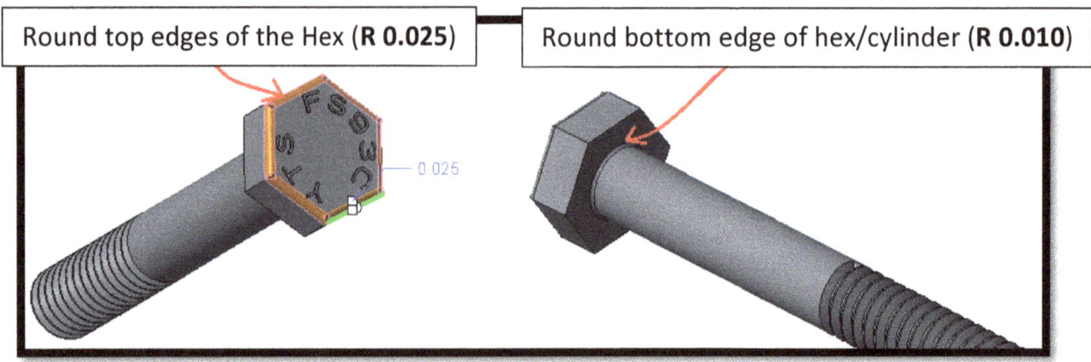

Step 9 – Save the part model and move on to the next part.

Part #3: Ram System U-Bolt

Step 1 – Set your **Working Directory** and create a **New Part** called *"Ram_Ubolt_L8"*

Step 2 –Set the material as **Steel**.

Step 3 – Adjust the options so that <u>3 decimal places</u> will be shown in sketcher, if not already done earlier.

Step 4 – 1 - Use the Sweep Tool (*not a* <u>*Helical Sweep*</u>) to create the U-bolt.

Step 4 – 2 - In the Sweep Toolbar switch back to the **Model Tab** and select the **Sketch Tool**. This will create a sketch that can be used as the trajectory for the Sweep.

Step 4 – 3 - Sketch on the Front datum plane, sketching a trajectory as shown.

1) Use the **Center & Ends Arc tool** to create a half circle at the origin.

2) Create two **vertical lines of equal length**.

3) Add a **Point** (*located below the centerline tool in the sketch toolbar*) to add a point to the bottom of the arc that can be used for dimensioning.

4) **Use the Dimension Tool** to set a dimension for the **overall height** between the top of the line and the sketch point on the arc **(2.500")**, as well as a **Diameter of the Arc (1.500")**.

5) Checkmark to accept the sketch.

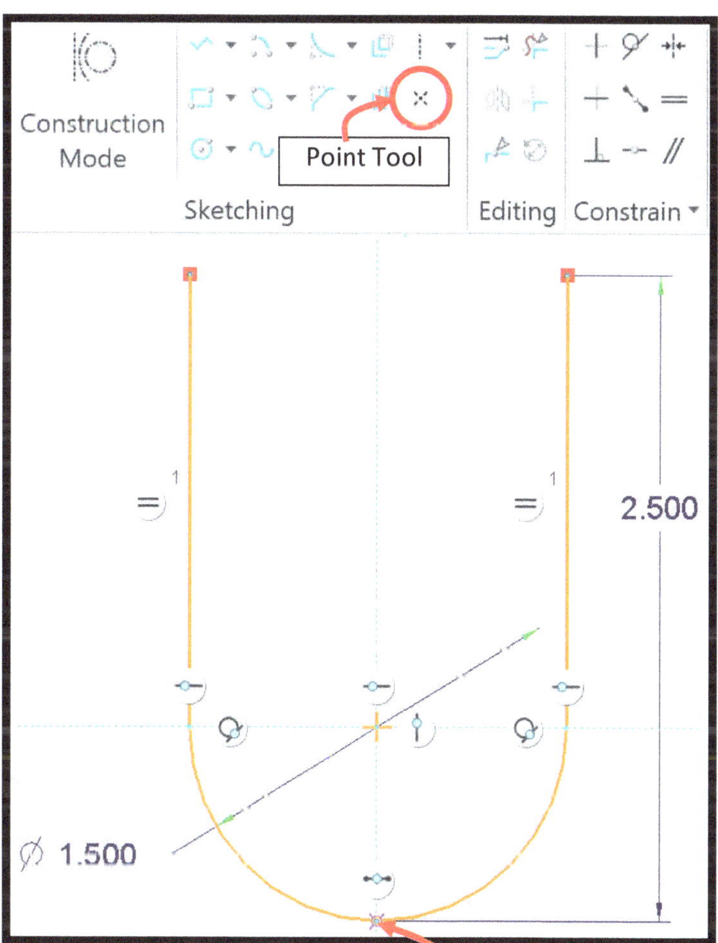

Step 4 – 4 - Toggle back to the Sweep tab in the top toolbar (*to get back to the sweep started earlier*). Open the **References** tab in the Sweep toolbar and select the *Sketch 1* as the trajectory by clicking on it, if not already done.

1) Press the "**Create or Edit Sweep Section" icon** to enter Sketch mode for the Sweep cross-section.
2) Press the **2D Sketch View** icon in the quick toolbar and **sketch a Circle** about the origin of the <u>trajectory.</u>
3) **Dimension** the Circle to have a **Diameter of 0.250"**.
4) Checkmark to accept the **Sweep Section** sketch

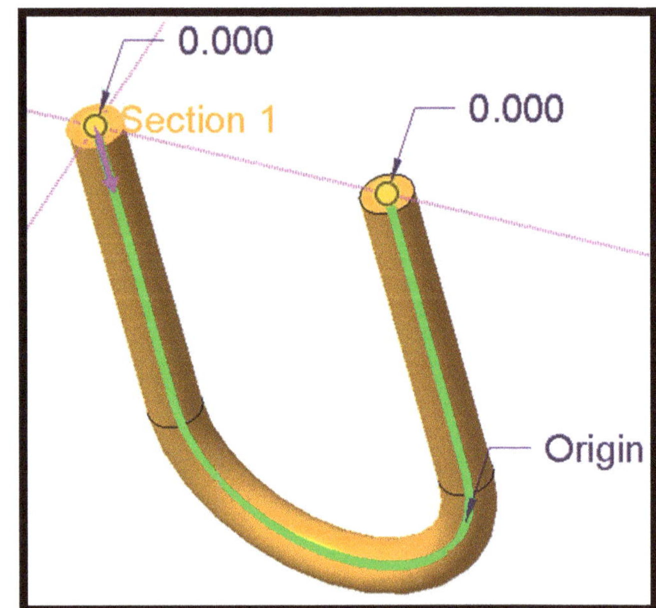

Step 4 – 5 - The Sweep should appear as a solid model on the screen. If not, check that the Trajectory (*Sketch 1*) and the Section were created correctly.

Tip – if you need to Edit Definition of Sketch 1, you can find it under the drop box of Sweep 1 in the model tree.

Step 4 – 6 - **Checkmark to accept the Sweep.**

Step 5 – 1 - Use the Cosmetic Thread tool *(under the Engineering drop box)*. This will add the details for standard bolt threads without modeling the threaded surfaces on the model, saving computer processing requirements on large assemblies.

Set the Cosmetic Thread options as listed below:

<u>Placement Tab</u>: LMB on the cylindrical (curved) surface of one vertical leg of the U-bolt to act as the threaded surface.

<u>Depth Tab:</u> LMB to select the top flat surface of the U-bolt leg with a **Depth of 1.500"** to define the threaded length.

<u>Toolbar Options</u>: Click to toggle on the **Standard Threads** option and specify the "**1/4 - 20" UNC** thread from the list.

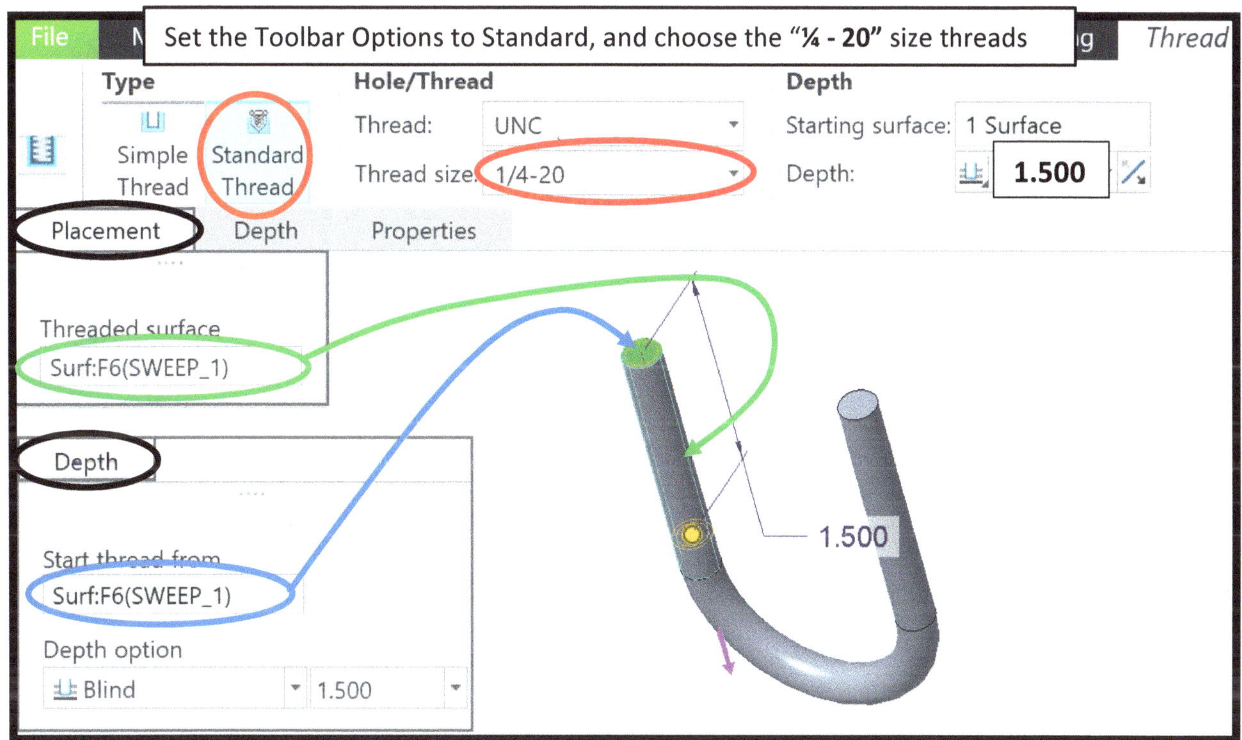

You can view details about the thread under the Properties tab, such as the thread diameter, class, etc. if desired

Step 5 – 2 - Press the Checkmark to accept the Cosmetic Thread.

Note that you will only see the outline of the Cosmetic thread in the graphics window if it is selected in the Model Tree. The threaded surfaces will not be visible on the part model as it is a "cosmetic" feature.

Step 6 – Select the Cosmetic Thread from the Model tree and use the **Mirror Tool** to create a copy of the threads onto the other leg of the U-bolt. Use the **Right datum** as the Mirror Plane.

Step 7 – Save your Part, and move onto the Drawing.

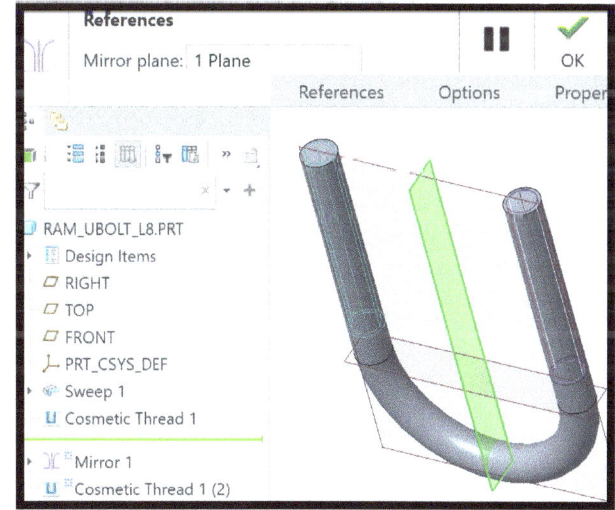

Lesson 8 Drawing:

Note that the Datums were **not** "Set" or Renamed on any of the parts for this lesson. Showing datums is not always required as it depends on the company standards or the need for geometric tolerances, which these parts do not need.

Step 1 – Create a New Drawing. **File – New - Drawing – "L8_Helicals"**

Step 2 – Setup the Drawing views to match the key of the first model only. Note that the drawing will start with the last open part model, so likely the U-Bolt. After getting the views for the first model complete then move on to the Step below to add the other models to the same drawing.

Step 3 – Add & Set the next part model to use on the drawing.

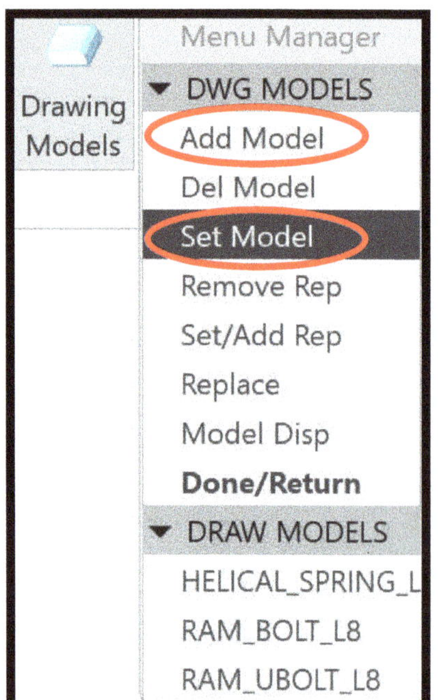

1) **Add the Models to the Memory**: Click on the ***Drawing Models*** icon in the top toolbar (*Layout Tab*) - click **Add Model** – find the Model of the *Ram_Bolt_L8* and **press OK**.
   Click Add again to find and add the *Helical_Spring_L8.*

2) **Set the Active Model**: In the ***Drawing Models*** tool, click **Set** and choose which model you want to set as the currently active model. CREO will use this active component when using the General View tool to add views. Close the menu by pressing **Done/Return**.

3) **Insert the Views of the Active Model**: use the **General tool** (*Layout Tab*) to insert a new view of the currently "Set" model. Set the View State to be the Front View (or whichever is needed to match the front view of the Key). Then use the **Projection View** tool (*Layout Tab*) to project the Top, Bottom views as necessary to match the key.

4) **Repeat the Steps for the next model:** "Set" the desired Model - Insert a new General view set to Front to match the key - then project different Views as needed using the Projection View tool.

Step 4 – Setup Custom Scales on the specific drawing views to match the key, as not every model will follow the Drawing Scale listed in the template box.

- **Double-click on the desired view – Scale Tab – Custom Scale – type in the value** as needed to match they key. This allows certain drawings views to differ from the drawing scale listed in the format box. The custom scale will be shown in a note next to the view to show which scale that specific view is using.

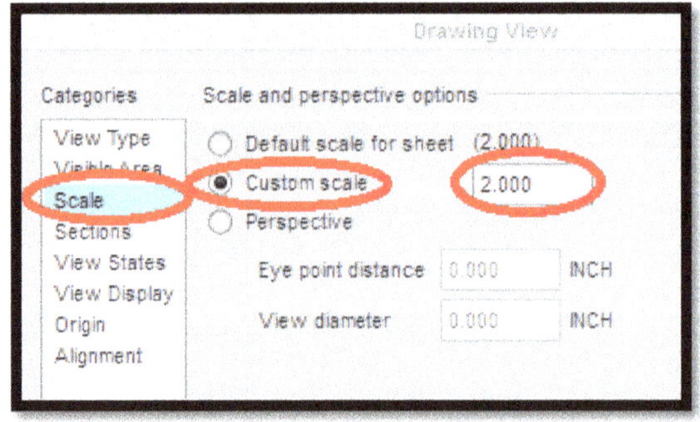

Step 5 – Continue to complete the Drawing details to match the drawing key, then save and submit your deliverables.

End of Lesson 8

Lesson 9 – Drafts & Renderings (Muffin Tin)

Lesson 9 is to create a muffin tin or baking pan, commonly used for baking delicious blueberry muffins. The Draft tool will be used to prepare the model for manufacturing, and Renderings will be used to make a realistic image. This will be the last lesson focused on Part Modeling, so users should be comfortable with the material by now and not need as detailed step-by-step instructions.

Drafts:

The Draft tool adds a small angle to desired surfaces of the model. Having a small angle on the side surfaces is useful during manufacturing as it allows part to be removed from a manufacturing mold easier, which is common for cast or injection molding methods. Commonly the angle on a draft is only a few degrees, enough to be effective for manufacturing but small enough not to interfere with the desired function of the modeled component.

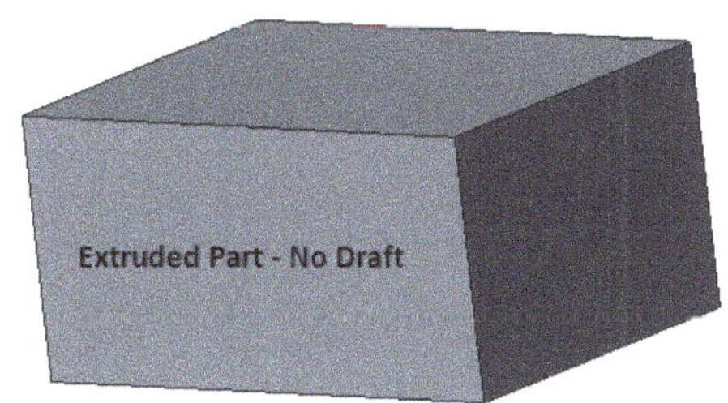

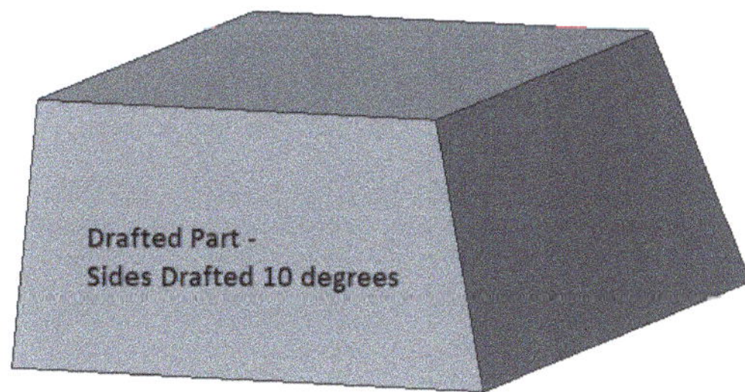

Renderings:

A realistic Rendering can be setup in CREO which produces a high definition image of the model with more realistic textures, color, lighting, and background. This is a useful option to prepare model images for a report cover sheet, presentation, or marketing.

Getting Started: Open the **CREO** *Parametric* software and follow the lesson steps carefully.

Step 1 – Set your **Working Directory** and create a **New Part** called "*MUFFIN_TIN_L9*".

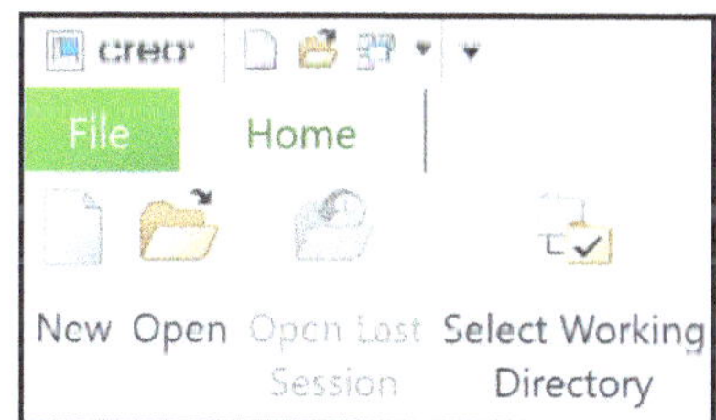

Step 2 – Change the material of the part to **Aluminum 6061**.

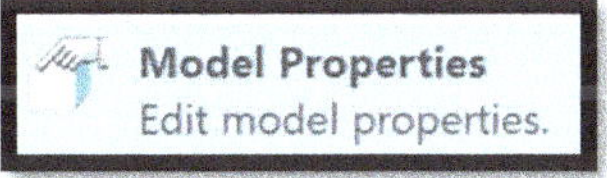

- **File- Prepare- Model Properties-** in the Material option select **change** – double click on **Legacy Materials** - <u>**double click**</u> on "**Al6061.mtl**" to set it as the material – press **OK** – press **Close.**

Step 3 – Use the **Extrude tool** to create a rectangular plate, placed on the **Top Datum**.

The rectangle should be **symmetric about the origin** and dimensioned to **11.00" (width) x 6.00" (height)**. The extrude depth setting should be left as the **default single direction option**, with an **Extrude Depth of 0.20"**.

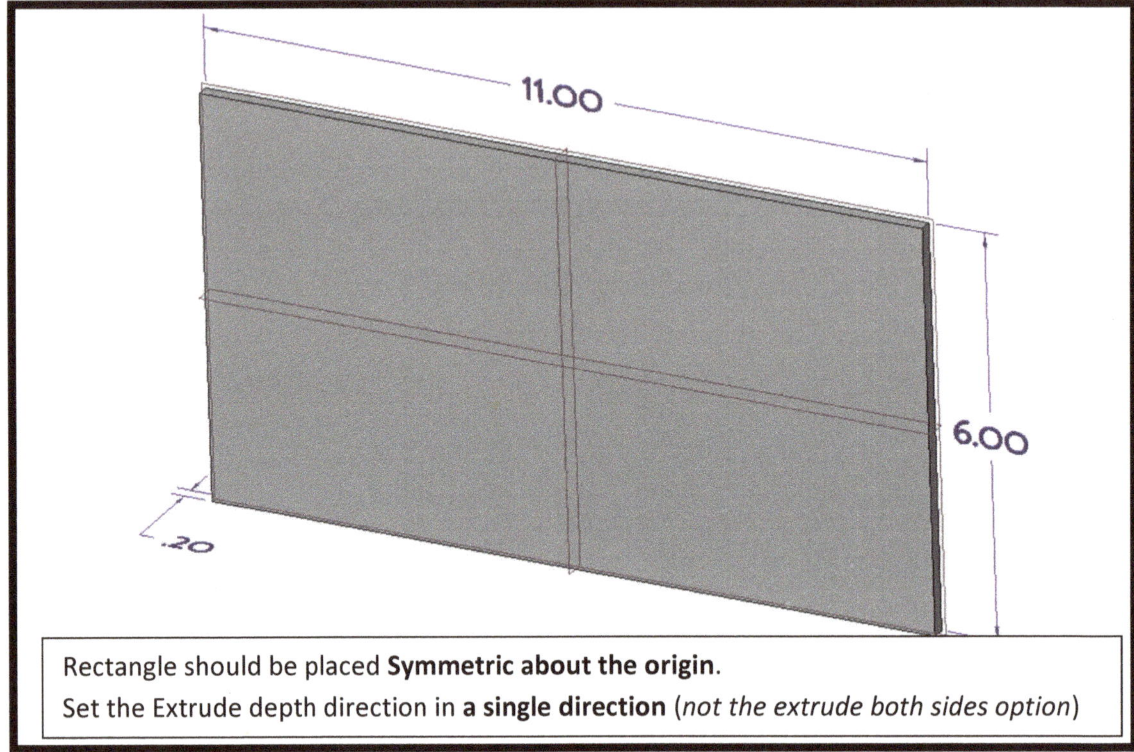

Rectangle should be placed **Symmetric about the origin**.

Set the Extrude depth direction in **a single direction** (*not the extrude both sides option*)

Step 4 – Create a second Extrude using the **Extrude tool** to create a cylinder placed on the **Top Datum** (which is the "bottom" surface of the rectangle).

- Sketch a **Circle** and set the placement dimensions from the outer corner edges as shown (**2.00"** from vertical edge & **1.50"** from the horizontal edge), with a cylinder diameter of **Ø2.50"**.

- **Flip the Extrude Direction Arrow** option so the cylinder extrudes **away** from the rectangular plate, as opposed to through it. *If the extrude incorrectly goes through the plate, the muffins will be out of spec as the height of the cylinder would be reduced by the thickness of the plate.*

- **Set the Extrude Depth to 1.75"**.

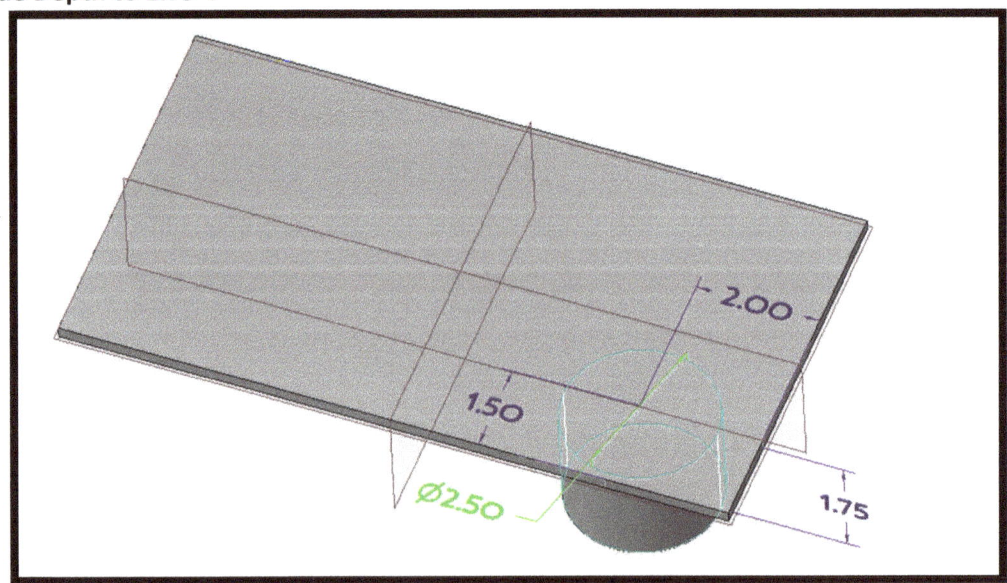

Your cylinder may end up located in a different corner than shown here, which is fine. Just make sure the cylinder is **not extruding through the plate** material, but **extends out from it** so it will be the correct height.

Step 5 – 2 - Press the Checkmark to accept the previous Extrude.

Step 5 – Use the **Draft Tool** to draft the **curved Cylindrical Surface** of the cylinder with a **Draft Angle** of **8.0 degrees**.

Open the **References Tab** to set the following references for the Draft:

1) Select the curved surface of the cylinder as the **Draft Surface**
2) Select the **planar surface (**not an edge!**)** of the plate (the side that the cylinder is on) as a **Draft Hinge** reference.
3) Set the **Draft angle to 8.0°** and toggle the **Pull Direction Arrow** in the top toolbar as needed so the cylinder is drafted to be narrower at the end, and remain the original diameter at the base.

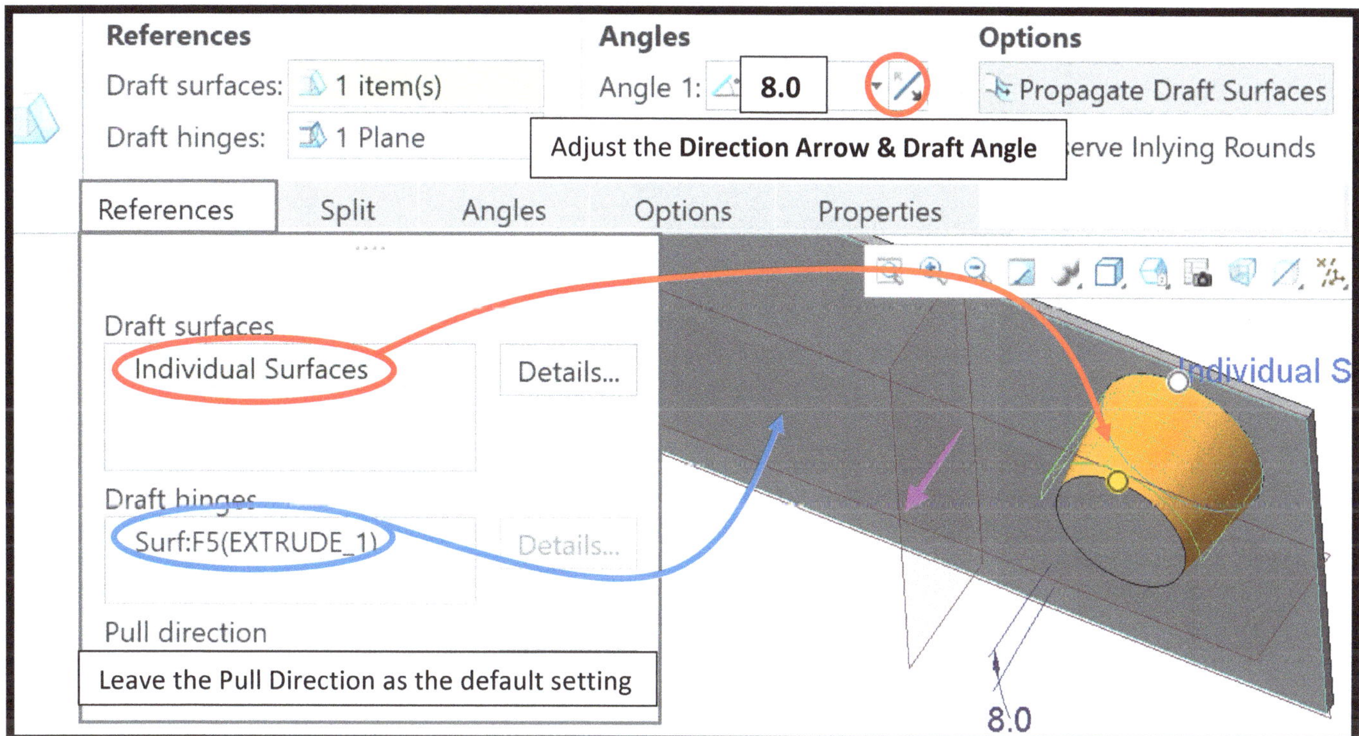

Step 6 – Group the Cylinder (*Extrude 2*) and the Draft feature together for patterning. **Do <u>not</u> add** Extrude 1 to the group!

- **Hold Control** on the keyboard and **LMB to select Extrude 2 and Draft 1** in the model tree – in the pop-up menu choose **the Group icon**.

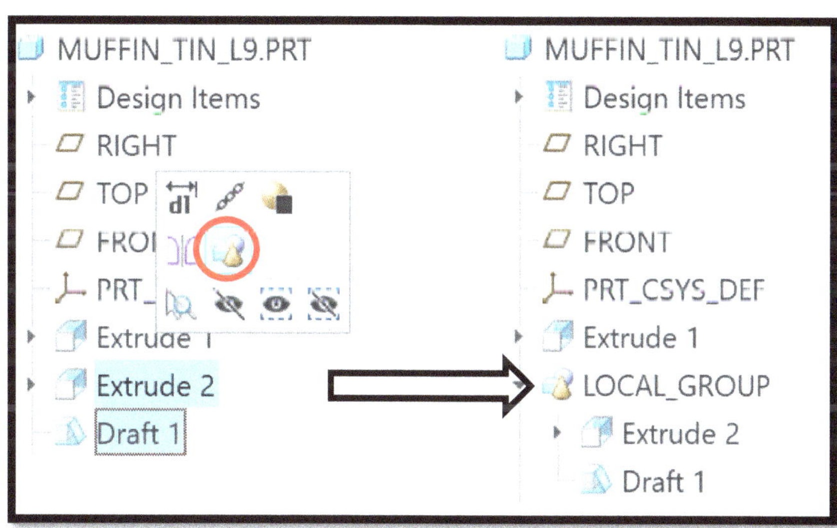

Step 7 – Pattern the Group by to create 6 muffin cups using the **Pattern Tool**.

-    Change the **Pattern Type to "Direction"** in the top toolbar

      - **Select the outer edges** of the plate as your direction references to pattern in two directions.

      - **Flip** the **Directions Arrows** as needed for each direction reference so the pattern goes onto the model instead of away from it.

      - Set the **# of instances** in each direction and the **Spacing** as specified on the **Drawing Key** dimensions (*look at the top view (shown below) to identify which dimensions specify the required pattern spacing values.*

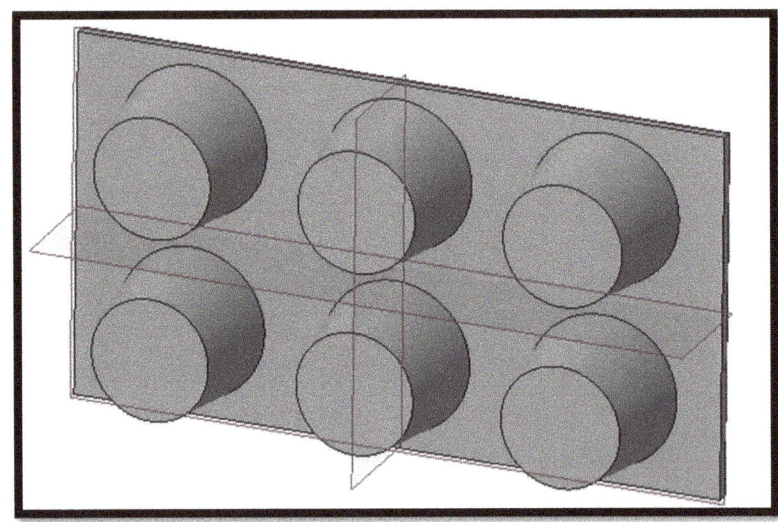

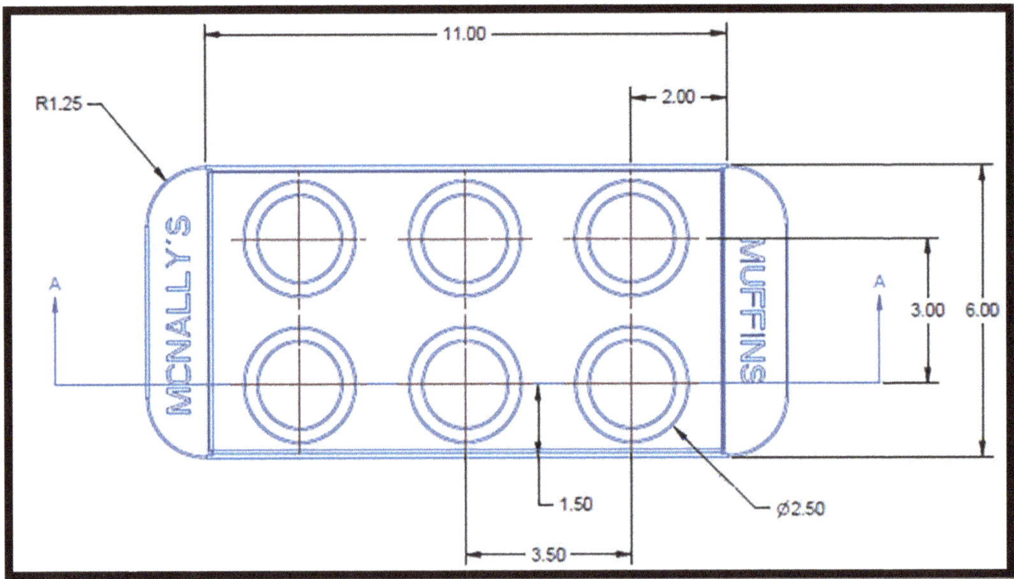

Drawing Key Dimensions for Reference: Look for the ones that specify the required pattern spacing

Tip: If your pattern fails or has errors it is possible you modeled the Draft or Extrude 1 improperly. You can also delete out the Draft, pattern the undrafted cylinders, and then draft all 6 cups all at once using the Draft tool (Hold control and pick all 6 cylindrical edges for the Draft Surface renderings).

Step 8 – Use the **Extrude Tool** to add the Handle shape to the model, sketching on the **Top Datum**.

- Use two **Center & Ends Arcs** to create the 90° arcs* as shown.
  - o   If the arc does not snap onto the corner, you can add them as Sketch References first (hold RMB – Refences tool - click the edges)
- Set an **Equal Lengths constraint** to set the arc Radii equal to each other, with **Radius of 1.25"**.
- Use the **Line tool** to create the straight lines to complete the closed-loop.

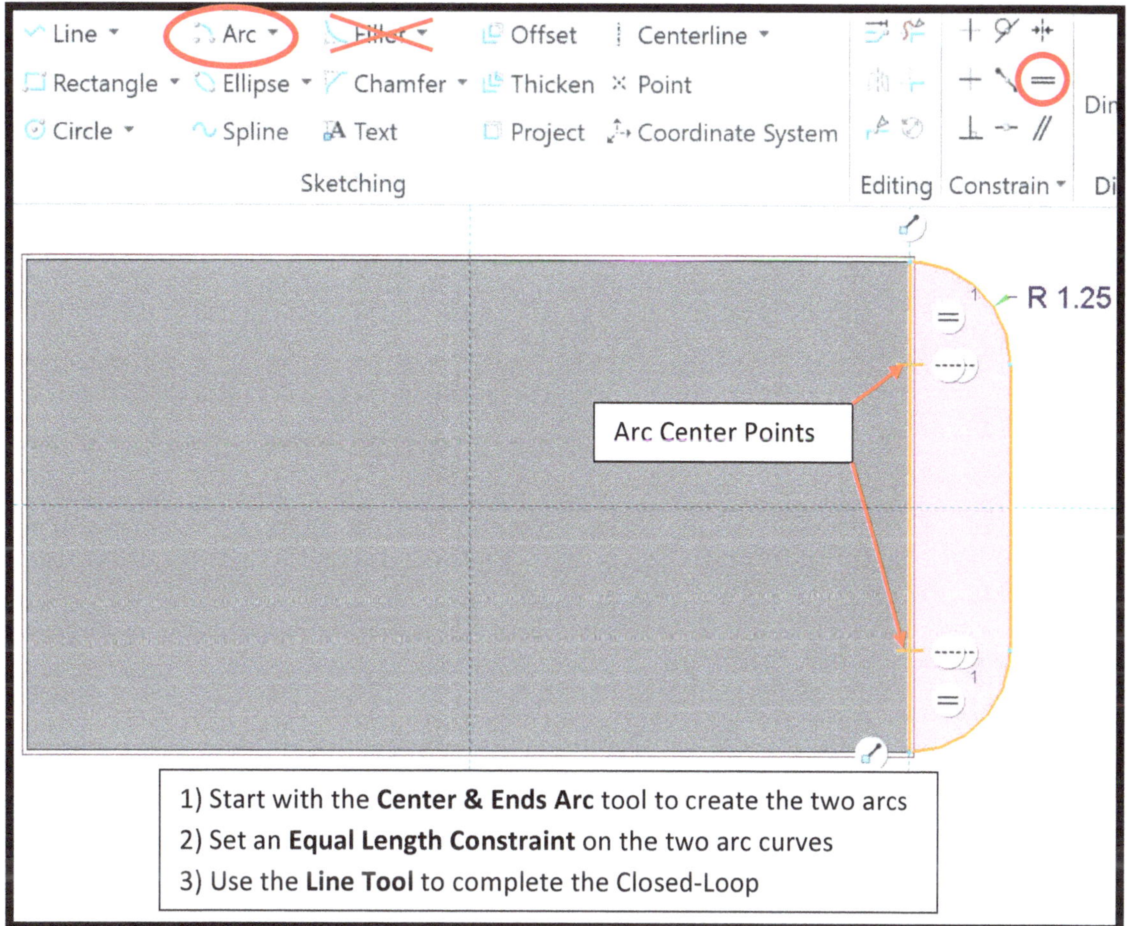

Note - *You <u>cannot</u> use the **Fillet tool** on a Rectangle to create the curve on the corners. The fillet tool could create a small radius, but will not be able to completely remove the horizontal lines as needed for this sketch.*

- Set the **Depth** setting to **"Extrude up to Selected"** and select the top **surface of Extrude 1** so that the extrude depth matches the depth of the Extrude 1.

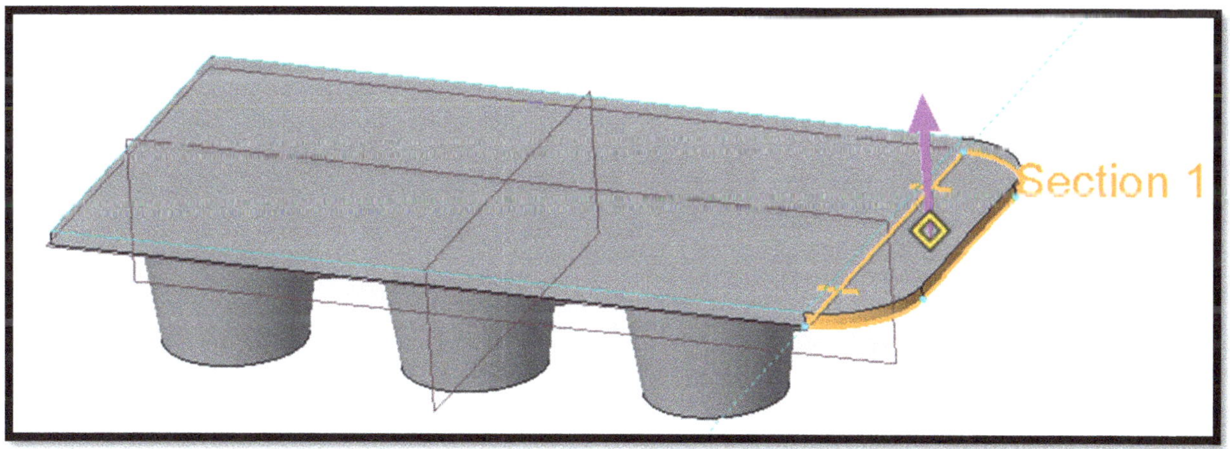

Step 9 – **Copy the Handle** (*Extrude 8*) to the opposite side of the part using the **Mirror Tool**.

If your Mirror fails: check that the sketch of Extrude 1 is **symmetric about the origin**. Also check that your Handle (*Extrude 8*) was sketched as a closed-loop properly, and does not have any extra weak dimensions or other errors on the sketch. If needed, model a second handle on the other side instead of using the mirror tool.

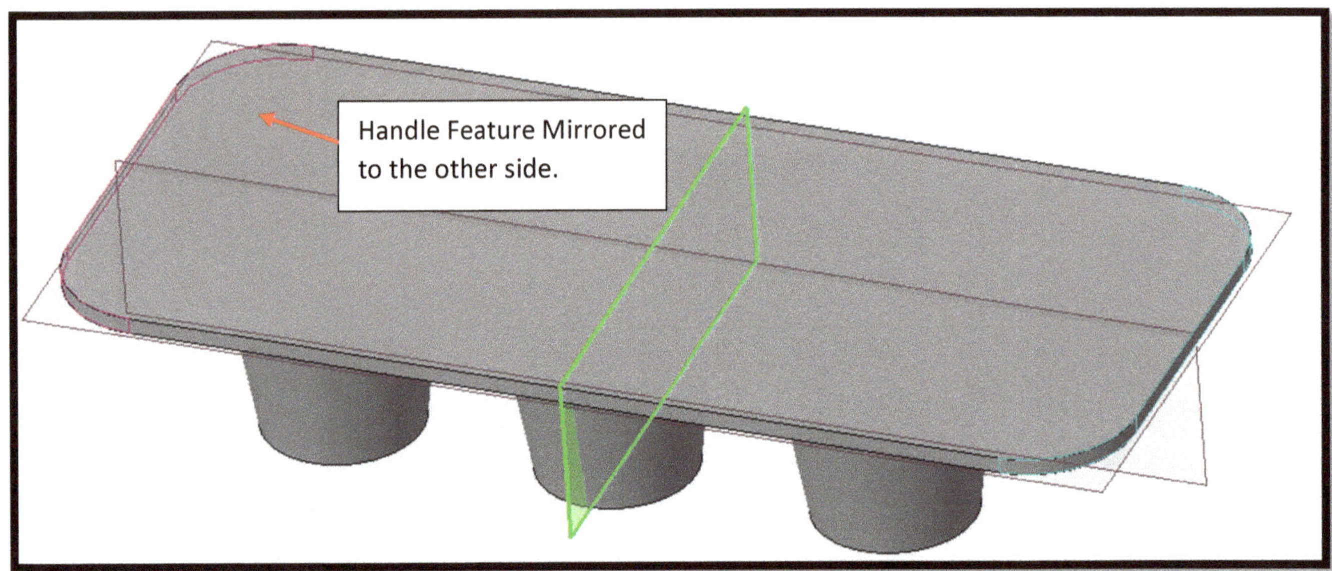

Step 10 – Use the **Shell Tool** to Shell the model to a **thickness of 0.10"**. Set the **Top surface** of the model as a **Removed Surface** in the References tab to remove the top surface of the model.

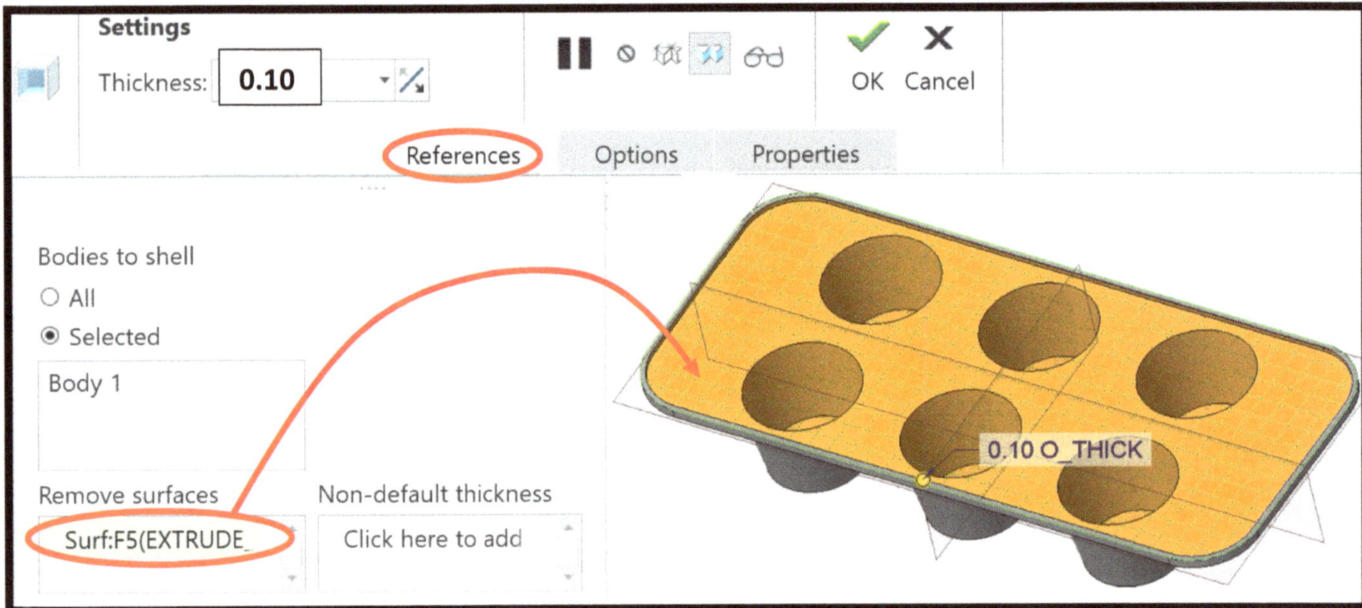

Design Change:

Your Muffin Tin should have "full thickness" handles to be more robust & luxurious for the wealthy engineers that will be using it. Modify the design such that the Handle Extrusion & Mirrored copy are excluded from the Shell.

Step 11 - Rearrange the Shell in the model tree so that the Shell comes before the handle extrusion and mirrored copy.

- **Click and hold LMB on Shell 1 in the model tree** to drag it to a new position before the handle Extrude 8. The Handles should now be unaffected by the shell.
- Sometimes the Shell will not reposition properly in the Model Tree, if so try the following;
    - o Drag the Shell on to the "Pattern 1" feature, and then it should be placed after that feature.
    - o Edit Definition of the shell and check that the Removed Surface is on the Extrude 1 surface, and not the surface of the Handle (Extrude 8), which would prevent the shell from coming first.
    - o Also try moving the Handles to come after the shell, instead of moving the Shell to come earlier.

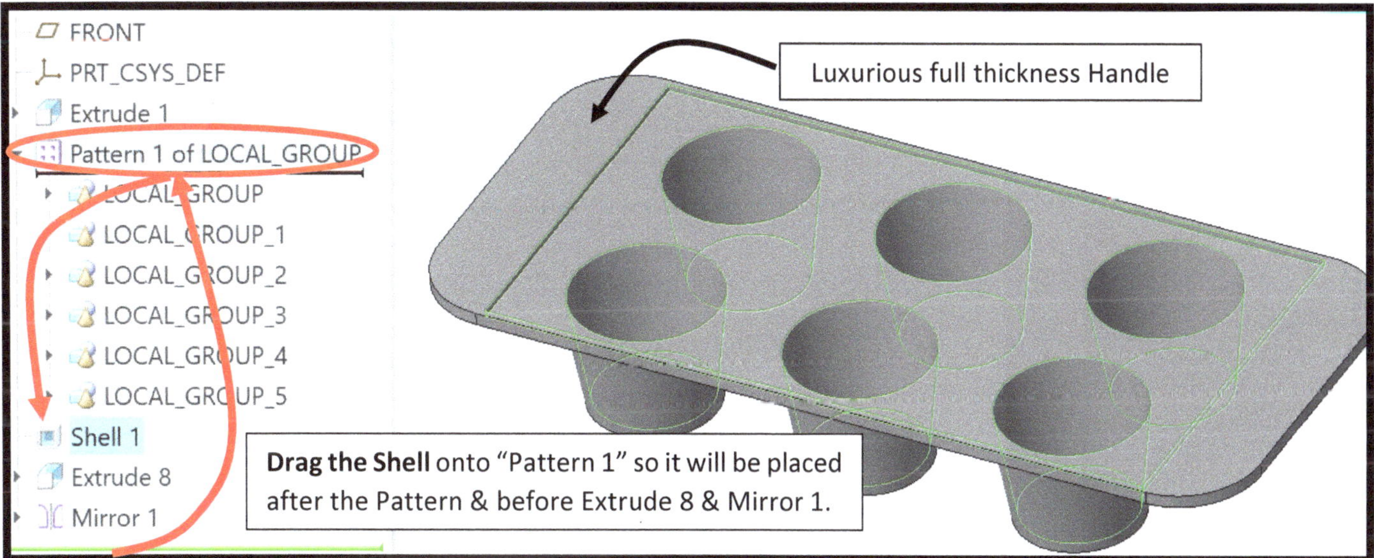

Step 12 - Add Text cutouts to the top of the Handles using the **Extrude tool**. Place the text extrusion on the **top** <u>surface</u> **of the handle.** Do not place the text on the underside of the handle, which would be a poor design choice for such an otherwise perfect baking pan.

- **You can use your own text letters, aspect ratio, font size and position**. A text height of **0.50" or similar** is recommended. You can change the Alignment to Center and place it on the centerline to center the text.
- **Tip**: Some settings may cause the sketch letters to intersect causing the sketch to fail, such as two "f" letters next to each other in the word muffin, depending on aspect ratio & font used.
- Set the Extrude options to **Remove Material**, a **Depth of 0.025",** and adjust the direction as needed.

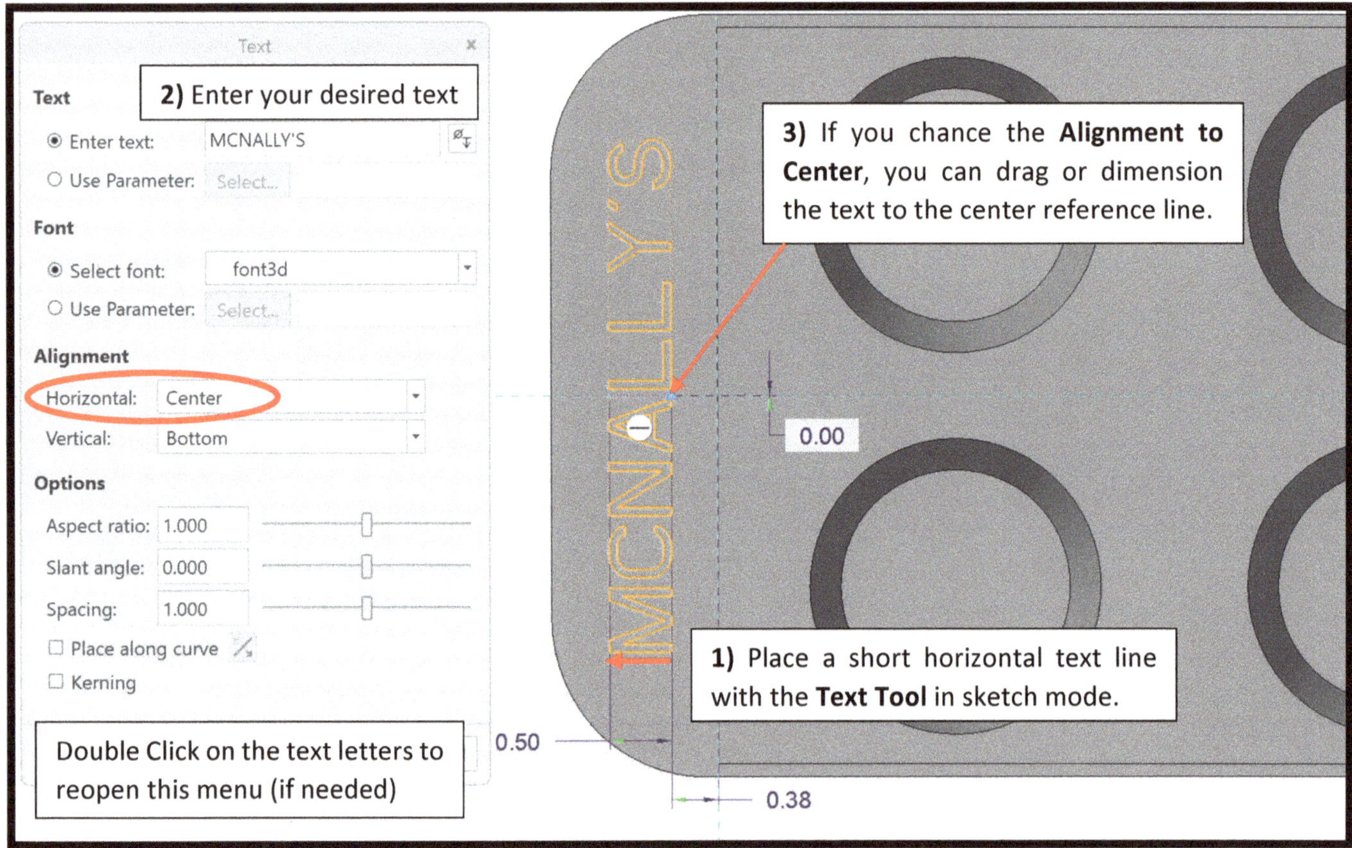

If the extrude fails: Check that you have the Extrude direction, depth, and remove material options set correctly to cut into the model and not away from it. You can also try adjusting the aspect ratio of the text, changing the font, changing the letters used, etc. in case the text geometry is causing an issue.

Step 13 – Add a plethora* of **Rounds** to this part to make it even more luxurious. (*If you don't use the word plethora in your daily vocabulary do you even deserve this high-quality muffin tin?)

- Use the **Auto-Round tool** *(expand the Round drop-box to find it)* with a **Radius of 0.05"**, which should round most of the edges except the text and some of the edges on the top of the model, which is OK.

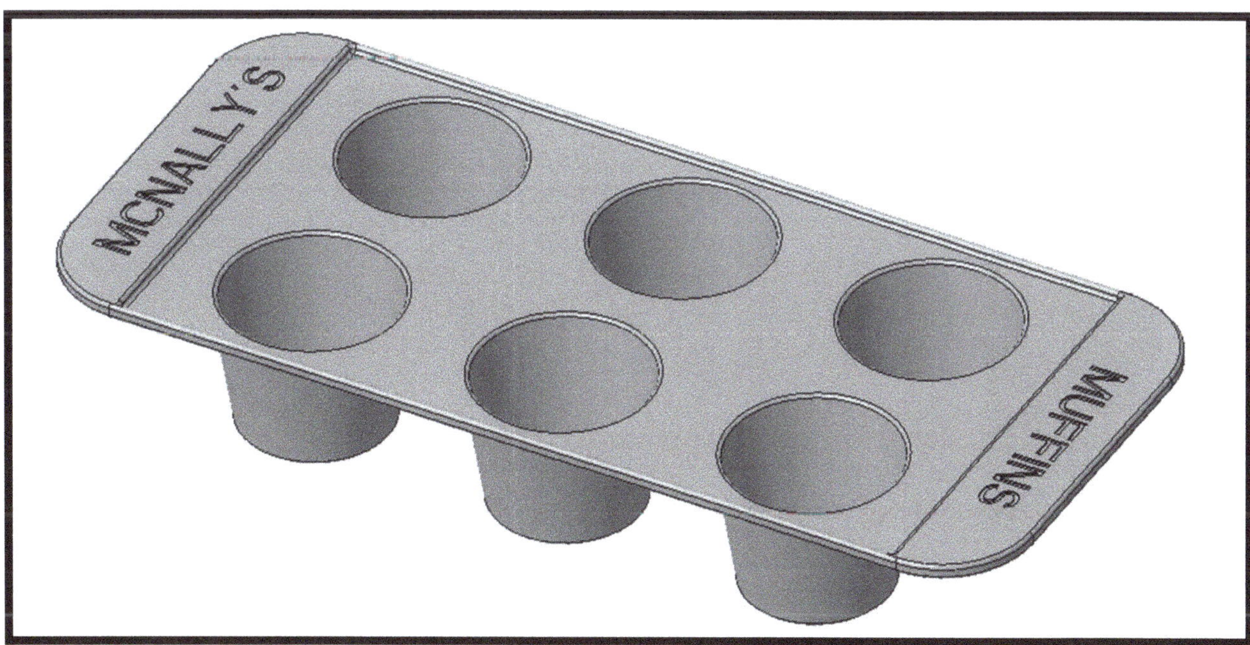

Step 14 – Use the **Plane Tool** to create a new **Datum Plane** to use for the drawing Cross-Section that is **Parallel to the Front Datum** that also runs through the **Axis of a Muffin Cup**. The Datum should run through a row of 3 muffin cups (either the top or bottom row is fine).

Plane

- **Datum Plane Tool – select an Axis** of a muffin cup as the first reference – **hold Control on the keyboard and select the Front Datum** as a second reference – **set the angle Offset to 0 degrees** – press **OK.** You should now have a datum running through a row of 3 muffin cups.

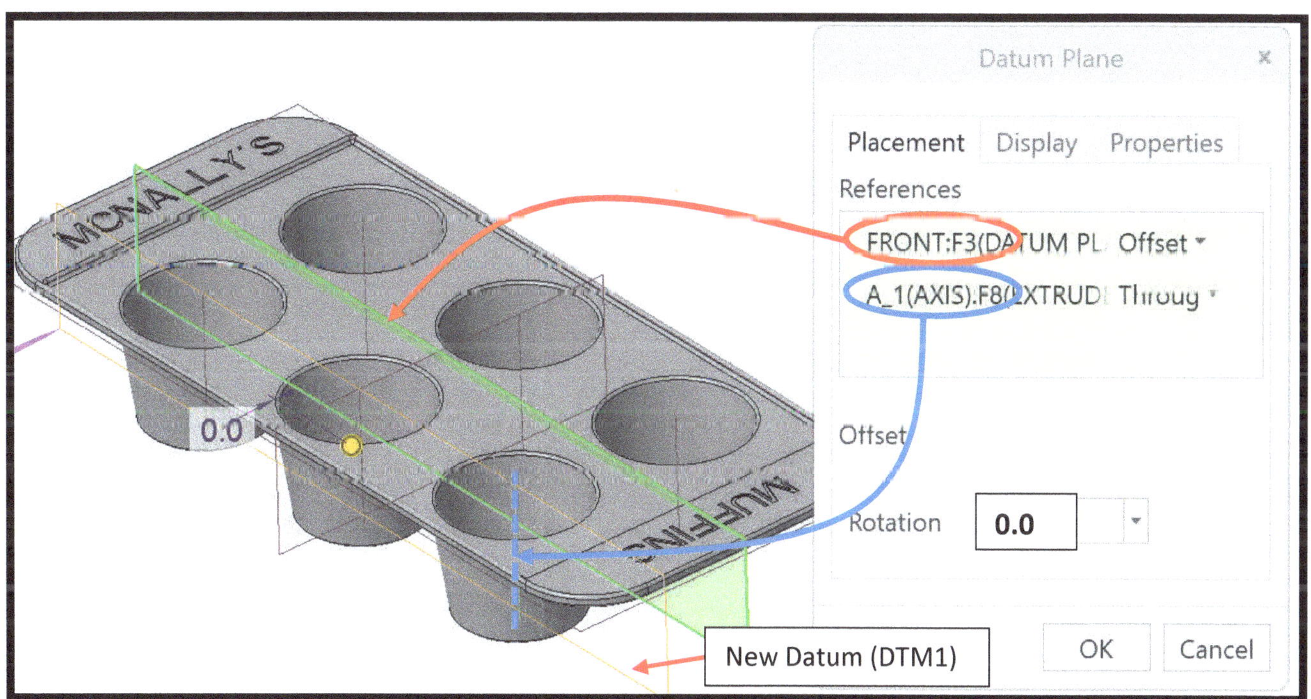

Step 15 – Adjust the Color/Texture of the Muffin Tin model. Use the **Appearance Gallery Tool** (View tab) and select a new Color Setting from the Library. You can choose whichever library or color option you desire.

- **View Tab** - **Appearances Tool** – click on a **Libraries option** – **browse for a metal library** (or other type if you want) – **select the color/texture** you want to use - a small **selection menu** box will appear in the upper right corner of your screen – **select the Part File name** from the Model tree – **press OK** in the small selection menu. This will change all the model material to the selected color.

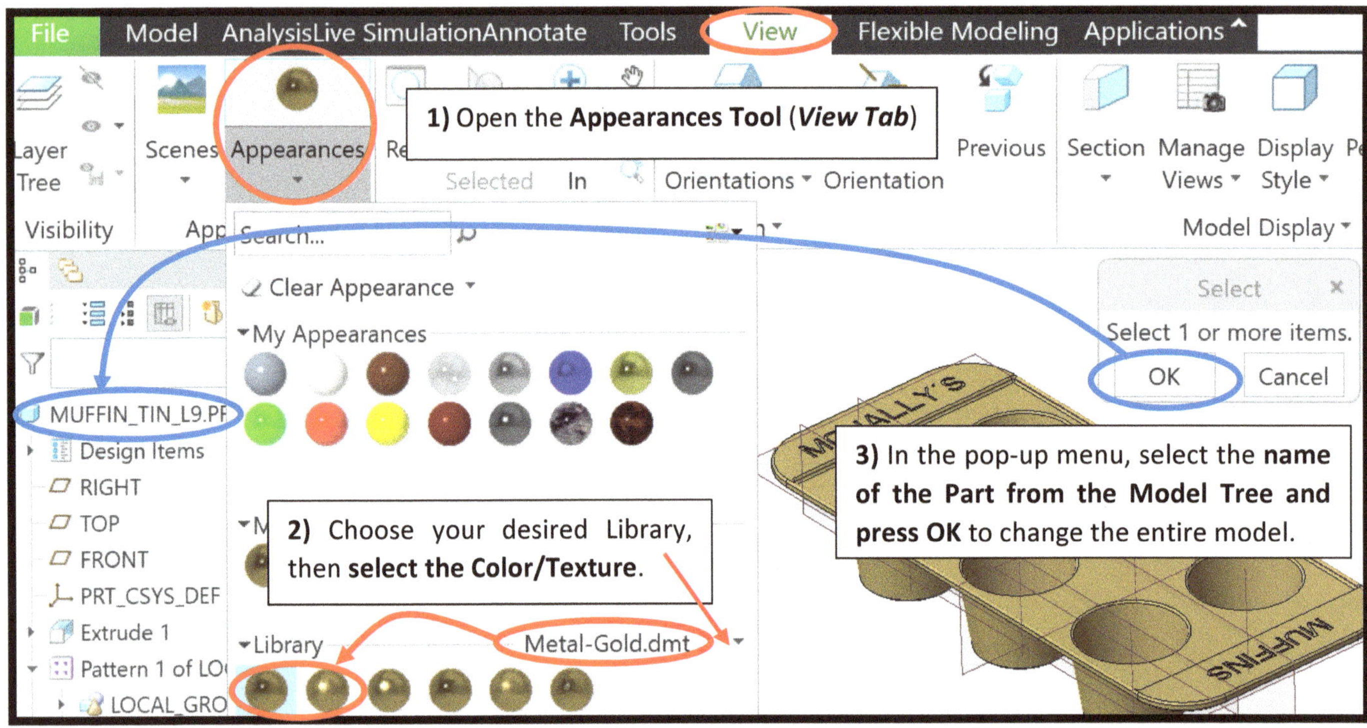

Step 16 – Create a Cross-Section on the Model to use on the Front View of the Drawing.

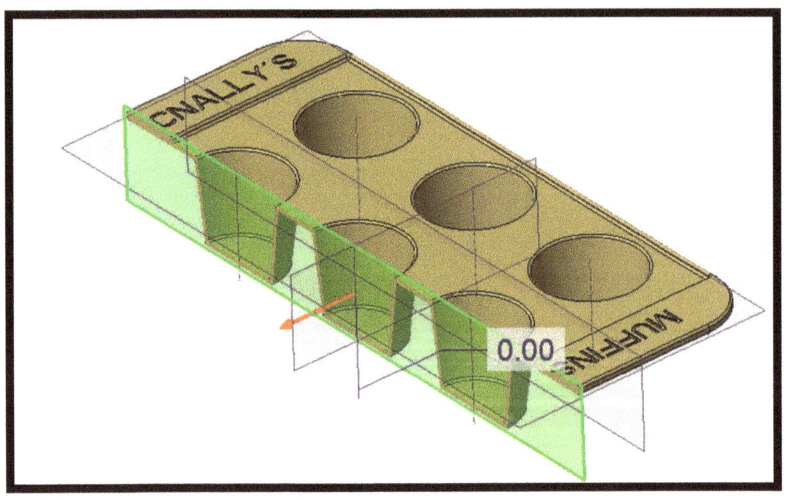

Create the Section: View Tab – Section Tool – **Select the Datum** you created as the reference (DTM1) – Properties Tab – rename the section to "A" – press the green checkmark.

Toggle off the section view: – LMB on Section A in the model tree –select the green **'Deactivate'** icon in the pop-up menu.

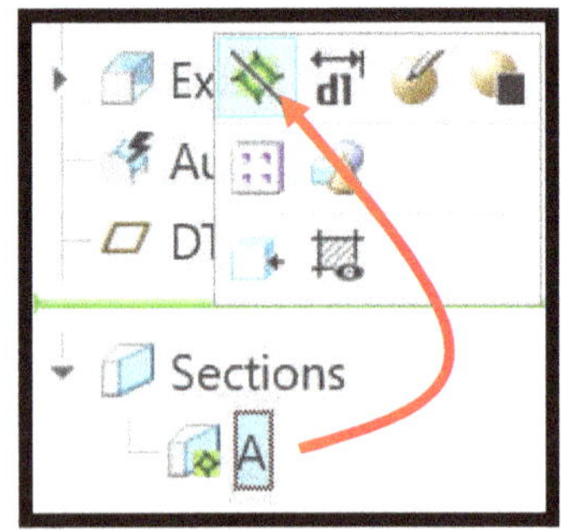

Step 17 – Save your part file and **attempt a Rendering**. You do not need to submit or save the rendered image.

Some graphics cards may not work properly for rendering, so if you have issues you can toggle off "Real Time Rendering" or skip this entire step. Also, some student home versions of CREO may not have the render studio installed, if so, you can skip this rendering step.

- **Applications Tab** in the top toolbar – **Render Studio** – **Scenes Tool** – **double click on a Scene** to add it as the background.
- Use the **Scenes tool – click on Edit Scenes – click on the Environment tab** to adjust the **"Floor Plane"** as needed: depending on your part model you can choose a different plane (Front, Bottom, Right, etc.) to orient the scene to your part model. **Press Close.**
- Toggle on **Perspective View** in the top toolbar to add depth to the model graphics window.
- **Press Render – Browse to choose a location** to output the image to and setup a max time of rendering.
- Use the **Screenshot Tool** (*or windows Snipping Tool*) to take a screenshot to use in marketing your muffin tin.

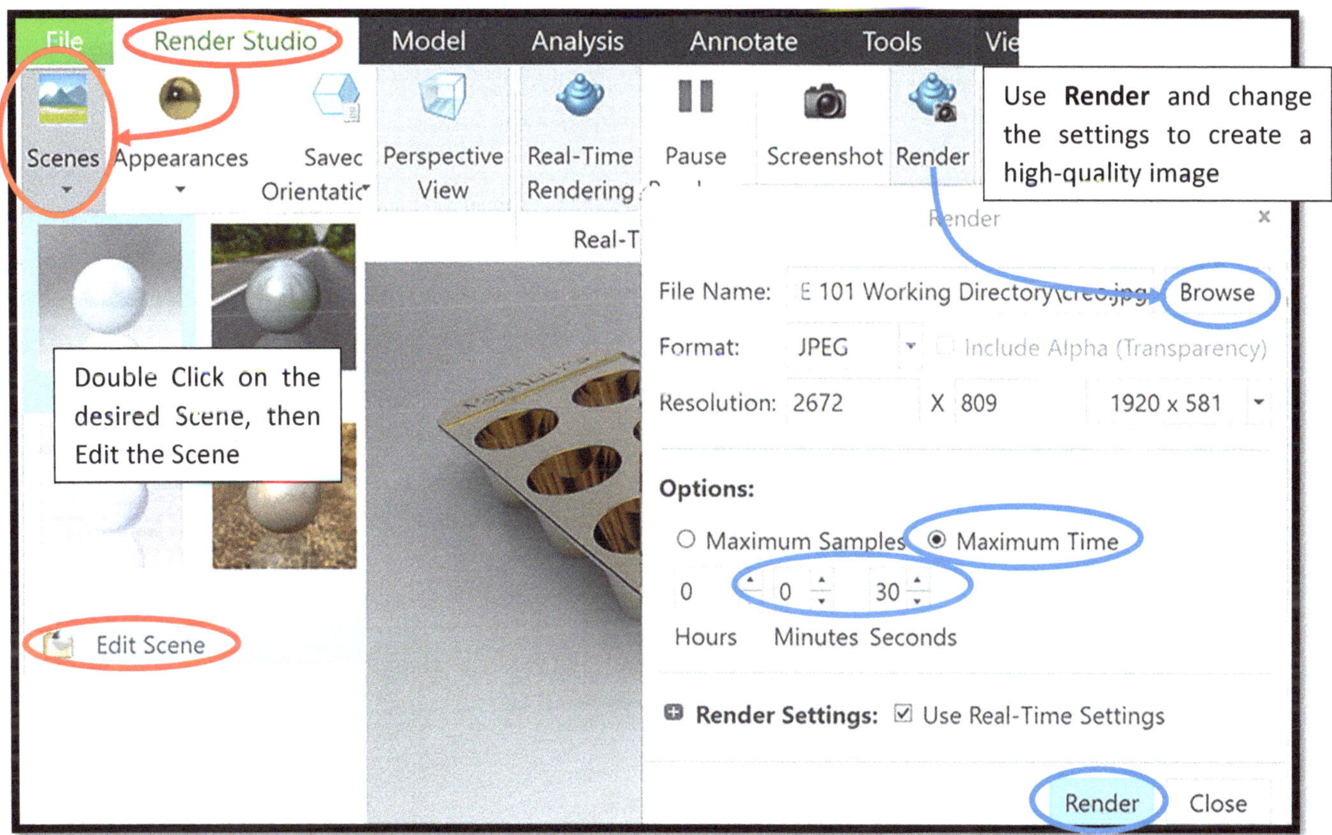

Step 18 – Create a drawing of the model named *"L9_Muffin_Tin"* that matches the key. You do not need to submit or save the rendered image.

End of Lesson 9

Lesson 10 – Assemblies 1(Gas Can Assembly)

Lesson 10 creates an assembly model of the Gas Can and its mounting components for the Rotopax© style system used for off-road vehicles. This lesson will cover Assembly mode and setup a basic Assembly drawing. One component will be created by you during this lesson but the other models you will need to download from Blackboard to use in this lesson.

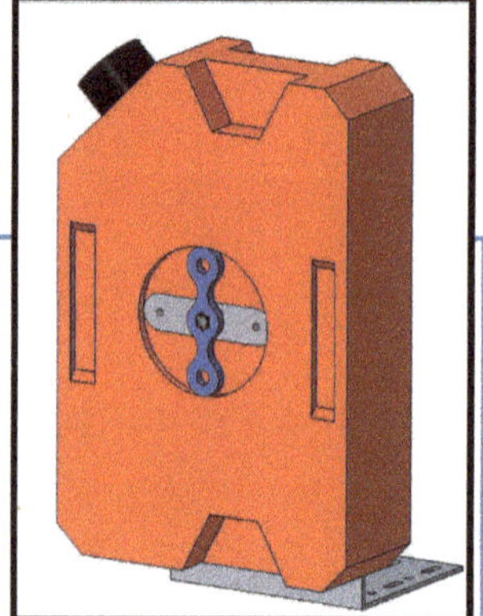

Assembly Mode Information:

An Assembly is a file type in CREO (.ASM) in which part models or other assemblies are added to the model space and have their position and placement defined by constraining them to other components in the assembly. Instead of using Features (such as Extrudes, Revolves, etc.), the Assembly will focus on Constraints and setting up references between part models or other assembly models, which are then known as a 'Sub-Assembly".

Assembly Methodology: There are two methods for creating assemblies;

Bottom-up Assembly: The practice of using **previously** modeled parts and adding them to the assembly with references to the assembly datums or other components in the assembly.

Top-Down Assembly: The practice of creating a **new part** in the assembly *"on the fly"*, using references from the assembled components to define the placement or size of the part model features of the new part. You must tell CREO how to locate the new part (i.e. aligning Coordinate systems), then will add features to the new part model with the benefit of being able see and reference other assembly parts as placement, sketch, or dimension references.

Assembly Constraints: Assembly mode is focused on the connection and placement of components. Use **Constraints** to define where in 3D space the components should be located and how they should be connected to the other parts or assemblies. The goal is to **Fully Constrain** every component to remove "Degrees of Freedom" or movement options of each component so it cannot not move out of position later.

Assembly Model Tree: By default, only Part Models, Sub-Assemblies, and the Assembly Datums are shown in top level hierarchy of the Model Tree. You can see Part Features and Part Datums by expanding the drop box of each assembled component in the Model Tree. In some versions of CREO (earlier than 7.0), you may also need to toggle on the display of Features in the Model Tree Settings - Tree Filter – make sure that Features is check marked.

Useful Assembly Tips:

- Always set a **'Default' constraint for the first component** of an assembly to make sure it is Fully Constrained.
- Turn off **display filters** in the quick toolbar as needed to hide datums, axis, etc. from the many parts from the graphic screen as needed to help selecting entities for constraints.
- **Fully Constrain** all components. If a component is not fully constrained it may "flex" any remaining degree of freedom later on unexpectedly. Use Edit Definition to bring back the Component Placement Toolbar if needed.
  o A component with a [] next to its name in the model tree is <u>not</u> Fully Constrained
  o [][] means that the component is dependent on an earlier unconstrained component, and therefore may not be constrained.
  o [·] signals that a "mechanism type" constraint is being used that has built in movement parameters.
- **Regenerate:** Use the Regenerate tool (or press CONTROL + G on the keyboard) to force an update to the model after making any changes. CREO may not allow you to save if you have made changes and have not "Regenerated" the assembly to force the update.

Getting Started: Open **CREO** *Parametric* and follow the lesson steps carefully.

Some new models are provided for use in this lesson that you will need to download.

Step 0 – Download, Unzip, and place the new files for this lesson in your working directory (not a sub folder). Some of the previous lesson's models are included in case you do not have your own. You can choose to either swap out your own model for the provided file for, or use your own.

Note: All files used in the Assembly must be **located in the same folder** (not a sub folder)! If they are not in the same folder, then the next time you open CREO the Assembly will not find the files, constraints may fail as they reference a missing part, and items will appear red colored in the Model Tree. You would need to find the missing files and move them to the same folder as the assembly and reopen CREO.

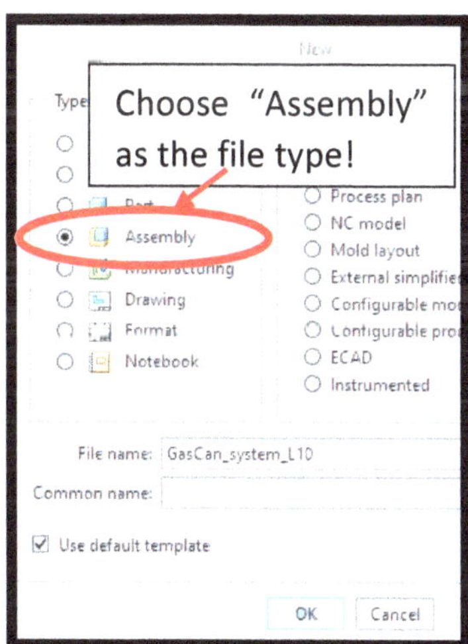

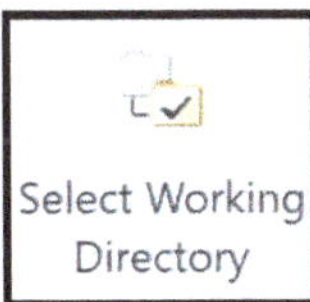

Step 1 – Set your **Working Directory** and create a **New Assembly** file type called "GasCan_System_L10".

File – New – Assembly type – type in the name - **OK.** If prompted, select the *"inlbs_asm_design_abs"* template.

Step 2 – 1 - Add the "main" component to the assembly using the **Assemble Tool** in the top Toolbar. For this lesson, **Assemble** the **GasCan_Bracket_L4** and press **Open.**

Step 2 – 2 - The bracket will appear on the screen in Purple and the constraint toolbar will be at the top of the screen, waiting to have constraints assigned to reduce the Degrees of Freedom (directions of unconstrained movement). For the first component of an assembly, use the **"Default"** constraint to fully constrain it to the Assembly Origin. If you do not set to *Default*, the model could be angled slightly which would be an issue later on.

-   **Select the constraint type as "Default"** to align the *GasCan_Bracket* with the Assembly coordinate system.

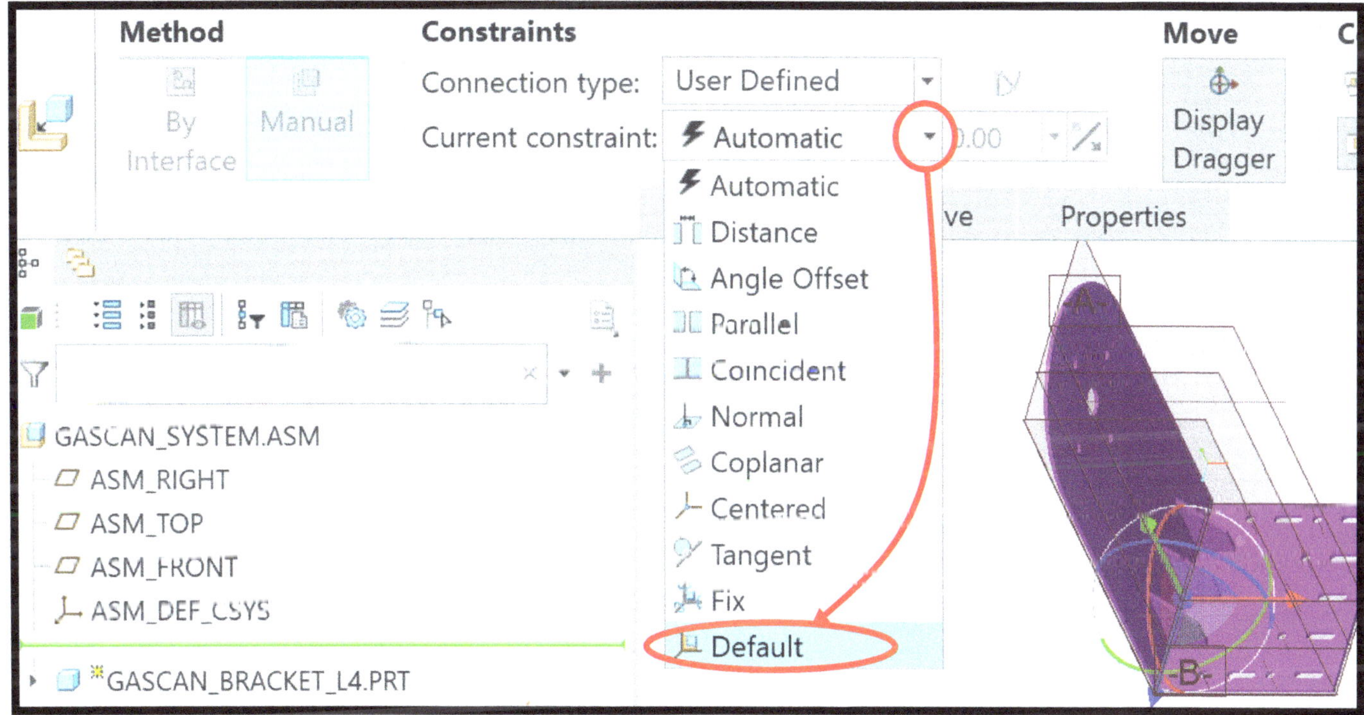

The Purple Color model above indicates it is not Fully Constrained. After choosing Default it will turn a gold color, indicating it is now Fully Constrained.

Step 2 – 3 - Press the checkmark to accept the component placement**. Then look at your Model Tree.**

- You should **not** have a [] symbol in front of the name of the *GasCan_Bracket*, as that would signal that it is not Fully Constrained. If needed, use **Edit Definition** and check that the component used a 'Default' constraint.

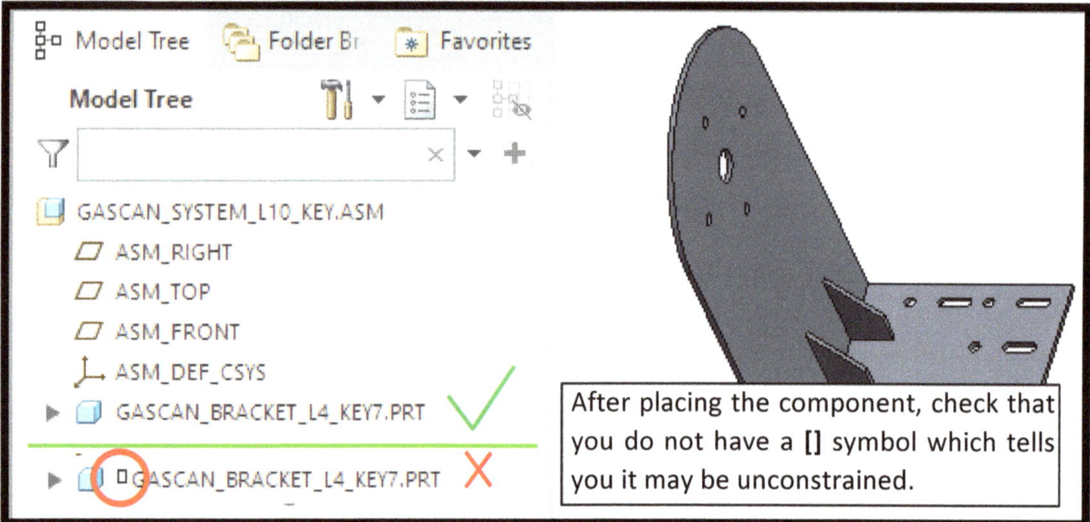

Assembly Tips:

You may want to **turn off the display of any Annotations/Notes, like a standard Hole Note** to declutter the graphic window. Use the quick toolbar to toggle off Annotations.

Also note that because these parts were modeled with their Datum properties **"Set"** as geometric tolerance datums they will not disappear when datums are turned off in the Display Filters. The datums can still be Hidden from the Model Tree if needed (Click and hold RMB on the datum - Hide) or by hiding datum layers (covered in Lesson 11).

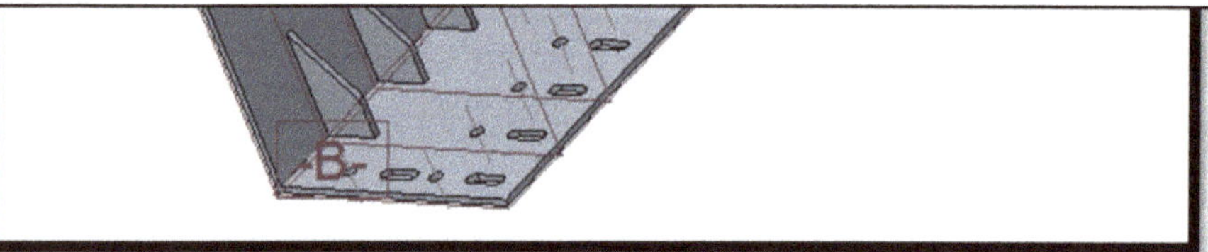
This Hole Note will be hidden after toggling off the **Annotations icon** in the quick toolbar.

The Datums are still visible even if the Display Filters are toggled off as they are "Set" datums from Lesson 4.

Step 3 – Assemble the Gas Can to the Bracket.

- **Assemble Tool**   select the ***GasCan_L2.prt*** from your working directory (or use the version provided in the Blackboard zip folder for this lesson)
- In the **Placement Tab** set the <u>**3 constraints**</u> listed below by selecting references from the Gas Can & the matching reference from the Bracket.
- **Tip: Be very careful where you are clicking.** If you add extra or failed constraints it may cause the component to fail or limit which type of constraint you can choose. You can remove a constraint or reference by clicking RMB on it in the Placement menu and choosing "Delete" or "Remove"

Step 3 - 1 - Constraint #1: (Surface to Surface)

Type: Coincident
Reference of Component: Select the front planar Surface of the Gas Can (side with the spout on the upper right corner)
Reference of Assembly: Select the rear Planar Surface of the Bracket (the side without the Ribs)

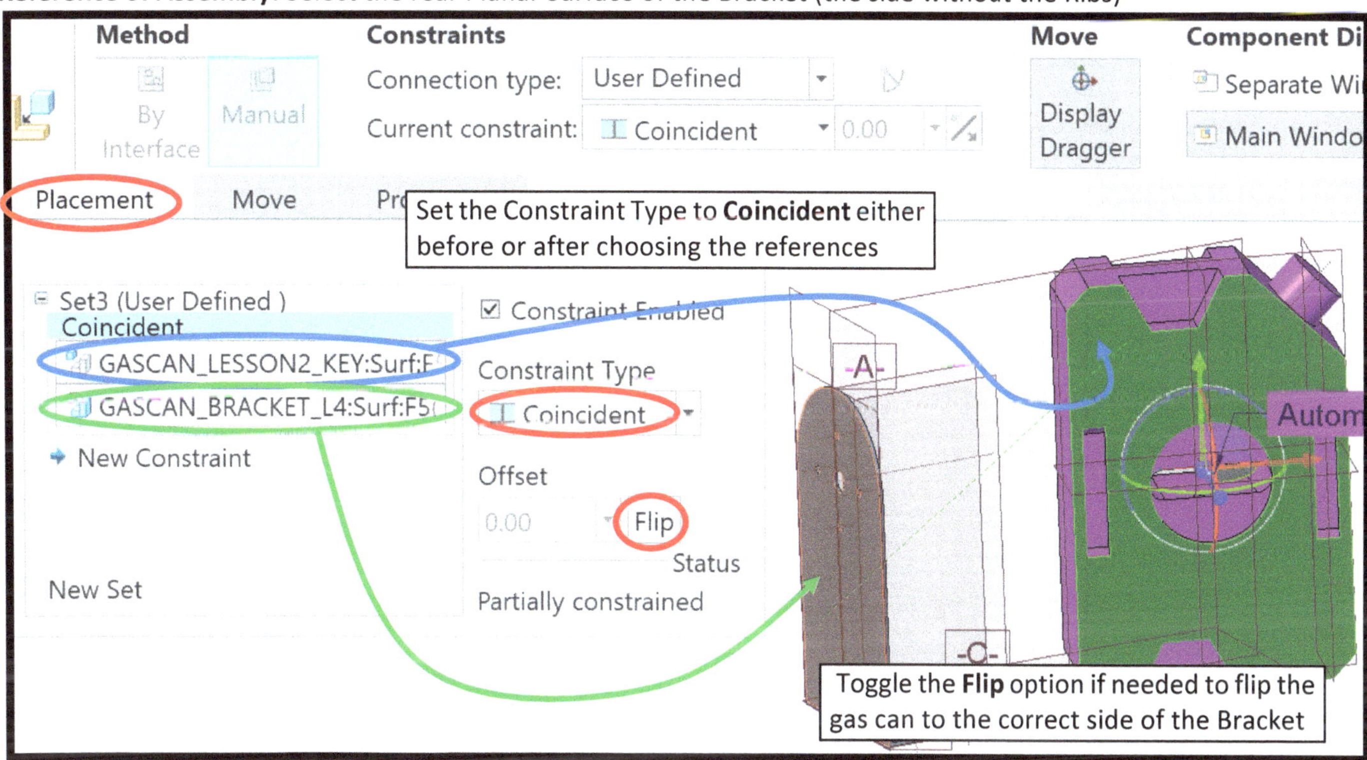

After setting Constraint #1 type to **Coincident** the Gas Can should move to the rear of the Bracket. You may need to toggle the "Flip" option under that Constraint if it is on the wrong side of the Bracket. It is OK if the Gas Can is upside down, as that can be fixed on constraint #3.

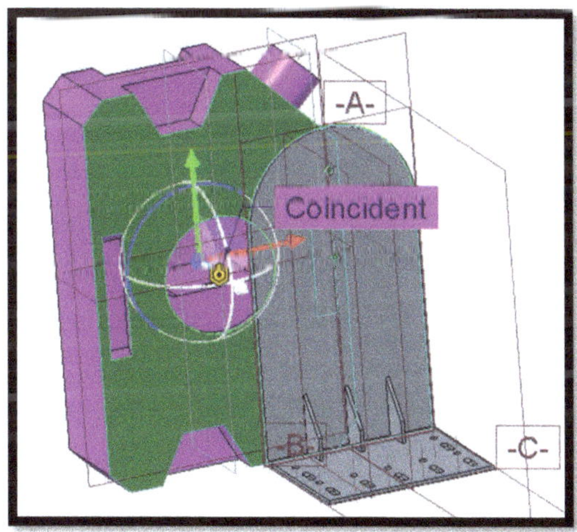

Step 3 – 2 - Select "New Constraint" in the Placement menu to begin the next constraint.

Note – if you add a constraint but select incorrect references or do not finish it, you should delete it from the Placement menu list!

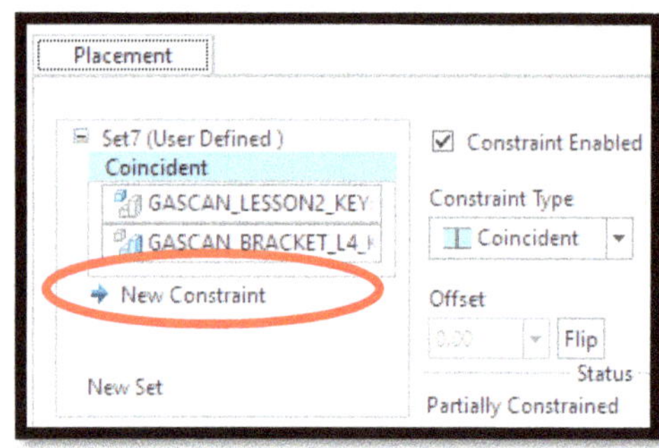

Step 3 – 3 - Constraint #2: (Axis to Axis)

- Make sure your **Axis Display Filters are toggled on** (*Quick Toolbar*). You will need to zoom in on the axis location for this constraint as the Axis have a very short line length.
- Toggle on the **'Show Component in a Separate Window'** option (shown in the image below at the top toolbar next to the checkmark) to view the incoming component in its own window.

Constraint 2:

Type: Coincident

Reference of Component: Select the Axis at the center of the Gas Can (Axis of the Circular Extrusion (**zoom in!**)

Reference of Assembly: Select the Axis of the large center Hole of the Bracket.

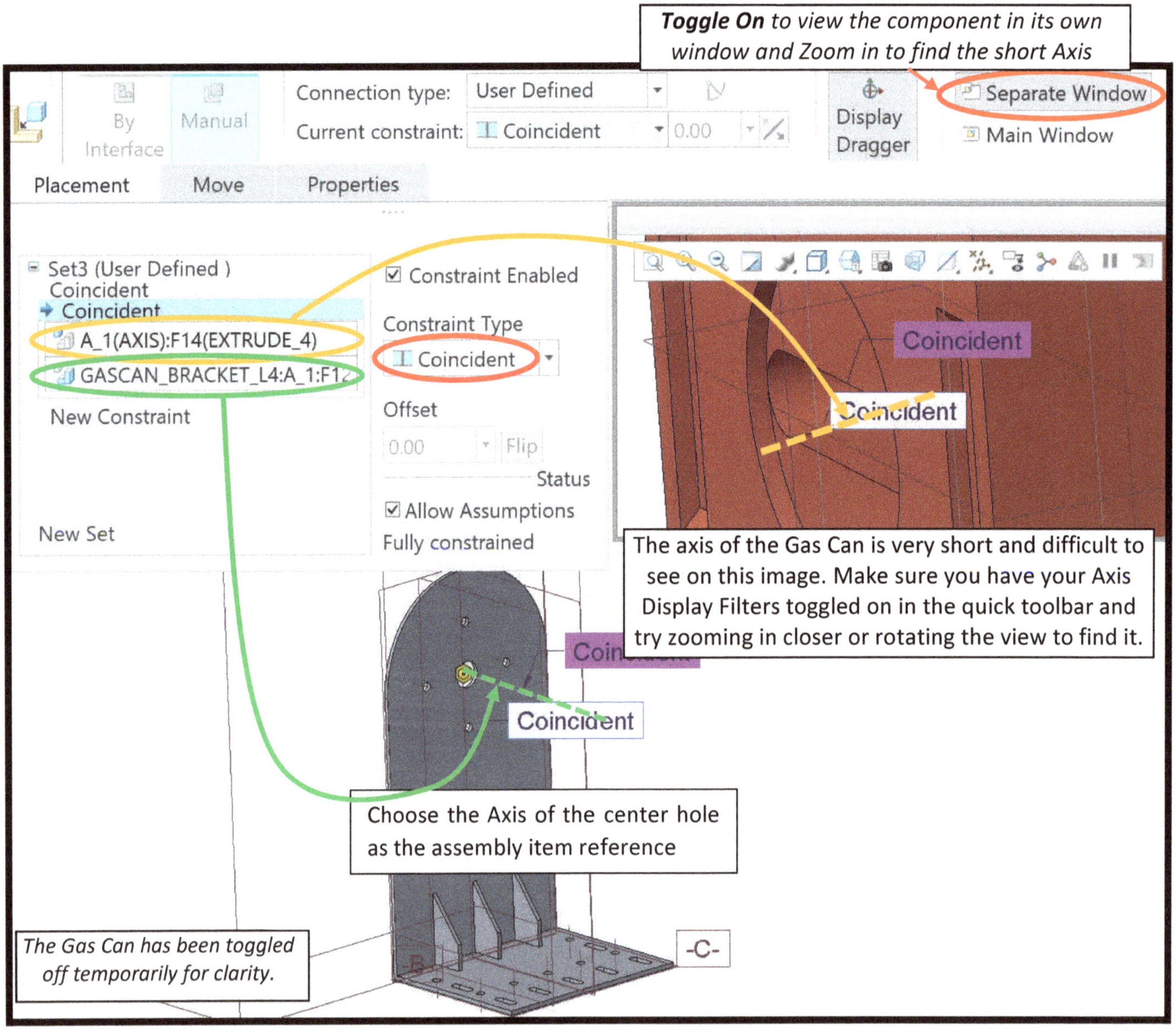

Step 3 – 4 - Notice the Gas Can now has a gold color (*fully placed*). CREO is assuming a constraint for the rotation of the gas can about the axis. To control the rotation a third constraint will need to be set.

- **Uncheck the "Allow Assumptions"** in the bottom of the placement menu
- Select **New Constraint** to add the Angle Offset constraint between datums.

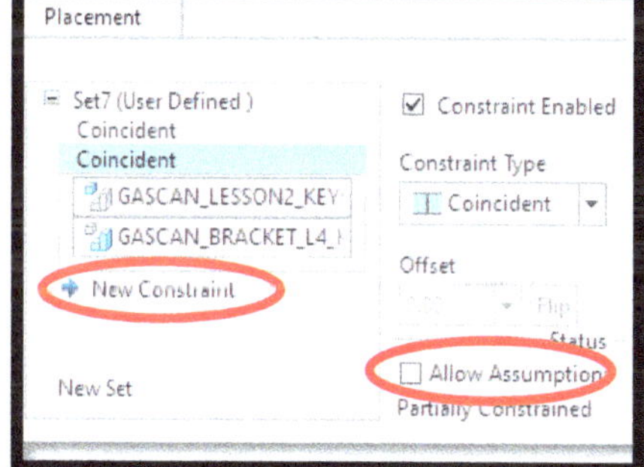

Step 3 – 5 - Constraint #3: (Datum to Datum, Angle Offset)

Constraint 3:

Type: Angle offset of either **0 or 90 degrees** (*as needed so can is upright*)

Reference of Component: Select the **Vertical Datum** of the Gas Can that runs the depth of the Gas Can

Reference of Assembly: Select the **Vertical Datum** of the Bracket

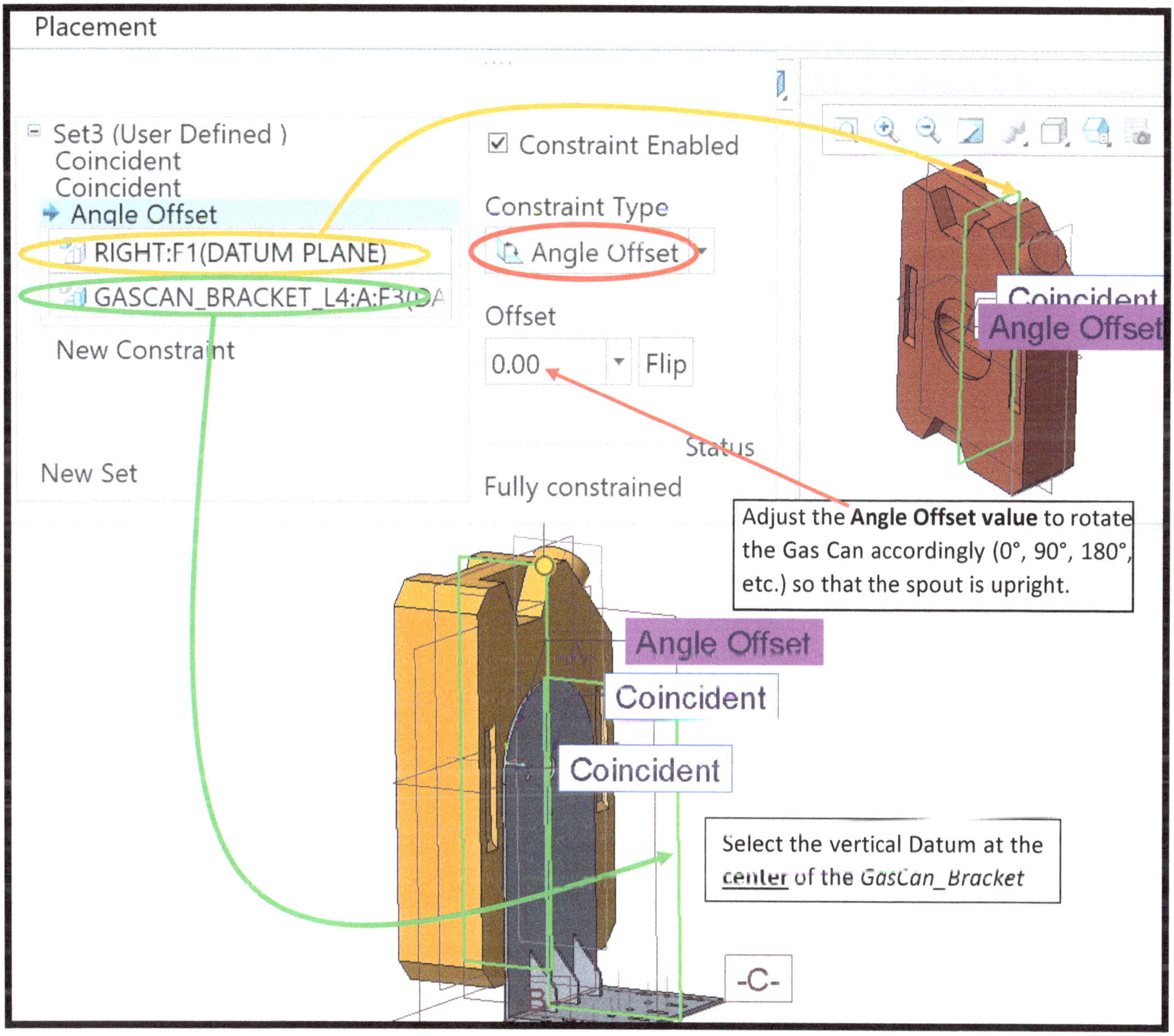

Adjust the **Angle Offset value** to rotate the Gas Can accordingly (0°, 90°, 180°, etc.) so that the spout is upright.

Select the vertical Datum at the center of the *GasCan_Bracket*

The Gas Can should now be **Fully Constrained**, with the Gas Can upright. Your Gas Can 'Spout' may be either in the upper right or left corner, depending on which surface of the gas can you used in Constraint #1.

If the Gas Can is not "Fully Constrained" (not a Gold Color), you likely have an error. Look at your Placement tab and check if you have extra constraints (should be 3 of them total), incorrect constraint types, or incorrect references. Delete out any extra constraints and delete any that may have issues and remake them.

Step 3 – 6 - Checkmark to accept the placement of the Gas Can to complete the Placement process. **If needed, you can use Edit Definition** to bring back the Placement Toolbar for the Gas Can and make any changes.

Bottom-Up Assembly (Creating the *GasCan_Mount* part model)

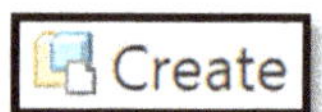

This next component to be added is the *GasCan_Mount*, which is not provided to you. You will make it "on the fly" using the "Bottom Up Assembly" methodology using the **Create Tool.**

Step 4 – Select Create from the top toolbar – select **Part** as the type and enter in **"GasCan_Mount"** for the name – **press OK** – select **"Locate Default Datums"** – choose the **"Align csys to csys"** option – press **OK** – CREO tells you it is waiting for you to select a CSYS in the very bottom of the screen* - **LMB** on the **Coordinate System of the Gas Can** (*PRT_CSYS_DEF*) on the graphics window (*you will need to toggle on the Csys Display Filters to make it visible*)

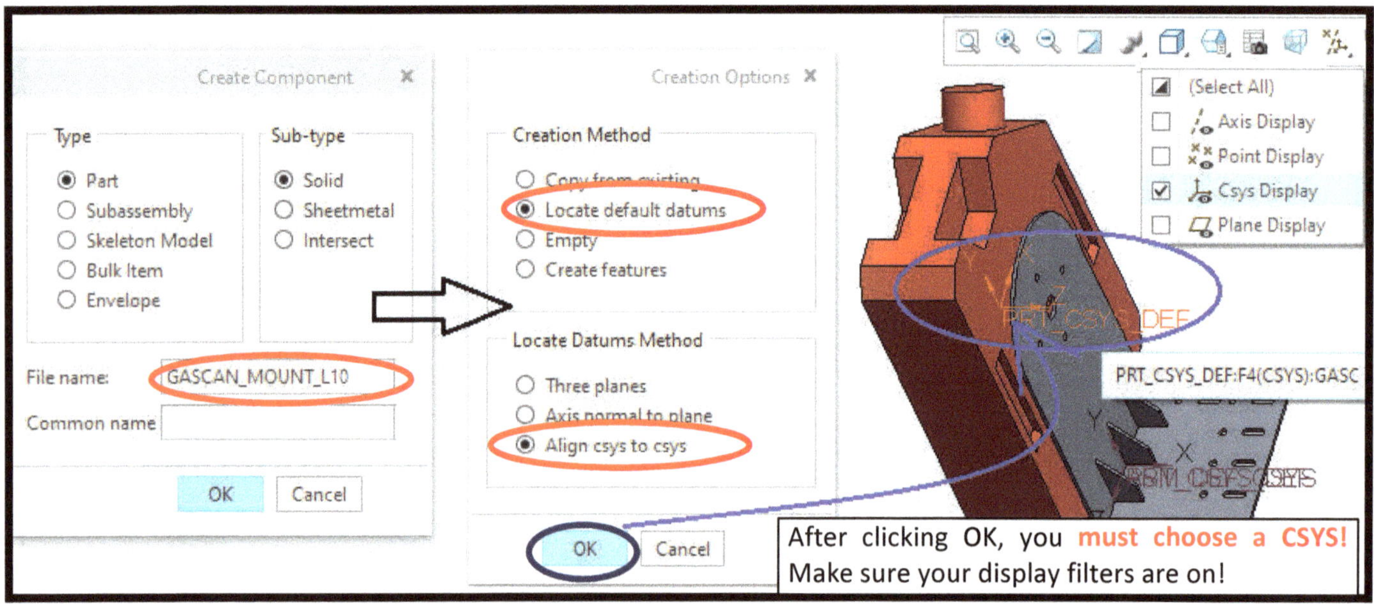

* Make sure you have selected a Coordinate System otherwise CREO will act frozen or locked up as it is waiting for you to finish the command properly!

Notice that the **Assembly components become inactive/greyed out** and the top toolbar is now in "Part Mode" modeling for the new part, which is just an empty Coordinate System and Datum planes as it has no other features yet.

There will be a green star next to the name in the model tree to signal that the new part is "Active".

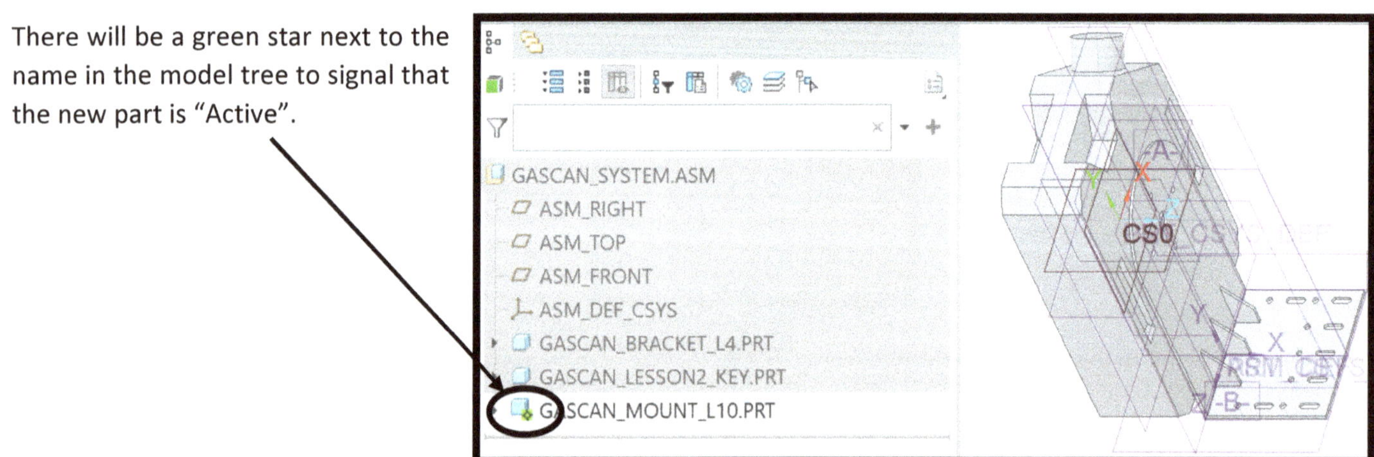

Step 5 – 1 - Create an **Extrude to add material to the new part**.

- Define the placement Sketch Plane as the rear planar surface of the *GasCan_Bracket* where the Gas Can is attached (the side without ribs).

- **Press Sketch** to enter sketch mode.

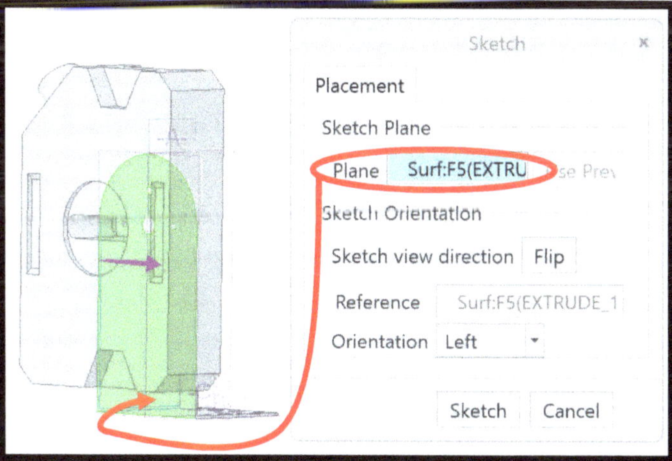

Step 5 – 2 - Use the **Project Tool** in sketch mode to create a closed-loop that matches the shape of the inside cutout of the gas can.

- In the Quick Toolbar: **Toggle off the Display Filters** and set the **Display Style** to **No Hidden** temporarily to declutter the graphics window for this step.
- **Click on the Project Tool** – with the project tool active and set to the Single option, **Hold Control & LMB on the six edges** of the center cutout shape of the Gas Can to create the closed loop shape (*Select the two horizontal lines and the four 90deg arcs of the ends*).

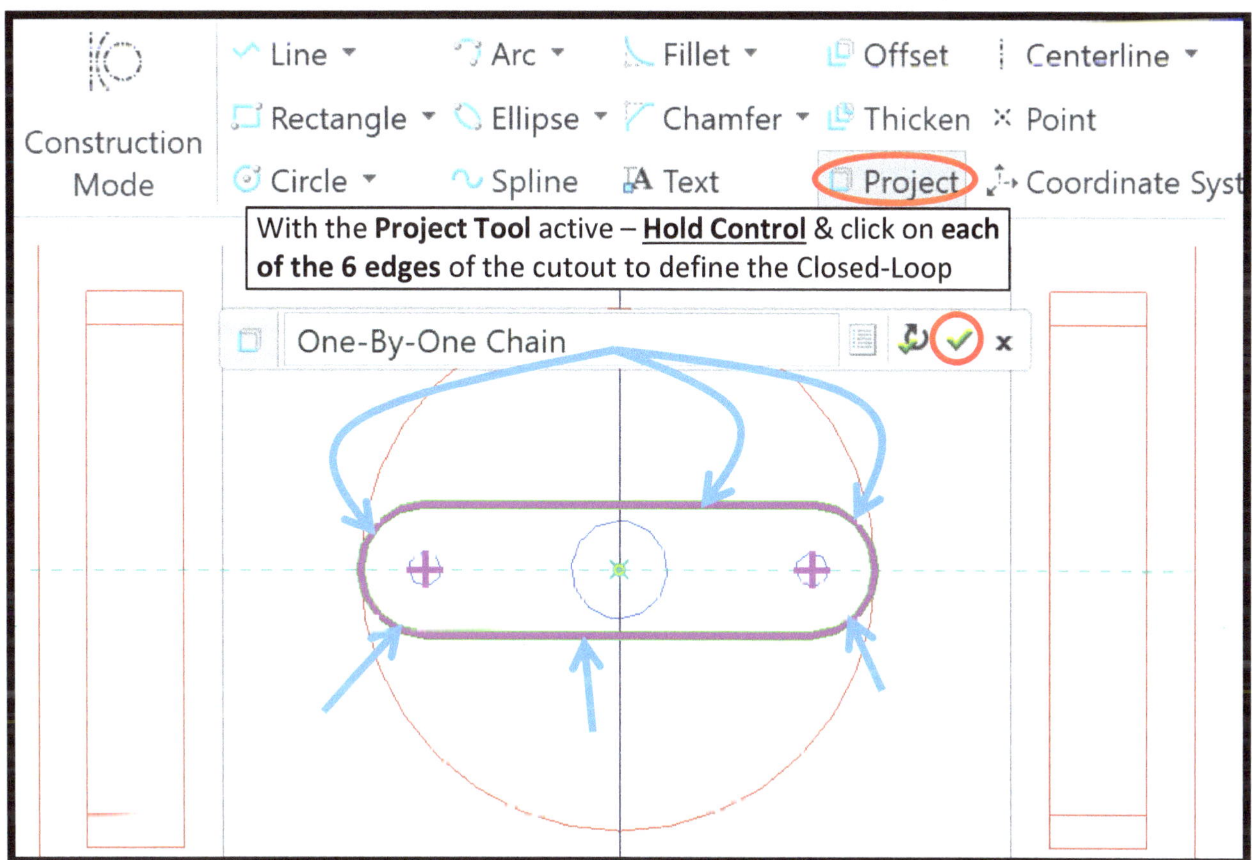

Step 5 – 3 - After check marking the Project tool you should see the shaded closed-loop shape.

- **Checkmark to accept** the Sketch. Change the **Display Style** in the Quick Toolbar back to **Shading with Edges**.

Step 5 – 4 - Set the **Extrude options to 'To Reference'** and select the flat surface of the circular cutout of the gas can to define the Extrude depth.

Error Notice: If you are having an issue where the Extrude will <u>only</u> remove material and the model is not "grey" that would signal you are making an *Assembly Extrude* feature and not making a *Part* feature. To do this properly, you must have the *GasCan_Mount* part file Activated and the rest of the assembly should be greyed out!

- **If needed** – cancel this Extrude and Activate the *GasCan_Mount* from the Model tree (Select it in the Model Tree - hold RMB – choose Activate (the green star icon)) and create the Extrude as a part feature while the Mount is activated. Or, if you do not have a *GasCan_Mount* in the Model tree, cancel this step and go back to Step 4.

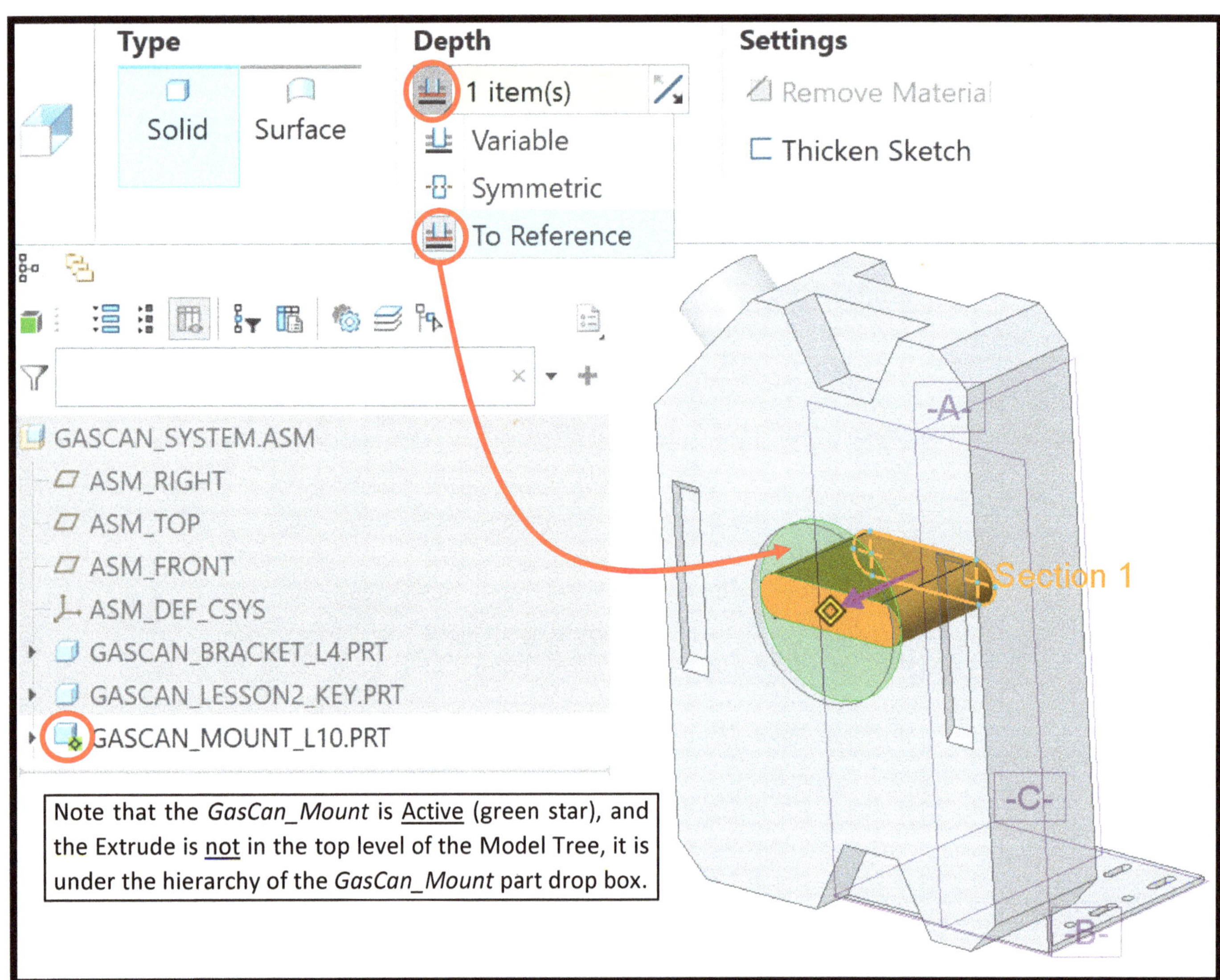

Note that the *GasCan_Mount* is <u>Active</u> (green star), and the Extrude is <u>not</u> in the top level of the Model Tree, it is under the hierarchy of the *GasCan_Mount* part drop box.

Step 5 – 5 - Checkmark to accept the Extrude.

Step 6 – With the Mount still active, **use the Hole Tool** to create a hole in the *GasCan_Mount* to line up with the left hole of the Bracket. Use the Axis of the *GasCan_Bracket* hole at the 9 O'clock position on the Bracket as the Coaxial Reference and the planar surface of the extrude from the previous step as the second primary reference. *Tip: Toggle on the Axis Display Filters if you turned them off earlier.*

Hole Type: **Standard**

Profile Options: **Straight, Tapped**

Shape Tab Options: Set the option to **Thru-Thread** depth

Toolbar Settings: **Screw Size: UNC "1 / 4 - 20", Hole Depth = Thru All**

Placement Type: **Coaxial**

Primary References: <u>Hold Control</u> on the keyboard when selecting the two primary references
- o   Axis of *GasCan_Bracket* Hole (the 9 O'clock or 3 O'clock position hole, either will work)
- o   Planar surface of the extrusion of the *GasCan_Mount*.

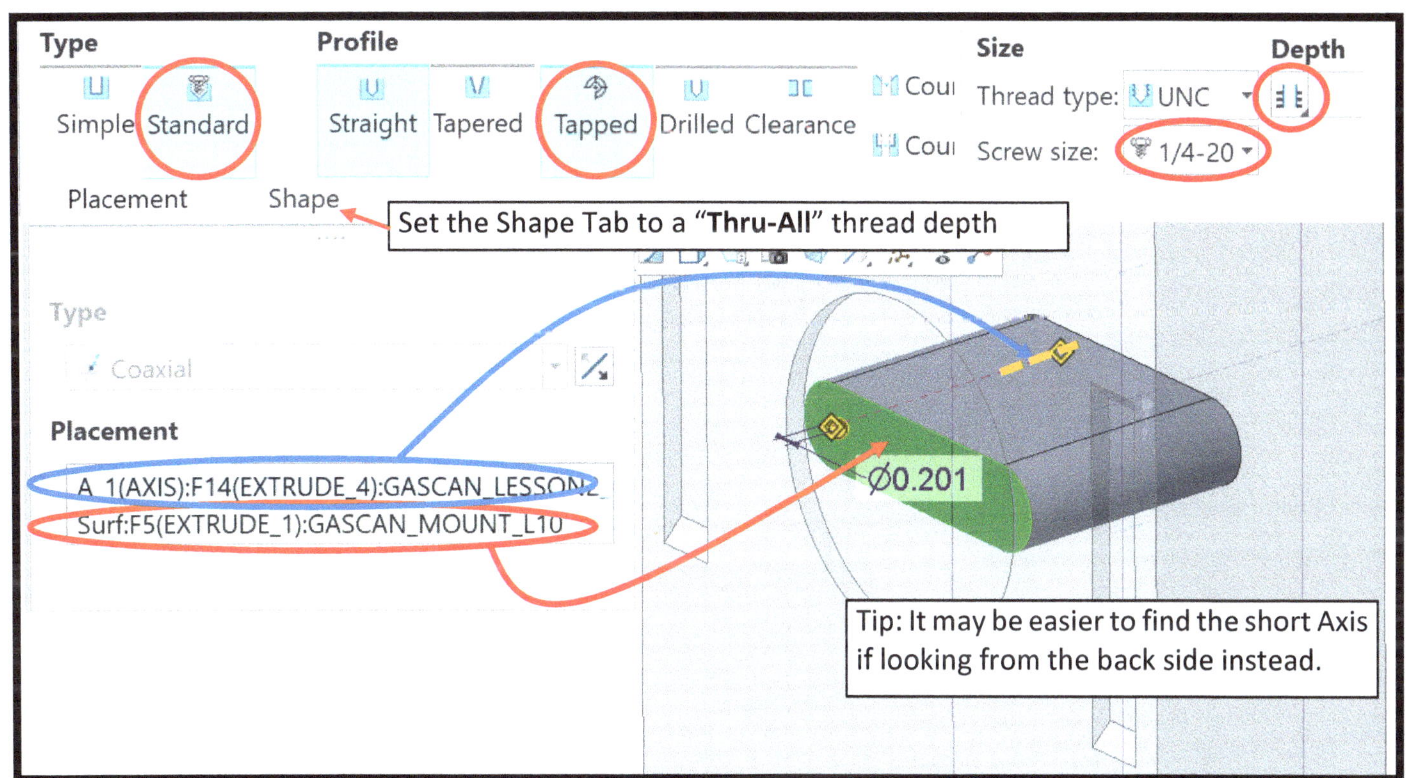

Step 6 – 2 - Checkmark to accept the Hole. While the Hole is selected you should see the outline of the threads.

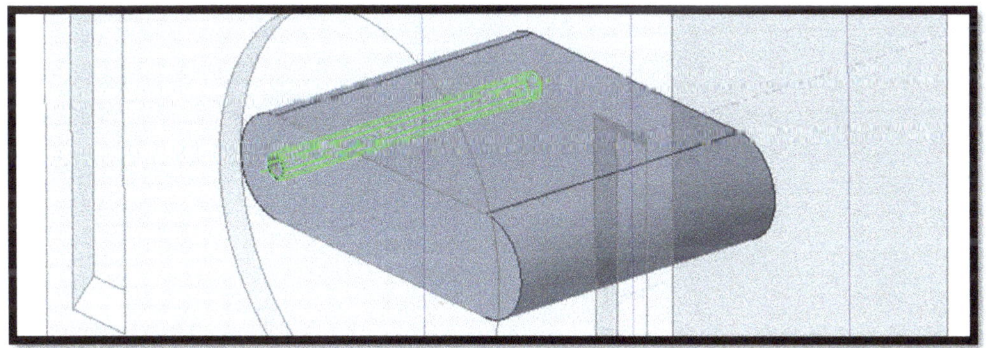

Step 7 – Mirror the Hole to line up with the opposite mounting hole of the bracket, using the **vertical Datum** of the *GasCan_Mount* as the mirror plane.

- **Expand the drop box** for the *GasCan_Mount* in the Model Tree to **find and click on Hole 1**.
- Use the **Mirror Tool** and in the **References tab** click on a suitable datum in the center of the Mount (toggle on Datum Display filters in the quick toolbar if needed)
- **Checkmark to accept** the Mirrored copy of the hole.

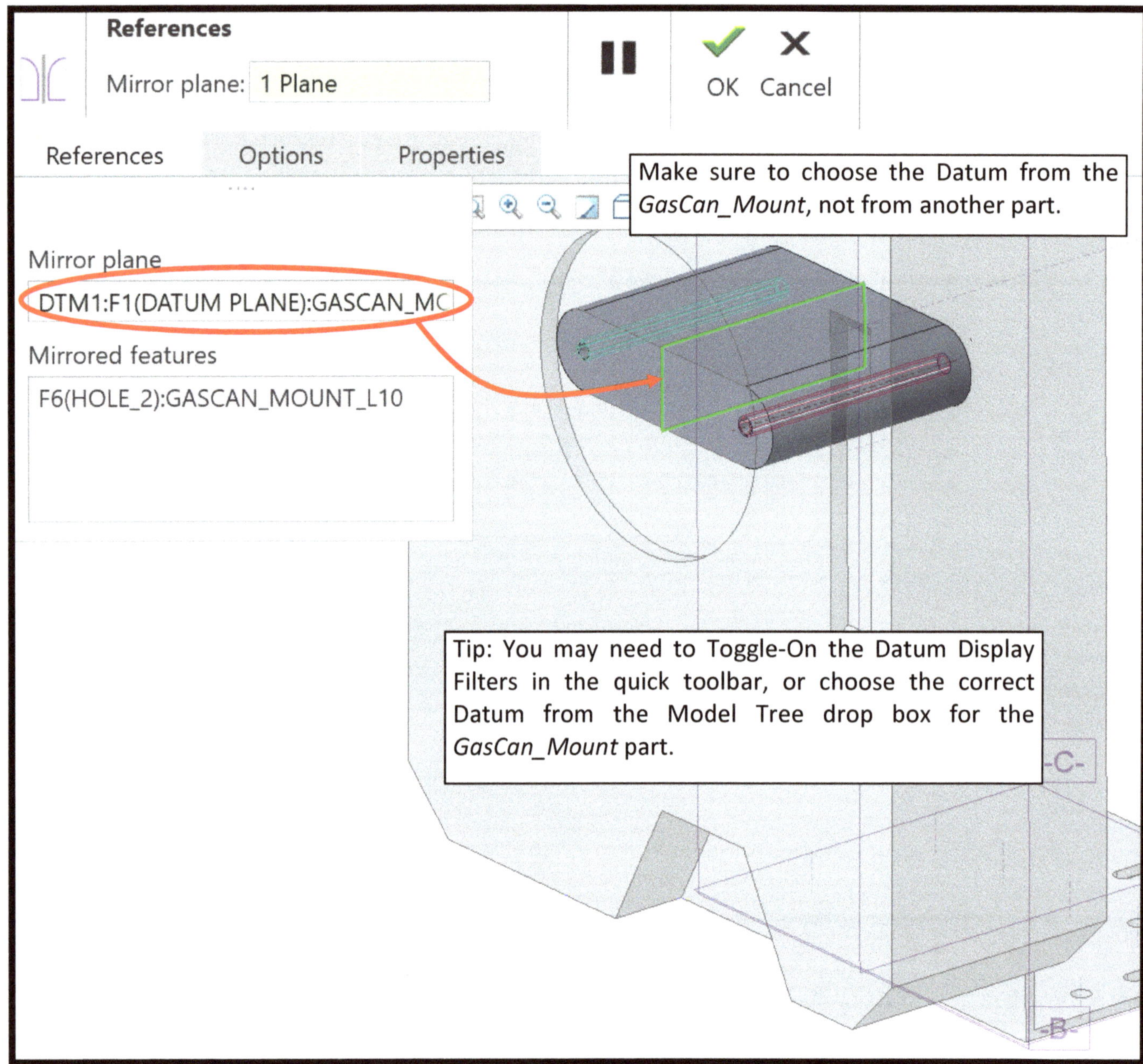

Note – if the Hole will not Mirror correctly you likely made an "**Assembly**" **Hole**, which is not a feature under the *GasCan_Mount* and will not Mirror correctly. To fix this: delete the Assembly Hole and then Activate the *GasCan_Mount* (RMB on it in the Model Tree – Activate) to be sure you are adding a Hole feature to the part, and not to the assembly.

Step 8 – Use the Hole Tool to create a hole through the center of the Mount, which will be used to thread in a handle later in the lesson. Use the center Axis of *the GasCan_L2* as the Coaxial Reference and the planar surface of the Mount as the second primary reference.

Tip: Toggle on the Axis Display Filters if you turned them off in the previous step.

Hole Type:	**Standard**
Profile Option:	**Straight, Tapped**
Shape Tab Options:	Set the option for "**Thru-Thread**"
Toolbar Settings:	Screw Size: UNC "**1 / 4 – 20**"
	Hole Depth = **Thru All**
Placement Type:	**Coaxial**

Primary References: Hold Control on the keyboard when selecting the two primary references
- center Axis of the *GasCan_L2*
- Planar surface of the extrusion of the *GasCan_Mount*.

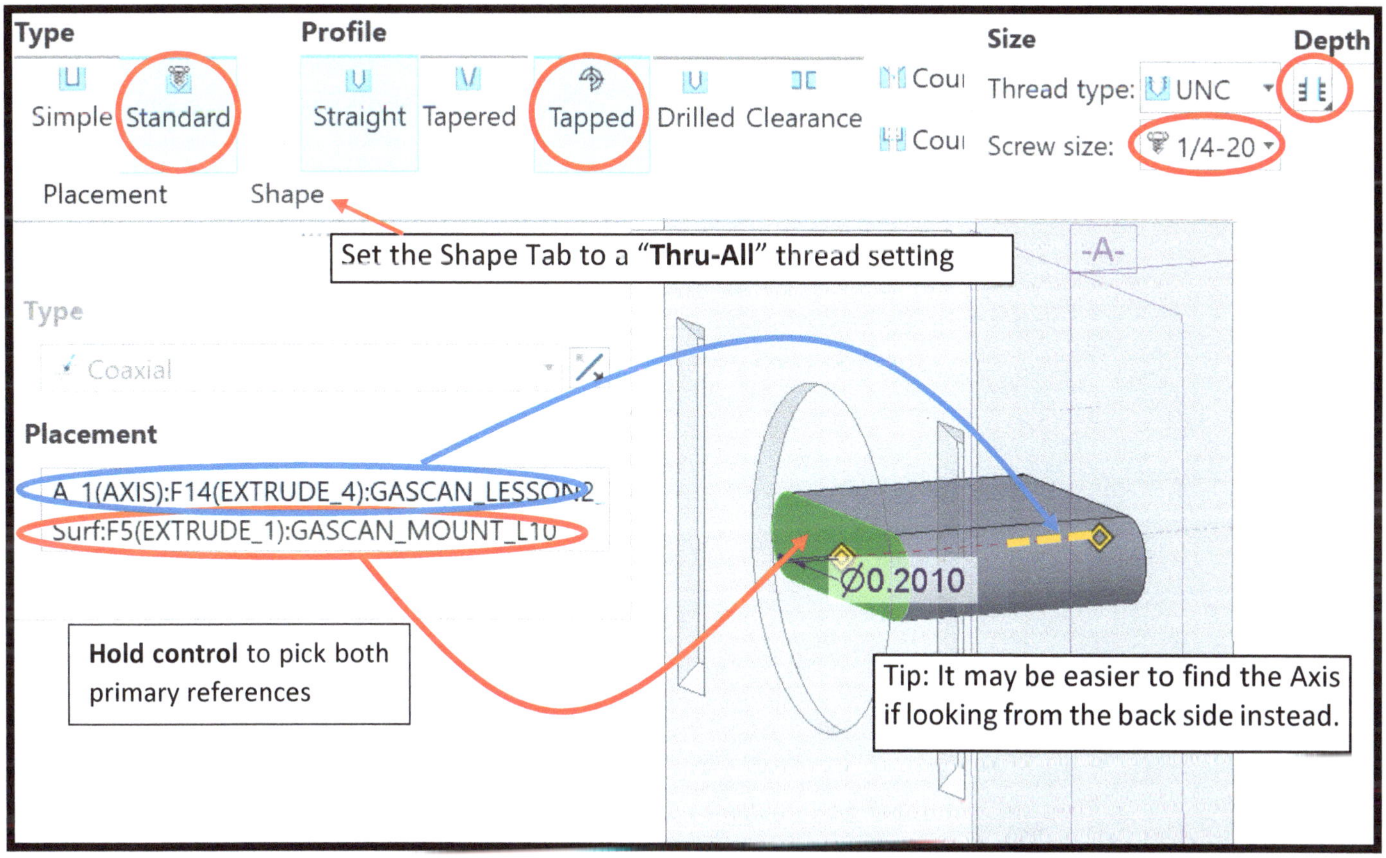

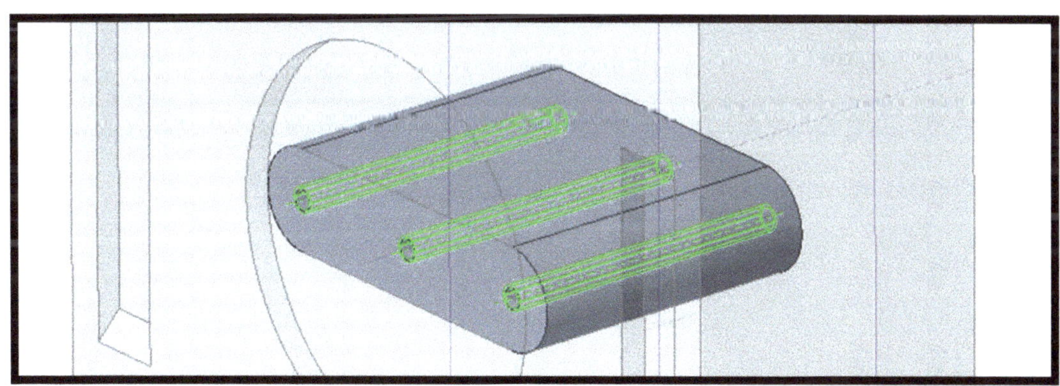

Step 8 - 2 – Checkmark to accept the Hole. You should now have 3 fully threaded holes.

Step 9 – Activate the Assembly to return to Assembly mode. This will end the "Part Mode" modeling of the new component. If you need to return to the *GasCan_Mount* part mode you can Activate it again to return to the part mode.

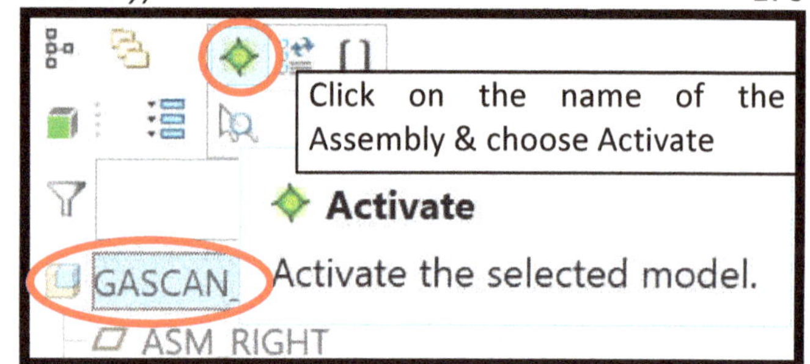

- **LMB** on the Assembly name in the Model Tree (*GASCAN_SYSTEM_L10*) – **choose "Activate".**

Step 10 – Assemble the next component in the Assembly, the ***Mount_Bolt_L10***. This part is provided in Blackboard and must be located in your Working Directory.

- **Assemble Tool** - select the ***Mount_Bolt_L10*** from your working directory – in the **Placement Tab** setup the constraints listed below by selecting references from the bolt & the matching reference from the Assembly.
- **Tip:** Toggle on the **Component Window** to see the component separate from the assembly, and turn on or off Display filters as needed to help you find the axis or datums easier. You may need try different view rotations or zoom levels to find the reference you want.

Step 10 – 1 - Constraint #1: (Axis to Axis)

Type: Coincident
Reference of Component: Select the center Axis of the Bolt
Reference of Assembly: Select the Axis of the left side Hole of the Bracket

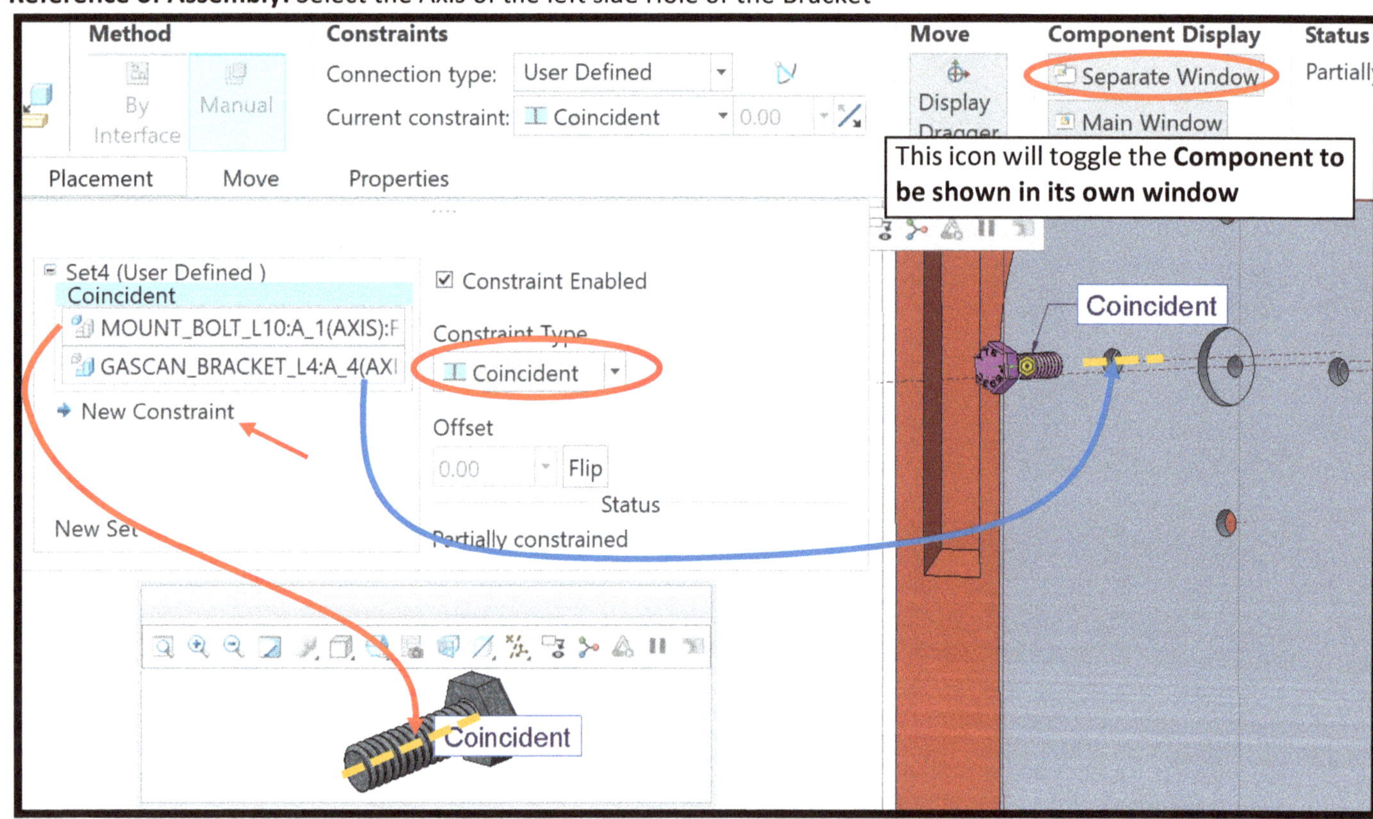

Tip: If the bolt is facing the wrong direction, click the Flip button under Constraint #1.

- **Then click "New Constraint"** and continue onto the next constraint setup.

Step 10 – 2 - Constraint #2: (Surface to Surface)

Type: Coincident
Reference of Component: Select the planar bottom surface of the bolt head
Reference of Assembly: Select the front planar surface of the Bracket

- You may need to **toggle the "Flip" option** so that the bolt is oriented through the Bracket into the Mount.
- By default, "Allow Assumptions" should be check marked and the component will be **Fully Constrained**. CREO will assume a default rotation of the bolt without needing a specific constraint setup, as the rotation is irrelevant to us.

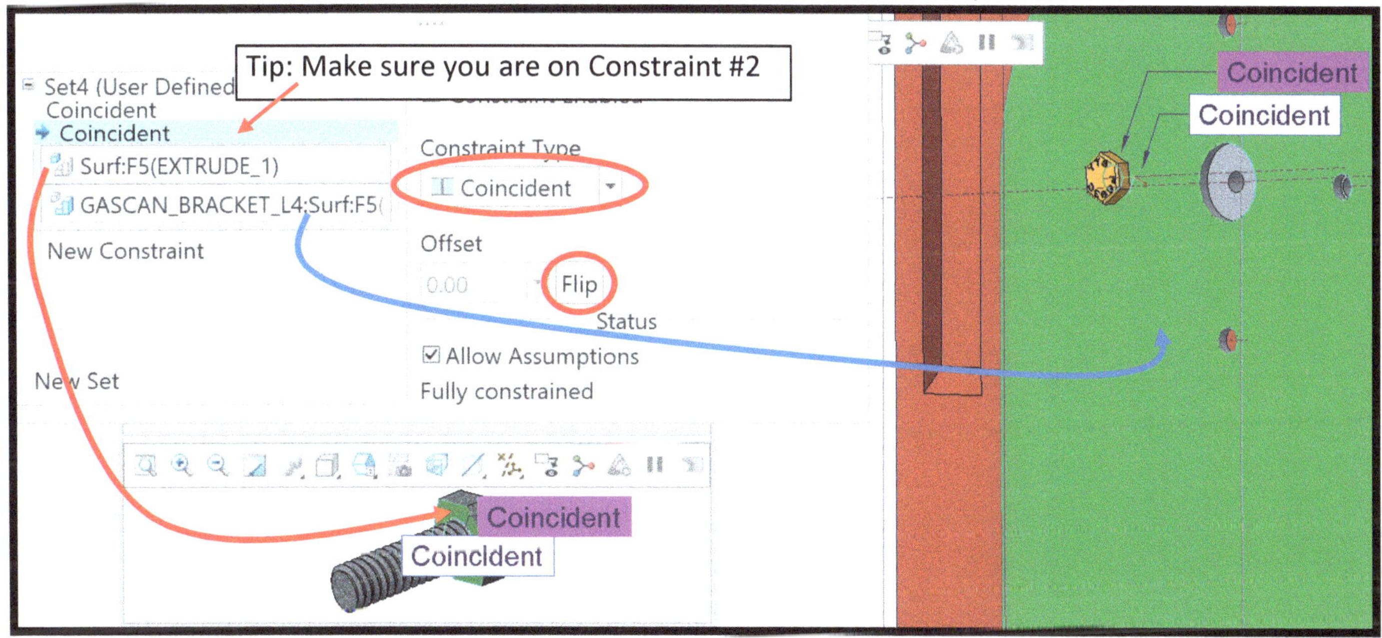

Step 10 – 3 - Check if your component is a gold color and says **Fully Constrained** in the Top Toolbar. If so, **Checkmark to accept** Bolt placement. If not, troubleshoot your constraints, remove any extra ones, or remake them as needed.

Step 11 – Use the Assemble Tool to place a **second *Mount_Bolt*** to the assembly. 

Follow the same steps and constraints used earlier but choose the Axis of the hole on the opposite side of the Bracket to place the 2nd bolt.

- **Checkmark to accept** the placement of the second bolt when it is Fully Constrained.

Note   It is also possible to create copies of components using the Pattern or Repeat tools, which is covered in Lesson 12. For now, you can just Assemble a second copy of the bolt.

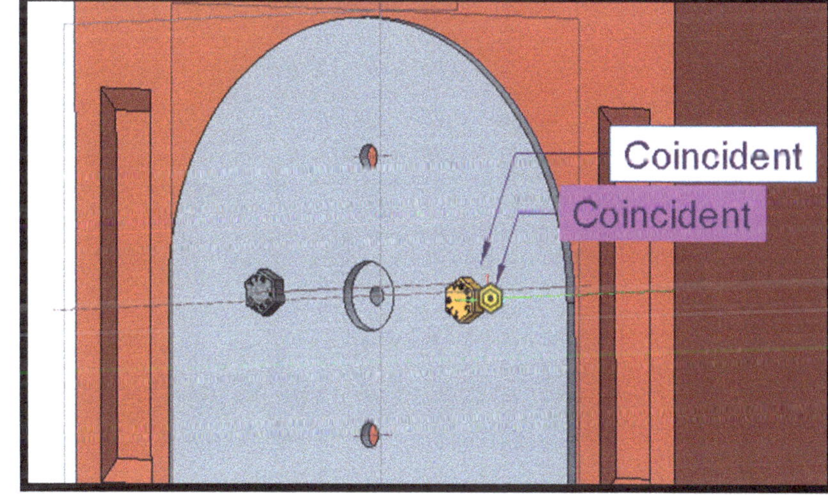

Step 12 – Design Update:

Modify the Hole placement diameter dimension on the *GasCan_Bracket_L4*. As the Hole position changes, the Bolt that is referenced to that hole should move to the new location after a regeneration process if the correct references were used.

- LMB to select the **GasCan_Bracket_L4** from the Model tree – choose **Activate** (*green star icon*) – expand the features list under the *GasCan_Bracket* in the Model tree – **Edit Definition of Hole 2** (expand the Pattern of the Hole to find the original Hole 2) – in the Placement tab **edit the Diameter reference to 2.50"**.

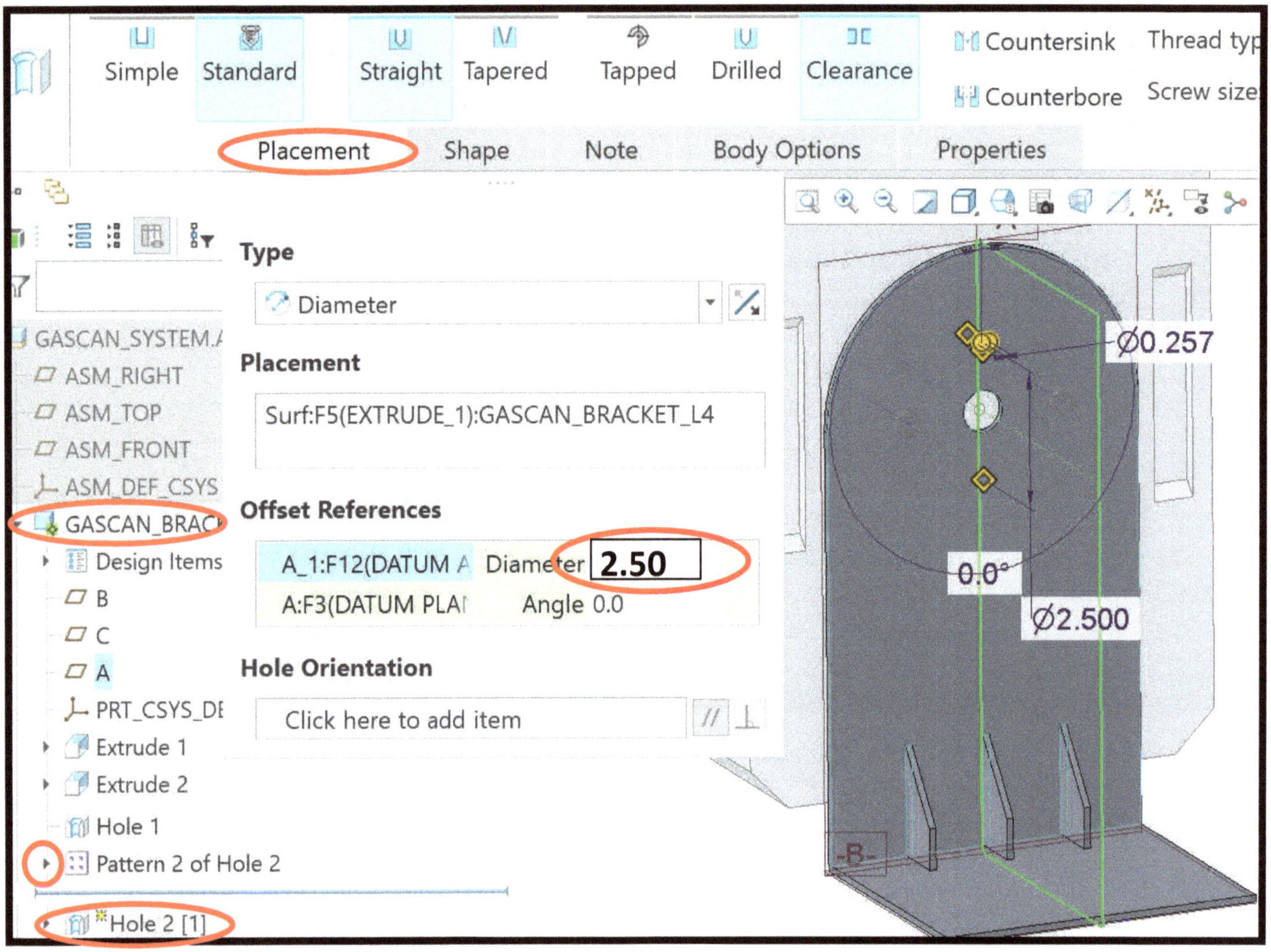

Step 12 - 1 – Checkmark to accept the changes to the Hole. You should notice the holes change position slightly.

Before changing the 3.00" diameter, and after changing to 2.50"

Step 13 – **Activate the Assembly** to return to Assembly mode**.**

- **LMB** on the **Assembly name** in the Model Tree (*GASCAN_SYSTEM_L10.ASM*) – in the pop-up menu **choose Activate** (green star icon).

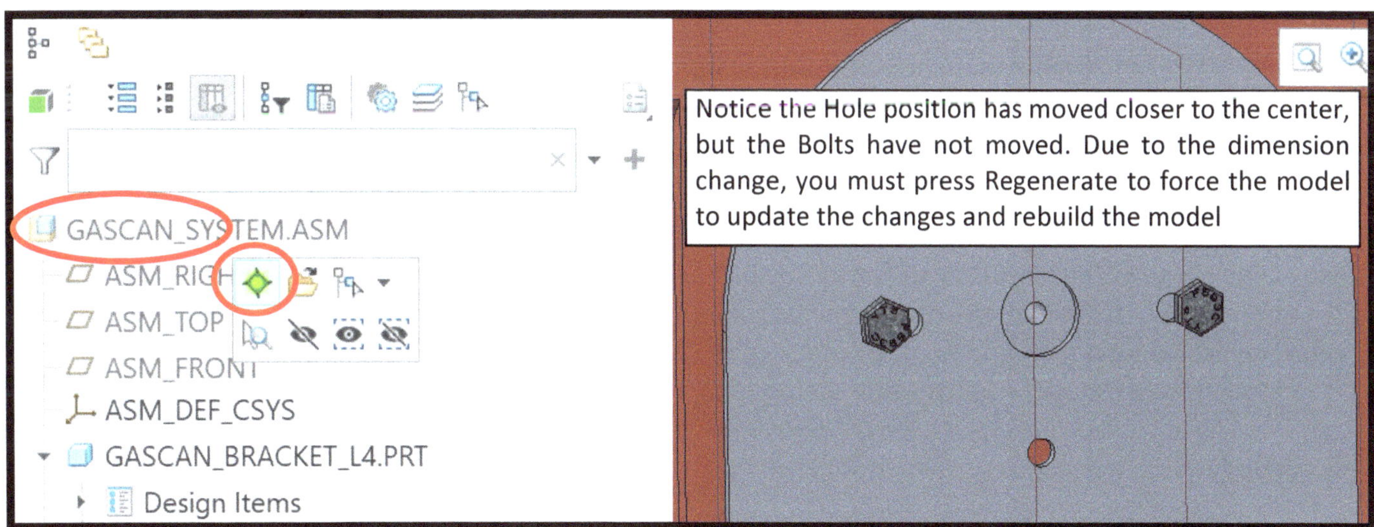

Step 14 – **Regenerate the Assembly**: Press **Control + G** or press the **Regenerate icon** (very top toolbar) to force the assembly to update with the recent changes. The Bolts should now relocate to the new location of the holes.

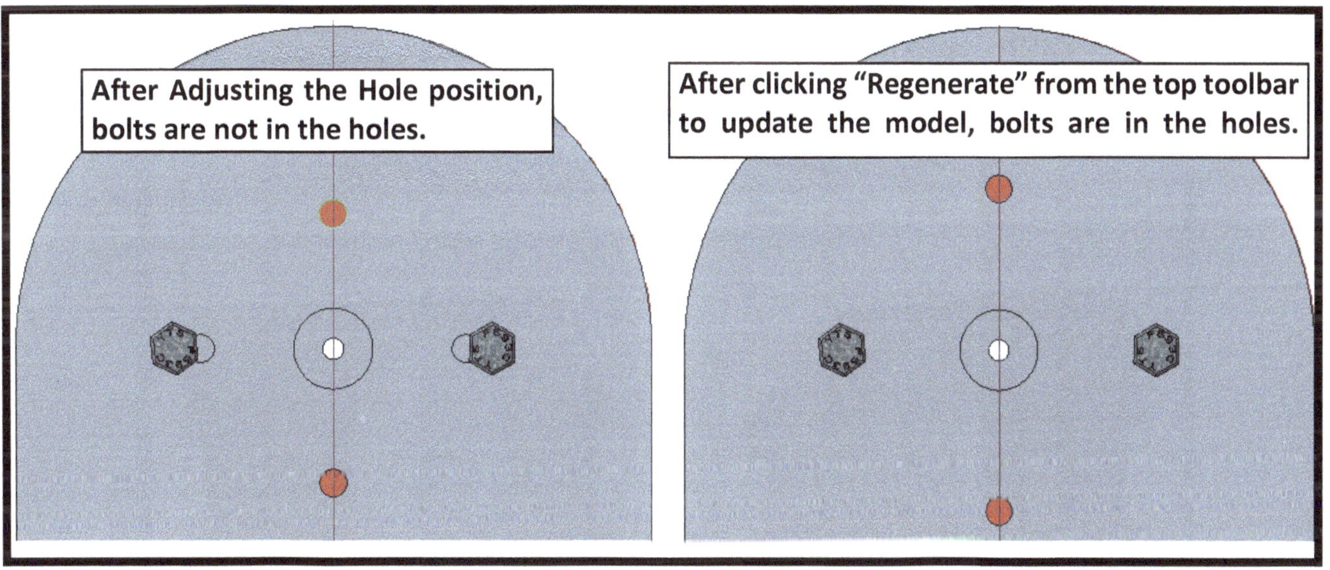

Tip: Often you may have to use Regenerate **before CREO will allow you to Save your work** after making changes to part features or constraints.

Flexing the Model (*Optional*)

Choosing **proper references for component placement** can create a more robust assembly that can handle some design changes, but there are limits. If the Holes are deleted, the Bolt will lose its reference and Fail (will be Red in the model tree). If the Gas Can rotates 90°, the holes in the Mount will likely not move to the correct position as the hole placement was referenced to the specific 9 O'clock and 3 O'clock Bracket holes. Try the following to **"Flex the Model"** and learn about how references affect the model (we will undo these changes at the end of this step).

- **Save your Assembly**.
- **Edit Definition** of the Gas Can – adjust the **Angle Offset constraint to 90°** to tip the can over - **press OK**.
- **Examine your model** to see if the Bolts and the Mount Holes have adjusted properly. You may notice the Mount is rotated properly, the holes look correct in the Mount, and the Bolts may have adjusted to their new position (results may vary depending on your chosen constraints). But …… the Model has not been regenerated yet so it may not be accurate!
- **Press Regenerate (Control + G)** to update the Model. Now you will notice the mount holes are in empty space, and the bolts have doubled up over the same hole (results may vary depending on your references).
- **Click <u>Undo</u>** in the top toolbar to undo the Angle Offset constraint and **return the model back to the original setup.**

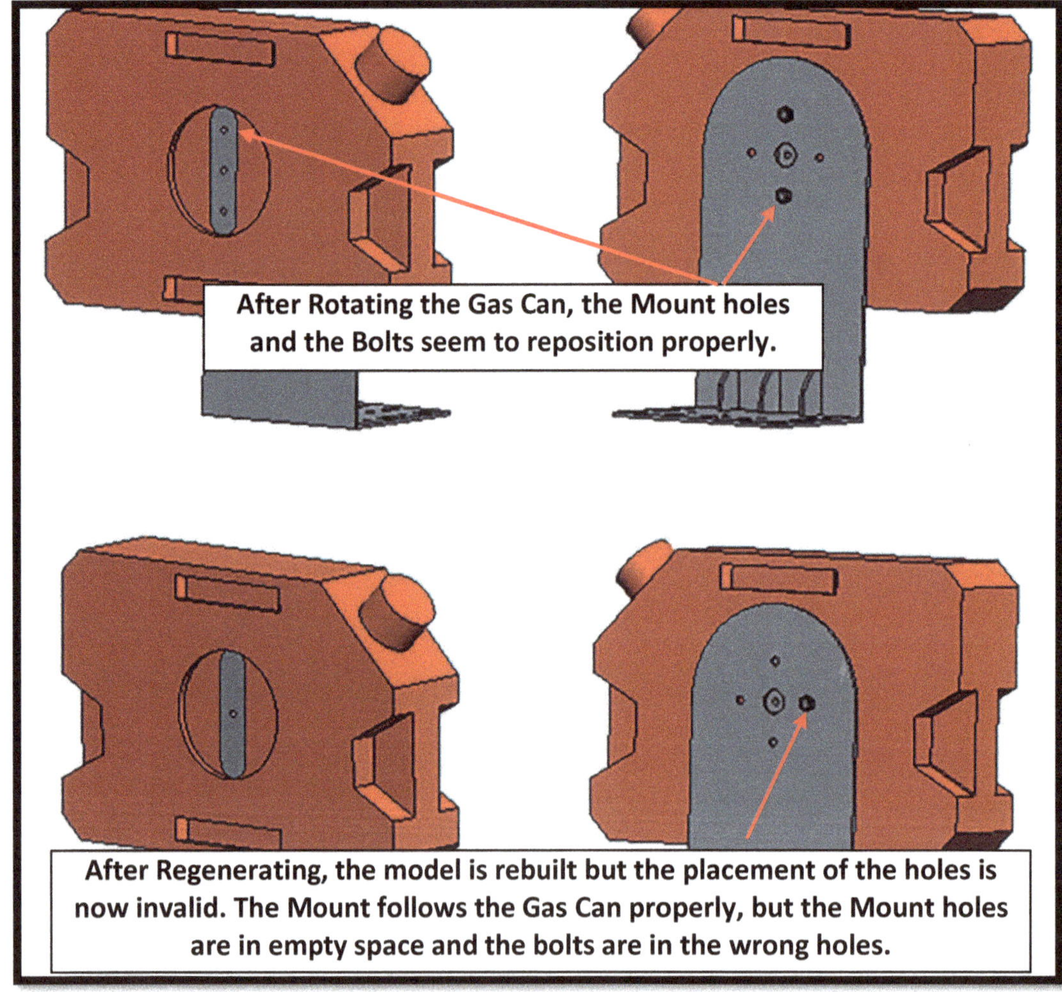

If you did want a more robust model, it would have been better to model the Bracket holes referencing the holes in the Mount, instead of vice versa. This is something that you will learn as you gain more experience on your own designs.

Step 15 – Use the **Assemble Tool** to add the ***Mount_Handle_L10*** to the assembly. This part is provided in Blackboard and must be located in your Working Directory.

Assemble

Step 15 – 1 - Constraint #1: (Axis to Axis)

Type: Coincident
Reference of Component: Center Axis of the Handle
Reference of Assembly: Center Hole Axis of the *GasCan_Mount_L2*

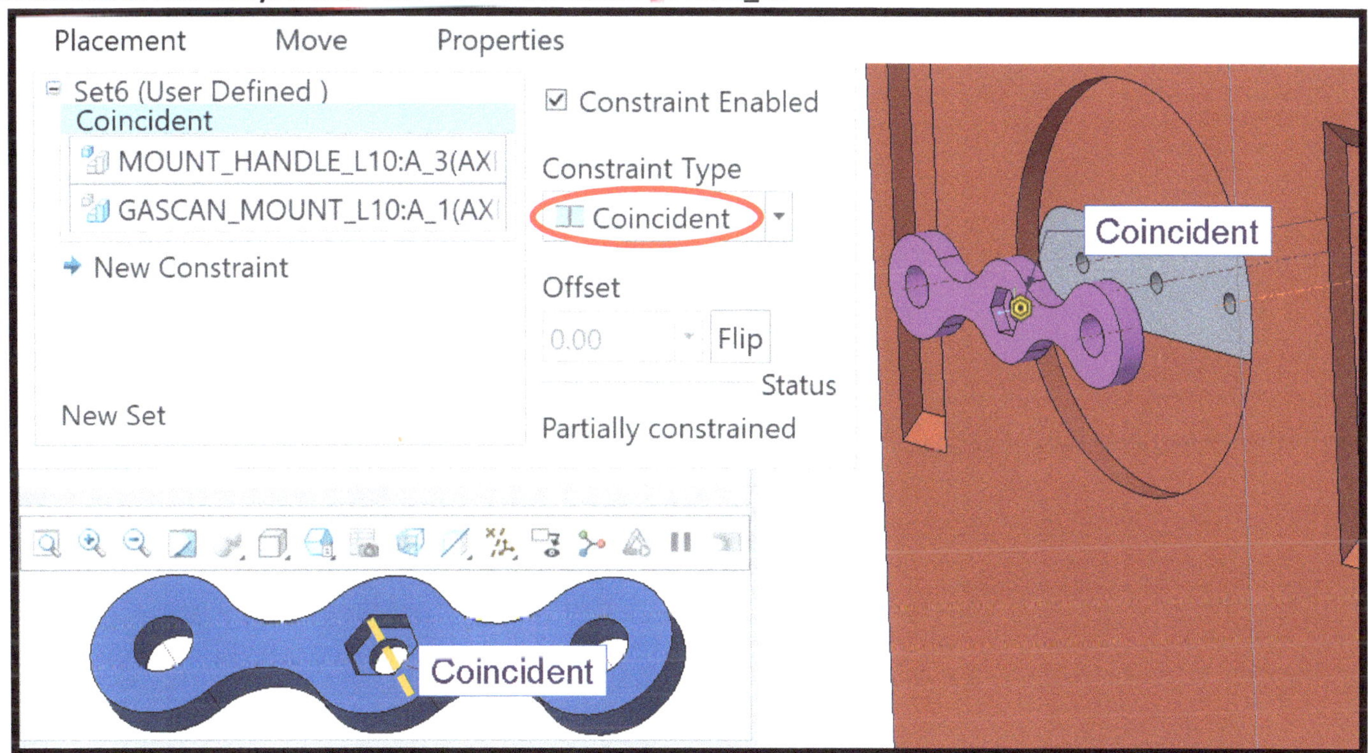

Step 15 - 2 – Constraint #2: (Surface to Surface)

Type: Coincident (toggle **Flip** if needed)
Reference of Component: Back Planar Surface of Handle (without the Hex cutout)
Reference of Assembly: Planar Surface of the *GasCan_Mount*

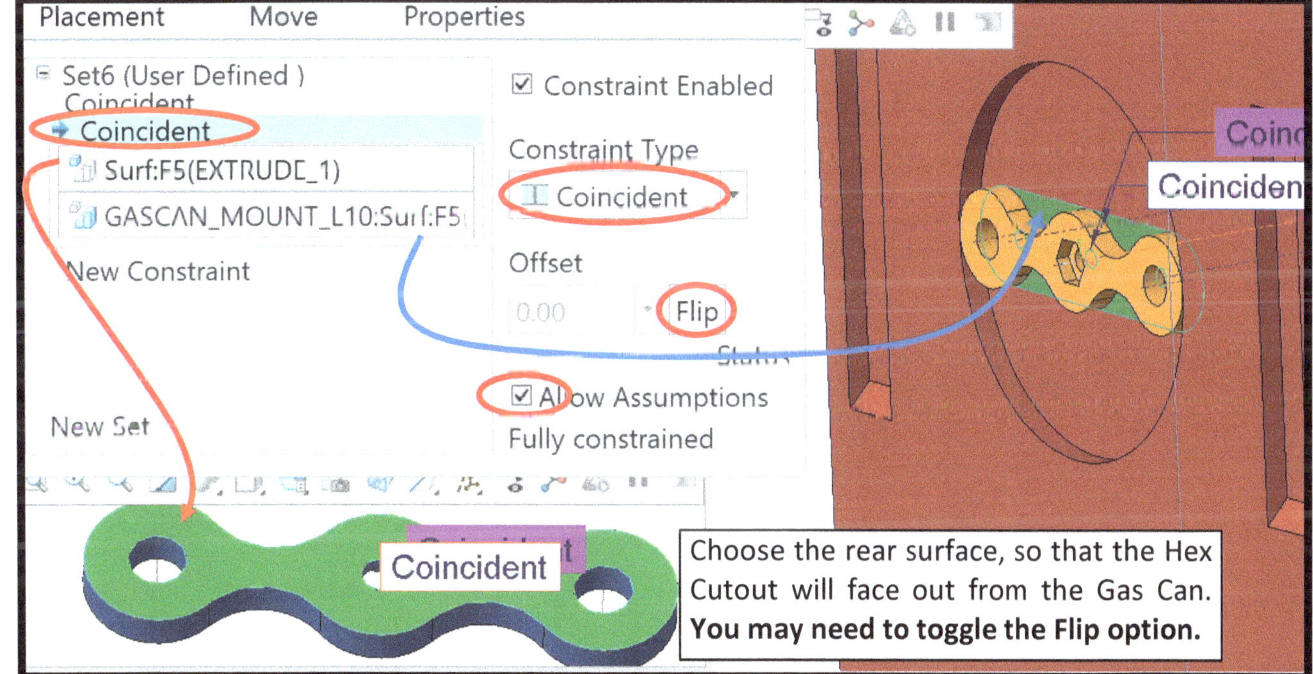

Step 15 – 3 - Constraint #3: (Datum to Datum)

- **Un-check the "Allow Assumption"** option in the Placement menu, and choose **"New Constraint".**

Type: Angle Offset with either **90° or 0°** (so that the handle is perpendicular to the mount)

Reference of Component: Datum of the Handle that sections through the Center Hole vertically.

Reference of Assembly: Vertical Datum of the *GasCan_Mount*

- Tip: Hover the mouse over a datum and the name will appear – choose the vertical datum of the **Mount part (not of the assembly or other part's datum)**

- Tip: **Tap RMB while keeping your mouse still** to cycle different selection options for where the mouse is currently located.

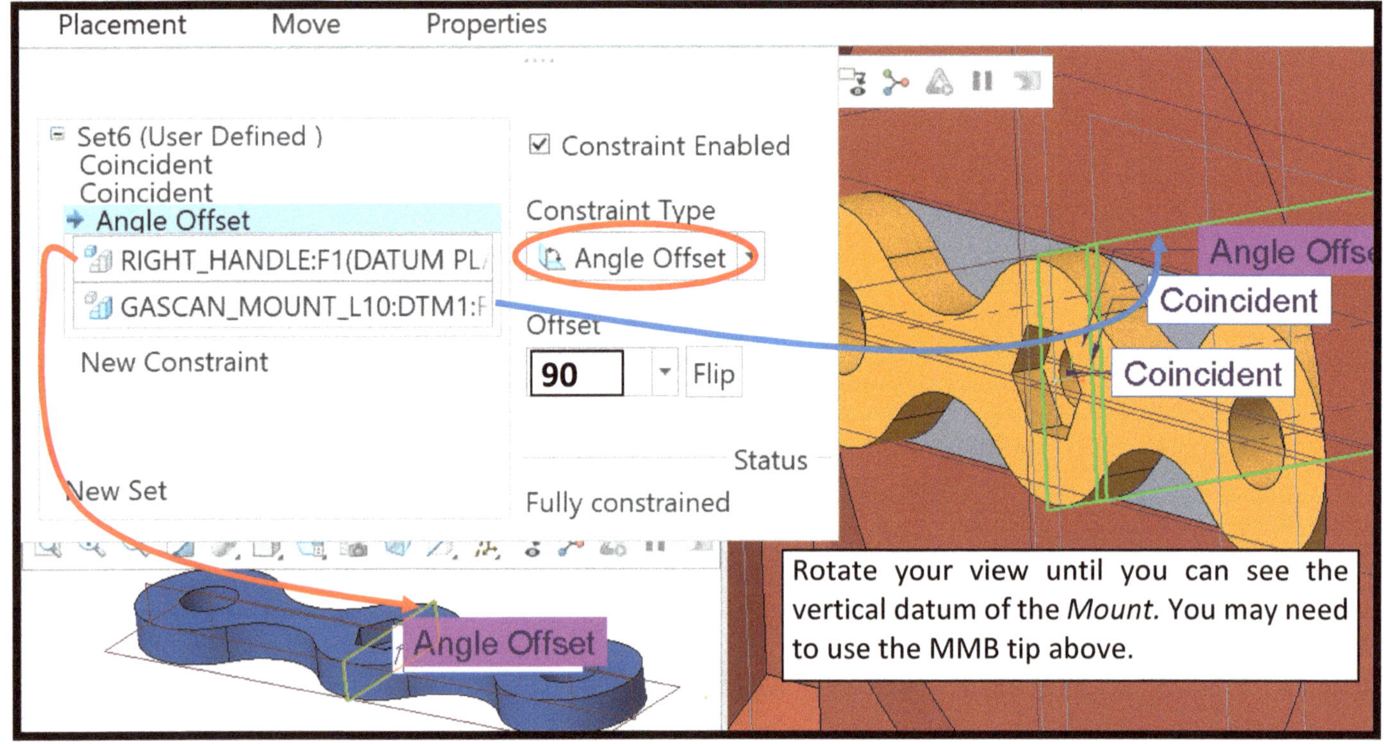

Step 15 – 4 - Checkmark to Accept the Component Placement.

The Handle should be vertical, which will prevent the gas can from sliding off the mount. If you need to change the constraints use **Edit Definition** to return to the constraint tool bar for the desired component.

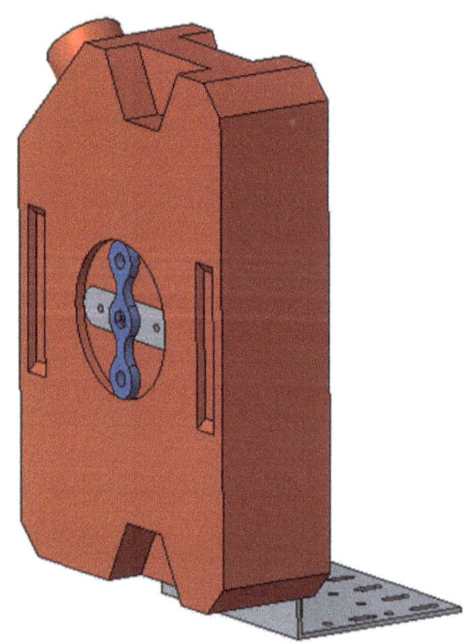

Continue on to the next page to add the final bolt to the connect the Handle to the Mount.

Step 16 – Use the Assemble Tool to add another instance of the ***Mount_Bolt_L10*** to the assembly, this time going through the Handle and the center hole of the Mount.

Assemble

Constraint #1: (Axis to Axis)

Type: Coincident

Reference of Component: Center Axis of the *Mount_Bolt_L10*

Reference of Assembly: Center Hole Axis of the *GasCan_Mount_L10*

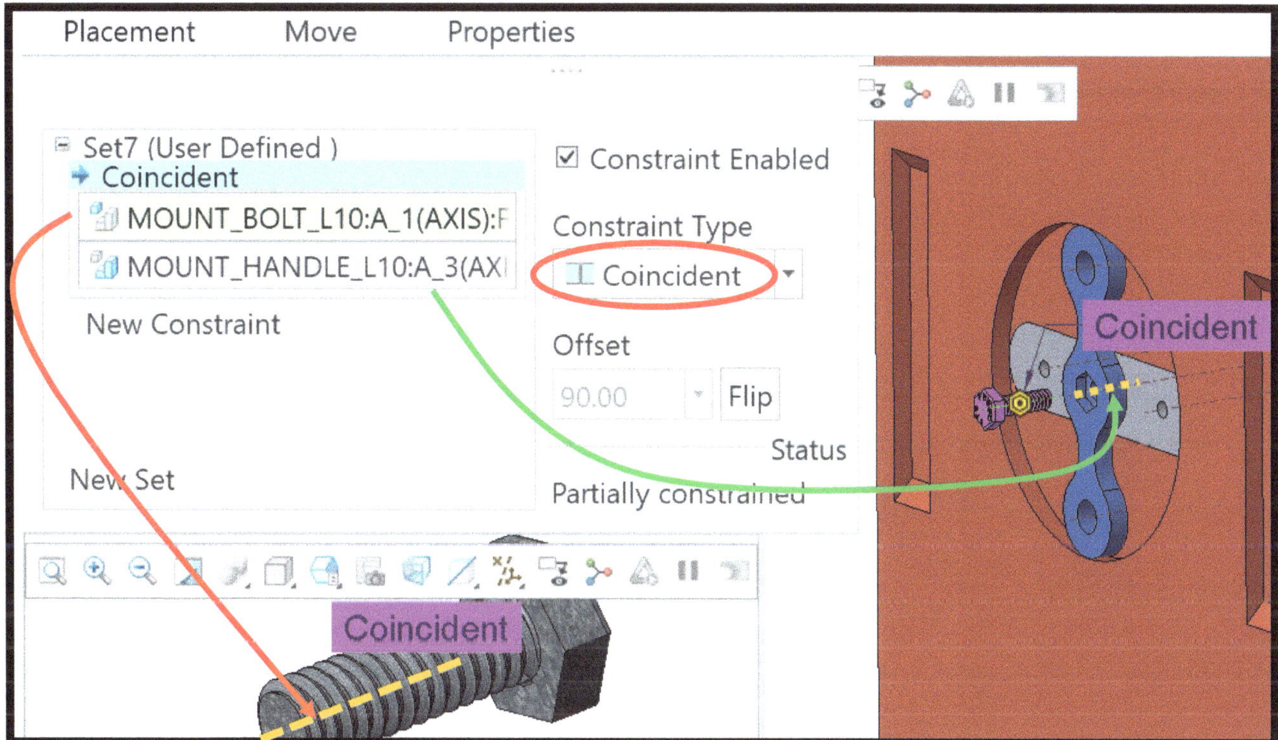

Constraint #2: (Surface to Surface)

Type: Coincident

Reference of Component: Bottom planar surface of the Bolt Head

Reference of Assembly: Planar surface at the base of the Hex Cutout on the Handle

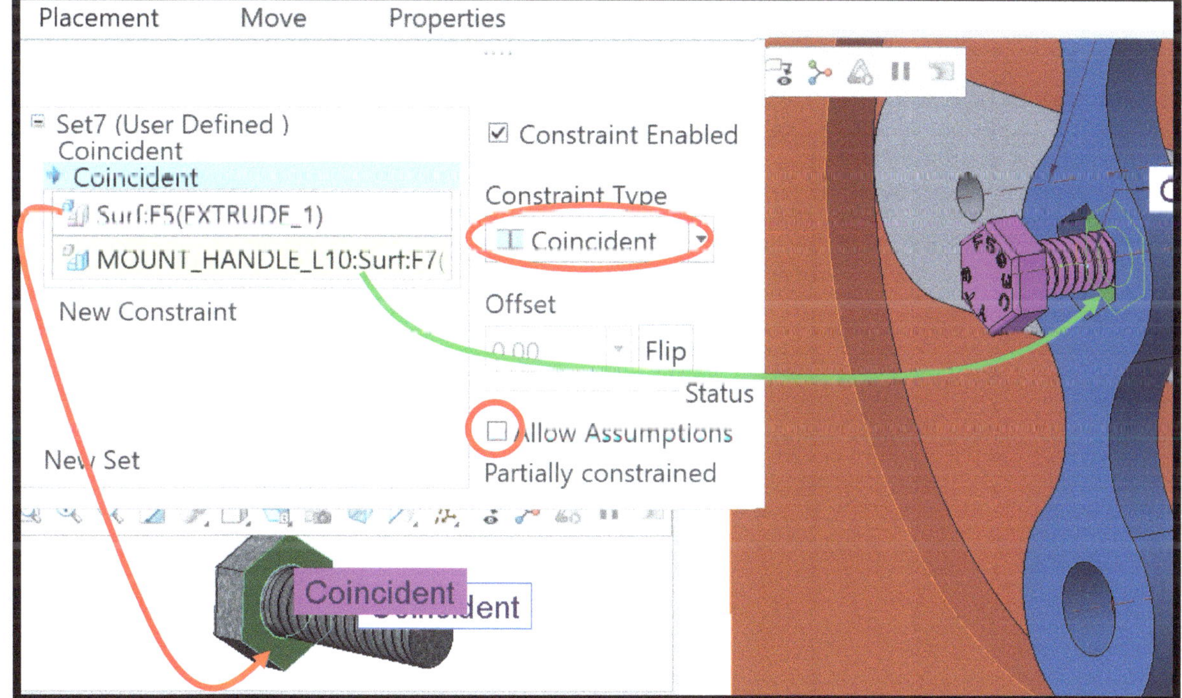

Constraint #3: Depending on your Model you may need to **Uncheck Allow Assumptions** and create a 3rd constraint. If the Bolt Hex does not fit properly in the Hex cutout add a 3rd constraint of an **Angle Offset between a vertical datum on the Bolt and a corresponding Datum on the Handle** with a degree offset of either **0° or 90°** as appropriate.

Tip: If can be difficult to find the desired Datum on the screen. Make sure Datum Display is on, then Zoom in and rotate the view Also try tapping RMB to cycle which entity under your mouse is highlighted, or temporarily change menu at the very bottom right corner of the screen from "All" to "Datum Plane" (but change it back to "All" when complete)

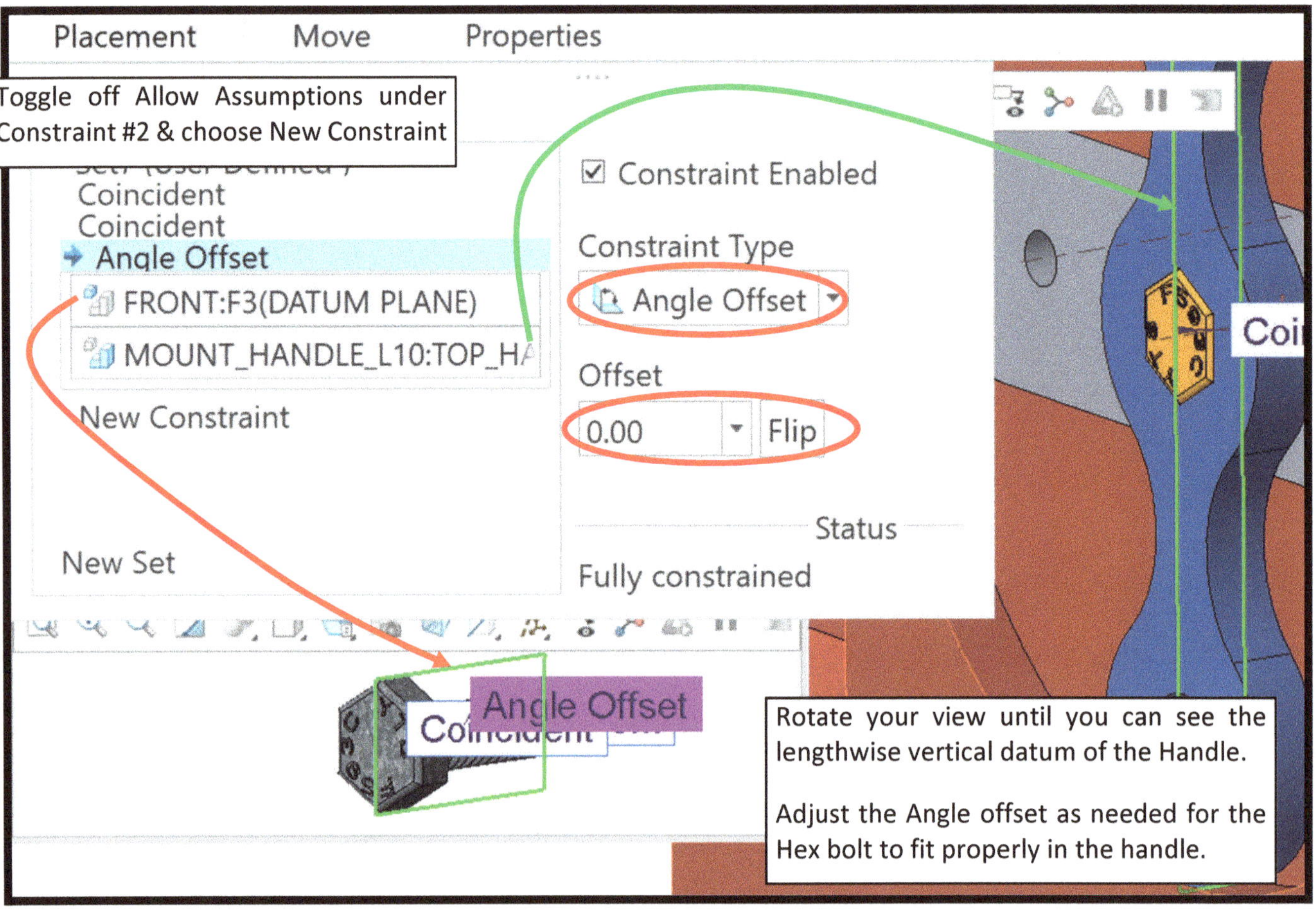

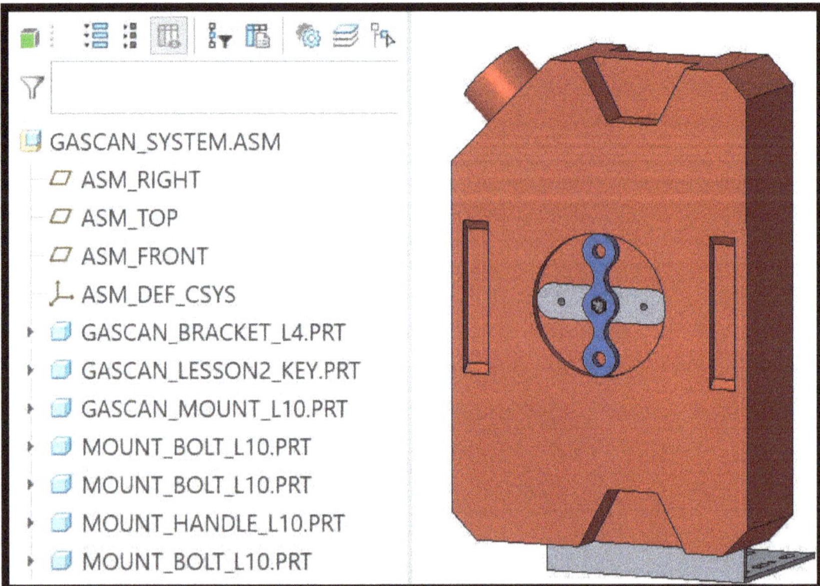

Step 17 – Checkmark to Accept the Bolt placement.

The model should look similar to the image shown (*Datums are hidden for clarity in the image shown at right*)

Step 18 – Open the *GasCan_Mount* so you can set the material properties for Aluminum. If you don't do this, the material will not populate the material name on the drawing Bill of Materials table.

- **Select it from the Model Tree – RMB – Open – File- Prepare- Model Properties-** in the Material option select **change** – double click on **Legacy Materials** - <u>double click</u> on "**Al6061.mtl**" to set it as the material – press **OK**

Step 19 – Save the *GasCan_Mount* part.

Step 20 – Return to the Assembly**, press Regenerate (Control + G) & press Save** to save your Assembly file.

It would be a good idea to check if **all the component files** (prt, .asm, etc.) used in this lesson are saved in your working directory and in the <u>same exact folder</u>. If they are not, you will have errors the next time you open the assembly as it will not be able to find the components and will give you red failed items in the Model Tree. Use Windows File Explorer to check that they are all located in the same folder. If needed, you can move them after you have closed down CREO (do not move them while CREO is open).

Step 21 – Start a new Drawing of the Assembly, name it "*L10_GasCan_System*". To save on time, a drawing of the newly created *GasCan_Mount* will not be required or expected for this lesson.

Step 22 – You will likely need to adjust the Parent View (Front View) by "Angles" to match the Key (with a Vertical rotation reference and 90 degrees rotation value)

Step 23 – Continue to setup the drawing for **Scale, UND Template, General View, & View Display Settings**.

Step 24 – Hide/Erase the Datums from the Bracket. These datums were "Set" during Lesson 4 so cannot be toggled off with the Display Filters like the other non-geometric tolerance datums can.

- **Annotate Tab** – in the Drawing tree on the left side of the screen **expand the "Datums" Item** for each View Template – hold control and select the datums to highlight them – choose the **Erase icon** in the pop-up menu. **Repeat for all 3 views**.

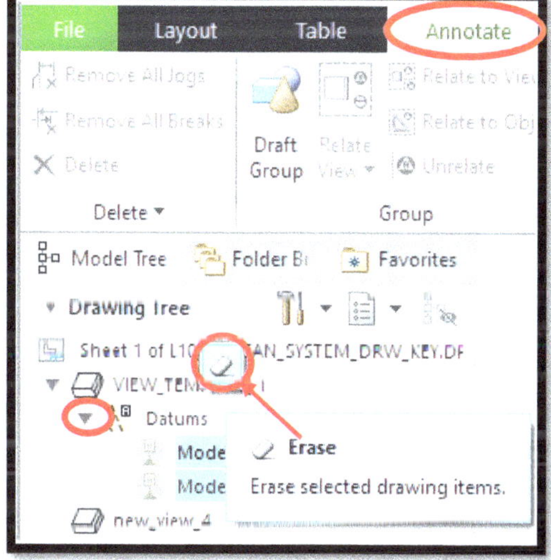

Step 25 – Follow Step #3 of the *Assembly Drawing Steps* (Appendix) to setup the BOM Table and Balloon notes and make a drawing to match the key. You will not complete the other *Assembly Drawing Steps* for this lesson, those will be covered in the next lesson.

Step 26 – Complete your drawing to match the key.

End of Lesson 10

Lesson 11 – Assemblies 2 (Ram Mount System)

Lesson 11 is to assemble the Ram Mount components from Lesson 3 as well as learn how to setup Assembly Cross-Sections and Exploded Views on the assembly drawing.

Assembly Dimensions: Notice that the dimensions on the Assembly drawing are used to detail the connections and placement of components, and **not to detail the <u>part features</u>** (e.g. hole sizes, part thickness, etc.) which can be seen on part drawings.

Dimensions that detail the placement of components on the assembly should be included, such as the distance a component is placed along a shaft, but components that are obviously mated or aligned but bolted holes do not need to have those dimensions in place.

Clearance dimensions (gap between a wheel and frame on a robot), Overall assembly size (Length, width, height of the entire assembly), and other reference type dimensions concerning the operation or movement of components may be added.

Assembly Cross-Sections: A Cross-Section can be made through all components on an Assembly, but it should be placed using an Assembly Datum (e.g. ADTM1), otherwise the section may be affected if a part is rotated or moved. Individual components should have different Hatch spacing and Angles to differentiate parts.

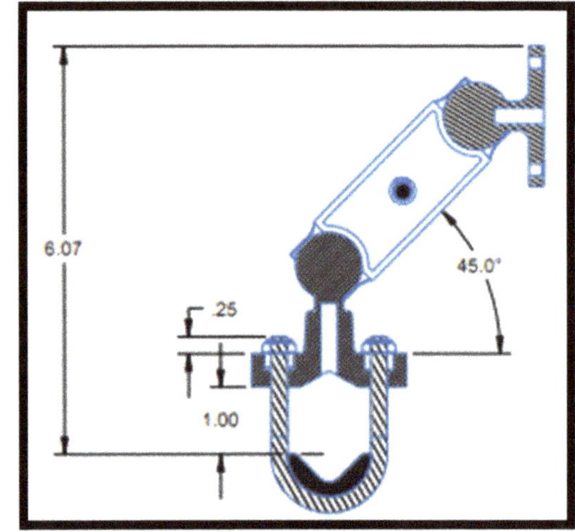

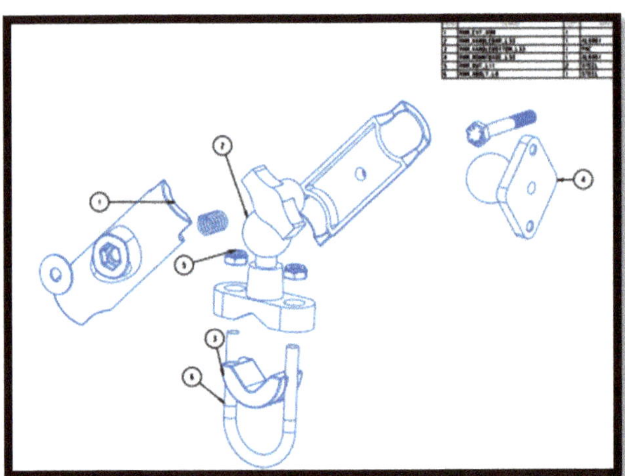

Exploded View States: Exploded Views are often used on Assemblies to show the individual components that make up an assembly. An Exploded View State can be created that allows for the Exploded state to be toggled on or off as needed and shown on the drawing.

Note – typically you do not explode every item in a large assembly. Instead you would only explode "main level" items from the model tree such that sub-assemblies and their parts would be kept together and exploded as a single unit on later assembly drawings.

Layers: The **Layer Tree** can be a very important tool for advanced users of CREO to affect groups of items all at once. Various layers are automatically created by CREO, such as a **Part Datums Layer**, that can be hidden to hide all part datums that make up the assembly all at once, useful for drawings that may have many geometric tolerance datums. User Defined Layers can be created to include all Holes or other groups of features that you want to adjust quickly. You can access the layer tree using the Show Items tool in the Model tree area, and opening the Layer Tree option. You can then switch back to the Model Tree using the same operation.

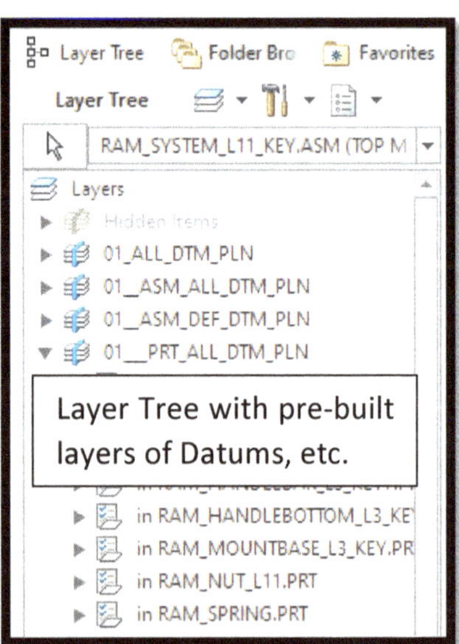

Layer Tree with pre-built layers of Datums, etc.

Assembly #1: Ram Mount System

Getting Started: Open **CREO** *Parametric* and follow the lesson steps carefully.

Step 1 – Set your **Working Directory**, and if needed download, extract, and place the components for this lesson in your directory.

Note: Make sure that **all CREO files** for this assembly are in your working directory. If you have to browse to find these components then the Assembly will not find them the next time you re-open the assembly and you will have red colored failed features in the model tree.

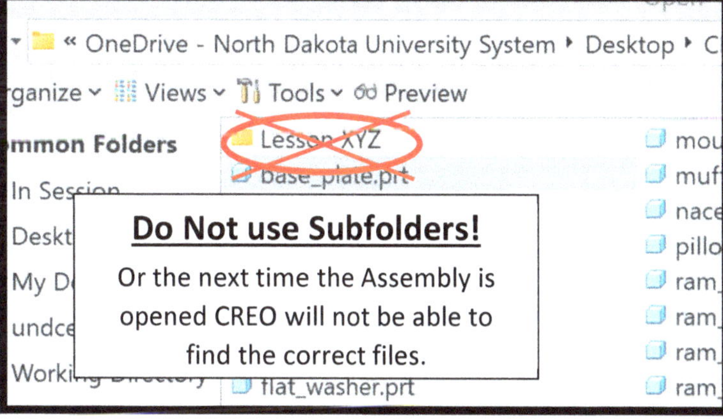

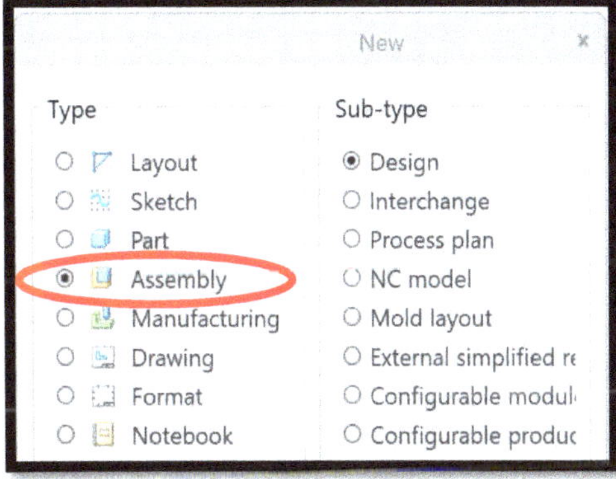

Step 2 – Create a new **Assembly file type** called *"Ram_System_L11"*.

- File – New – select **Assembly** as the file type – enter in the desired filename - OK.

Step 3 – 1 - Add the main component to the assembly using the **Assemble Tool** in the top Toolbar. For this lesson, **Assemble** the *Ram_Handlebar_L3* part first.

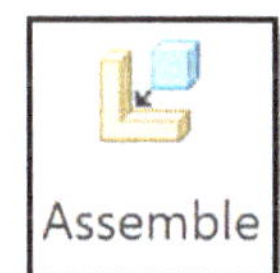

Step 3 – 2 - Set the "Current Constraint" type as "Default". Tip: **Always use "Default"** for the first component instead of leaving it in place unconstrained otherwise it may later rotate or move in undesired ways.

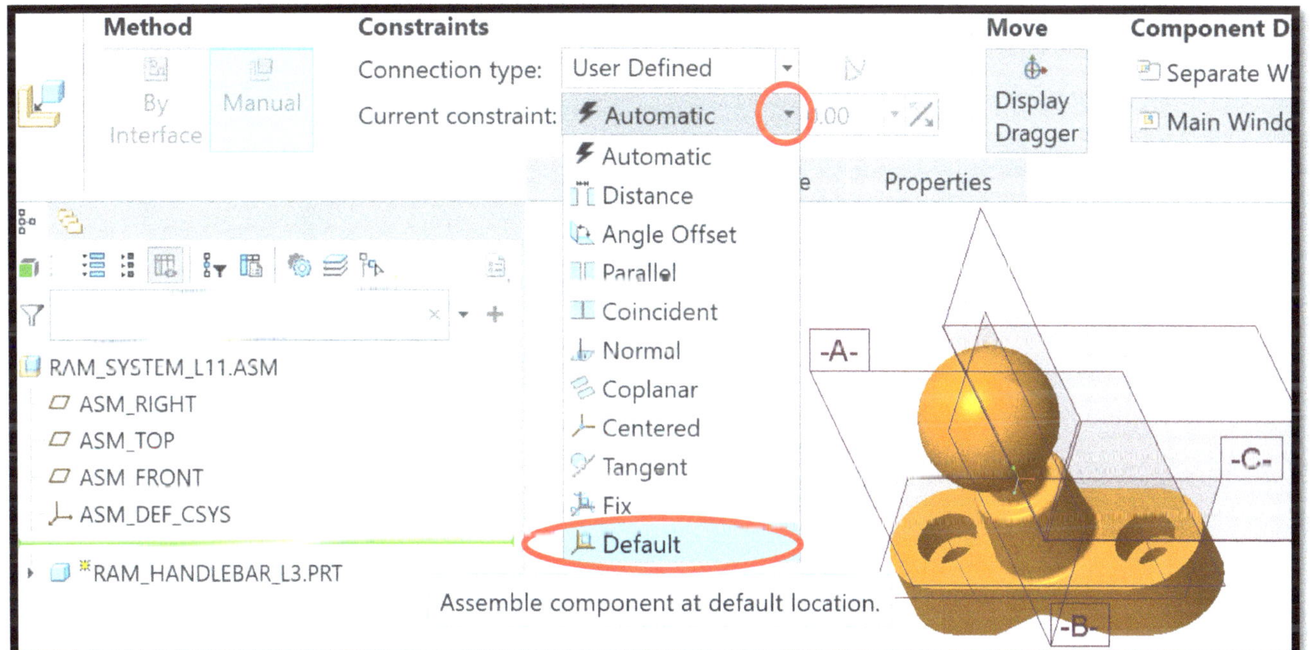

Step 3 – 3 - Once the model is constrained (*gold color*) **press the checkmark** to accept the component placement.

Step 4 – Assemble the ***Ram_Ubolt_L8*** to the assembly using the Assemble tool in the top toolbar.

Constraint #1: (Axis to Axis)

Type: Coincident
Reference of Component: Axis of the left "U" leg of the U-bolt
Reference of Assembly: Axis of the left Hole on Ram Part

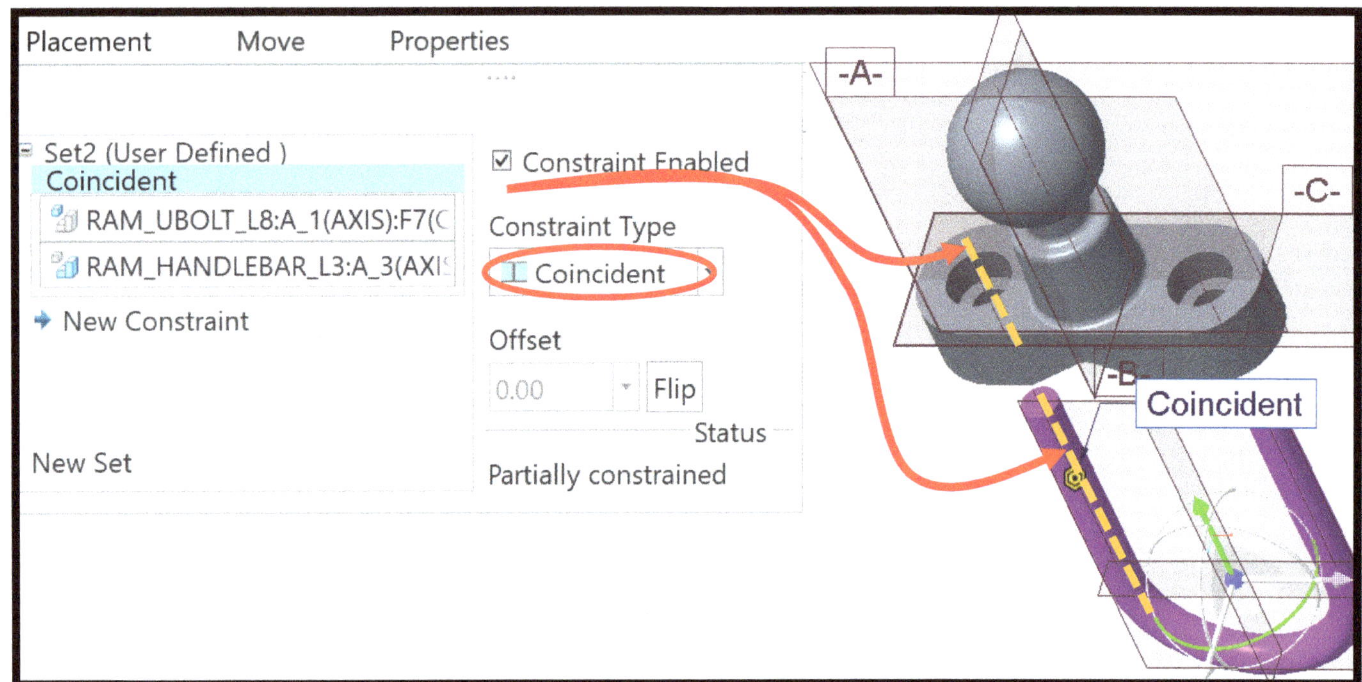

Although the U-bolt may appear to be aligned with both holes you still need to provide a constraint to ensure the U-bolt cannot twist out of alignment. **Click on New Constraint** to begin Constraint #2.

Constraint #2: (Datum to Datum)

Type: Coincident
Reference of Component: Front Datum of U-Bolt
Reference of Assembly: Front Datum (A) of Ram Part

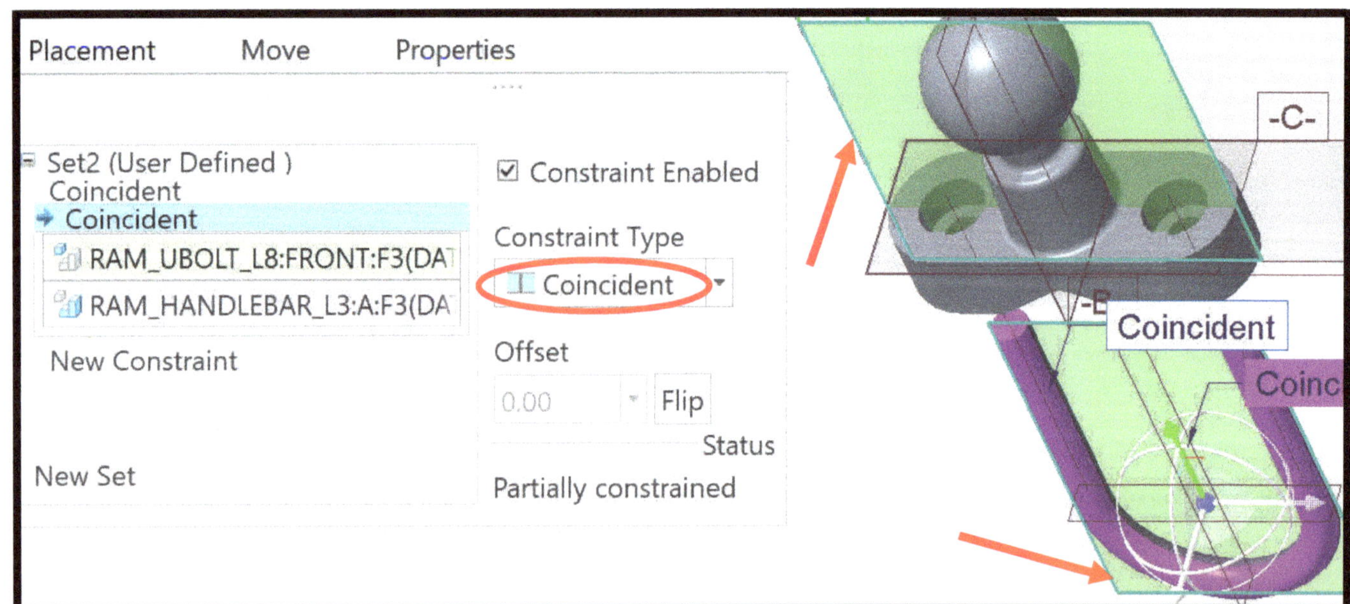

Now set a "Distance" constraint to control how far the U-bolt is inserted through the hole.

Constraint #3: (Surface to Surface)

Type: Distance (offset value of **0.25" above the surface**)
Reference of Component: Planar Surface of U-Bolt End
Reference of Assembly: Top Planar Surface of the Ram_Handlebar Part

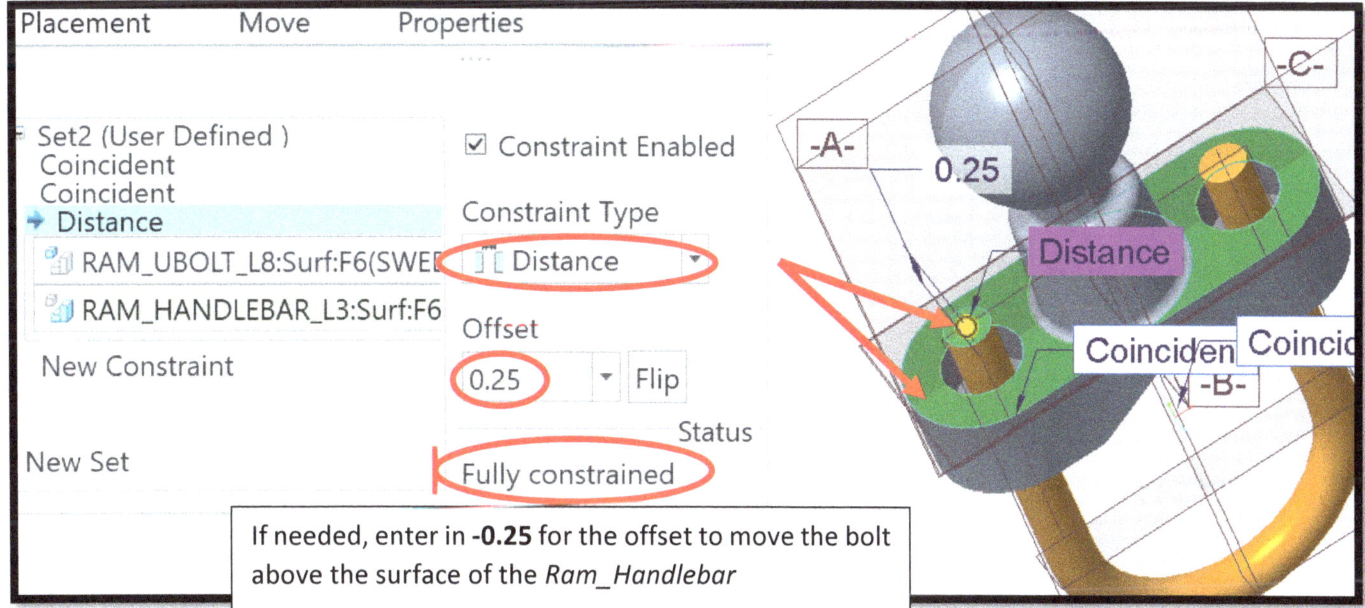

Step 4 – 2 - Accept the Placement of the U-Bolt. Tip: If you need to adjust a constraint use Edit Definition to bring back the placement toolbar.

Step 5 – Assemble the *Ram_Nut_L11* onto the left leg of the U-bolt using the constraints shown below. Detailed images of each constraint setup are not provided for this step, refer back to previous steps if needed to review.

Constraint #1: Axis to Axis (Coincident)

Constraint #2: Bottom surface of Nut to top surface of the Ram Part (Coincident)

Make sure "Allow Assumptions" is checked so that the part is Fully Constrained. This reduces the need to specify the rotation of the nut as it is irrelevant for this part.

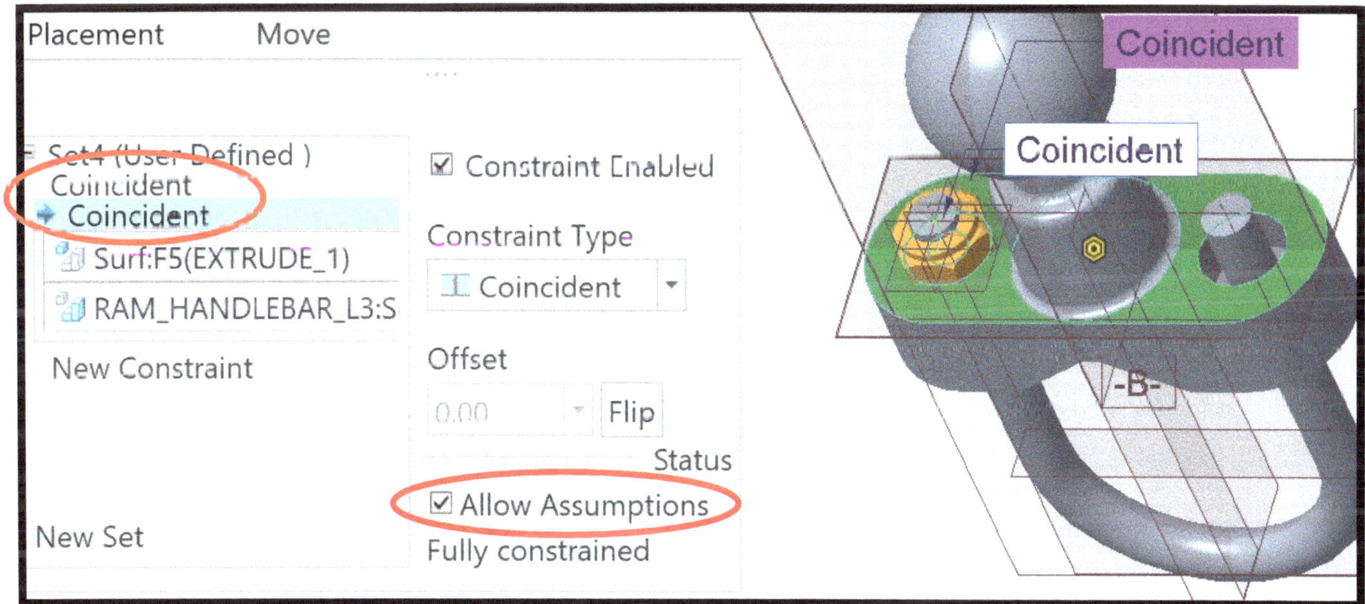

Step 5 – 1 - Accept the placement of the *Ram_Nut*.

Step 6 – Add an additional *Ram_Nut_L11* component to the opposite side of the U-bolt using the Assemble Component tool.

Assemble

Constraint #1: Axis to Axis (Coincident)

Constraint #2: Bottom surface of Nut to top surface of the Ram Part (Coincident)

Make sure Allow Assumptions is checked so that the part is Fully Constrained.

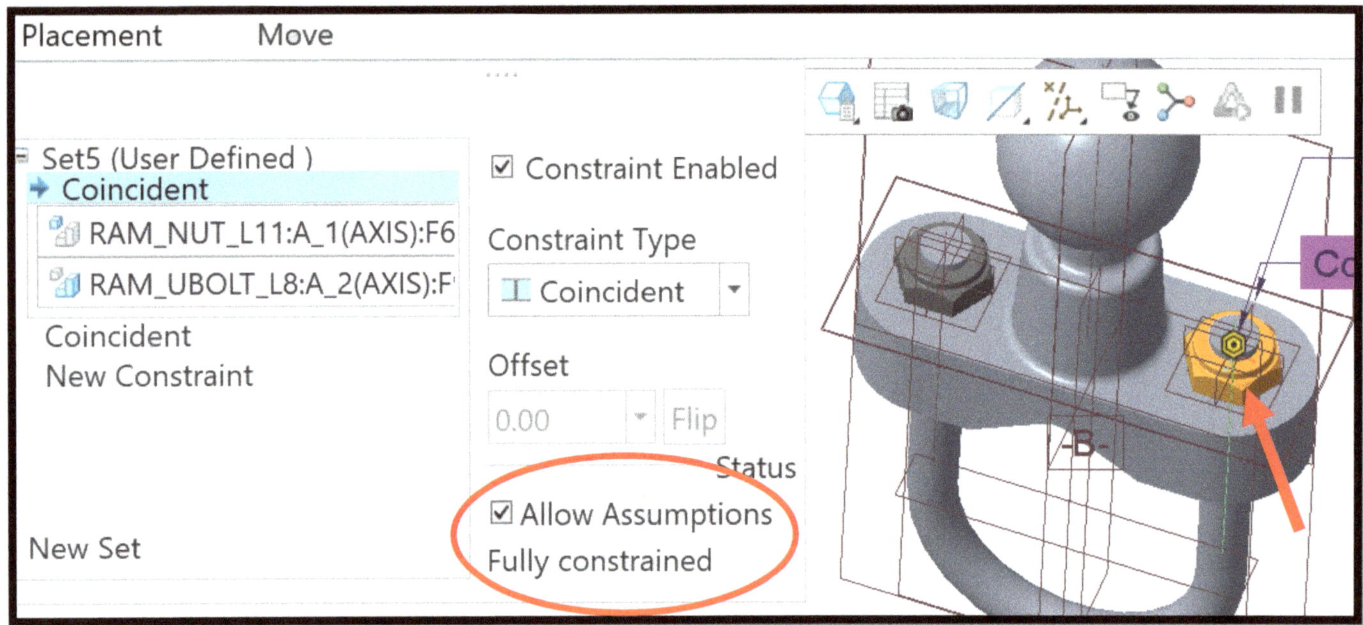

Step 7 – Assemble the *Ram_HandleBottom_L3* part using the constraints shown below

Constraint #1 (Surface to Surface):

Assemble

Type: Coincident

Reference of Component: "U" Cut out <u>surface</u>

Reference of Assembly: curved cylindrical surface of the bottom part of the U-Bolt

Tip: Note that the surfaces may not align as you would expect as it is still free to rotate out of the plane of the U-bolt. Disregard any issues with this constraint until the other constraints are added in.

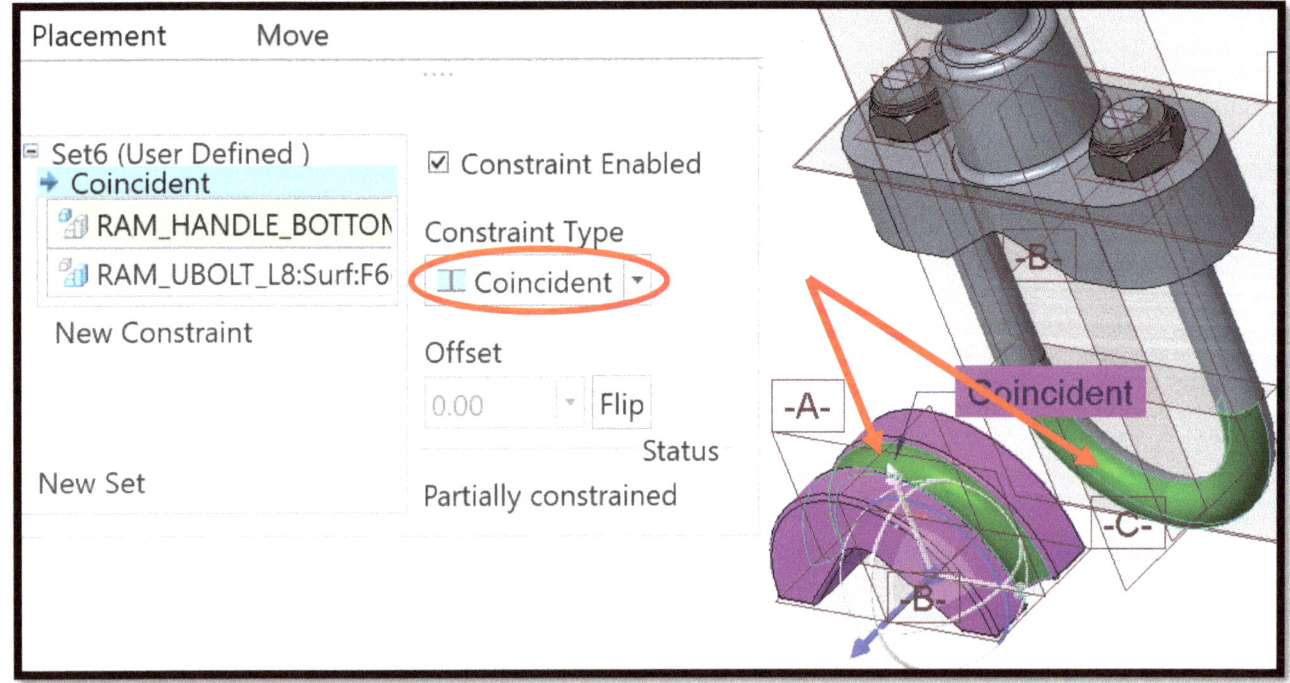

Constraint #2 (Datum to Datum):

Type: Coincident

Reference of Component: Datum A (Front) of *Ram_HandleBottom*

Reference of Assembly: Front Datum of U-Bolt. **Note:** there may be other part datums that would work but reference the **datum from the Ubolt** as that is the part we want this part attached to.

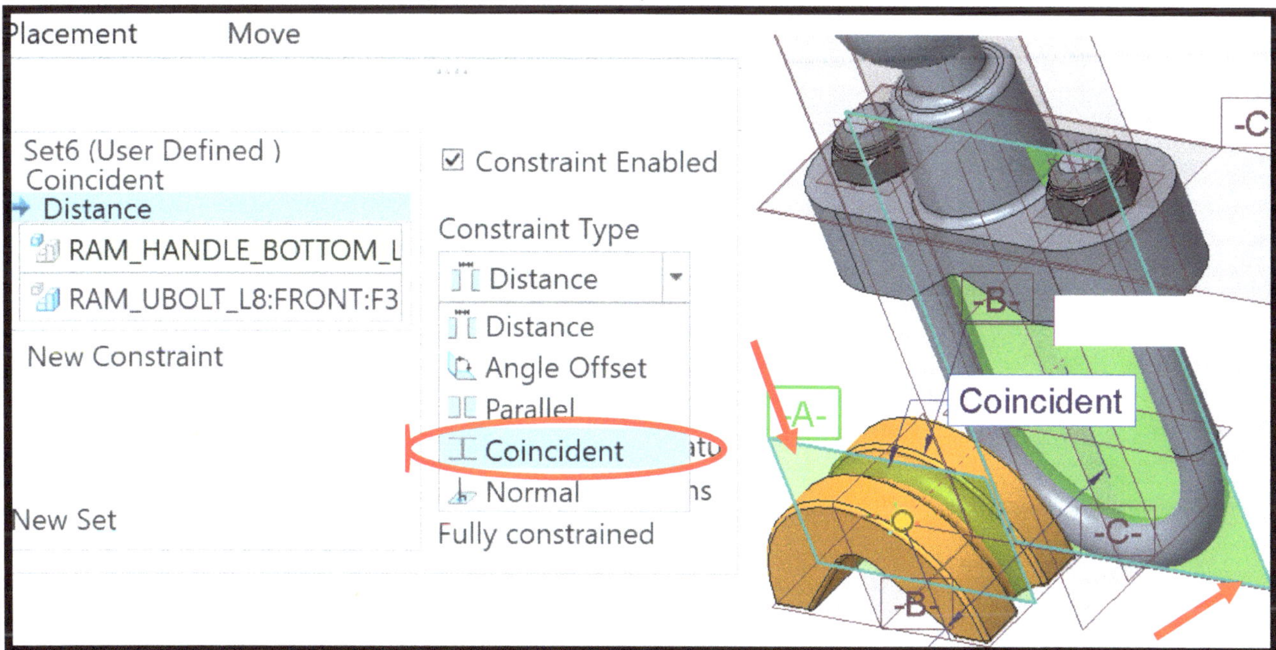

Constraint #3 (Parallel Surface to Surface):

Type: Parallel

Reference of Component: Planar top surface of the *Ram_HandleBottom*

Reference of Assembly: Planar bottom surface of the *Ram_HandleBar*

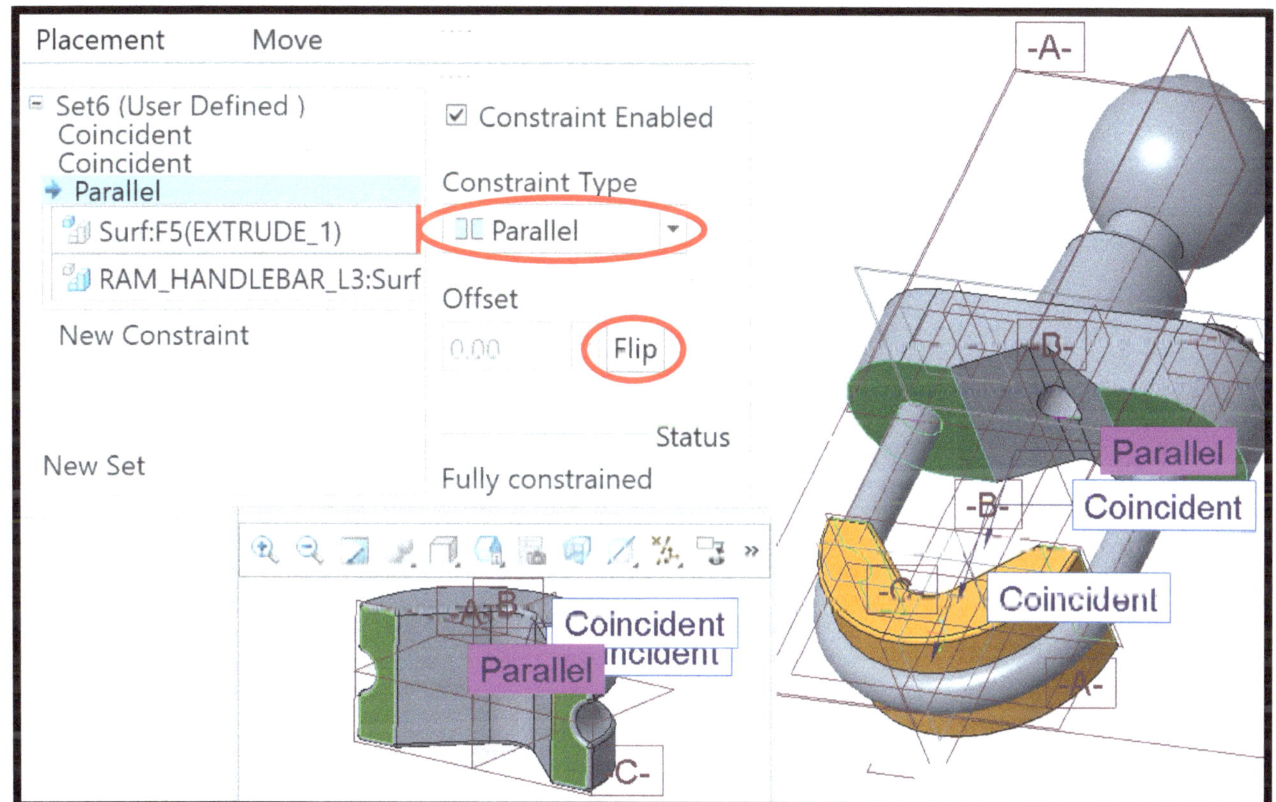

Step 7 – 2 - Checkmark to Accept the Fully Constrained placement of the component & Save the Assembly.

Step 8 – Open the *Ram_Extension_L5* part in its own Part window (not the assembly): **File - Open – *Ram_Extension_L5*.**

This part must be modified to remove a strip of material from the center of the part. The part will also have one "side" removed so that the part acts as one half of an assembly, which matches how this component is built in real life.

Modification 1: Extrude a cut through the middle of the part lengthwise placed on Datum C to remove material **symmetrically** from both sides of the Datum. This will remove a small thickness of material near the center of the model to provide space for both halves to pivot and clamp on the spherical ball mount later on.

- **Extrude Tool** – placement on **Datum C** - sketch a closed-loop rectangle larger than the part (a 5"x 2" rectangle)
- Set the Extrude Options to **Remove Material, Extrude Symmetrically on Both Sides*, and a Depth of 0.10".**
**Tip: Make sure you are extruding symmetrically so the center of the part is removed properly on each half.*

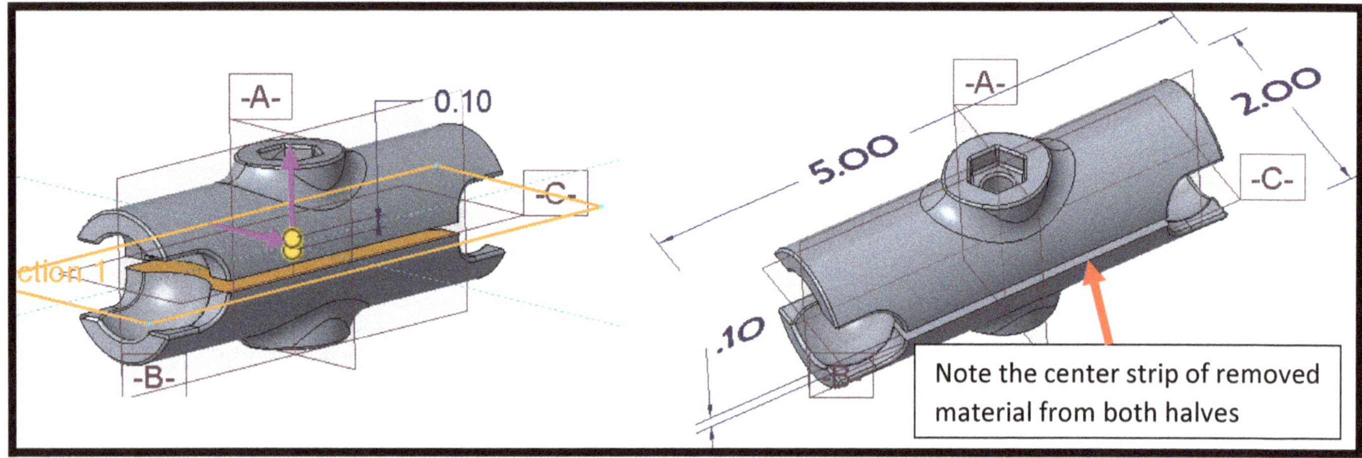

Modification 2: Cut away an entire half of the part so it is no longer two disconnected halves. The final product will be assembled later on in Assembly mode using two copies of the same side to match how it would be fabricated and assembled in reality.

- Use the **Extrude tool** placed on Datum C and sketch a closed-loop larger than the part.
- Set the Extrude Options to **Remove material, Direction should be a single direction** *(towards the screen),* **and depth of Thru-All.** This should remove one half of the part model without having to change any of the features.

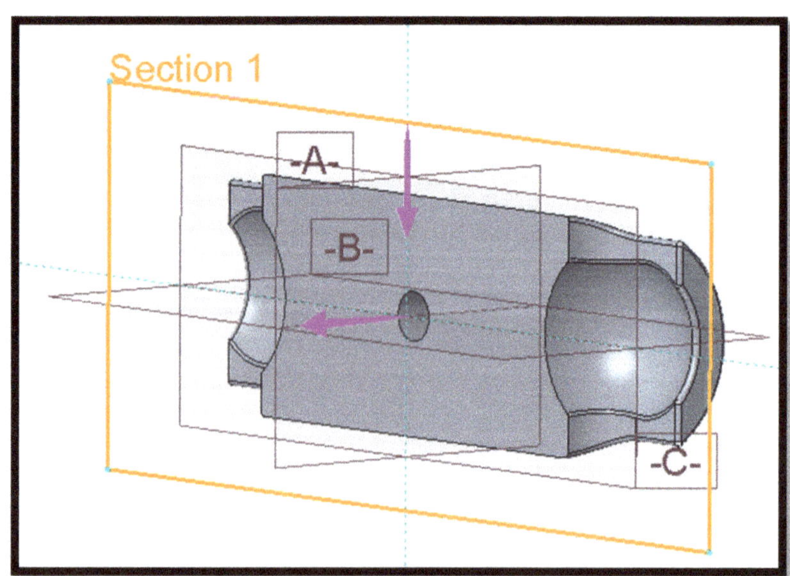

Modification 3: Remove material to hollow out the part using an Extrude. A Shell will not work due to the geometry of the Sphere and Hex walls, so use the **Extrude Tool placed on Datum C** to hollow out some of the model.

1) Place the Extrude on the center **Datum C** * (*not the top surface or the extrude will be too deep!*)

2) In Sketch mode use the **Offset Tool** as shown in the image below to create a closed loop **offset -0.10"** from the outer edges of model. This tool is not as intuitive as the Project tool as you cannot just Control click multiple lines.

- While in Sketch Mode – **Offset Tool** – disregard the message box – click on the **Edit Chain Properties Icon** – select **Rules Based** – click on an **edge** of the planar surface to use as the "anchor" – with the Rule set by default to **Tangent** it should show a purple offset Closed Loop on the sketch.

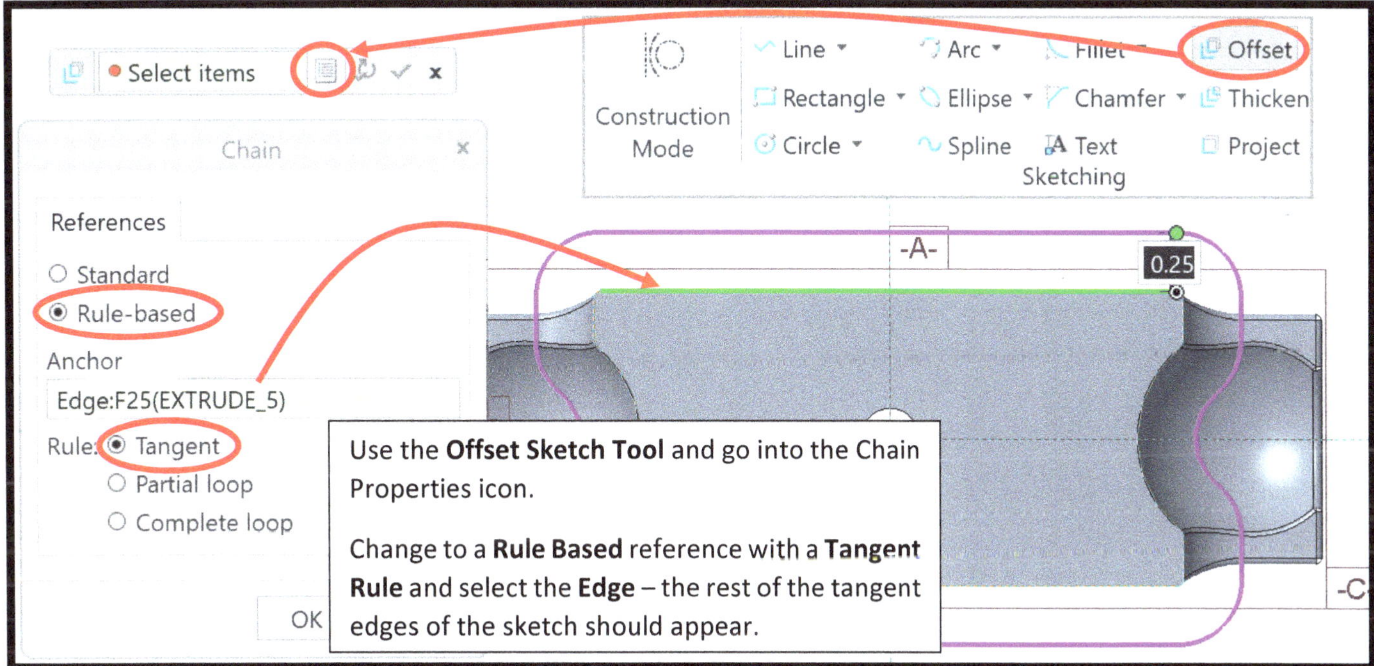

Tip: If the entire closed loop does not appear as shown above you can instead choose the "Complete Loop" Rule – click on the edge line – **in the Loop Reference box** click on the planar surface and the shape should appear.

3) Click in the value box and **type in 0.10 or -0.10** to offset the loop to inside the model. The (-) will change the direction of the offset to the opposite side of the anchor line.

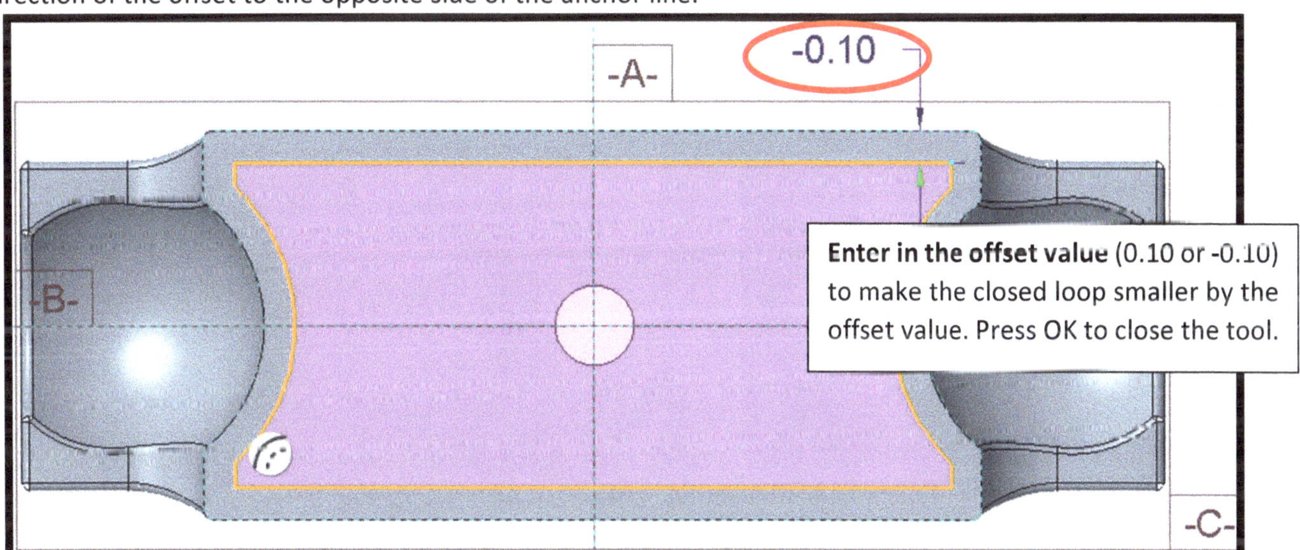

4) **Press OK in the Chain Menu** then press the **Checkmark icon** in the Offset Toolbar (if you don't it will not save the shape you just created) and finally Checkmark the top toolbar to accept the Sketch.

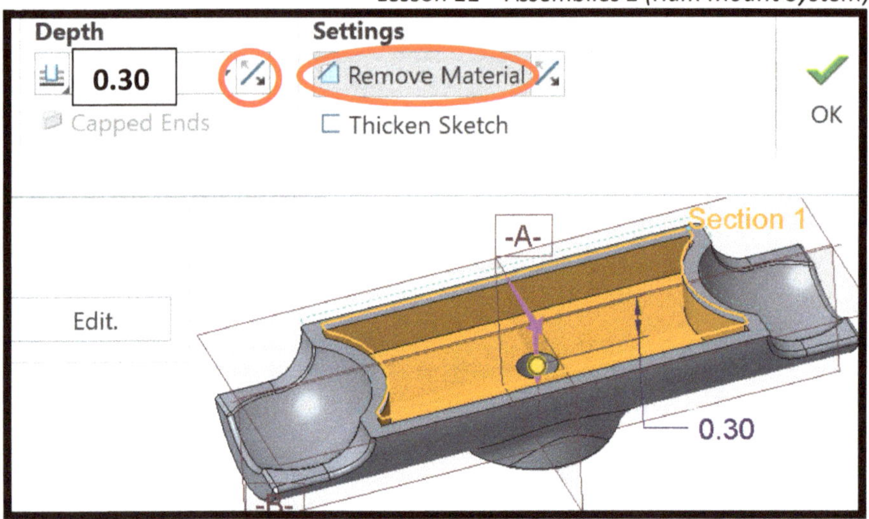

5) Set the Extrude Toolbar Options to **Remove Material, Depth of 0.30", Flip the Direction Arrow** as needed to cut into the model.

- **Tip:** If your Extrude appears too deep and is cutting through the model wall you likely did not place the Extrude on the center Datum C, or you did not extrude on Both Sides in Modification #1.

Step 9 – Save a Copy of the Part, called ***Ram_Extension_L11*** and then **Close the file**. *This should save to your working directory, but check to make sure as it must be in the same location as the other files.*

Step 10 – 1 - Start a **new __Assembly__ file** called "***RAM_EXT_ASM_L11***". This will assemble the components of the Ram Extension product in its own sub-assembly that will be added to the main assembly later.

Step 10 – 2 - Assemble the *Ram_Extension_L11*, using a **Default Constraint**.

Tip: Make sure you have created a new __Assembly__ (not a part file), and are __not__ working in the previous assembly. This should be its own new assembly file.

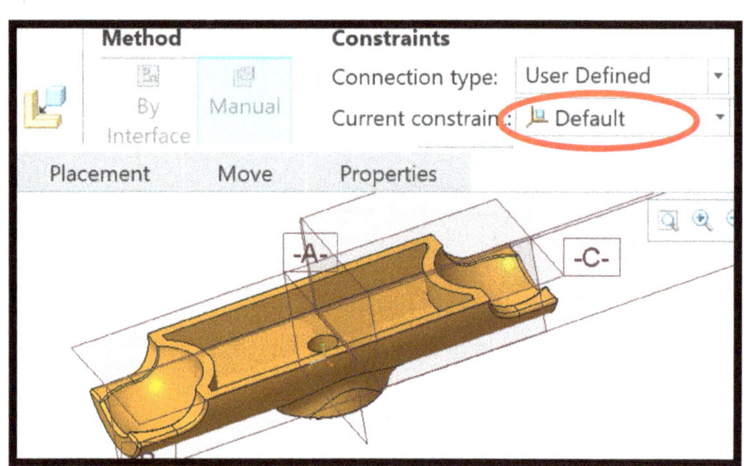

Step 10 – 3 - Assemble a second copy of the ***Ram_Extension_L11***.

Constraint #1: (Align the Horizontal Datums)

Type: Coincident (**Flip** as needed)
Reference of Component: Datum C
Reference of Assembly: Datum C

Tip: Use the red **Drag Arrow** to slide the copy further away so you can select its Datum C.

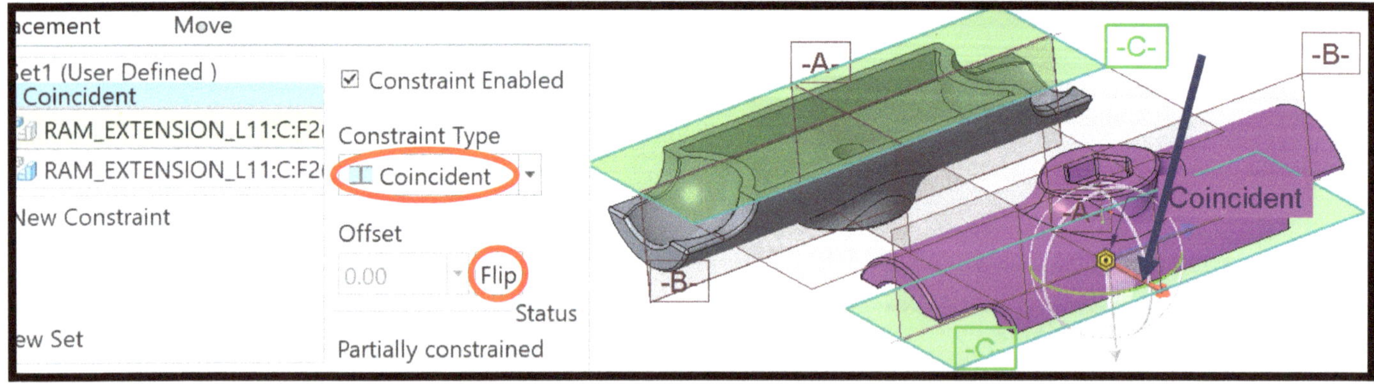

Tip: You can hover over a datum and tap RMB to cycle which item under the mouse should be highlighted.

Constraint #2: (Align the Vertical Datums)

Type: Coincident
Reference of Component: Datum B
Reference of Assembly: Datum B

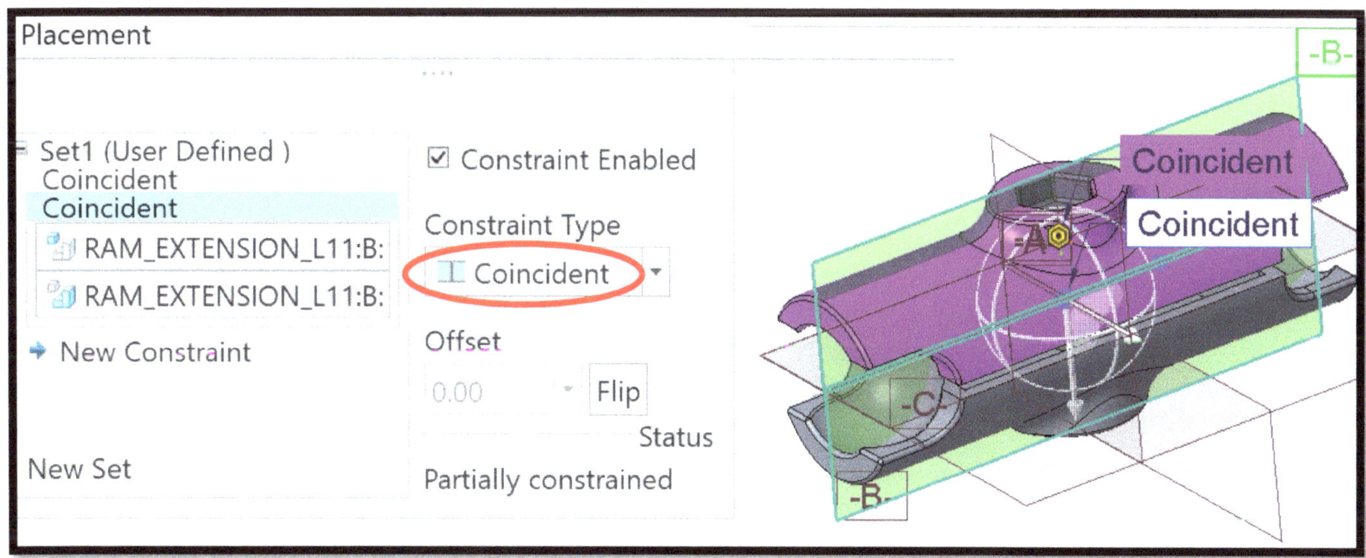

Constraint #3: (Align second set of Vertical Datums)

Type: Coincident
Reference of Component: Datum A
Reference of Assembly: Datum A

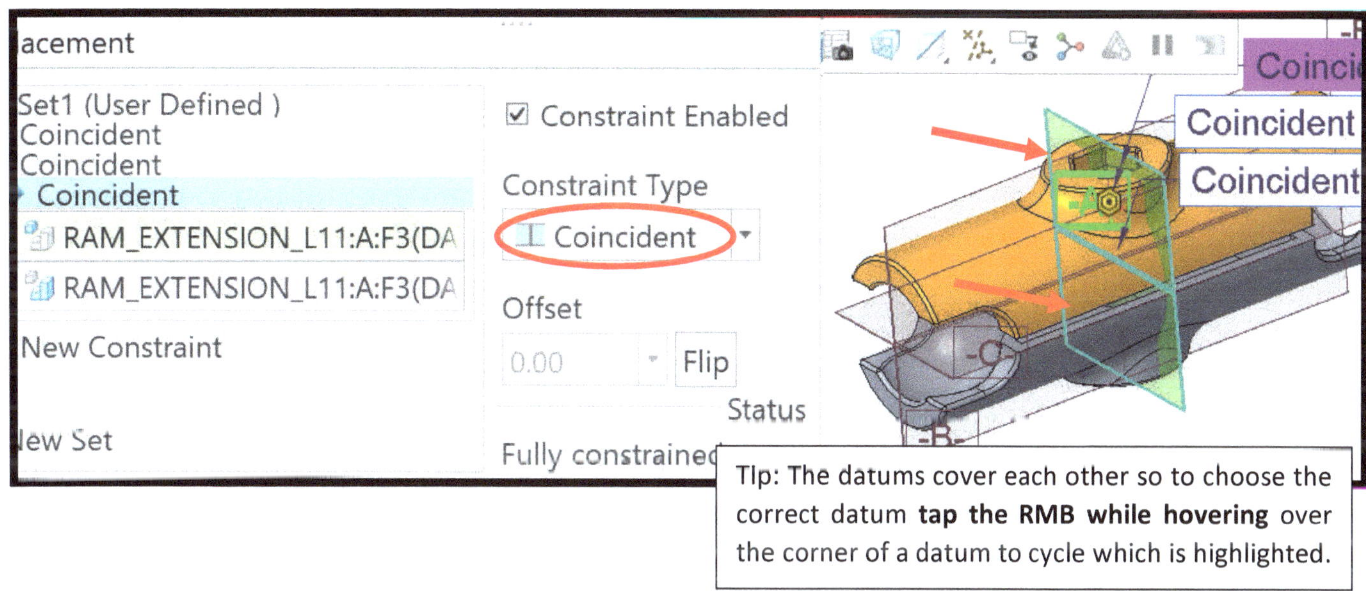

Step 10 – 4 - Accept the Placement of the Component once it is Fully Constrained.

Step 11 - Assemble the *Ram_Bolt_L8* through the Hex hole on the second instance of the *Ram_Extension*.

> **Constraint #1:** Axis to Axis (**Coincident**)
>
> **Constraint #2:** Bottom surface of Hex Head to planar bottom surface of the Hex cutout (**Coincident**)
>
> **Constraint #3:** Side Planar Surface of the Hex Bolt Head to side Planar Surface of Hex Cutout (**Parallel**), so that the bolt fits in the hex cutout properly without interference.

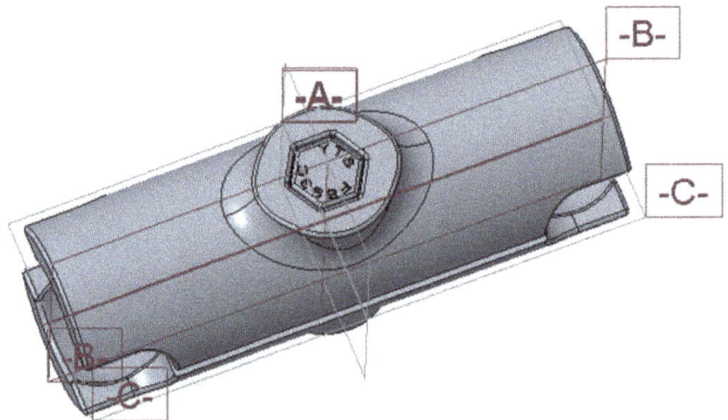

Step 12 - Assemble the *Ram_Washer_L11* part onto the other end of the Bolt.

> **Constraint #1:** Axis to Axis (**Coincident**)
>
> **Constraint #2:** Bottom surface of Washer to top surface of the Hex cutout (**Coincident**). Use Flip if needed.
>
> **Allow Assumptions** to fully place the washer.

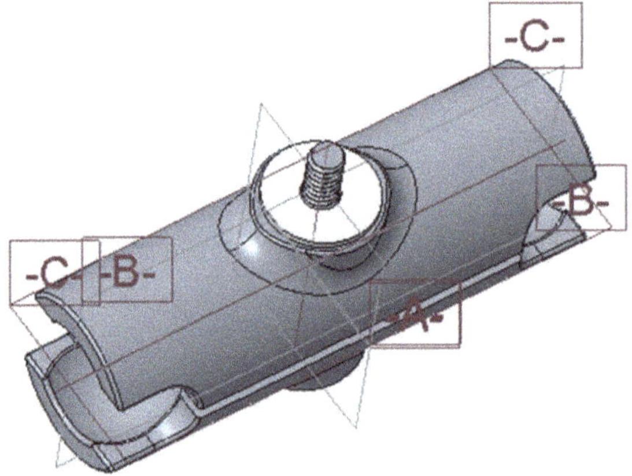

Step 13 - Assemble the *Ram_Wingnut_L11* component on top of the Washer, with similar constraints to the Washer placement, but sitting on top of the washer.

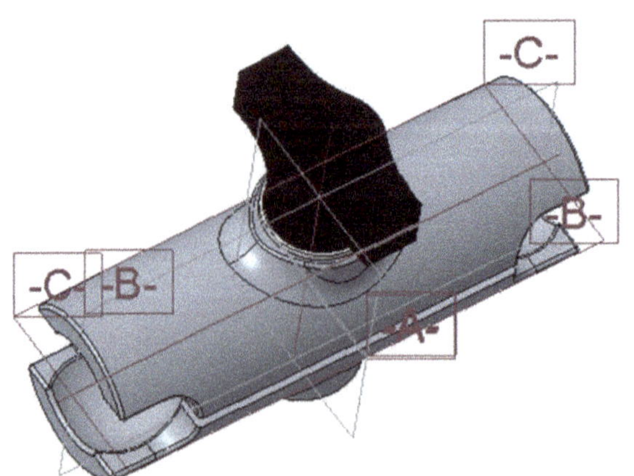

Step 14 – 1 - Hide the *Ram_Extension_L11* part and **Assemble** the ***Ram_Spring_L11.***

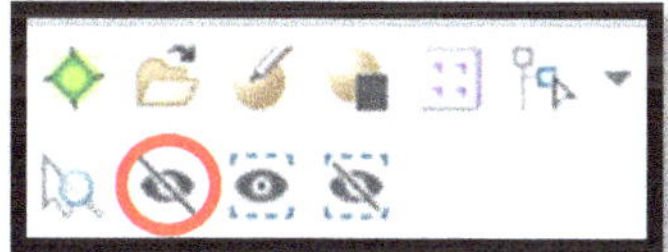

Hide the Part: **Select** the upper *Ram_Extension_L11* part from the model tree - select the **Hide** from the pop-up menu.

> **Constraint #1:** Axis to Axis (**Coincident**).
>
> **Constraint #2:** Flat end Surface of the Spring to the inner flat surface of the *Ram_Extension_L11* (**Coincident**)
> **Allow Assumptions** to fully place the Spring.

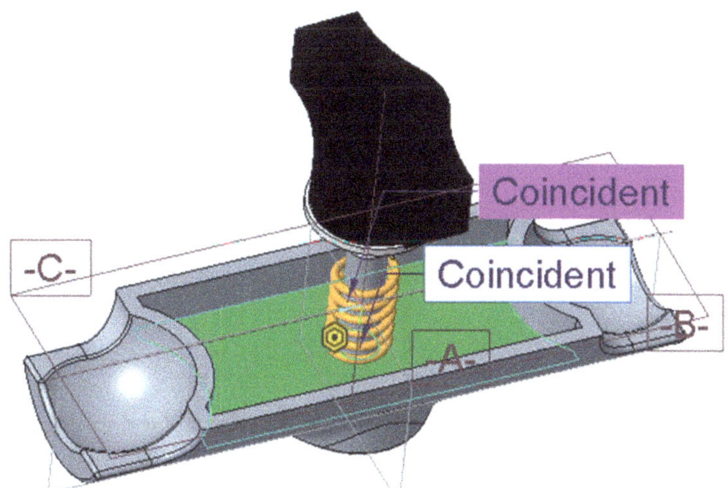

Step 14 – 2 - Accept the placement of the Spring and **Show** the *Ram_Extension_L11*.

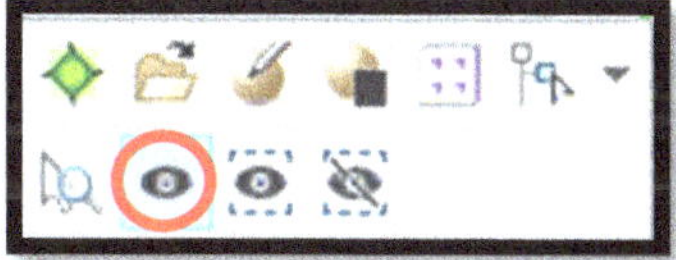

- **Select** the hidden *Ram_Extension_L11* from the Model Tree – **select the Show** icon from the pop-up menu

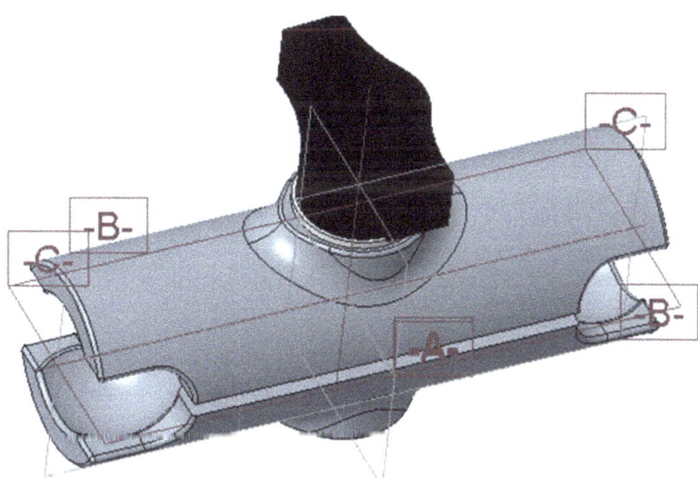

Step 15 – Save the *Ram_Ext_ASM* Assembly. If CREO will not let you save, press **Regenerate** (Control + G) to regenerate and force any updates on the model if needed before you can save.

Step 16 – Go back to the ***Ram_System_L11*** **Assembly**, and **Assemble** the ***Ram_Ext_Asm.ASM*** to the spherical ball.

Constraint #1: (Center to Center of Spheres)

Type: **Centered** *(change to the Centered type instead of Coincident!)*
Reference of Component: Inner Spherical Surface on left end of the *Ram_Extension_L11*
Reference of Assembly: Spherical Surface of the Ball on the *Ram_Handlebar*

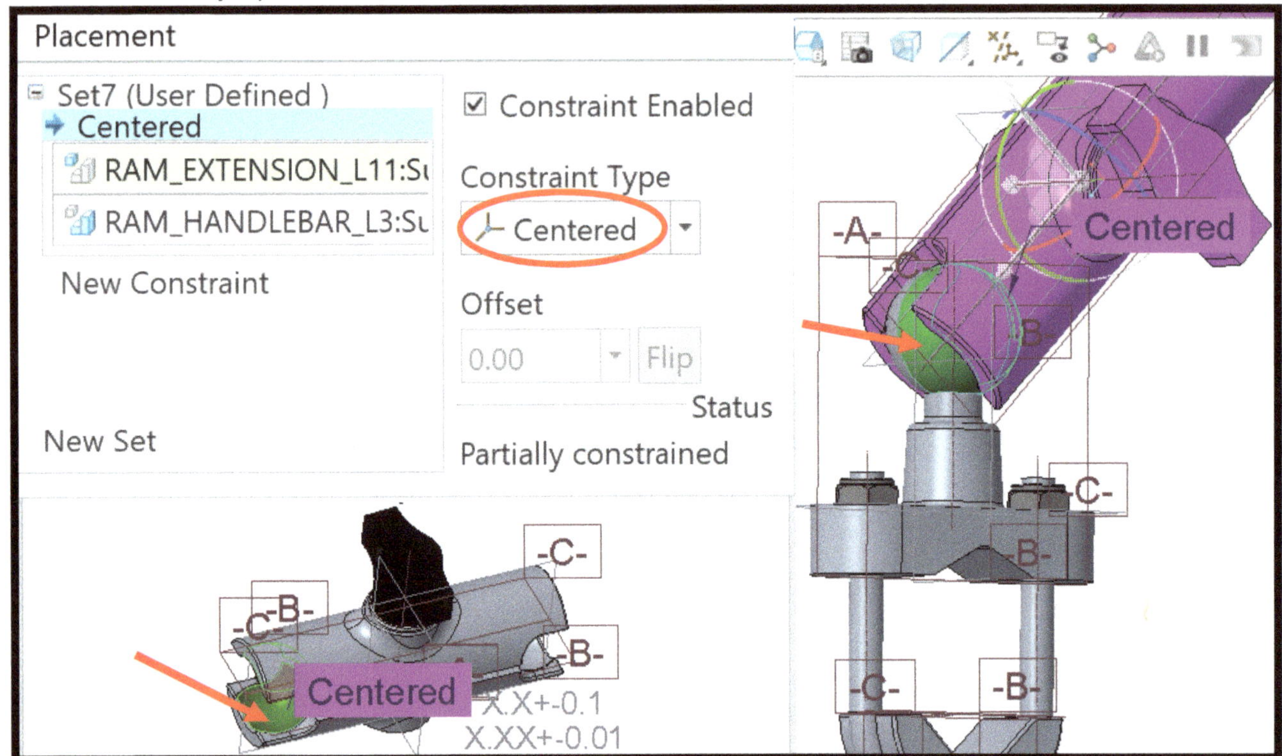

Constraint #2: (Datum to Datum)

Type: Coincident
Reference of Component: Datum C of *Ram_Extension_L11*
Reference of Assembly: Datum A of the *Ram_HandleBar* component
Tip: Use the Circular Dragger handle (green arc on this example) to rotate the Extension in plane for visibility.

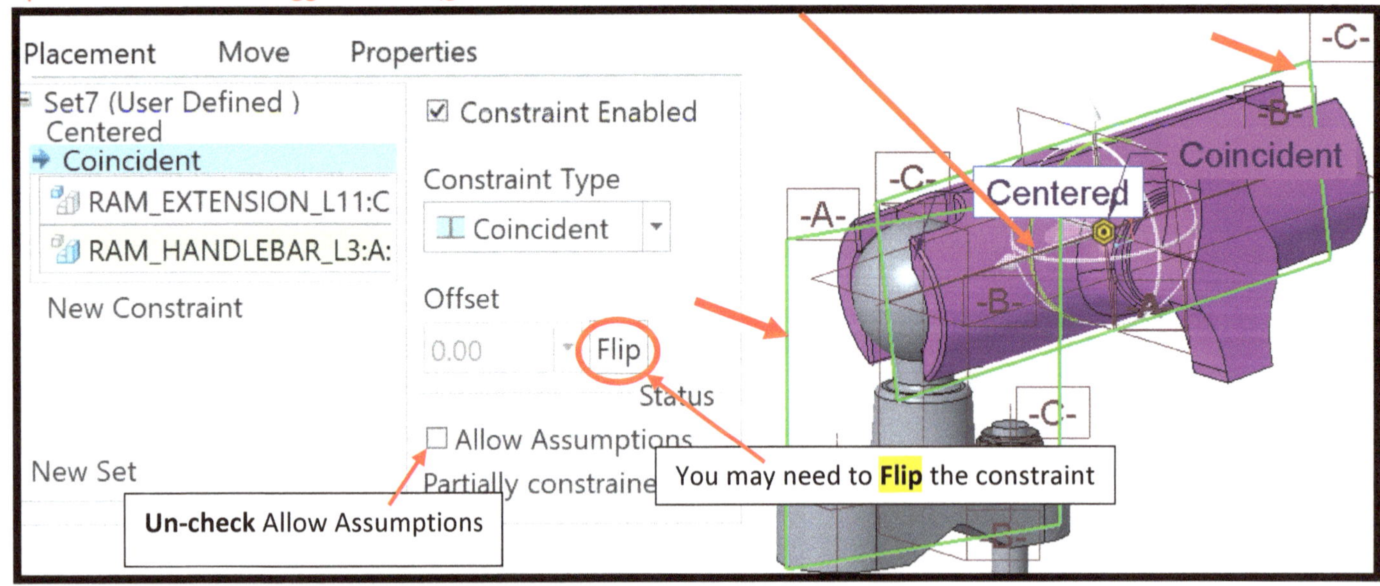

Note – Setup this constraint so the **Wingnut faces the screen or Front of the assembly**. Do not choose a datum that rotates the extension so the wing nut is on the top, back, or bottom of the ram-extension when viewed from the front.

Constraint #3 (Angle Offset Datums)

Setup this constraint so that the angle of the *Ram_Ext_ASM* is at 45 degrees up and to the right of the Ball.

Type: Angle Offset of 45 degrees (or similar, you may need to use a multiple of 90 such as -45, 135, 225, etc.)
Reference of Component: **Datum B of Ram_Extension_L11**
Reference of Assembly: Top Planar Surface of *Ram_Handlebar* component

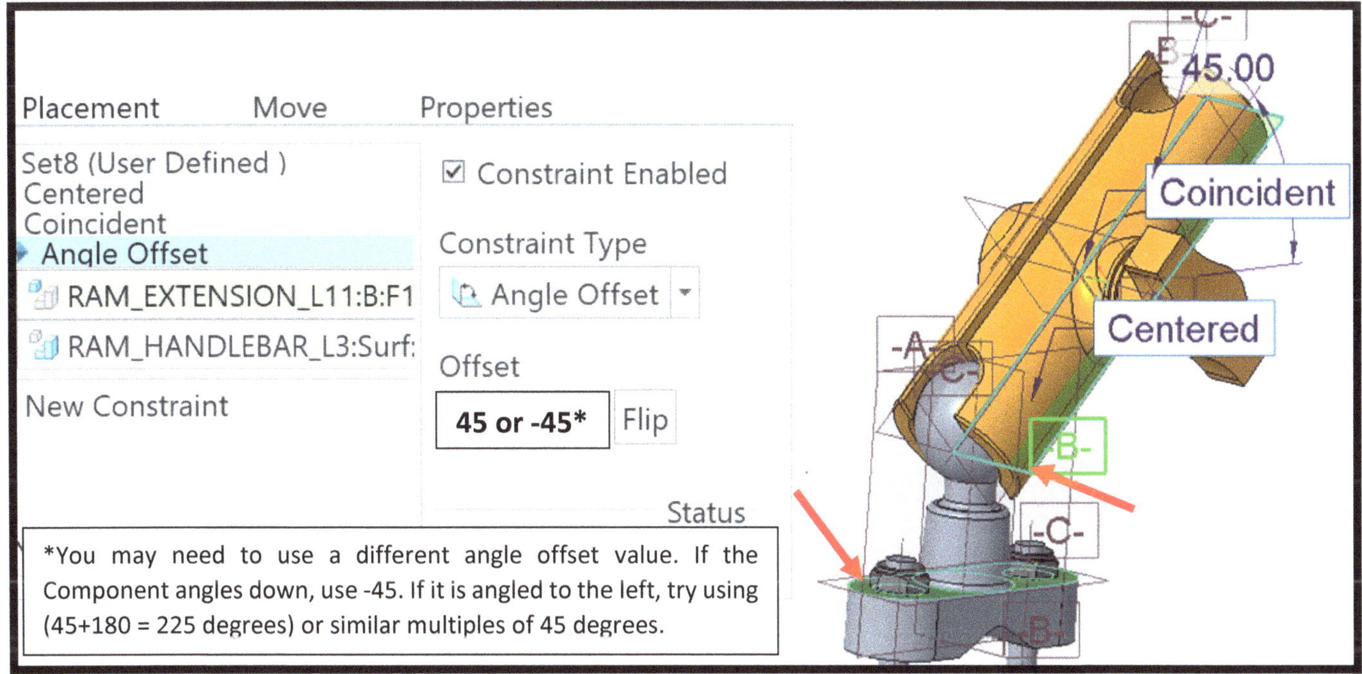

Step 16 – 2 - Accept the placement of the fully constrained sub-assembly.

Step 16 – 3 - Check that your wingnut is visible from the Front view and that the Ram Extension ASM is angled properly. **Select the Front View option in the Quick Toolbar** to check how it will look on the Drawing which will need to match the key. If not, use Edit Definition of the *Ram_Ext_ASM* and change or remake the constraints as shown earlier.

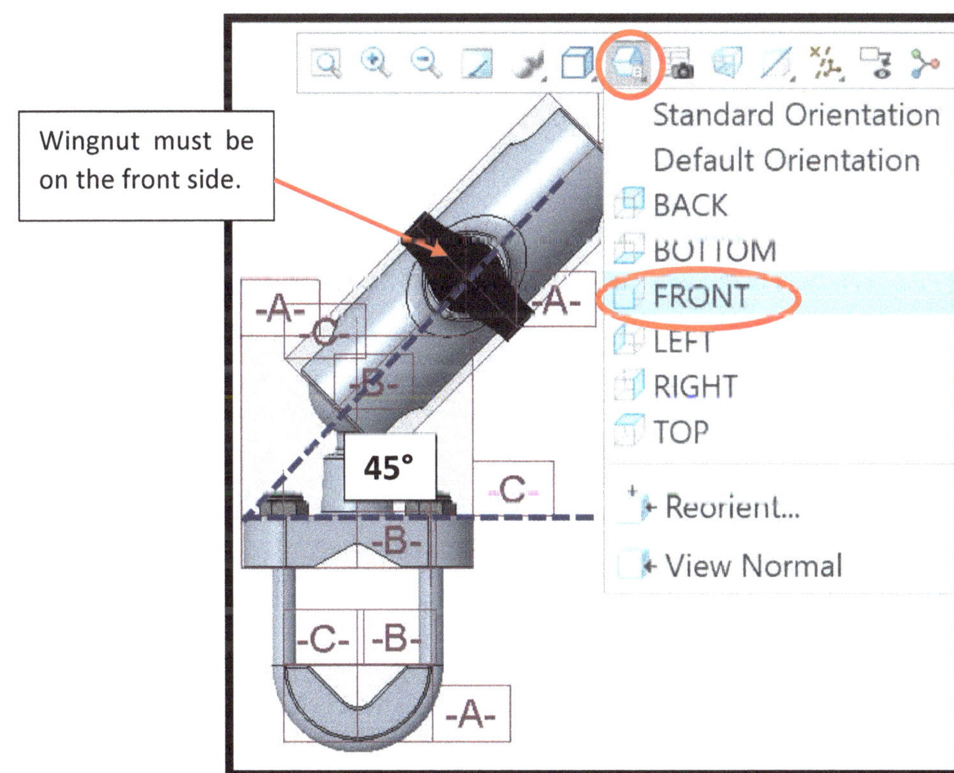

Step 17 – Assemble the *Ram_MountBase_L3*. This component is provided to you and must be extracted and saved to your working directory.

Use similar constraints to the previous step, but have the *Ram_MountBase_L3* component point to the right, with the rotation setup so that the holes are positioned above and below the Ball, not side to side.

Constraint #1: Spherical surface of the ball end to the Spherical cutout of the Ram Extension (**Centered)**

Constraint #2: vertical Datum to Datum **(Coincident)** to keep the part in line with the extension, with holes positioned properly at the top and bottom and not side to side.

Constraint #3: Bottom of *Ram_MountBase* to a Parallel Datum of the *Ram_Handlebar* component (**Angle Offset)**
         Set a 0°, 45°, or 135° degrees offset as needed so the flat surfaces is vertical & facing to the right.

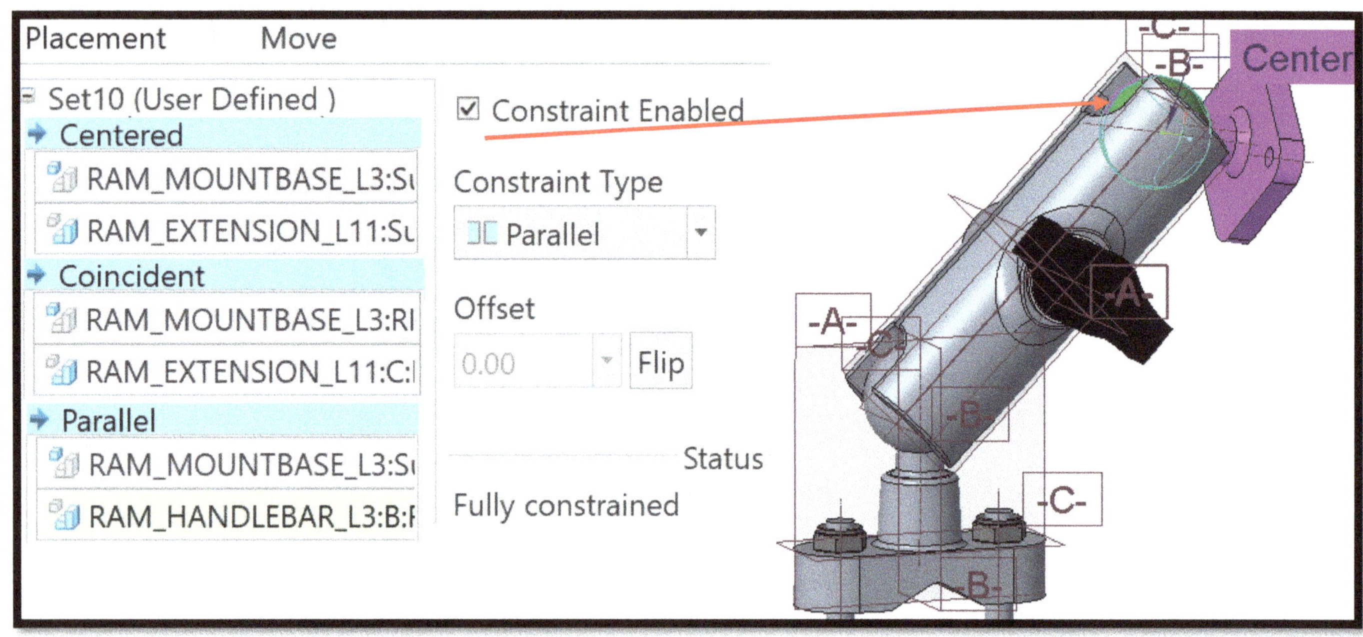

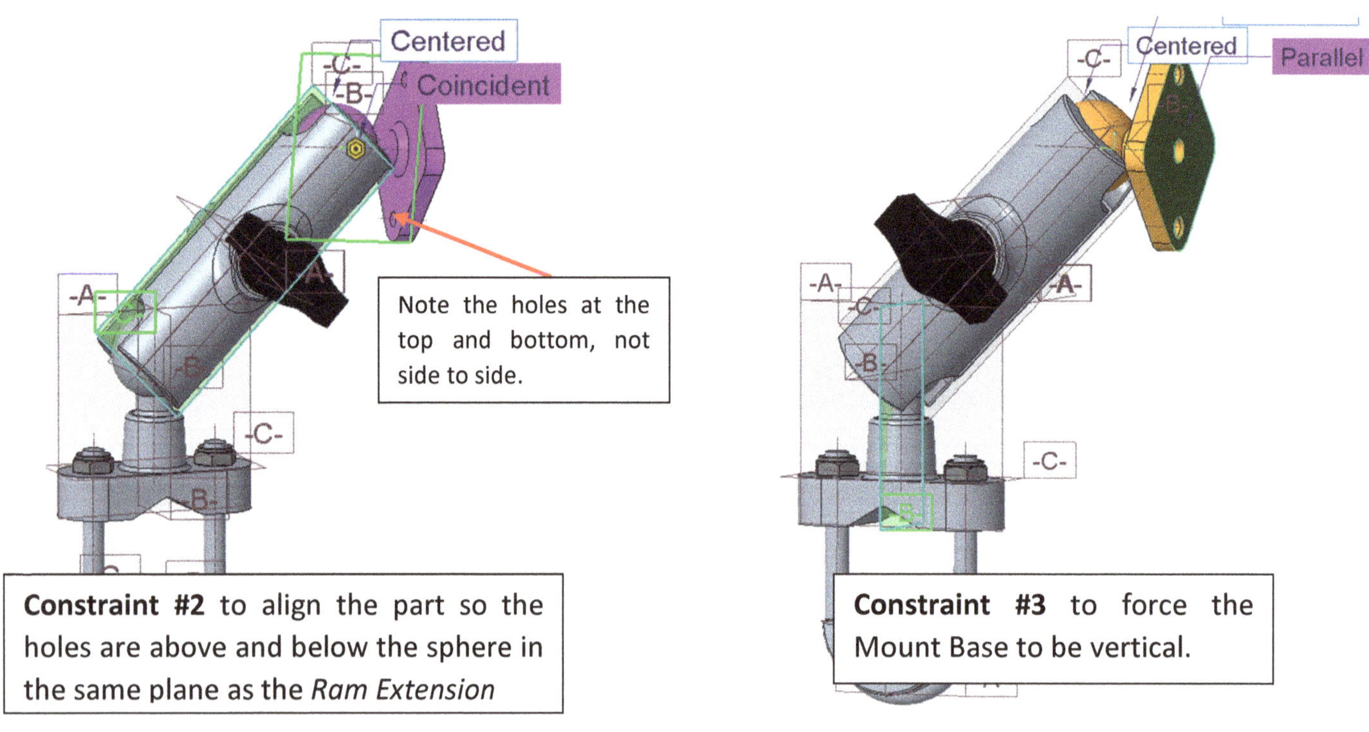

Constraint #2 to align the part so the holes are above and below the sphere in the same plane as the *Ram Extension*

Constraint #3 to force the Mount Base to be vertical.

Step 17 – 2 - Checkmark to accept the component placement.

Step 18 - Constraint Check: Any components that are not Fully Constrained will have a **[]** or similar symbol in front of the item In the Model Tree. You can use Edit Definition on that item or any parts that they reference to make sure all items are Fully Constrained.

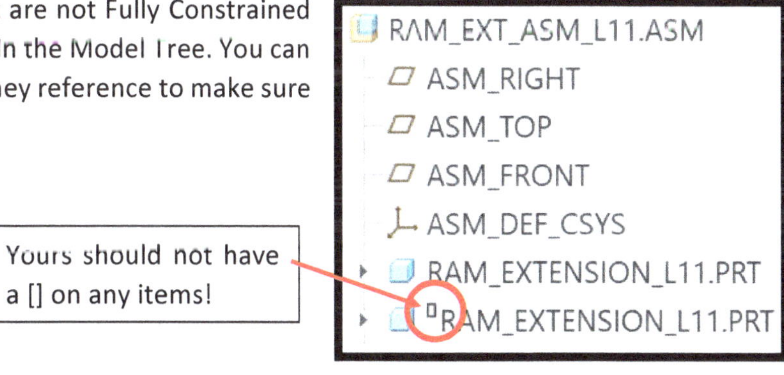

Step 19 - Create a new Assembly Datum (ADTM1) that will be used to place a Cross-Section on for the drawing. If you are making a Cross-Section from Drawing Mode you **must use an Assembly Datum** so it interacts with all components.

   - Plane Tool – Select the **ASM_FRONT datum** in the Model Tree as the reference – set the **offset of 0 – press OK**.

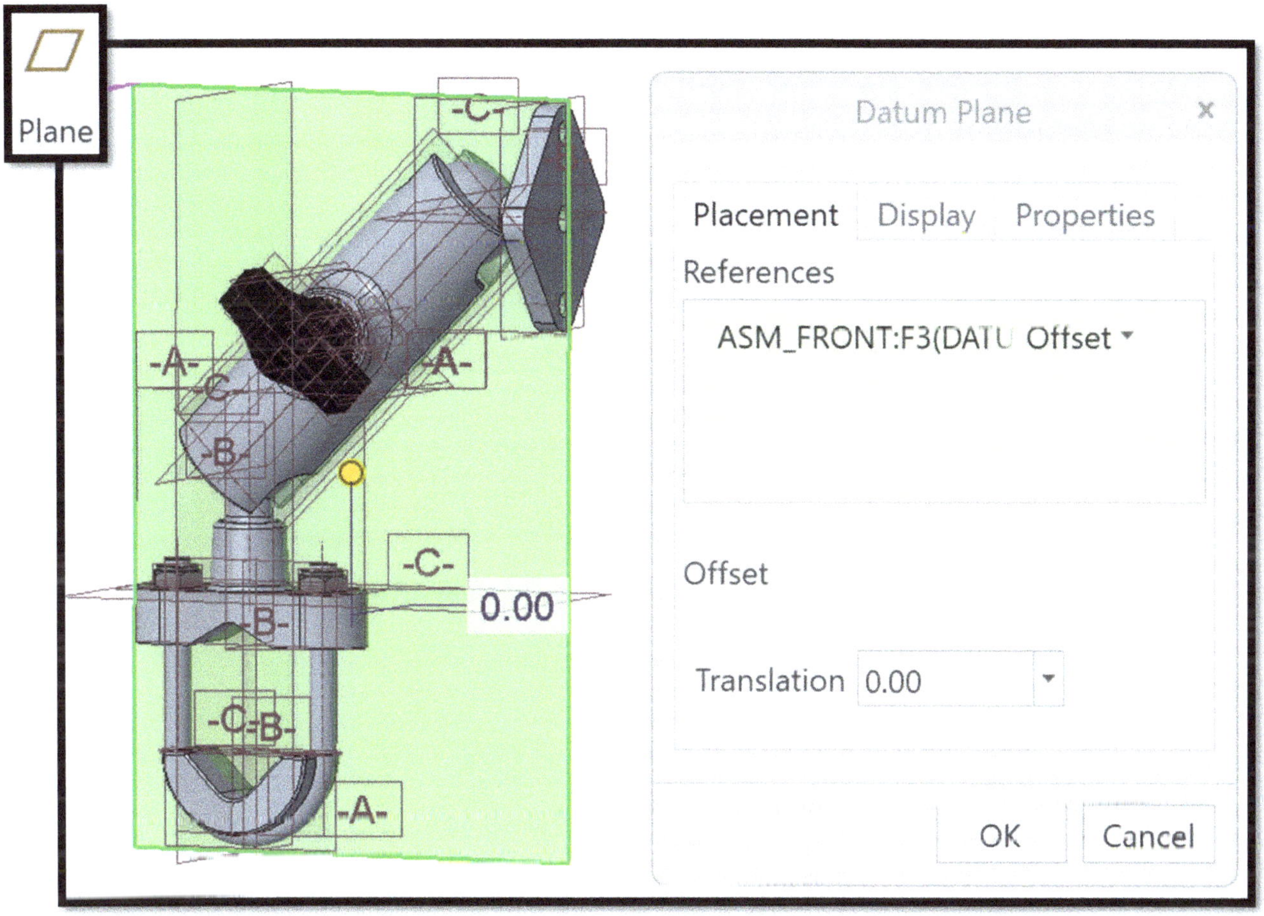

You should now have a new Assembly Datum ADTM1 in the Model Tree.

Step 20 – Regenerate (press Control + G or use the icon in the top toolbar) and **Save the model.**

Step 21 – Start a new Drawing of the *Ram_System_L11.ASM*. Then follow the steps on the next page to Hide or Erase the excessive amount of Datums to declutter the drawing easily.

Step 22 – Hide all the part datums on the drawing. You cannot toggle them off using the Display Filters as many of them were "Set" Geometric Datums that were setup in earlier lessons. There are 2 methods to hide these datums easily:

> **Option 1)** While in Drawing Mode, hide/erase the individual datums from each View Template while using the Annotate tab (**shown in Step 2 of the *Detail Drawing Steps***).
>
> > **or**
>
> **Option 2) Use the <u>Layer Tree</u>** to hide all Part Datums all at once from the model (shown below).

Using the Layer Tree to hide Datums:

Layers can be used to interact with all types of entities at once on a model (all Holes, all Part Datums, etc.). Using Layers is a quick and easy way to hide all part datums at once but if you want to show them again you need to **unhide** the Layer. CAD experts often use Layers for other operations such as creating custom layers for specific types of parts or features to organize more complex models or handle variations in a design that they want to interact with.

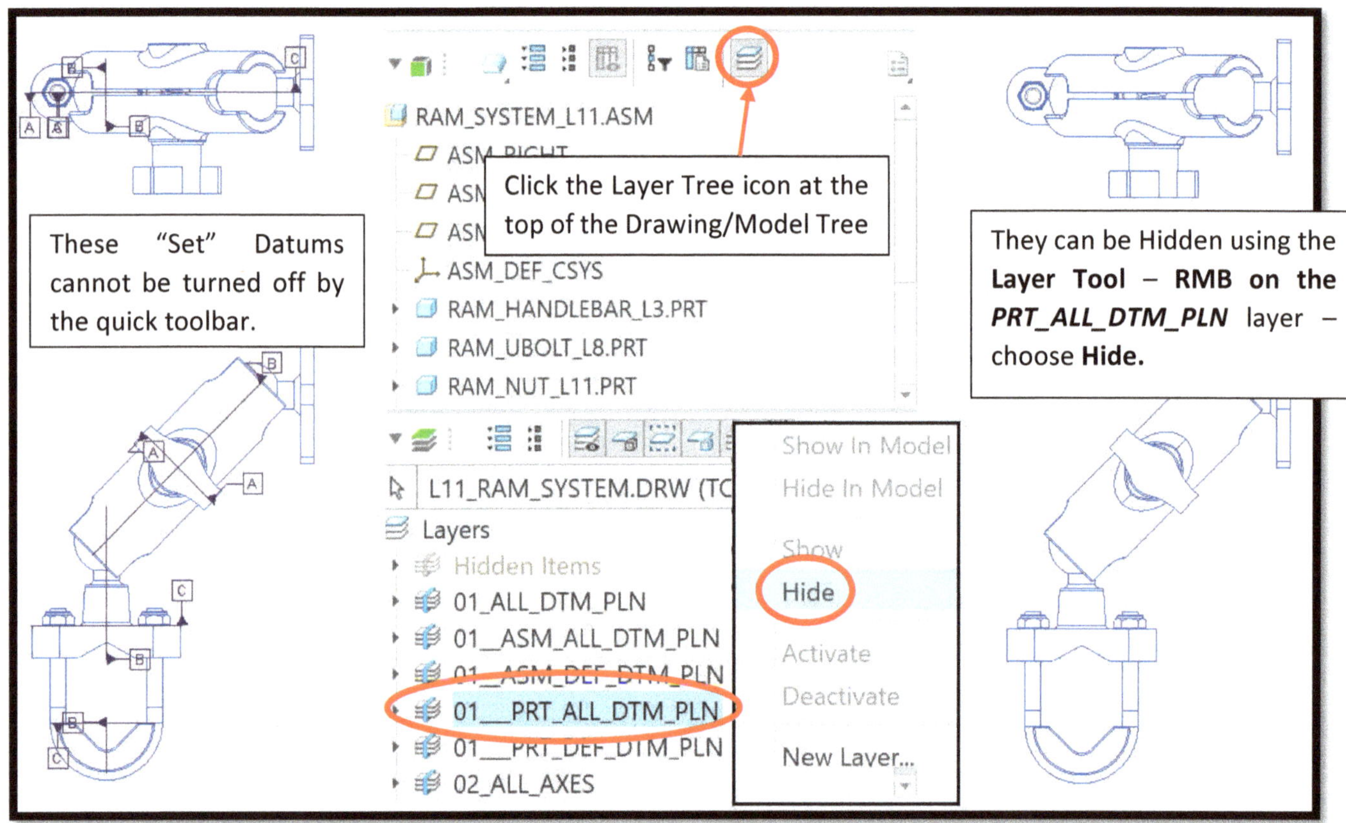

Step 23 - Complete the remainder of the Drawing by following the steps listed in the *Assembly Drawing Steps* (Appendix) as needed to match the drawing key. You will not create a drawing of the *Ram_Extension_L11* subassembly to save time on this lesson, although typically you would create an assembly drawing of sub-assemblies as well.

End of Lesson 11

Lesson 12 –Test Rig Base & Tower Assemblies

Lesson 12 is to create two assemblies to build the "Test Rig Stand" used for testing small wind turbines. This test rig stand is mounted in front of a Fluid Testing Station, which includes an air blower capable of producing 40mph winds. Students will later design a wind turbine platform that attaches to this Test Rig Stand for testing their own wind turbine designs.

This lesson has fewer step-by-step instructions for completing each step of the model. The goal and settings for each step are provided but students may need to refer back to earlier lessons to review how to operate a feature or complete a constraint placement.

Repeat Tool:

You can use the **Repeat tool** to assemble multiple copies of a component quickly by selecting which constraint reference will be replaced for each repeated copy. For example, a bolt assembled in a hole can be repeated to switch out the axis constraint reference to the axis of the next hole. This is faster than manually assembling a second copy and is more robust than using the Pattern tool, as the Pattern copies will not update the position of the bolt if hole changes to a new location.

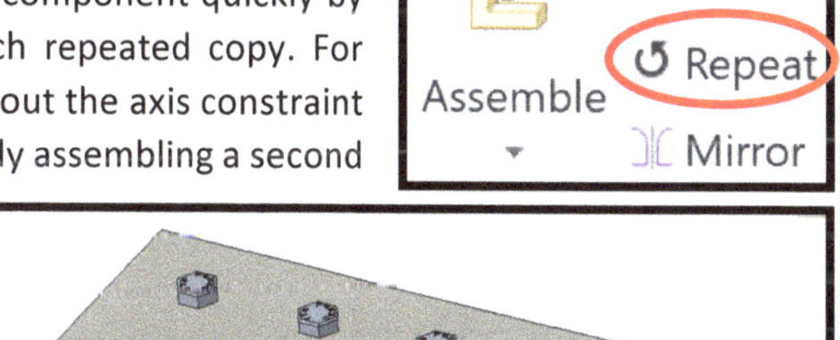

Note: If the Repeated component is not appearing in the correct orientation, you may need to Edit Definition and toggle the "Flip" button under a constraint (if that option is available), or use the Pattern tool or Assemble another instance instead of using Repeat.

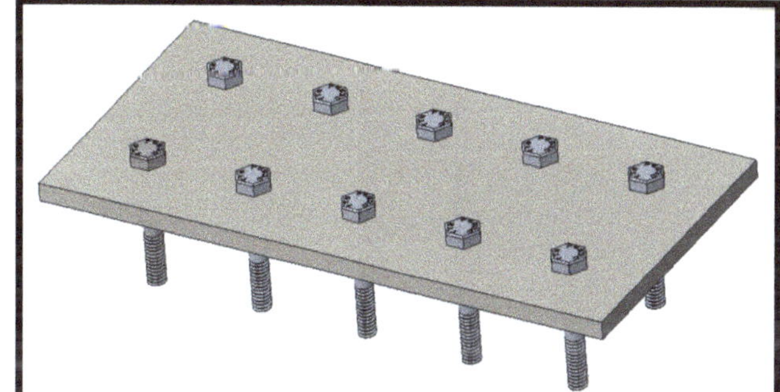

Useful Assembly Tips:

- Turn off **Display Filters** in the quick toolbar as needed to hide datums, axis, etc. from the many parts from the graphic screen as needed to help selecting entities for constraints. "Set" datums will not hide using Filters.
- **Fully Constrain** all components. If a component is not fully constrained it may unexpectedly "flex" any remaining degree of freedom later on. Use Edit Definition to bring back the Component Placement toolbar to make the component Fully Constrained.
    - A component with a [] next to its name in the model tree is not Fully Constrained
    - [][] means that the component references an earlier unconstrained component
    - [*] signals a mechanism type constraint that has built in movement parameters.
- **Regenerate:** Use the Regenerate tool (or press CONTROL + G on the keyboard) to force an update to the model after making any changes. CREO may not allow you to save if you have made changes and have not "Regenerated" to force the changes to update.
- **Tangent Constraints –** A tangent constraint places a surface against a curved surface. This can sometimes be useful in CREO but can also prevent which options you may want to use for secondary constraints. If you need to use a Tangent Constraint, **suggested practice is to place the Tangent constraint last**.

Assembly #1 - Base Stand Assembly

Getting Started: Open **CREO** *Parametric* and follow the lesson steps.

Step 1 - Make sure that all files for this assembly are in your Working Directory.

- **Download, extract/unzip, and place the files from Blackboard** for this assignment to your working directory. Save all files in the same working directory folder or CREO will not find them next time the assembly is opened.

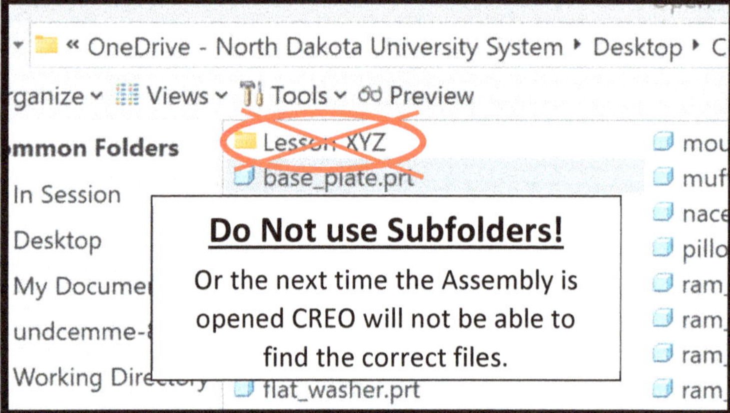

Step 2 – Set your **Working Directory** and create a **new Assembly** called "*Base_Stand_ASM*".

Step 3 – **Assemble** the *Base_Plate* component using the Assemble Tool with a "**Default**" constraint type. This will "Fully Constrain" the new component to the Assembly coordinate system.

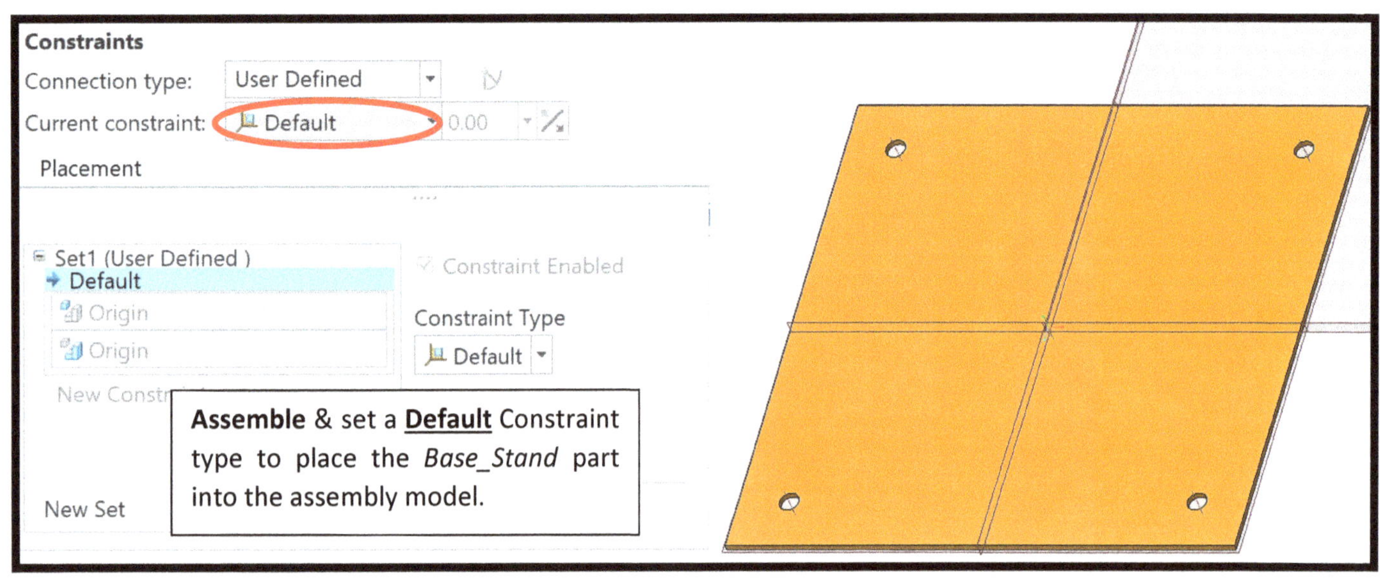

Step 3b – **Checkmark** to accept the placement of the *Base_Plate,* then **check your Model Tree**.

- You should **not** have a [] symbol in front of the name of the *Base_Plate*, as that would signal that it is not Fully Constrained. If needed, use Edit Definition and check that the 'Default' constraint is setup correctly.

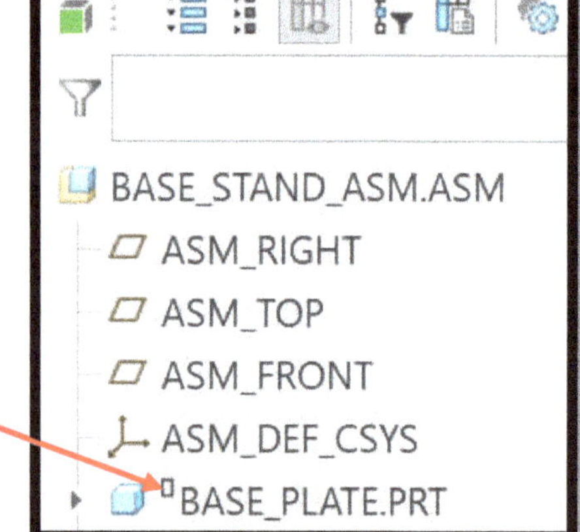

If you see this then your component is not Fully Constrained!

Step 4 – Assemble the *Base_Stand_Support* part to the assembly using the Assemble tool. **Fully Constrain** the component by setting constraints that will place the component on the top surface of the Base Plate and centered in the middle of the plate *(detailed instructions are not provided).*

Assemble

- **The side hole in the *Base_Stand_Support* should be near the top and also facing or perpendicular the ASM_Front datum in order to match the key.**

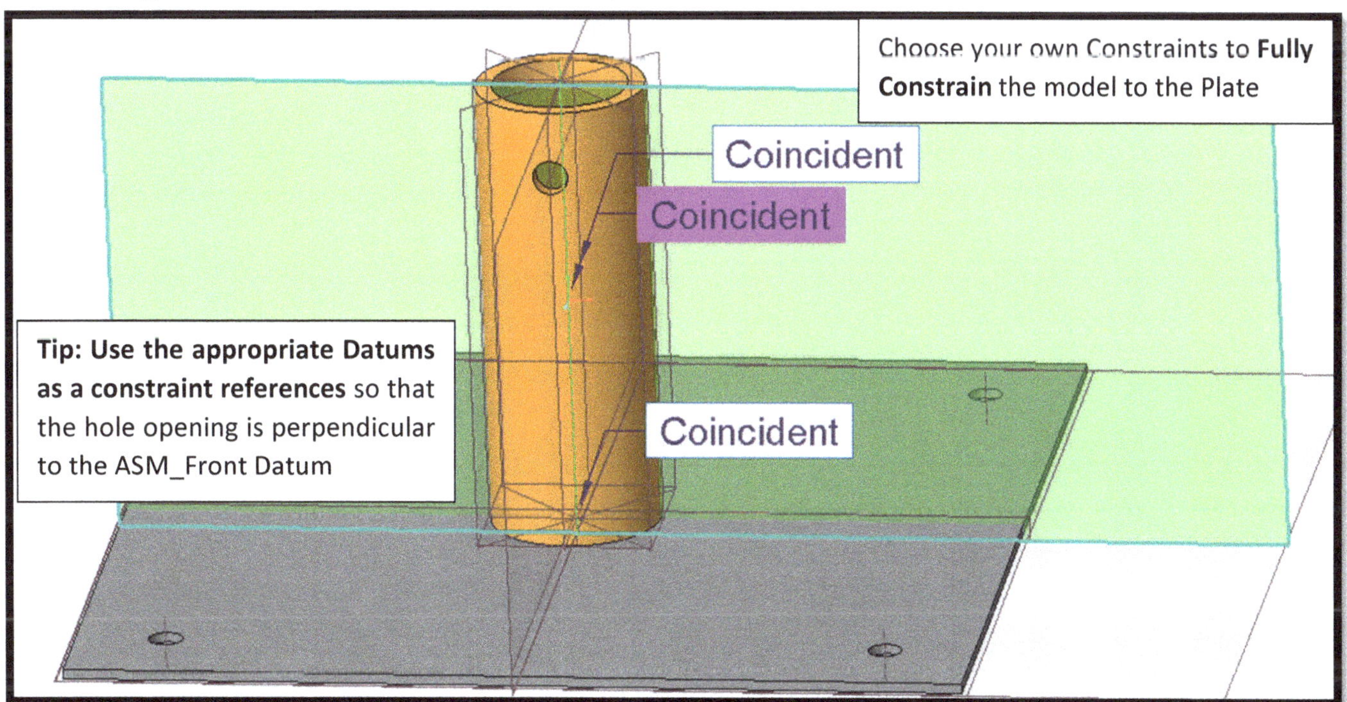

Step 5 – Assemble the *Base_Stand_Gusset* component to the assembly using the **Assemble tool**, following the proper **Constraint order listed below**:

1) Planar Surface of the "shorter" bottom side of the gusset **Coincident** to the top surface of the base plate.

2) Vertical Datum (Top) of the Gusset **Coincident** to the 45° Vertical Datum (DTM2) of the *Base_Stand_Support**.

3) Planar surface of the "long" side of the Gusset set to be **Tangent** to the outer curved surface of the cylinder. It is best to set this Tangent constraint after setting the other constraints.

* If your Gusset is on the wrong side of the cylinder, try the **'Flip' option under Constraint #2**, and make sure you used the Datum of the "Base_Stand_Support" and not an Assembly Datum or *Base_Plate* datum.

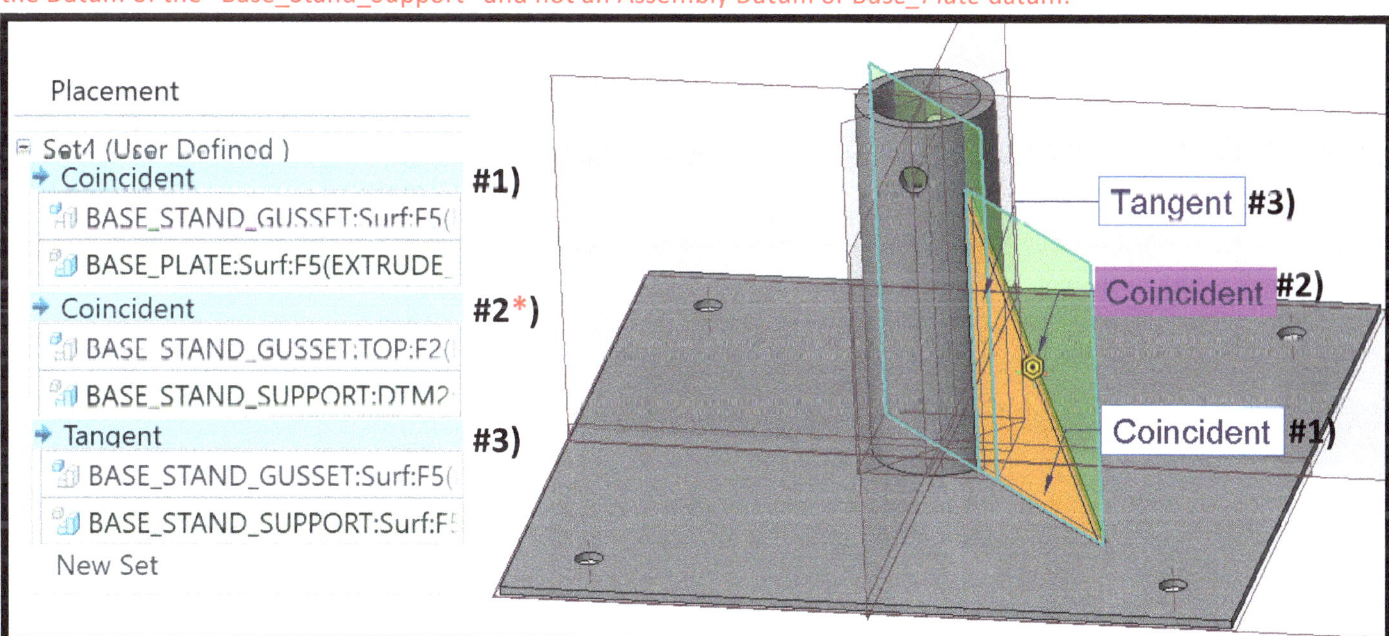

Step 6 – Use the **Repeat tool**** to repeat the placement of the ***Base_Stand_Gusset*** pieces around the cylinder. The repeat tool creates copies of a component by replacing one of the constraint references to a different entity to control the placement of the new copies.

1) Select the ***Base_Stand_Gusset*** in the Model Tree, then select the **Repeat** tool in the top toolbar.

2) Select the vertical datum constraint reference to replace during the repeat operation: **Click on the** *"Coincident - Top - DTM2"* (or similar) constraint in the menu and **click Add**.

3) Now choose the new reference for the "Added" Constraint. **Click the vertical datum of the cylinder** to use for the placement of next gusset to be 45° apart from the original. You should see the copied Gusset added to the model.

4) **Click on the next vertical datum** to place the final copy of the Gusset as shown below.

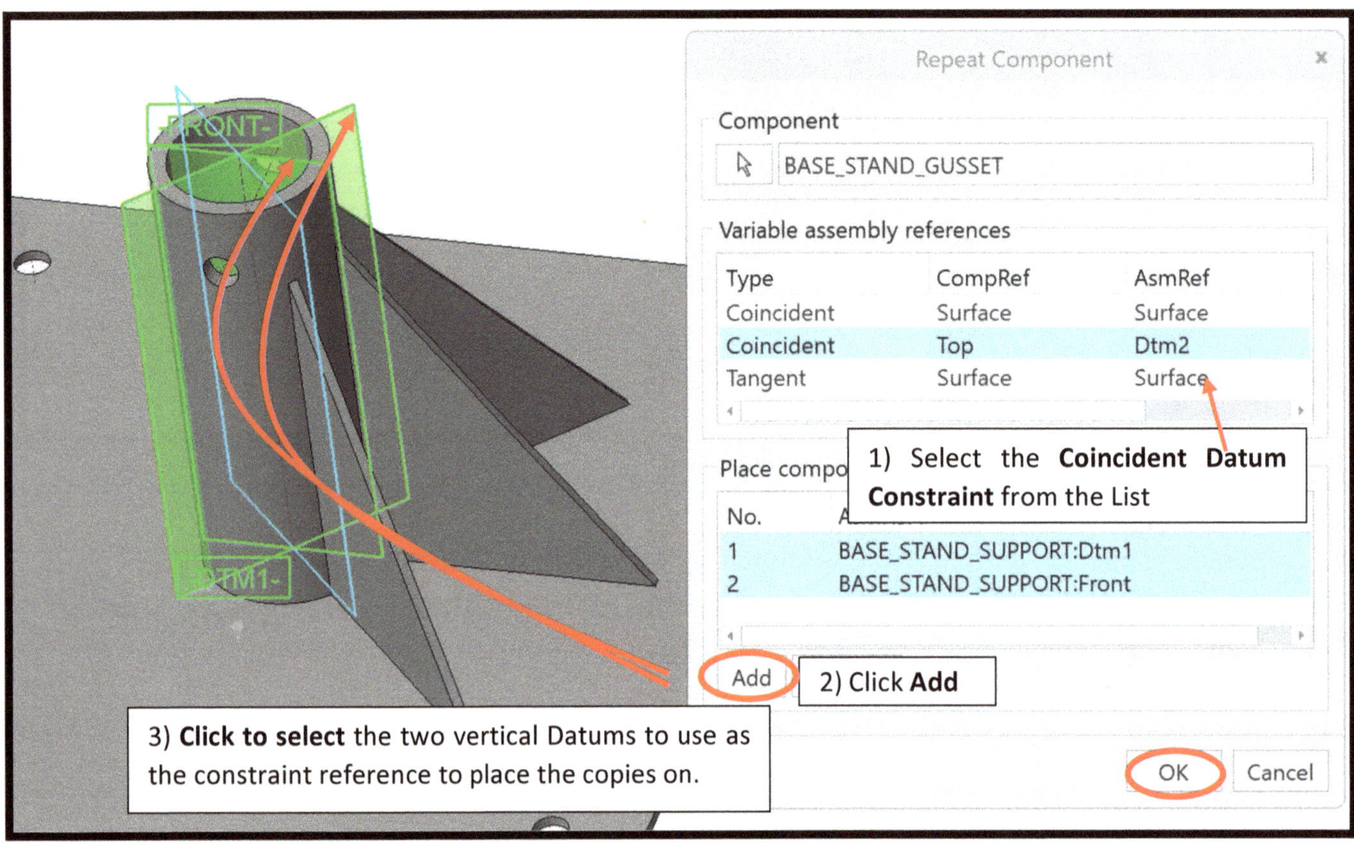

****Repeat Tool not working?** If the Gusset repeats are not working properly it could be due to the type of constraints used for the copies or original part (invalid constraints, not fully constrained, different types of constraints, etc.) or choosing the wrong constraint to swap out.

If the Repeat command does not work for you in this step you can instead Assemble another 2 copies of the Gusset manually using the Assemble Tool, or use a Pattern (Axis Type, 45° spacing, & may need to do an 8 instance pattern and toggle off the ones that are not desired) to complete the assembly similar to the image above.

Step 7 – Save the assembly before moving on the next assembly.

Assembly #2: Stand & Tower Assembly

Step 1 – If not already done earlier; download, unzip, and copy/paste the files from Blackboard for this assignment to your working directory.

Step 2 – Create a **new <u>Assembly</u>** file called "***Stand_And_Tower_ASM***". If prompted, select the "*inlbs_asm_design_abs*" template.

Step 3 – Assemble the ***Base_Stand_ASM*** using the **Assemble Tool** with a **"Default" constraint type.** This will "Fully Constrain" the sub-assembly to the Assembly coordinate system.

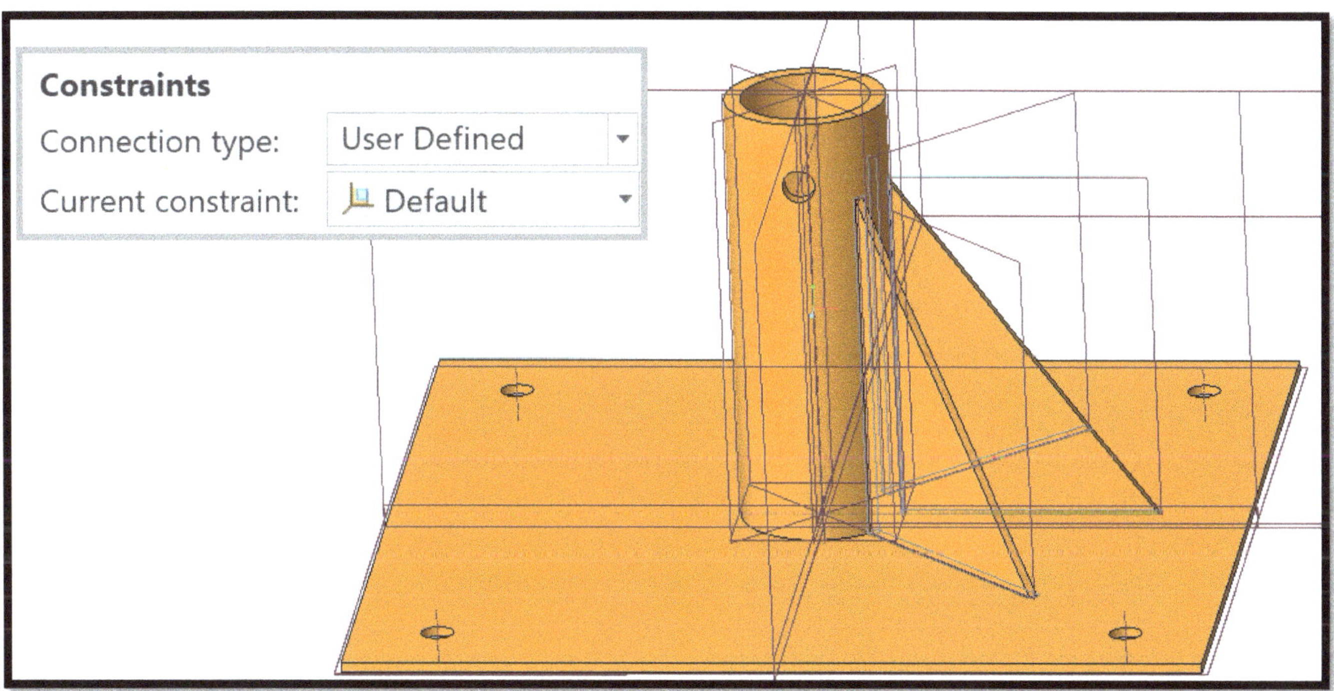

Step 4 – Assemble the ***Base_Post*** to the assembly using the Assemble tool in the top toolbar.

-   **Fully Constrain** the component by setting constraints that will place the component inside the *Base_Stand_Support* cylinder & be coincident with the top surface of the *Base_Plate*.

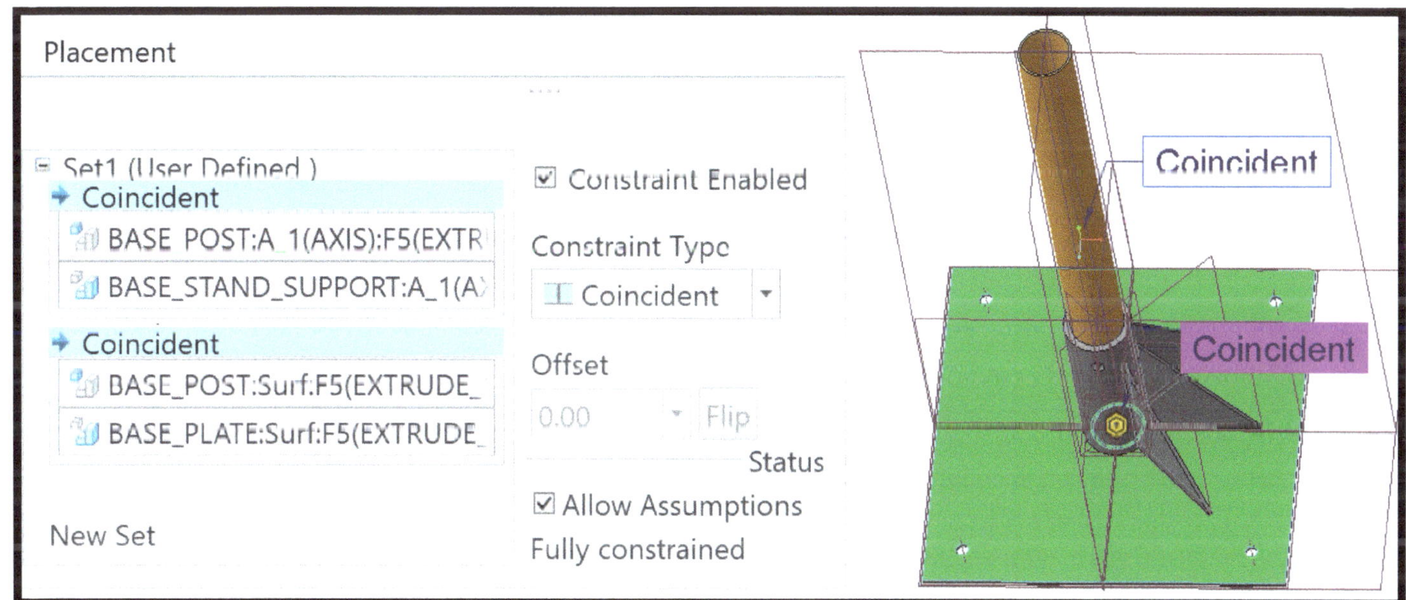

-   **Checkmark to accept the component placement** and **check your Model Tree** to make sure you do not have an unconstrained icon [] in front of the name.

Step 5 – Make changes to the *Base_Post* part file while in Assembly Mode using Activate:

The *Base_Post* requires a design update to include a new hole that will align with the hole in the *Base_Stand_Support* part, as well as removing a length of material from the bottom of the *Base_Post.* This will create a gap at the bottom of the cylinder to account for any welds, burrs, or other production effects that could cause the post from bottoming out early and prevent the holes from aligning properly in the physical prototype.

Design Change #1: Add a Hole to the Post

Activate the *Base_Post.PRT* and create a **Coaxial Hole** that aligns with the hole in the *Base_Stand_Support* so that a bolt can be added later.

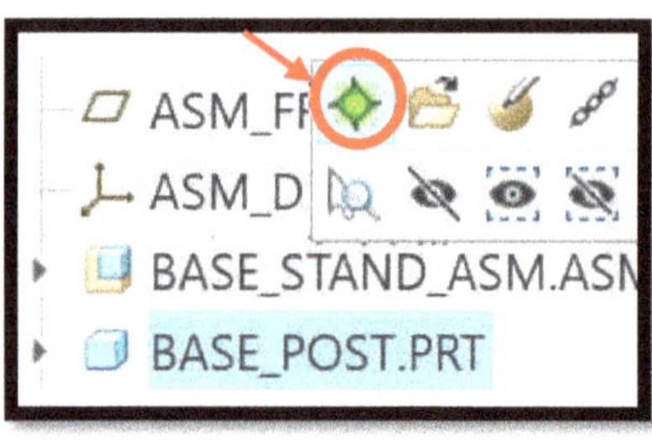

- **Activate the Post: Click on the *Base_Post.PRT* in the model tree** – choose **Activate.**
- After activating, the Post should become the only active component (everything else will be greyed out) which effectively brings you to part mode for the Post.
- **Use the Hole Tool** with the settings shown below

Hole Placement: For a Coaxial Hole Placement, **you must hold CTRL on the keyboard** to select the two primary references of an **Axis** and a **start Plane**, which should be the <u>center vertical datum</u> instead of the curved surface.

Hole Settings: Simple Type, 0.38" Diameter, Symmetrical with 5.00" Depth (*enough to go through both sides*).

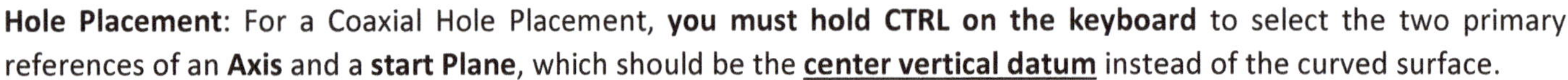

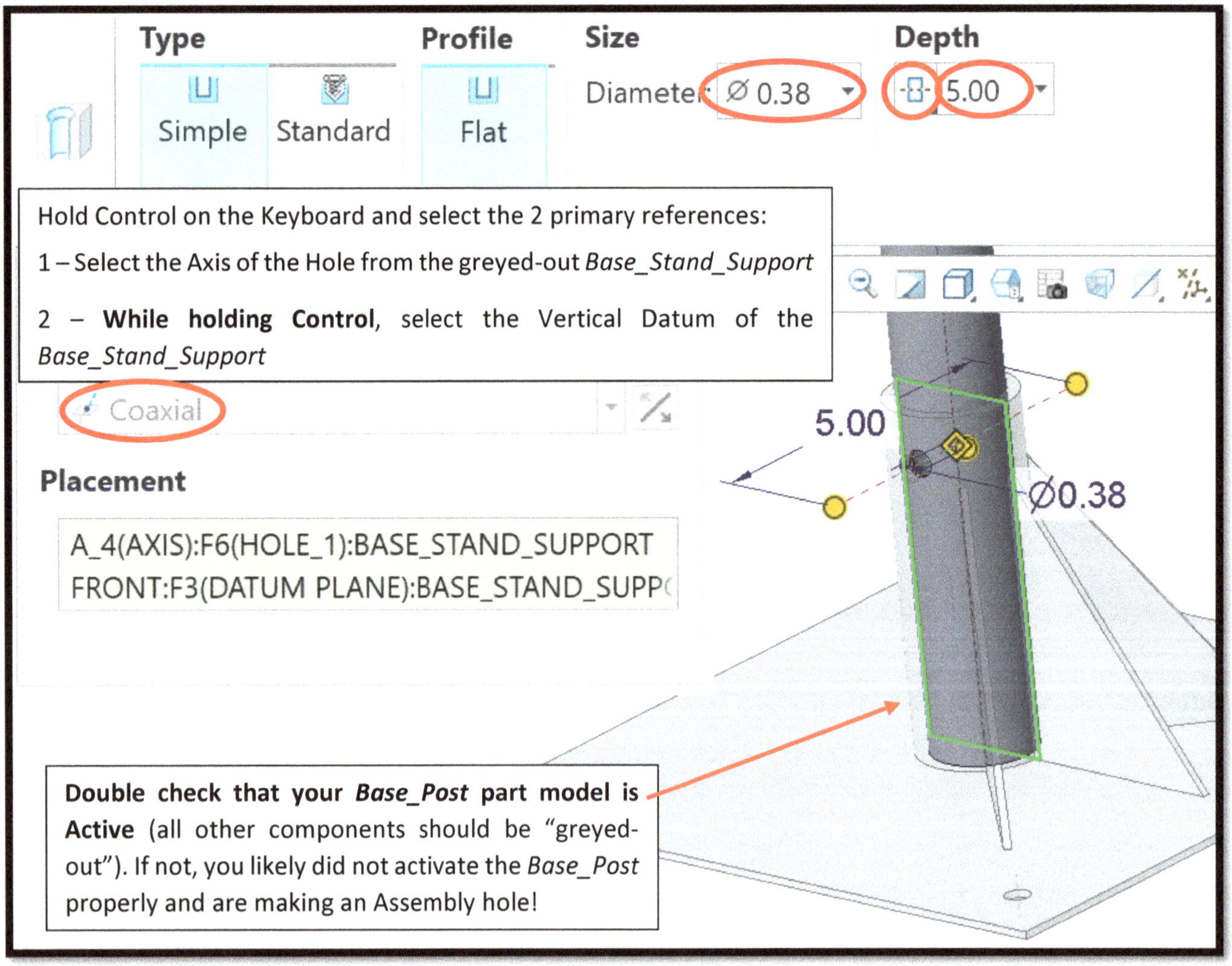

Design Change #2: Remove the bottom section of the Post

Activate the *Base_Post.PRT* (if not already active) and add an **Extrude** to remove the bottom **1.25"** of material, effectively shortening the length of the Post without affecting the position of the Hole.

Extrude Settings: Use the **bottom planar surface of the pipe as the sketch plane*** **and sketch a shape larger than the pipe diameter.** Set the **Extrude Depth to 1.25",** toggle on Remove Material, and toggle the direction as needed.

*Tip: You can **tap RMB** to cycle which entities under your mouse are being selected to choose the correct sketch plane, as the *Base_Plate* will be blocking the view (or temporarily "Hide" the *Base_Plate* in the model tree). You could also use the top surface of the *Base_Plate* as the sketch plane.

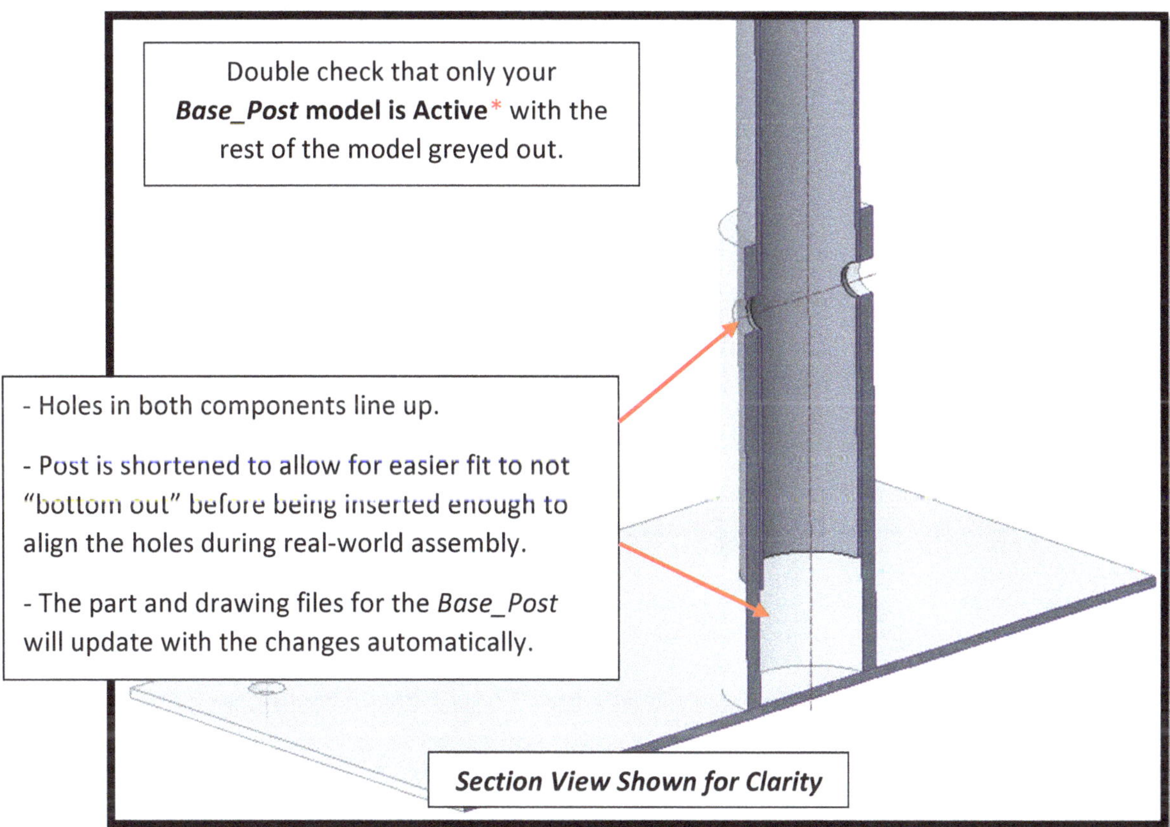

* Note: **You must have had the *Base_Post* "active"** otherwise you will be making an "Assembly" feature. Assembly features act differently as they can only remove material, do not update the part files, and will interact with all components within its volume instead of just a single part.

Note the image at right which shows the "assembly" Extrude which interacts with the entire assembly incorrectly and is listed as a main level item instead of within the *Base_Post.PRT*. This will occur if you did not Activate the Part.

Step 5 – 2 - Save the Part & then Activate the Assembly in the Model Tree (**RMB on the Assembly name – Activate.)** *Or press CTRL+A on your keyboard to activate the current window.*

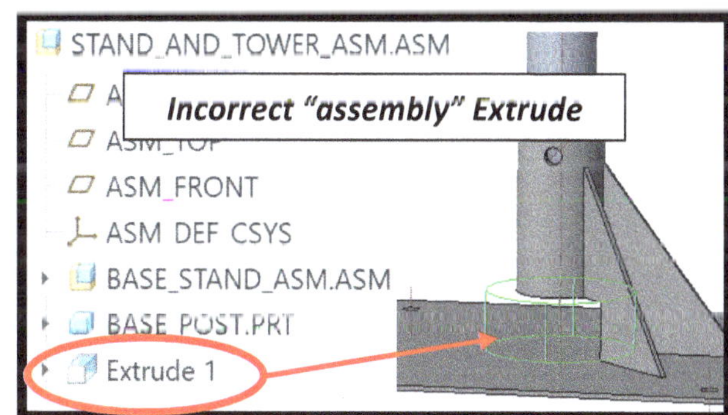

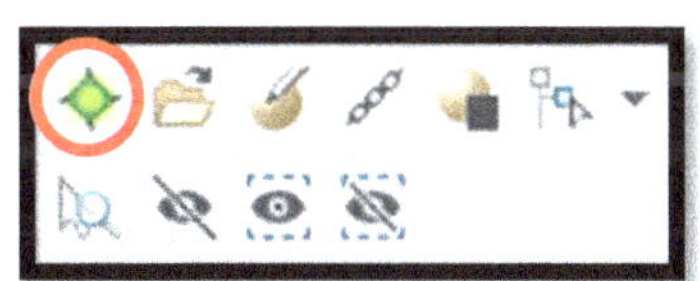

Step 6 – Use the website www.mcmaster.com to download and import the two parts (a bearing and a pin) for use in the Assembly. These files are also included in the lesson's Zip File if you cannot access the McMaster website.

6 – 1 - Search for the Part # of each item listed below on **McMaster.com**. Under the details for the item change the CAD file type to **3D Step File** and press **Download** (*or RMB on the Download button and choose Save Link As*).

- Save these files to a known location and then move to your working directory.

Desired Components: 6383K244 and 98416A018

Note: CREO can open 3D Solidworks files from McMaster, but it is recommended to use the STEP file option as there may be issues that prevents saving or modification of the Solidworks files from McMaster CREO.

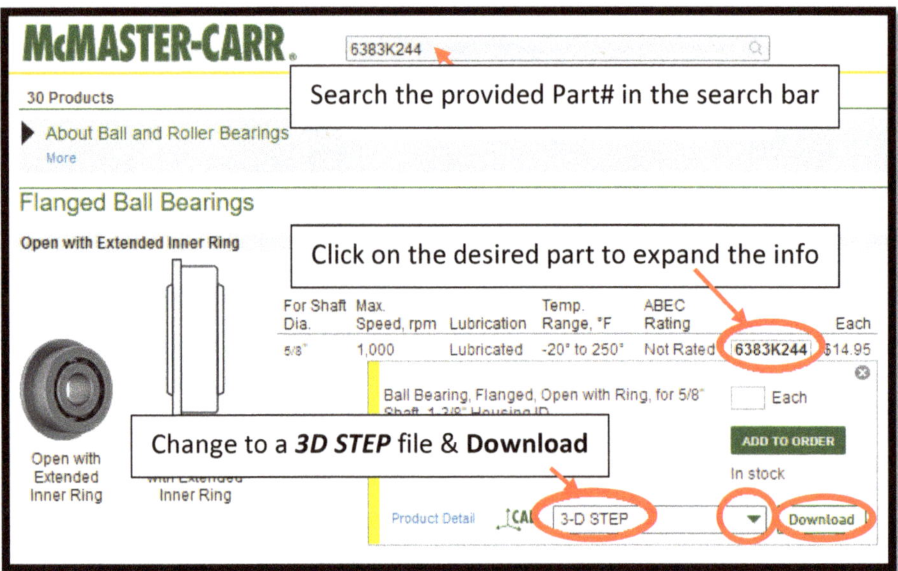

6 – 2 - In CREO; use **File – Open – change the File type to ".STEP" or "All Files"** - locate the downloaded STEP File - **press Import.** *By default, CREO only looks for CREO files so you must change the Type setting to "All Files" or ".STEP".*

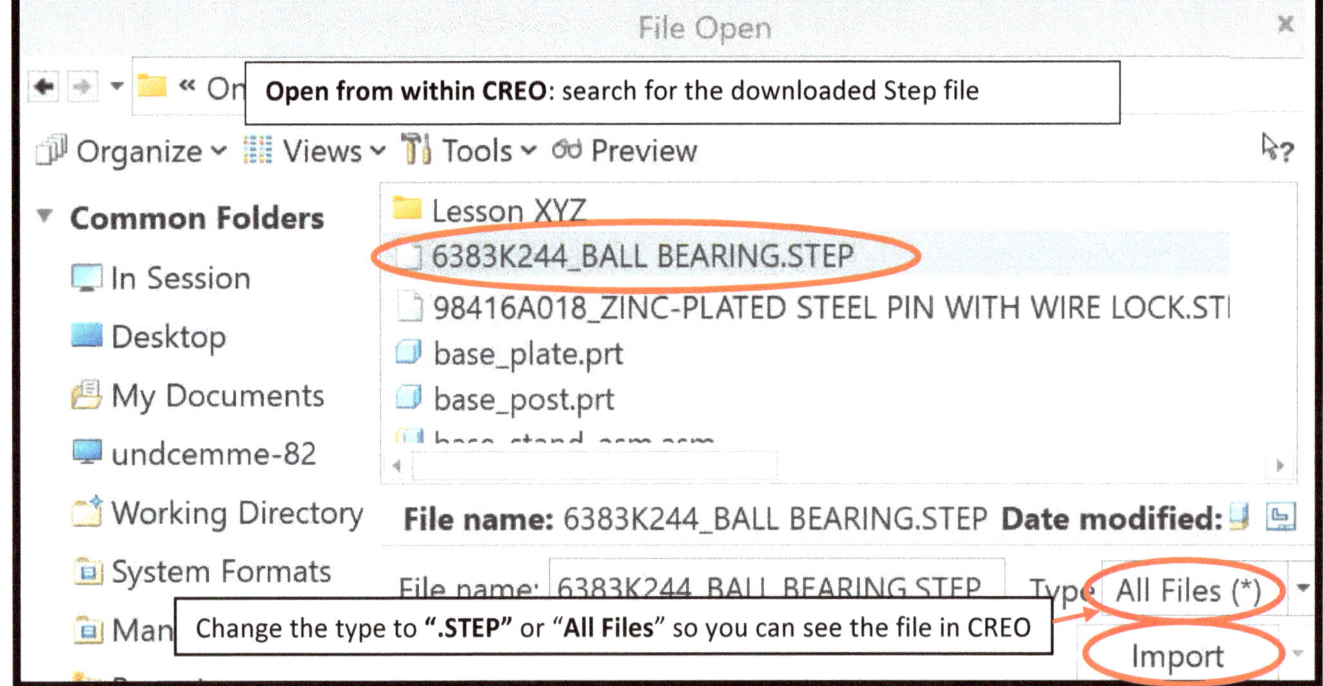

Tip: If you cannot find the file: make sure it is not compressed/zipped, that you are allowing CREO to look for STEP or All Files, and that you are looking in the correct folder for your downloads.

6 – 3 - The **File Import window** should open and you need to choose if it is a Part or an Assembly. Even though the Bearing & Pin are assemblies, we want them to behave as a single unit for this model so will **choose to import them as Part files.**

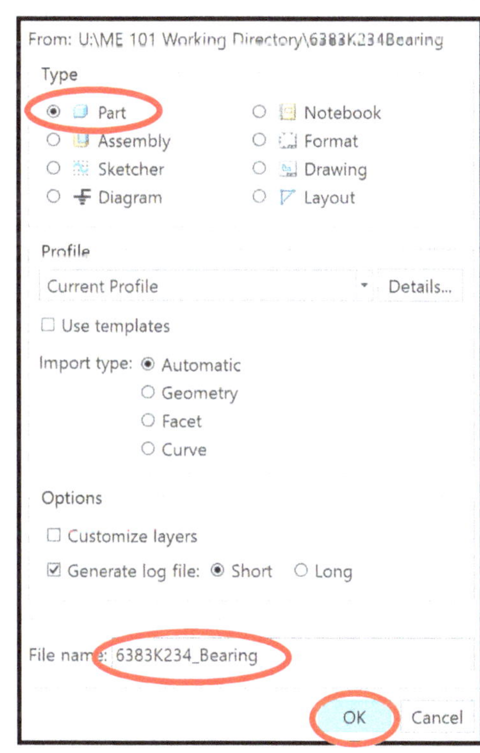

- Choose "Part" as the component type, **provide a proper name** (*use the McMaster Part# with some description added),* then press **OK**. The Part should open in a new Part window. You may get an error message about the names of some features being modified; disregard these errors and press close to continue on.

6 – 4 - Repeat the process to download and import the Pin component (**98416A018**) as a Part file into CREO as well.

Step 7– After importing both the parts into CREO you will need to **create a Datum Axis and Datum Planes** to use as Assembly Constraint references. By default, the imported parts do not have Front/Top/Right views or Axis on holes which means you cannot easily make drawings of the part or assemble it until you complete some more steps.

Drawings of these parts are not needed for this assignment. But if they were you would download the 2D Drawing PDF from McMaster to include in a drawing package instead of creating your own. If the imported parts will be modified and require drawings you can refer to the Appendix to learn how to setup the required datums as the Front/Top/Right for drawings views so that a drawing can be created of an imported part.

Bearing Part Setup: After importing you need to __add an Axis__ to the center hole & set the **Material** and __color__.

7 – 1 - Open the Bearing part file (if not already open) – use the **Axis tool** from the top toolbar – LMB to select the outer edge of the hole or cylinder as a reference – press OK. The Axis is now added to the part Model.

7 – 2 - Set the **Material properties (Steel)**

7 – 3 - Set the color of the bearing with a Blue or similar color to provide contrast to the other components. Use the **Color Appearance tool (View tab).**

7 – 4 - Press Save.

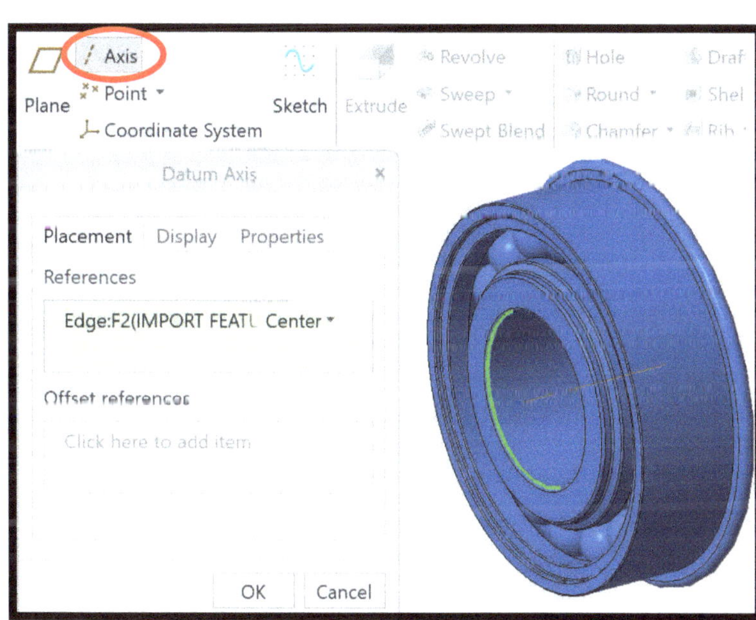

Pin Setup:
After importing you need to **add an Axis** as well as a **horizontal Datum** to the model that can be used for Assembly constraint references.

7 – 5 - Open the Pin part file (if not open already) – use the **Axis tool** from the top toolbar – LMB to select the outer edge of the cylinder as a reference – **press OK**. The axis should now be added.

7 – 6 - Use the **Datum tool** to add a Datum – select the center Axis as the first reference – while holding **Control on the keyboard** select **a vertex point on the edge of the bolt circle** (the point should appear around the 3 O'clock or 6 O'clock positions of the circular edge of the face.)

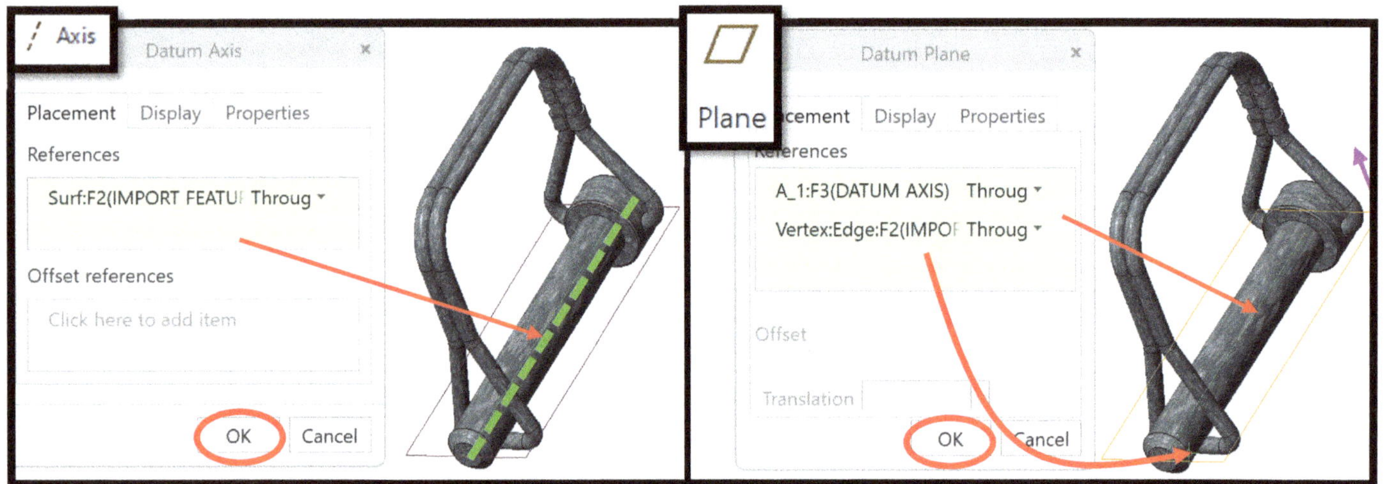

7 – 7 - Set the Material Properties (Steel) and **color** of the Pin with some version of a Steel color or similar to provide contrast to the other components. Use the **Color Appearance tool (View tab).**

Note - Having the Material properties set for these parts is required for the cell to autofill in the BOM table for the assembly drawing.

Step 8 - Save the component. Make sure that both of these new CREO Part files **are located in your Working Directory.** You can remove the McMaster STEP files from your directory if you want as they are no longer needed.

Step 9 – Assemble the Bearing in the top of the ***Stand_And_Tower_ASM*** using the Assemble tool. *Make sure to assemble the CREO file for the bearing and not the STEP file.*

- **Fully Constrain** the bearing by choosing constraint types & references that will place the lip of the Bearing on top of the *Base_Post* with the majority of the bearing body pressed into the pipe.

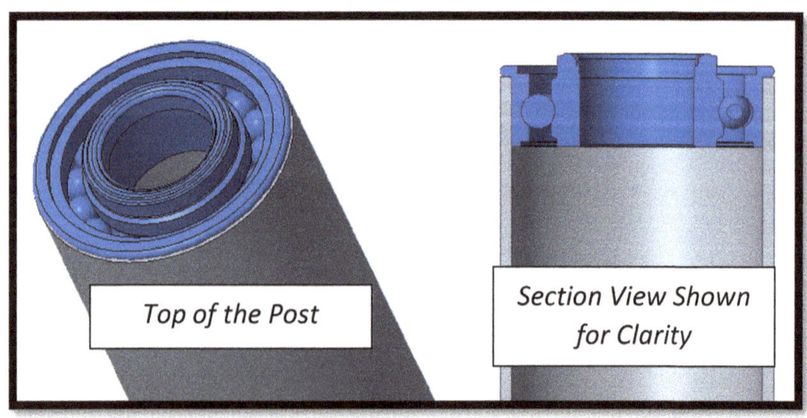

Step 10 – Assemble the Pin to the assembly using the Assemble tool in the top toolbar. **Fully Constrain** the component by choosing constraint types & references that will place the Pin in the holes of the Support and the Post.

- The Pin "handle" should be to be horizontal and positioned over the Gussets to match the key. Use a **Parallel** or **Angle Offset Constraint** between the Datum on the Pin and the appropriate vertical Datum on the assembly to control the handle angle. You may need to toggle off "Allow Assumptions" to add the 3rd constraint.
- Use either a **Tangent** or a **Distance Offset** to position the head of the bolt near the cylinder surface. It does not matter which hole the Pin head is near as it can be inserted either way.

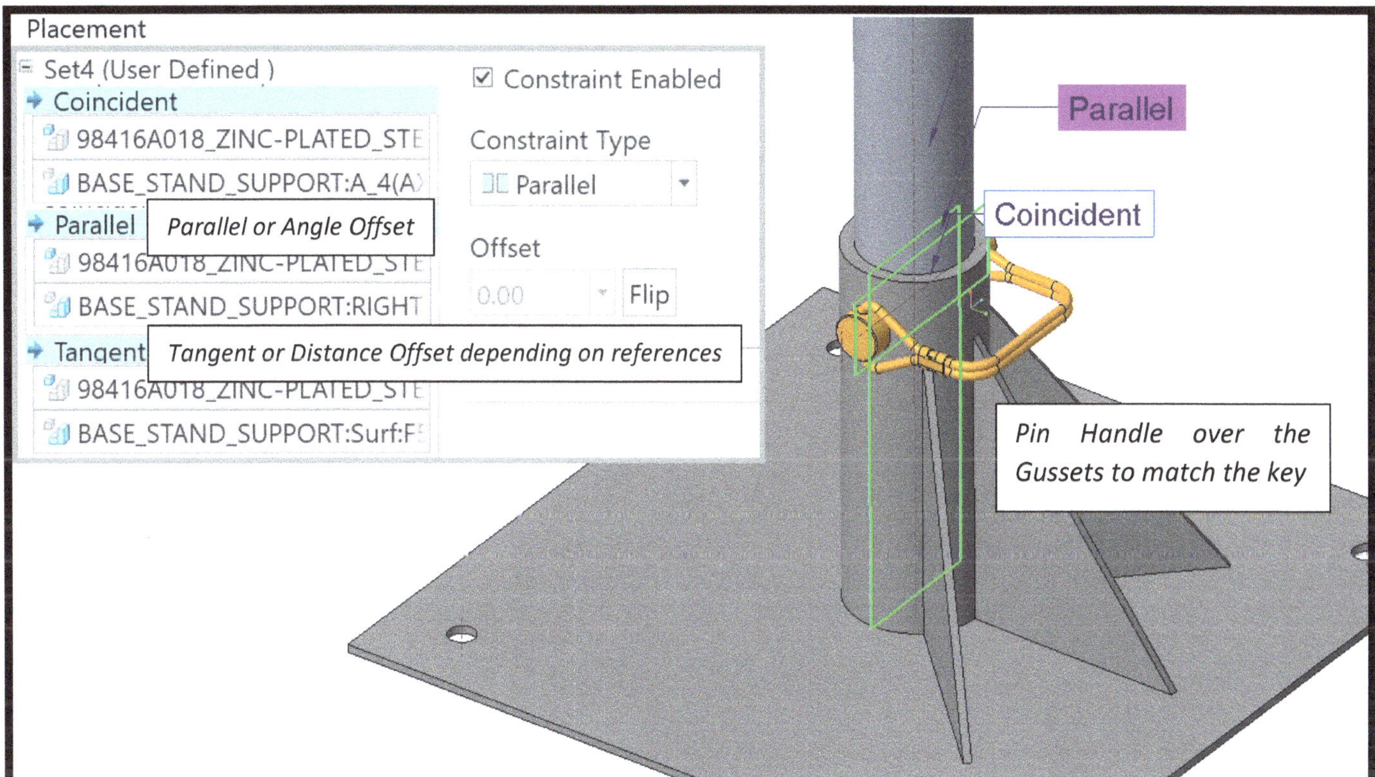

Step 11 – Save the assemblies and create two assembly drawings to match the keys. Save all these Parts, Assemblies, and Drawings for future use on the project.

Note that all files must be in the same directory or you will have missing components when you next re-open the assembly after closing this session of CREO! Check through Windows File Explorer that your CREO versions of the Bearing and Pin are actually saved to the correct directory with the rest of the Assembly files.

End of Lesson 12

Lesson 13 – Nacelle Platform Assembly & Post Insert

This lesson will complete the assembly of the Nacelle Platform for the wind turbine project. You will be specifying your own hardware from McMaster to use to assemble the components together properly as well as modeling and designing your own "Post Insert" component to connect the platform to the bearing from the previous lesson.

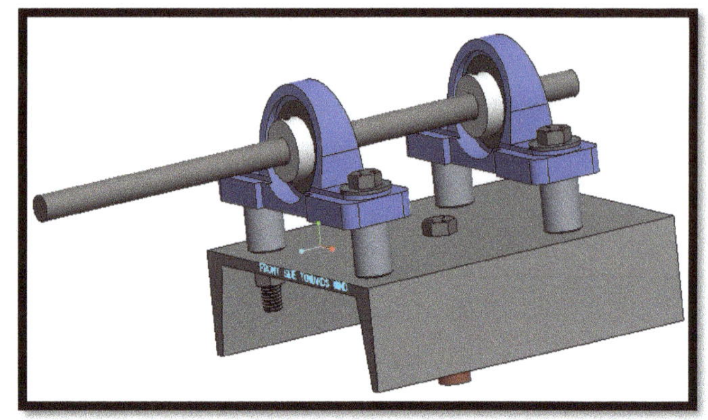

The Nacelle Platform is designed act as the core component for the rest of the wind turbine design, such as the blades, generator, and tail fin that will be created in future lessons. This platform was designed so that it can quickly attach & detach from the base stand to minimize down time between testing of each prototype for the campus students, as well as serve as a mounting area for the main shaft and other components that will be added later.

Assembly #1 – Nacelle Platform Assembly

Step 0 – Make sure that all files for this assembly are in your working directory, including the assembly and part files from Lesson 12.

- **Download, extract/unzip, and copy/paste the files from Blackboard** for this assignment to your working directory. Save all files in the same working directory folder or CREO will not find them next time it is opened.

Step 1 – Set your **Working Directory** and create a **new <u>Assembly</u>** called "**NACELLE_PLATFORM_ASM**".

Step 2 – Assemble the **Nacelle_Base. PRT component** using the Assemble Tool with a **"Default" constraint type.** This will "Fully Constrain" the new component to the Assembly coordinate system.

Assemble

Note that the *Nacelle_Base* has a **blue Sketch of text** on the Front surface of the part to identify which face will face the wind. This is not a functional feature in the part and is only there for your reference.

After completing this lesson, you should **delete the Sketch feature** from the model tree so that it does not appear on the drawings or future models.

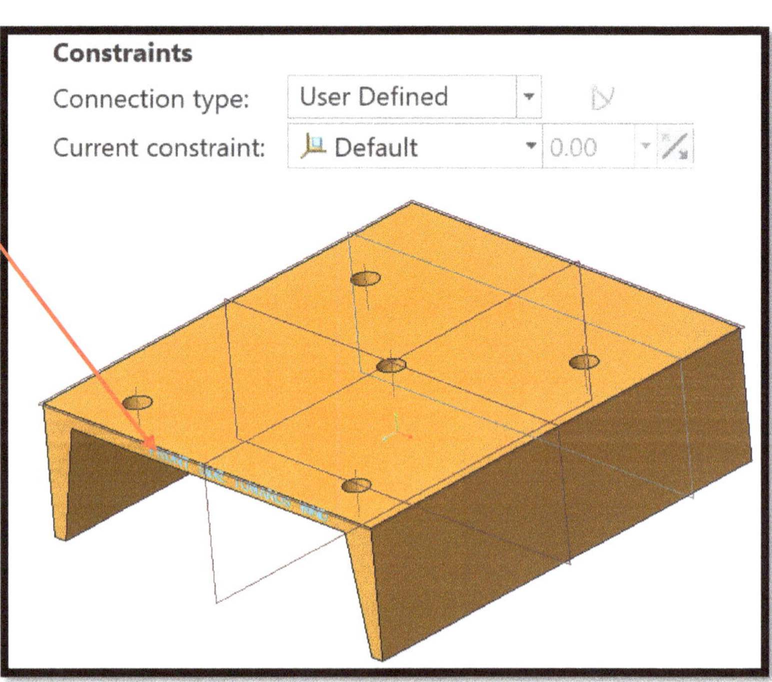

Tip: Toggle off the standard Hole Note Annotations in the quicktoolbar to declutter the screen.

Step 3 – Assemble the *Pillow_Block_Bearing_1_2_inch.PRT* to the assembly using the **Assemble tool** in the top toolbar.

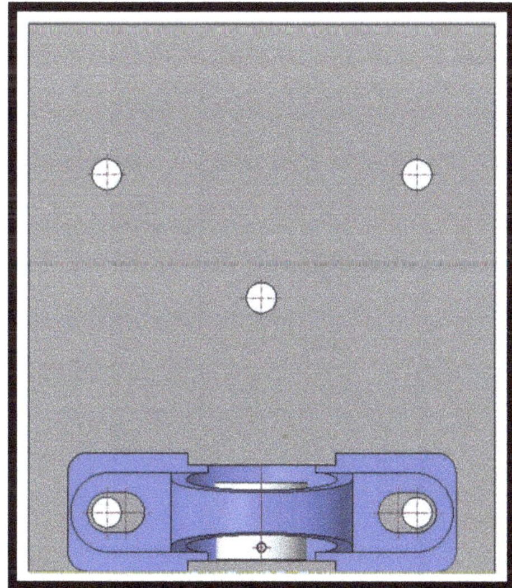

- **Fully Constrain** the component by choosing constraint types & references that will place the component on the top surface of the Nacelle Base, lined up with the mounting holes, and **have the longer/protruding cylindrical "raceway" of the bearing facing towards the front**.

Note that the bearing has a slotted hole for mounting, which has two axes on each slot. You will need to select the correctly spaced axis for a constraint to be sure the Bearing lines up with the opposite mounting hole on the Base.

Step 4 – Assemble a 2nd copy of the Bearing to the *Nacelle_Base* that lines up with the rear pair of mounting holes. Use either the **Assemble Tool** or the **Repeat Tool** to place the second bearing.

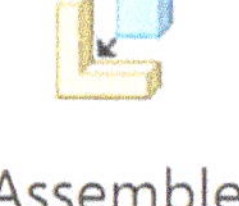

Assemble

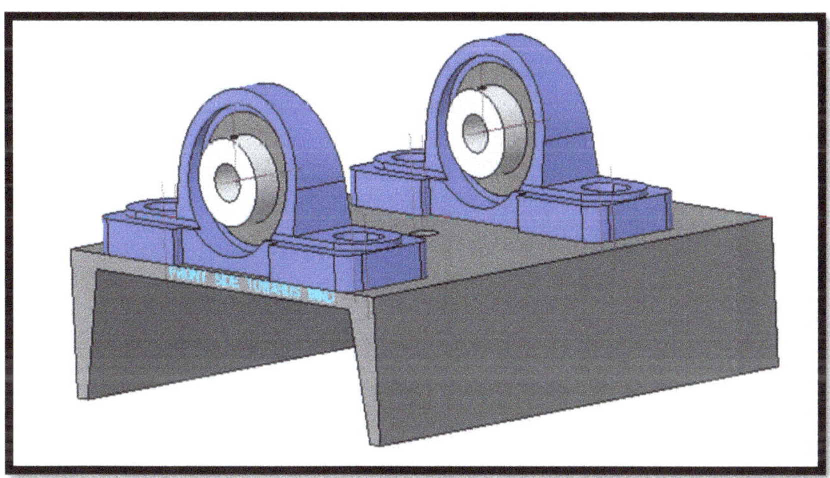

Repeat Tool Option: Depending on how you constrained the original copy you may be able to **Repeat** *(covered in Lesson 12)* and swap out both Axis references at once by holding Control on the keyboard in the Repeat menu when choosing which references to replace. Otherwise, just assemble another copy manually making sure it is fully placed

Note: A **Pattern** is not ideal for this step. A Directional type pattern could be used, but would not automatically update if the hole position on the base changed later on. A Reference type pattern will attempt to add extra copies of the bearing on each axis of each hole (which adds too many copies) and the Point type Pattern uses the center of the Bearing as a reference, which doesn't line up with the hole axis.

Step 5 – Assemble the ***Flat_Washer_3_8.PRT*** to the assembly. **Fully Constrain** the component by choosing constraint types & references that will place the Flat Washer on top of the bearing, aligned over the appropriate mounting hole axis.

Tip: The center Axis will be very small and difficult to select since the washer is very thin. Toggle on the option in the top toolbar to **show component in a Separate Window** to better zoom in and find the Axis of the Washer.

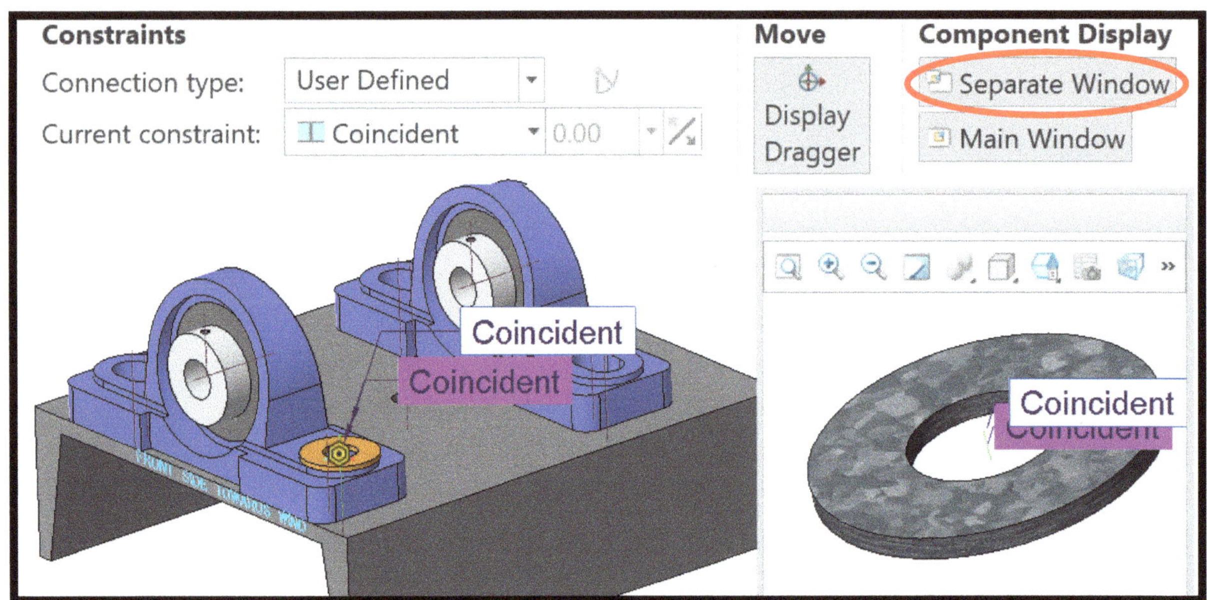

Step 6 – Repeat to add more copies of the ***Flat_Washer*** onto the top of the other mounting holes for both bearings**.**

- Select the Flat Washer in the Model tree – **Repeat Tool** – choose the reference to swap out -**press Add** – click the new axis reference for each new copy.

Note that this lesson will have you Repeat the Bolt, Washers, & Nuts individually, as grouping items does not work with the Repeat tool. Other methods could also work, such as Group and Pattern (Point type pattern) or assembling all hardware together as a sub-assembly and then repeating the sub-assembly.

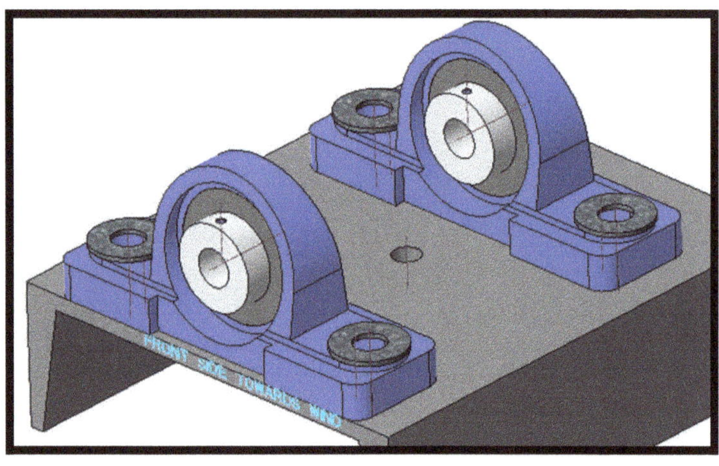

Step 7 – Assemble the ***Lock_Washer*** to the underside of the *Nacelle_Base* aligned with the base mounting holes.

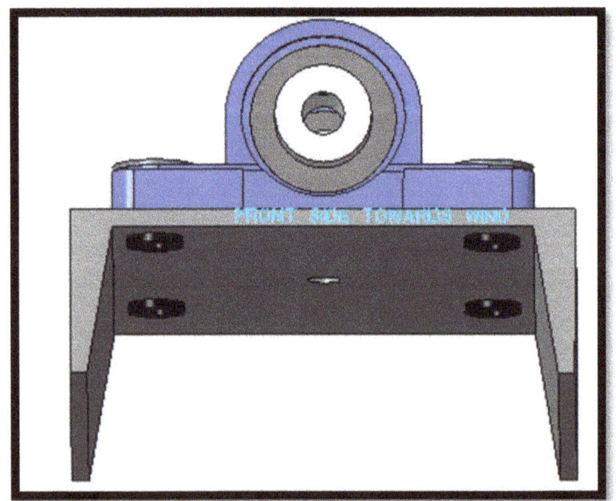

Step 7 – 2 - Use the Repeat Tool to add the Lock Washer to all four mounting holes. (Select the Lock Washer in the Model Tree - Repeat Tool – select the Axis constraint - Add - click the desired axis of the holes to add the copies to.)

Hardware Specification: "Spec" or find a Bolt from McMaster to meet the requirements.

A bolt needs to be sourced from McMaster that can be used to physically attach the Bearings to the Nacelle Base. To do so we first need to measure the max diameter and determine the length of bolt that will work for the design.

Step 8 – 1 - Use the **Measure Tool** (on the **Analysis Tab**) to measure the **Distance the bolt will need to pass through** as well as the **Diameter of the hole** on the bottom of the *Nacelle_Base*. These will be used to choose a bolt size.

When using the **Measure Tool** you must first choose <u>which type of measurement</u> you want in the toolbar (see below).

- For a **Diameter** you should click on the curved edge or cylinder of the hole on the bottom of the *Nacelle_Base* (the smallest hole the bolt will need to pass through).
- For the **Distance** measurement, **you must hold Control on the keyboard** to select the two <u>planar surfaces</u> as references to measure between the top of the flat washer and the bottom of the lock washer. *Tip: Do not select edges as it will measure at an angle between 2 points which is not the same as the perpendicular distance between two surfaces.*

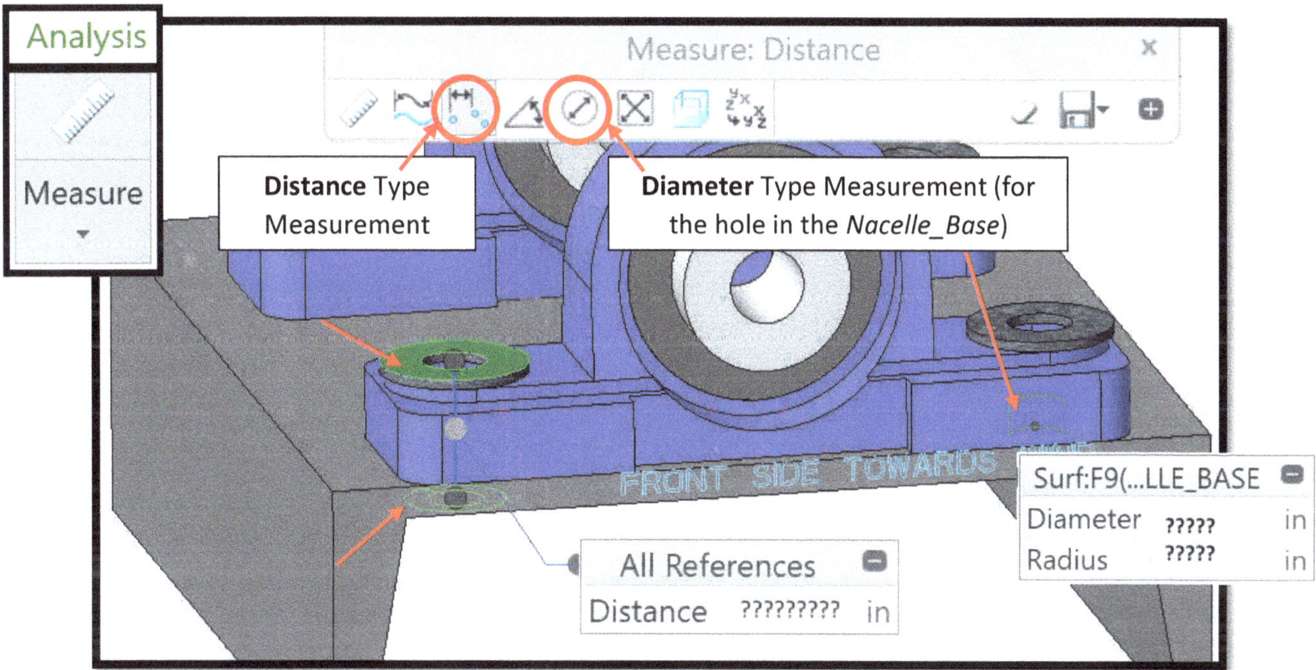

Step 8 – 2 - Now choose a "standard" (non-metric) **Bolt Size** to fit within the measured hole size. Bolt sizes are specified in fractional sizes as well as the option of "Fine" or "Course" threads, with Course having fewer threads per inch.

For example, a ¼" bolt comes in either a course ¼-20 size with 20 threads per inch or a fine ¼-28 size with 28 threads per inch. Typical options for the course bolt sizes would be (**1/4"- 20**), (**5/16"- 18**), (**3/8"- 16**), (**7/16"- 14**), (**1/2" - 13**). <u>Choose the size from the list above that fits best.</u>

Step 8 – 3 - Determine a **Bolt Length**. It must be long enough to protrude out the bottom enough to fit a nut as well as a future Bearing Spacer that will be added to the model that will raise the Bearing from the platform. Choose a length that accounts for this by adding 3/8" for the nut as well as 1.25" to 1.75" extra length.

Bolt Length = (Measured Distance between washers) + 3/8" (Nut Thickness) + (1.25" to 1.75" Extra Length)

Bolt Length Range = Min: ____________ Max: ____________

Step 8 – 4 - Other Bolt Requirements:

Other specifications that must be considered when choosing a bolt are the Bolt Type are listed below.

Partial vs Full Threads: Either will work for this project. Typically, in higher stress applications the non-threaded section of the bolt is beneficial to provide more strength (it has more material than the threaded cross-section.)

Bolt Head/Type: Any non-metric bolt with a flat-bottomed head (i.e. **Hex**, Socket, Flange) will work fine.

Material & Strength: Bolt material can be any material which provides a **Tensile Strength greater than 30,000 PSI** (typically **Steel**). This value is provided in the details of the specific catalog pages.

Coatings: Black Oxide is the lowest cost but prone to rust. Zinc has better corrosion resistance but Galvanized has the long-term resistance to moisture if exposed outdoors. But since this is to be used indoors it doesn't matter.

Cost: Choose a Bolt & Nut that will combined **cost < $30** (disregard shipping charges), with at least **4 quantity**.

Nut: Also find a nut that matches the bolt size and material.

Step 8 – 5 - Find your bolt:

Go to **McMaster.com** – click on the "**Screws & Bolts**" Category – use the filters on the left menu to set the **Size/Thread (diameter), and Head Type** and then start browsing the remaining options in the main window. Leave Length unfiltered so you can find items in the desired range instead of a specific length.

Note that the more settings you add to the filter the less options will be available so don't try and set all filters at once or you may end up with a specialty/expensive bolt option. It is suggested to set a few filters at first and then browse before adding more.

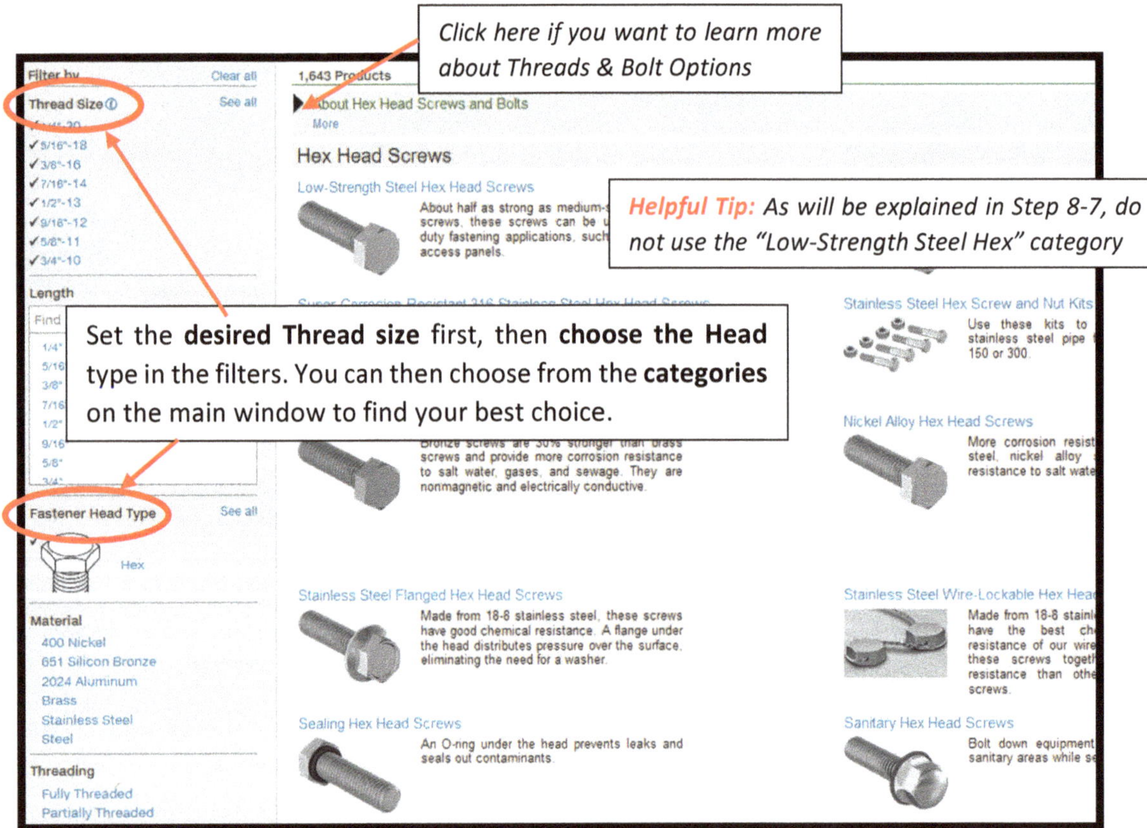

This Step Continued on Next Page:

When you get to the actual catalog page for the chosen item you will see a column for the **Tensile Strength, Length, Cost**, etc. so that you can choose a bolt that fits the requirements.

Step 8 – 6 - After finding a suitable bolt click on the **Part #** to expand the details. Download both the **2D PDF** (to save for later in the final report) and the **3D STEP file** to import into CREO. **You will not create Drawings of these models**. Instead download the *2D PDF* provided by McMaster to save for later.

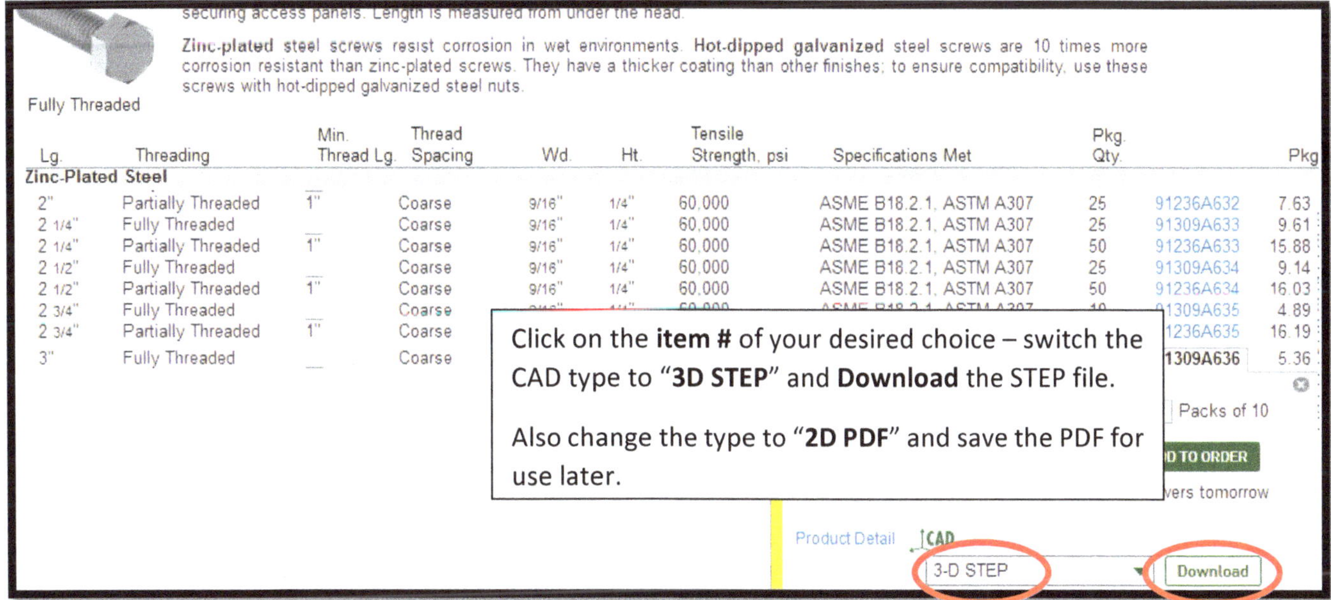

Fully Threaded

Lg.	Threading	Min. Thread Lg.	Thread Spacing	Wd.	Ht.	Tensile Strength, psi	Specifications Met	Pkg. Qty.		Pkg
Zinc-Plated Steel										
2"	Partially Threaded	1"	Coarse	9/16"	1/4"	60,000	ASME B18.2.1, ASTM A307	25	91236A632	7.63
2 1/4"	Fully Threaded	—	Coarse	9/16"	1/4"	60,000	ASME B18.2.1, ASTM A307	25	91309A633	9.61
2 1/4"	Partially Threaded	1"	Coarse	9/16"	1/4"	60,000	ASME B18.2.1, ASTM A307	50	91236A633	15.88
2 1/2"	Fully Threaded	—	Coarse	9/16"	1/4"	60,000	ASME B18.2.1, ASTM A307	25	91309A634	9.14
2 1/2"	Partially Threaded	1"	Coarse	9/16"	1/4"	60,000	ASME B18.2.1, ASTM A307	50	91236A634	16.03
2 3/4"	Fully Threaded	—	Coarse			60,000	ASME B18.2.1, ASTM A307	10	91309A635	4.89
2 3/4"	Partially Threaded	1"	Coarse						91236A635	16.19
3"	Fully Threaded	—	Coarse						91309A636	5.36

Step 8 – 7 - Import the STEP files as <u>Parts</u> into CREO. Refer to the steps in the Appendix for **"Importing Vendor Files"** for instructions on importing STEP files.

Important Note: Some CAD files from McMaster use an outdated Surface Model (hollow instead of a solid model). The **"Low Strength" Hex Bolt bolt chosen above** is one of these that will **<u>not work</u>**. The solution is to choose a different bolt category which probably does have updated models, or on many bolts there is the option to Download the 3D Step **"<u>No Threads</u>"** version which has a solid model that works fine.

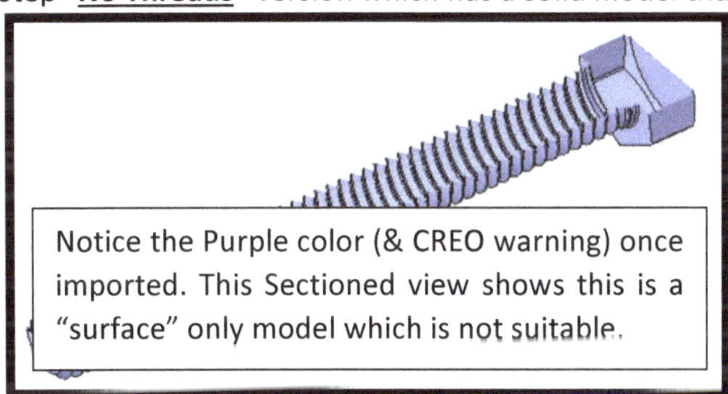

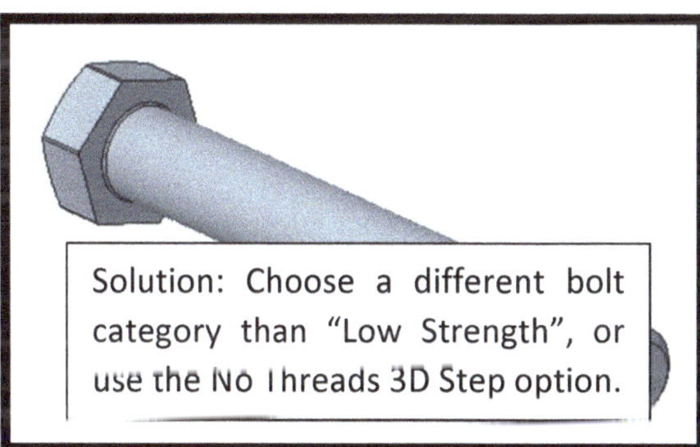

Step 8 – 8 - Setup the **Material Properties, set a Color, and add a Datum Axis** to use as an Assembly references.

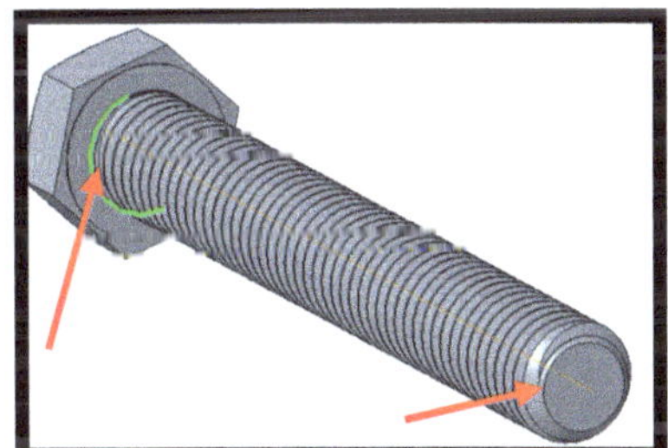

Tip: Adding an Axis to a Fully Threaded Bolt:

To add an Axis to a fully-threaded bolt you should use the circular edge at the very bottom or top of the threaded length as a reference. The edges on the threads themselves are angled and may not create an axis in the center of the model.

Step 8 – 9 - Go back to the McMaster home page and **repeat the above steps** to find a suitable **Nut** for the bolt and import it into CREO.

Step 9 – Assemble the **Bolt and the Nut** to the assembly, and then **Repeat** each of them onto the other 3 locations.

Note- You should use an Axis to Axis constraint type when assembling the bolts & nut into the holes so that the Repeat tool can swap out the axis reference when making copies. You likely already did but if it doesn't work that could be why.

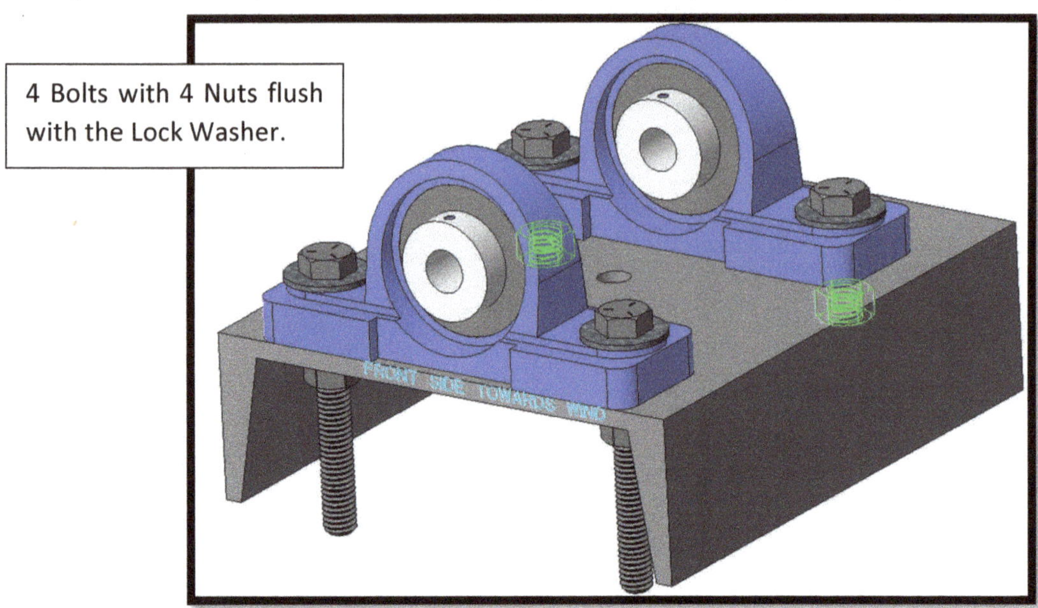

Step 10 – Clean up the Model tree by **adding all the mounting Washers, Bolts, & Nuts to a Group** and rename the Group as ***Bearing_Mounting_Hardware***. Note that this group will not affect the BOM tables on the drawings but it does make the model tree easier to read.

To Create a Group: <u>Hold Control</u> on the keyboard and select all the desired hardware items (they must be next to each other in the model tree without gaps) – in the pop-up menu choose the **Group icon** (you may need to hold down RMB on the item to have the menu appear)

To Rename a Group: LMB on the group – hold RMB - Rename.

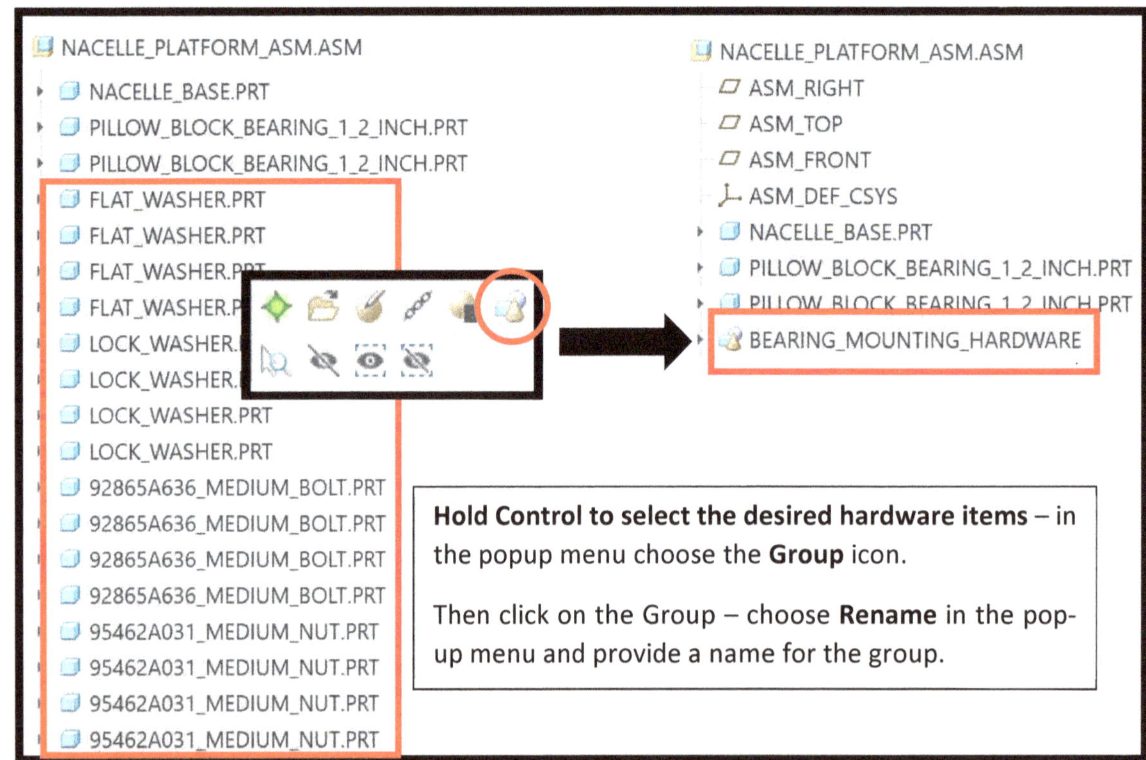

Step 11 – Use a **Pre-Defined (*Mechanism Type*) Constraint** to Assemble the ***Shaft_Threaded_1_2.PRT***.

These Pre-defined constraints allow a **Range of Motion & Limits** to be set to control or limit the part movement while still being constrained in the other degrees of freedom. For this part, use a **Cylinder** predefined constraint to allow rotation of the shaft and apply limits on the linear translation of the shaft in and out of the bearing hole. These types of constraints are required to be able to use the **Mechanism Application** in CREO (not covered here).

Assemble

1) **Assemble the Shaft Part** – select the drop-box for the "User Defined" option – **select "Cylinder"** – open the **Placement Tab.**
2) **Setup the "Axis Alignment"** Axis to Axis constraint between the shaft and bearing.
3) Select the **"Translation constraint"** between the rear planar surface of the shaft and the rear vertical planar surface of the rear bearing
4) In the **Position options** for the Translation Constraint, **set the Minimum and Maximum Limits** to allow for the shaft to slide from a **minimum of 0.00" to a maximum of 2.00"** away from the bearing surface. If your Shaft movement is on the front side of the bearing, set the Min Limit to -2.00" and the max to 0" to reverse the direction of movement. Disregard the "Current Position" setting.
5) The **Rotation Axis constraint** does not need to be set for this constraint as the shaft should be allowed to spin all the way around without limits.

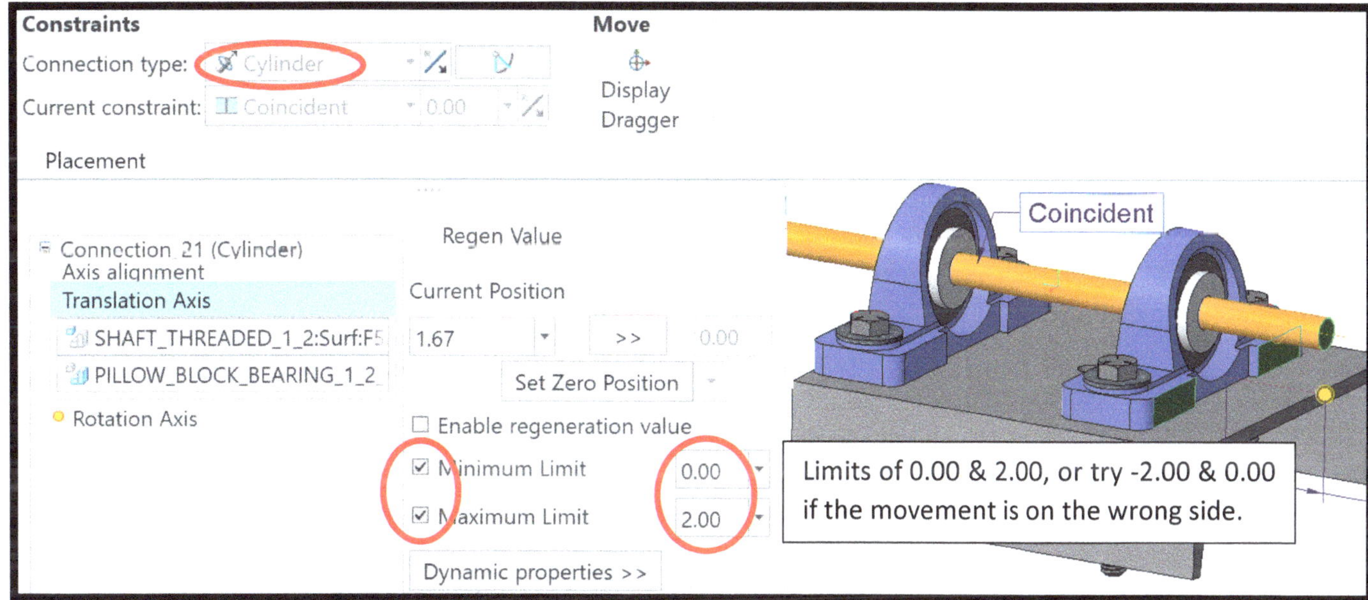

Step 11 – 2 - Check marking the component placement, then test out the movement of the shaft within its range limit by using the **Drag Component tool** in the top toolbar. While using the Drag tool – LMB to select the shaft   move the mouse and the Shaft will move within the range limits – LMB to end the Drag operation. You can leave the shaft anywhere within its range for now as the precise position does not matter.

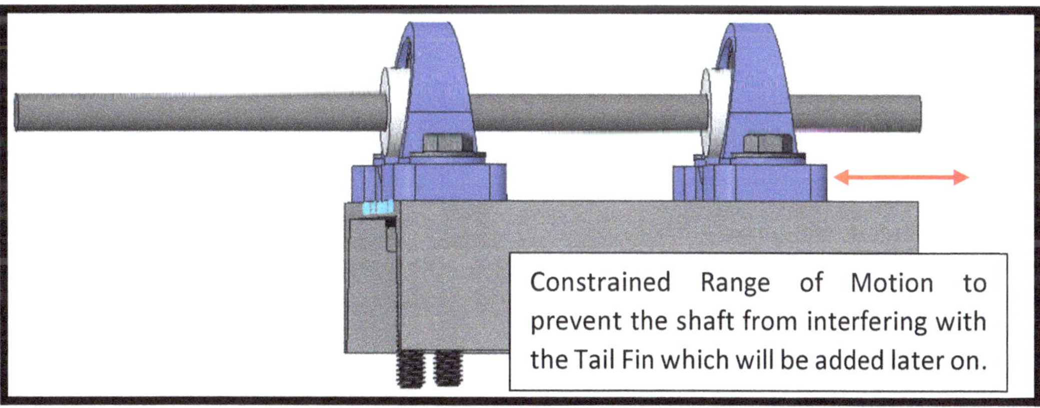

Step 12 – Design Update: There is not enough room between the *Shaft* and the *Nacelle_Base* to fit a 32 Tooth generator sprocket to the model. You will need to add in the ***Bearing_Spacer.PRT*** to the model underneath the pillow block bearings. This can be accomplished by assembling the Spacer component onto the *Nacelle_Base* (without referencing any entities on from Bearing!), moving it earlier in the Model Tree sequence, and then using Edit Definition of the *Pillow_Block_Bearing* to change the constraint reference to the *Bearing_Spacer* surface instead of the *Nacelle_Base*.

Step 12 – 1 - Assemble the *Bearing_Spacer.PRT* to the model. You can temporarily **Hide** the other parts from the model to assist in placing the Spacer on the *Nacelle_Base*.

1) **Hold Control & LMB on the Bearings, Shaft, and Hardware in the Model Tree – hold RMB** – select **Hide** (the Eye symbol). *To show any hidden items you would repeat these steps but choose "Show".*

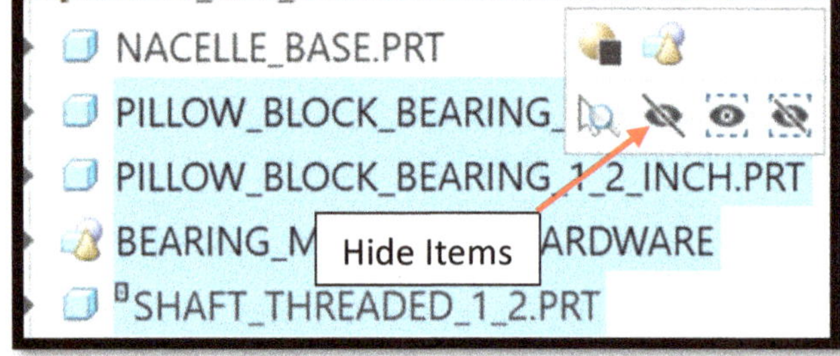

2) **Assemble** the ***Bearing_Spacer.PRT*** as shown on a hole, then **use the Repeat Tool** to add a Spacer to all 4 holes on the *Nacelle_Base*.

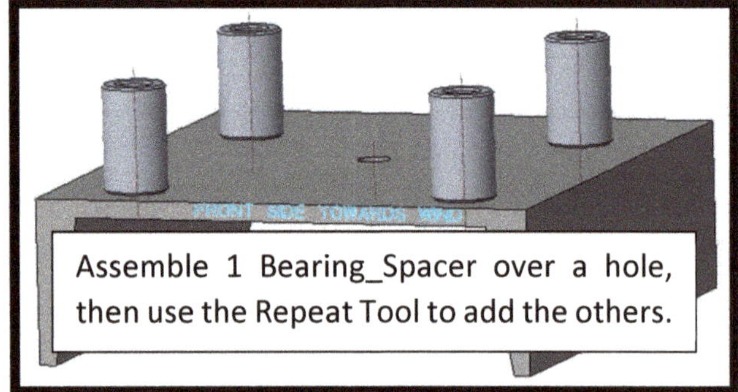

3) **Select all 4 *Bearing_Spacers*** in the Model Tree and Drag them to reorder their sequence in the Model Tree to come <u>before</u> the Bearings in the sequence of the Model Tree.
   Tip: Your Spacers must not use Placement or Repeat references from the Bearings or Hardware axis or surfaces, as that would prevent you from being able to re-order them to come before those items in the Model Tree.

4) Select the Hidden items from the Model Tree and **Show them again** using the **Show icon** in the pop-up menu.

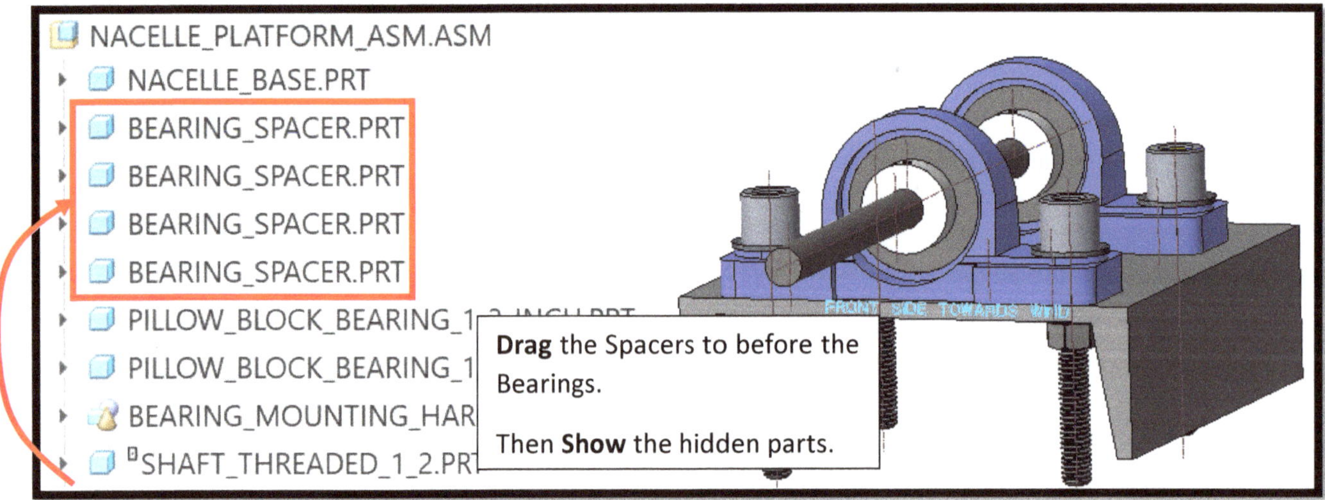

The Bearings will now be interfering with the Spacers and **will require a change of constraint references** to position the Bearing on top of the Spacer.

Step 12 – 2 - Edit Definition of the first *Pillow_Block_Bearing.PRT* in the Model Tree. Find the Coincident Constraint used to place the Bearing on the *Nacelle_Base* surface- then **hold RMB on the reference for the *Nacelle_Base* surface** – choose **Remove –** then **click on the top surface of the Spacer** to add it as the reference. Or you can delete out the entire Constraint and remake it.

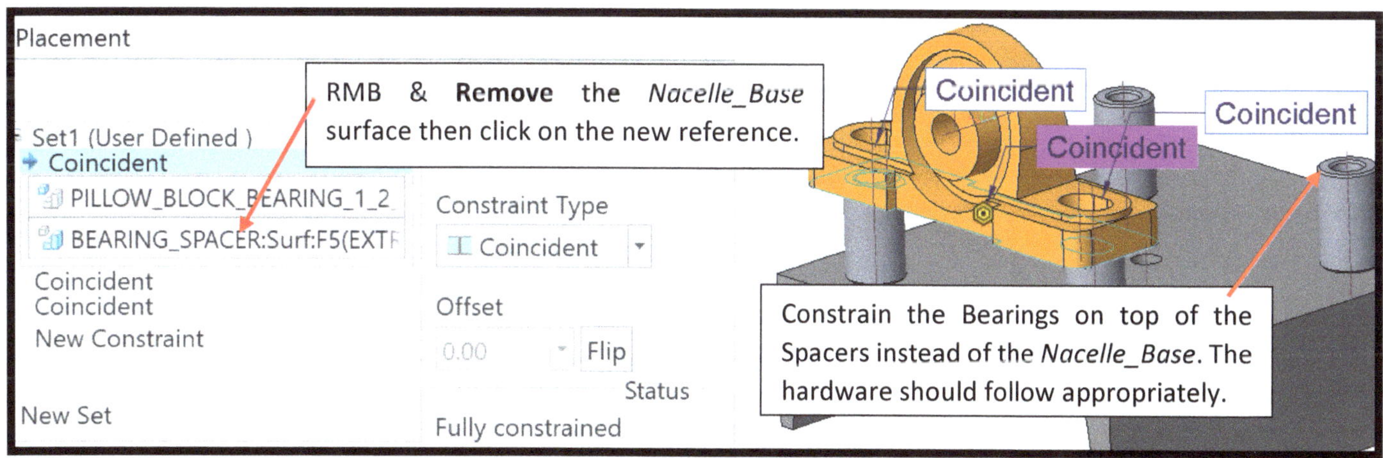

Step 12 – 3 - Repeat for the 2nd bearing. After making the changes the Bolts, Washers, & Nuts should move with the bearings to their proper positions, if they were constrained properly to the correct references. If you notice any red items in the Model tree, those features have failed and may need to have their constraints fixed using Edit Definition.

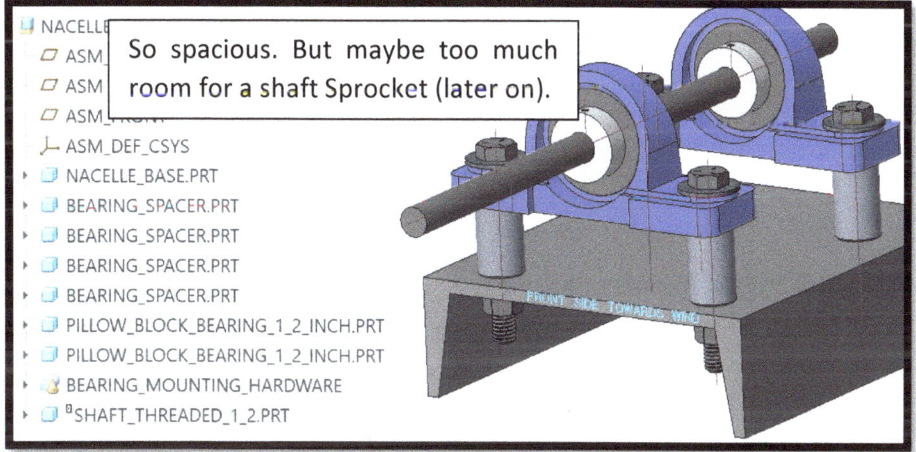

The Model should now have the Bearings on top of the spacers with the bolts and hardware properly placed.

Check that the Model Tree does not have any unconstrained items. The [•] in front of the Shaft signals that it is a Mechanism Constraint that allows movement, so counts as Fully Constrained.

Step 13 – Reduce the Height of the Bearing Spacer as that is too much height, then **Regenerate to update the change.**

- **Activate** the *Bearing_Spacer* in the Model tree (RMB – Activate icon). Then Edit or Edit Definition of **Extrude 1 –** change the height dimension **to 1.00".**

- Regenerate the change by pressing **Control + G on** the keyboard (or the Regenerate icon at the very top toolbar.)

- **Activate the Assembly** itself or press Control + A. The change should now be updated in the model as well as on the Part and Drawing for the *Bearing_Spacer*.

- **Save** the Assembly.

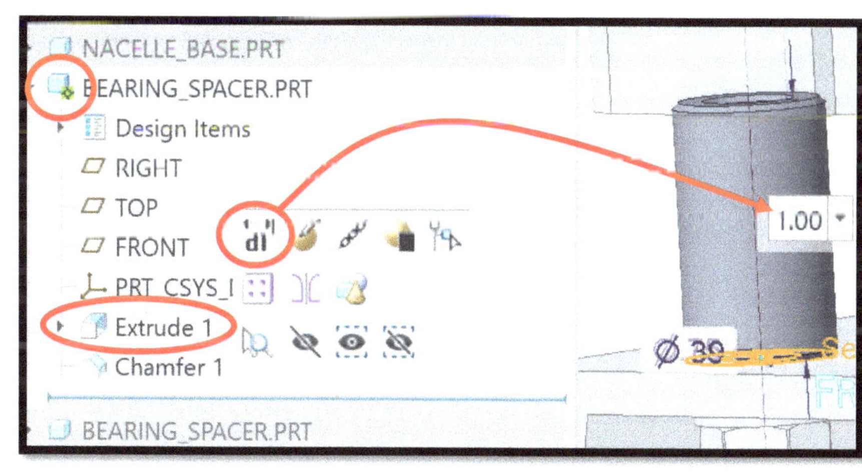

Design a New Part (*Nacelle Post Insert*)

Step 14 – Design a new Part, named *Nacelle_Post_Insert*, to connect the *Nacelle_Base* to the bearing in the top of the Post from Lesson 12 according to the requirements below. **Design this part as if you will be taking your CAD Drawing to a machine shop to fabricate the part.** As such, you will need to search McMaster.com for a suitable "stock" material that can be used to fabricate the design in a machine shop. After finding a suitable material size you also need to check that your design stays within that size to keep costs low. (e.g. If using 2" round stock, don't design a part > 2" wide).

Step 14 – 1 - To get started, look at your CAD model for the *Test Stand Assembly* from Lesson 12. **Hand sketch an idea** for the *Nacelle_Post_Insert* part that will be mounted to the bottom of the *Nacelle_Base* and can then be inserted into the Bearing opening. Consider the dimensions, hole size, etc. that may work before modeling in CREO. **Read the design requirements** below.

Tip: This part does not need to be complex. The simpler it is, the easier it is to fabricate (i.e. cut to length, turn down (adjust the diameter profile), add holes, etc. are all "easy" operations for a machinist. But adding unnecessary smooth curves, spheres, interior geometry, etc. are a sure way to make a bad impression with your machinist.

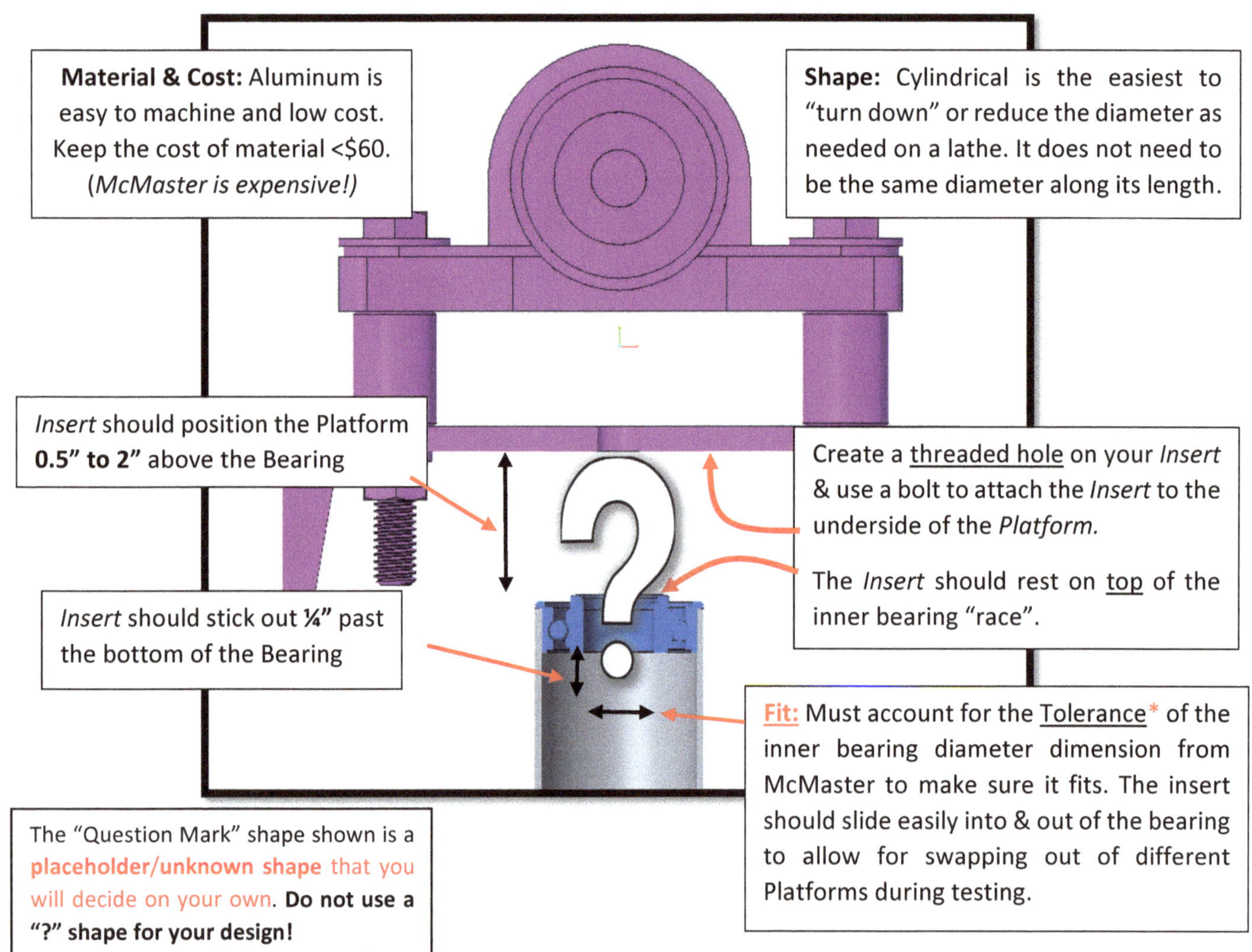

***Bearing vs Insert Diameter Sizing:** Recall that tolerance on a dimension will determine what actual dimension the Bearing from McMaster may have for the inner diameter. To ensure the Insert fits into the bearing easily, you must calculate an Insert diameter dimension with a tolerance that after fabrication is not too large to fit in the bearing. **Follow the steps on the next page** to determine the correct size.

Step 14 – 2 - Refer to the detail drawing from McMaster (shown below) and **calculate what the smallest Bearing inner diameter** could be from the manufacturer using its tolerance range.

Then **calculate the diameter** of the bottom section of your Insert to ensure it will fit in the bearing. Keep in mind that our class drawings use the number of decimal places to dictate the tolerance (e.g. X.XXX = + - .001).

Does the max actual size of the Insert fit within the smallest actual diameter of the Bearing? Keep this in mind as you create the Detail Drawing for your part as you will want show the correct tolerance (decimal places) on that dimension.

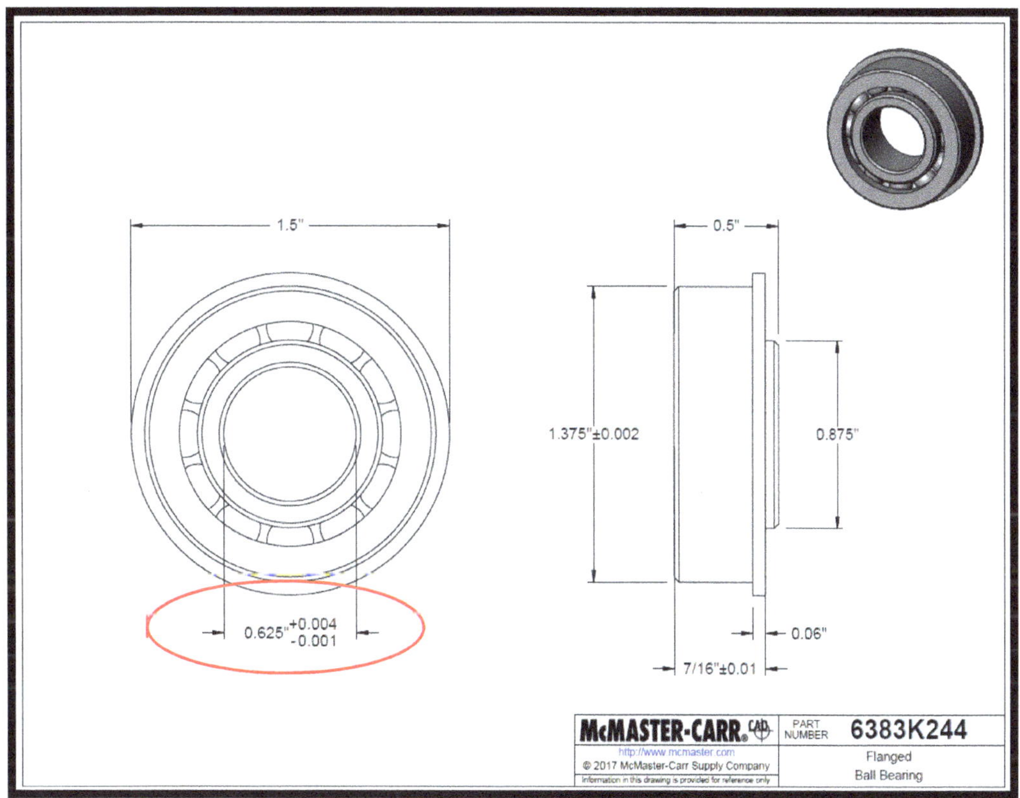

Step 14 – 3 - Now that you have an idea for the design, you need to **source a "stock" material** that a machinist would use to work with. Go to **McMaster.com** & search within the **Categories** such as **"Round Stock", "Shafts", or "Metal Rod"** to find a material and shape that will work (aluminum is ideal, mild steel could also work).

Is the Stock material close to the dimensions you want to use (disregarding length)? If so add to your records what the McMaster part # is for the stock material to include in the final report Bill of Materials (later on).

Note: Do **not import** the McMaster CAD file into CREO for modeling; design and model your own part file from scratch that could be custom manufactured from the stock material you have chosen Typically, you **do not import** a raw material STEP file and whittle it away, instead you should model knowing what material it will be fabricated from.

Step 14 – 4 - Now model your *Nacelle_Post_Insert* from scratch as a new Part file. Also make sure to **set the Material, Color & Texture**, etc. Again, do NOT import a STEP file from McMaster to work with, model your own design like has been done through the book.

Step 15 – Specify a bolt for mounting the ***Nacelle_Post_Insert*** to the ***Nacelle_Base*** through the center hole. This can be a shorter version of the bolt you specified earlier, or perhaps even the same bolt if it works for your design.

Tip: Make sure the "Threaded" Length of the mounting hole in your *Nacelle_Post_Insert* is long enough to thread the bolt into the *Nacelle_Post_Insert.*

By default, Standard holes are set to an arbitrary threaded depth, but there is an option to adjust the threaded length or set to "Through Thread" or a specified Distance of Threads.

- Edit Definition of your Hole - Shape Tab – check your Thread Depth value – change to Through Thread (only an option if you have a Thru-All hole Depth).

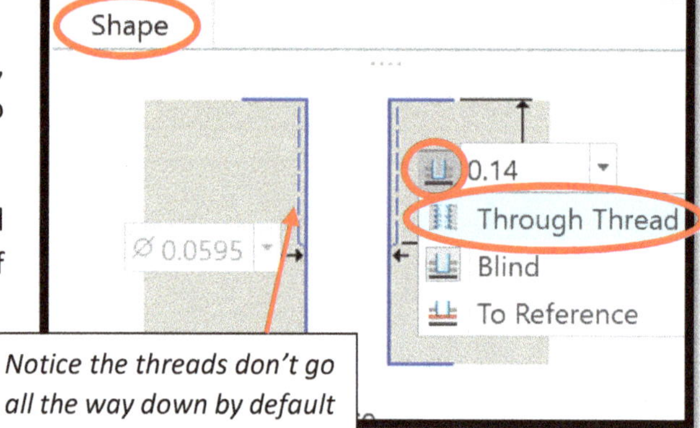

Step 17 – Assemble your *Nacelle_Post_Insert* and the **Bolt** to the main assembly file**,** using the center hole on the *Nacelle_Base***.**

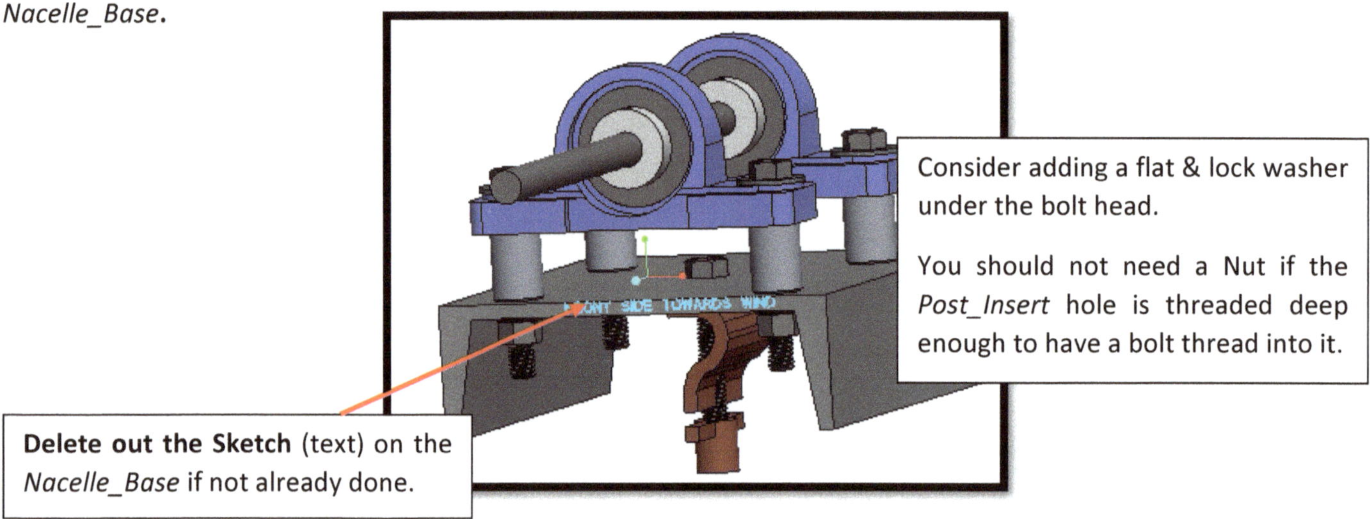

Consider adding a flat & lock washer under the bolt head.

You should not need a Nut if the *Post_Insert* hole is threaded deep enough to have a bolt thread into it.

Delete out the Sketch (text) on the *Nacelle_Base* if not already done.

Step 17 – Save the Assembly. Regenerate the model if needed. Also **Save all the parts** that may be open. To open this assembly again in the future, keep in mind that all the imported Hardware part files, the *Post_Insert*, *Bearing*, *Shaft*, *Base*, etc. must all be in the same directory, and not in a sub-folder!

Step 18 – Create an **Assembly Drawing** similar to the key (note there are 2 sheets on that drawing), as well as a **Part Drawing** for your design of the ***Nacelle_Post_Insert*** to class standards for a detail part drawing.

Your Part drawing should **include a Cross-section**, centerlines, dimensions, a note for the threaded hole, and check that the tolerance on the diameter dimension for the section that goes into the bearing fits the requirement.

<h2 align="center">End of Lesson 13</h2>

"Basic" Drawing Steps

Follow the steps below on each of your drawings to setup a basic drawing. This is a generic guideline, you may need to apply the step settings differently depending on the drawing keys for each lesson.

Overview of Steps:

1) Insert the General Isometric View
2) Turn Off Display Filters
3) Use Sheet Setup to add the UND template
4) Adjust the Orientation or Angle of the Parent Views to match the key
5) Adjust the Scale
6) Adjust the View Positions by turning off Lock View Movement
7) Set the View Display for each view
8) If needed, create and show Cross-Sections (also adjust the Section Hatching and Add Arrows)
9) Peel back the Datum Tag lines *(starting in Lesson 4)*
10) Show Centerlines and Notes *(starting in Lesson 3)*
11) Save, Export as PDF, & Submit for grading.

Step 0 – Start a new Drawing file: - **File – New –** choose **Drawing** as the file type – provide a **Name** to match the key - **OK** - choose the **desired default model** that you want a drawing of - leave the Template options of *"C_drawing"* – press **OK.**   Note: *You do NOT choose the UND C size format in this step!*

Step 1 – Insert the General (Isometric) View.

- Make sure you are on the **Layout Tab – General View– press OK -** if a "Select Combined State" menu pops up choose *"No Combined State"* and press OK – **LMB** on the screen to set where the view location – in the *Drawing View* pop-up menu choose the **View Type** category - select **Standard Orientation** – set the Default Orientation to **Isometric – OK.**

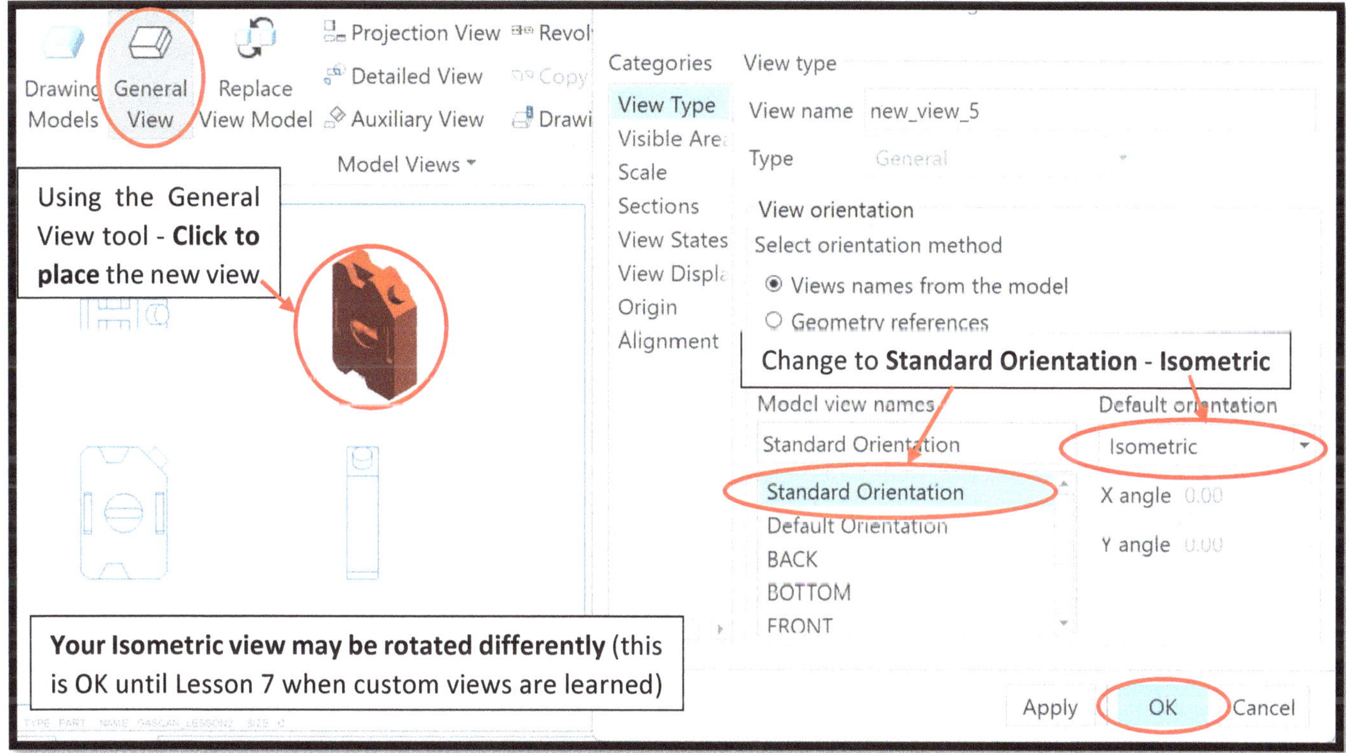

Tip: If you need to get back to the menu above, double click on the desired View.

Step 2 – Toggle-Off the Display Filters in the Quick Toolbar

By default, CREO will display the **Datums, Axis, & Coordinate Systems** on the views unless the **Display Filters** are unchecked in the quick toolbar. While these are useful in Part Mode, they clutter the drawing and must be turned off before you finalize the PDF. Note that these Display Filters will be toggled on if you reopen the drawing later on.

- In the **Quick Toolbar – Display Filters icon – uncheck all –** click **Repaint** ("wiper" button) in the quick toolbar.

***Tip:** These Display Filters are useful during modeling in Part mode, so if you go back to Part mode be sure to toggle them back on.

Note: Display Filters cannot remove "Set" Datums from the model (such as when the datums are renamed with Set toggled on)

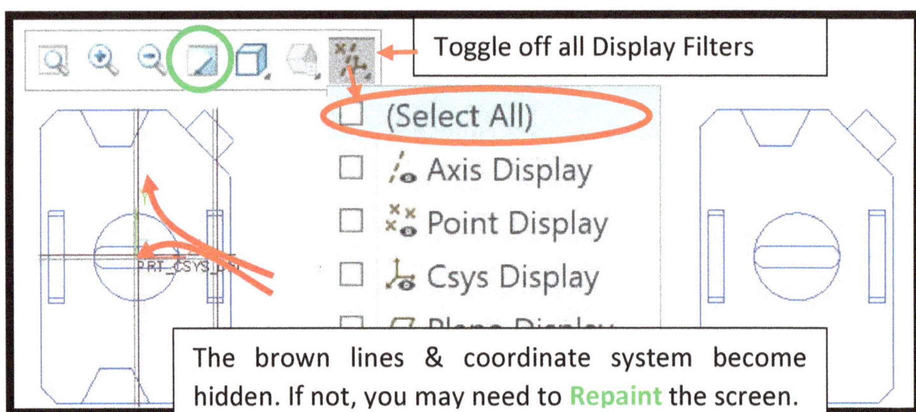

Step 3 – Add the UND C_Size Template (*Sheet Setup*).

**Make sure that the C_Size_UND.frm template from Blackboard has been saved to the same working directory on your computer that the part and drawing files are located in!*

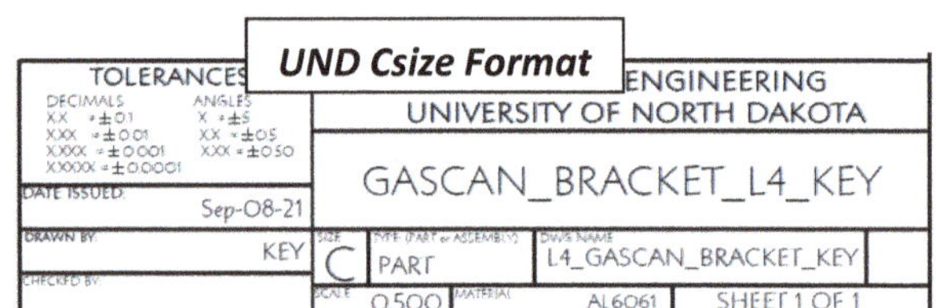

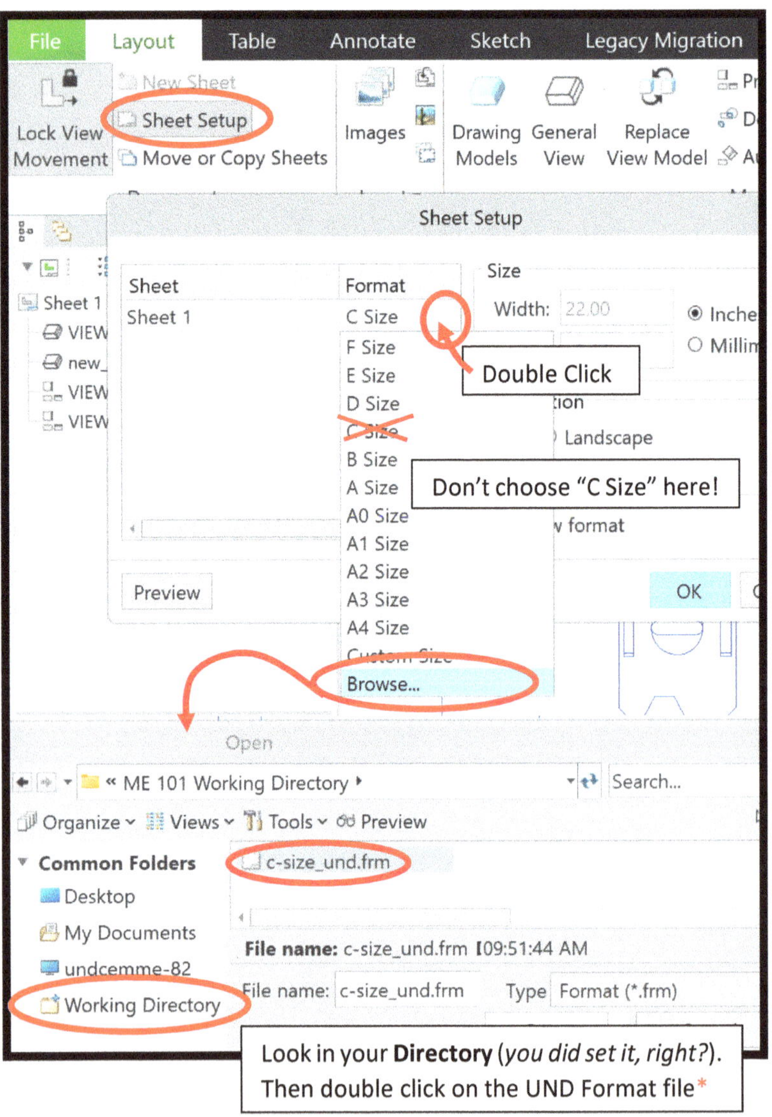

Look in your **Directory** (*you did set it, right?*). Then double click on the UND Format file*

1) Click on the **Layout Tab** – choose **Sheet Setup** – double click on the *Format size box* – **Browse** – Navigate to your **Working Directory** where all your CREO files are saved to – **select "c-size_und.frm"** – click **Open** – press OK

2) A menu box will appear asking you for your Name - enter in your Name and **press Enter**.

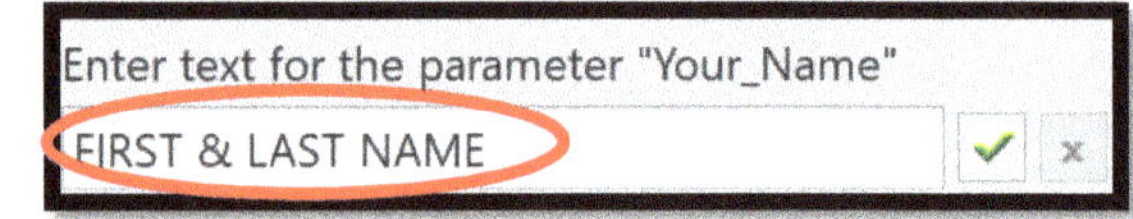

Note: CREO will not let you continue until you have answered its request for **Your Name!**

Step 3 – Specify the Material *(if needed)*

The Material should come in automatically, *but only if you set your Material in the Model Properties!* If not, go back to the part model and set the material. (*or use the tip below to type it in*)

*Tip: You may need to **Adjust the Font Size** of some words in the Template cells. You can change the font size or words of a cell by clicking on the Annotate tab – double click on a cell – then change the font size or text as desired.*

Step 4 – Adjust the View Orientations

If your view orientations do not match the key you must adjust the orientation of the **Parent View** *(bottom left view)*. Changing this Parent View will also rotate the Top and Right views since they are "children" or projections from the Front parent view. Note that the Isometric General view will not follow the parent view.

You can adjust a view **_by Angles_** to rotate the view a specific number of degrees about the chosen Horizontal, Vertical, or Normal (Perpendicular to the screen) axis.

- **Layout Tab – double click LMB** the _Parent View_ (*i.e. the Front or lower left view*) – **View Type** category - select **Angles** – click the **drop down box** of the Rotation Reference – **choose a reference** to rotate the view about (**horizontal, vertical, normal**) – **type in 90** degrees for an angle value – **Apply** – (*choose "Yes" to modifying the children views if a pop-up appears*) - all views except the general view should now have rotated. **Repeat as necessary – press OK** to close the menu.

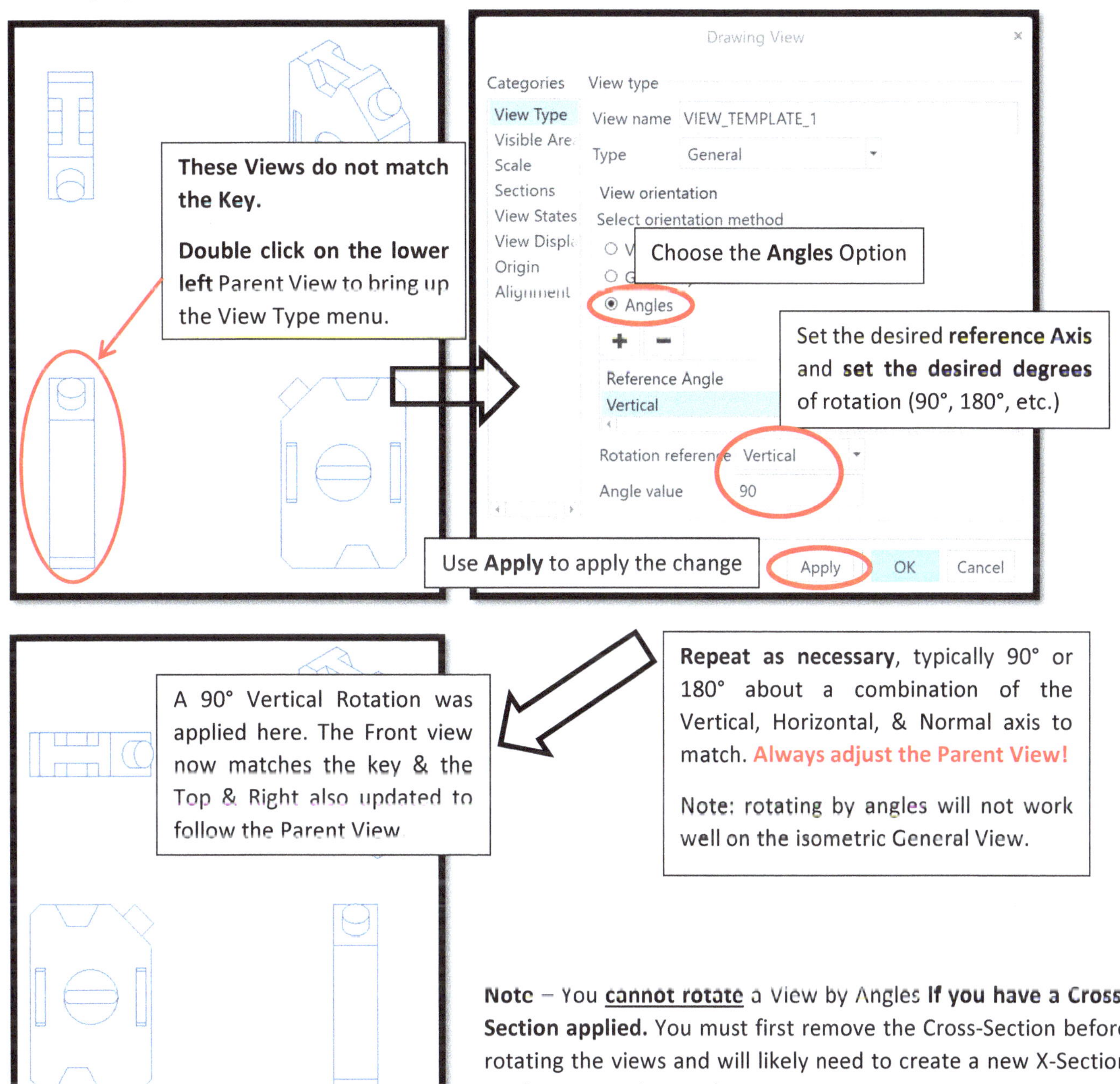

Note – You **cannot rotate** a View by Angles **If you have a Cross-Section applied.** You must first remove the Cross-Section before rotating the views and will likely need to create a new X-Section on the correct datum plane.

Step 5 – Adjust the Drawing Scale to match the key

- **Double click LMB** on the **Scale** in the bottom left of the screen – type in the desired **Scale Value** (*Check the Lesson Key for which value to use*) – **press Enter** to accept it.

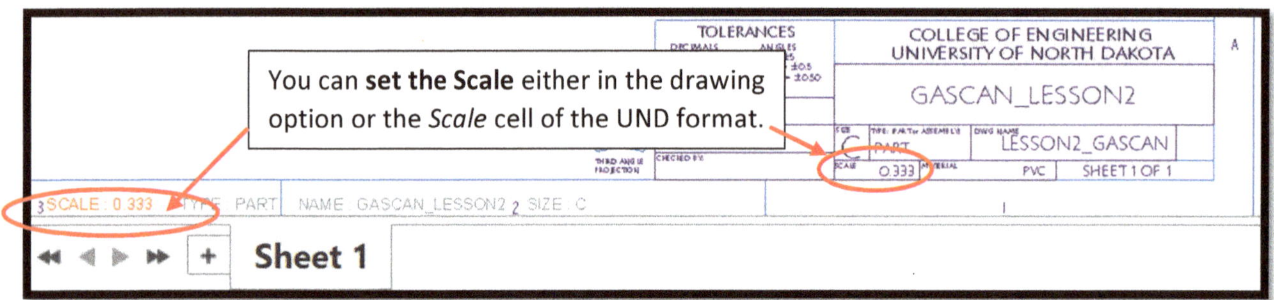

Tip – if your scale seems off (i.e. a very small number is needed to match the key) you likely have a unit issue, such as measuring in cm but modeling in inches by forgetting to change the model properties to cm units!

Step 6 – Adjust the View Positions

Properly locate the position of all views so that the views are not intersecting the edges of the UND Template or the boundaries of the drawing sheet. Spacing & Position should generally match the key, but does not need to be exact.

- Layout Tab – LMB click on the **Lock View Movement** button to toggle off the lock setting – **LMB click on a view to select it – click and hold on the view to drag** the view to a new position*.

*The Top and Right views are projections of the Front View. You must move the Front (Parent) View in order to move the other views in certain directions. The main views must remain **aligned** (not off center from each other).

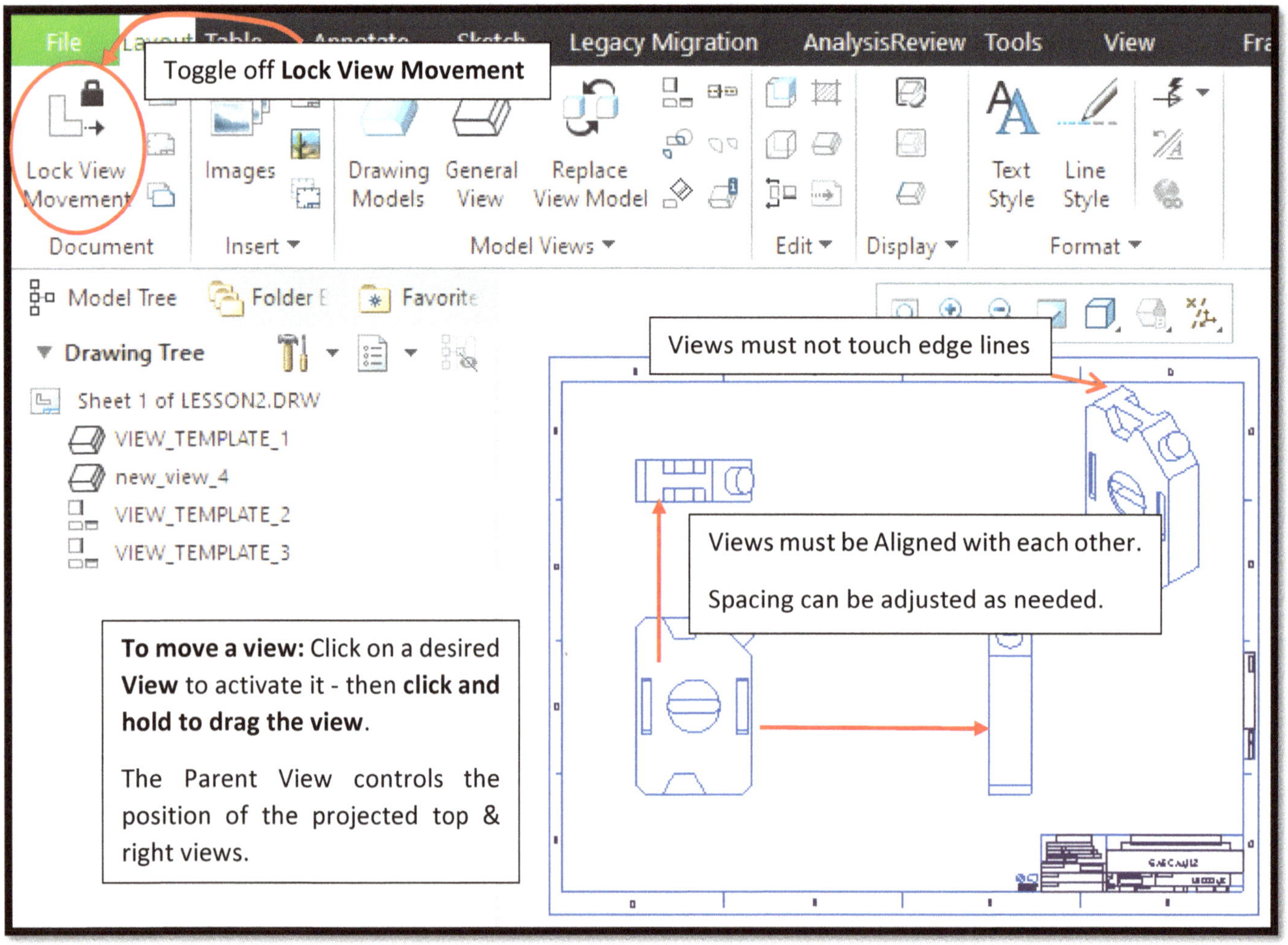

Step 7 – Set the View Display for each view

The View Display of each view must be setup to match the key. This will commonly be a *"No Hidden"* style so that interior edges and geometry will not appear on the drawing which provides a cleaner view. Cross-Section views should always be No Hidden!

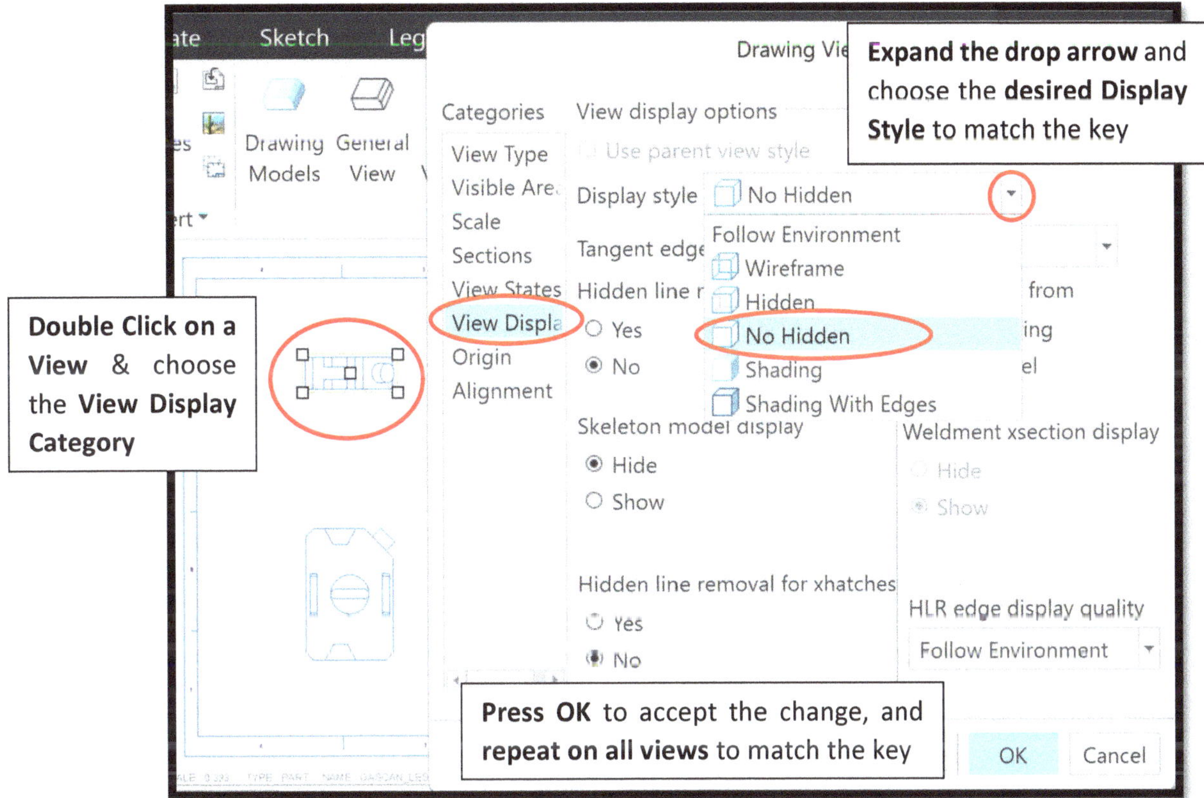

- **Double click LMB** on a specific View to bring up the view's *Properties Menu* – select the **View Display category** – expand the drop box for **Display Style** – select **No Hidden** *(according to the key)* – click the **drop-down box** of the **Tangent Edge Display Style** option – select **Default/None/Solid** *(according to the key)* – **OK.**
- **Repeat for each View** to match the lesson key instructions. Not all views will be the same on each drawing.

You must repeat this step on each view to set the specified View Display!

Step 8 – Create or Show any Cross-Sections (if needed)

1) While on the **Layout tab - double click on the desired View*** to display a Cross-Section (*cut-away view*) on.
   - * *The view that you will be able to see the "cut" surface and Hatch lines on.*
2) Choose the **Section tab** of the *Drawing View* properties box.
3) Select the **(+)** sign. Then choose the desired Cross-Section to show (if previously made) or click Create New.

Option 1: <u>Showing a Previously Created Cross-Section</u>: If you already created the cross-section, like was done in Lesson 3 for the *Ram_HandleBar* in Part mode, you can **select a previously created Cross-Section** from the list and **press OK**. It must have a green check mark next to the name to be valid on the chosen view. A **red X** by the name indicates the view cannot display the Section properly (i.e. the cut surface is not visible on this view).

Option 2: <u>Create a New Cross-Section</u>: To create a new Cross-section choose **Create New** from the menu.

a. After selecting *Create New*, leave the default options of *Planar* & *Single* and **choose Done** menu

b. A Name Menu will appear near the quick toolbar: **Type in the Name of the section** using a single Capital Letter (A, B, C, D, etc. or if you already have a failed A you can use AA as the name) and **press Enter**.

c. **Select the desired datum plane:** Choose a datum from one of the <u>other</u> views that will allow the model to be sliced so the cut surface & hatch lines will be visible on view you are adding the Cross-section to.

   Tip: You may need to temporarily toggle on "Display Filters" in the Quick Toolbar to see the datums.

d. **Press Apply –** the section view should be visible with a green checkmark in the list. A **red X** means you chose the wrong datum and must create a new section again (and must use a different Section name, like AA)

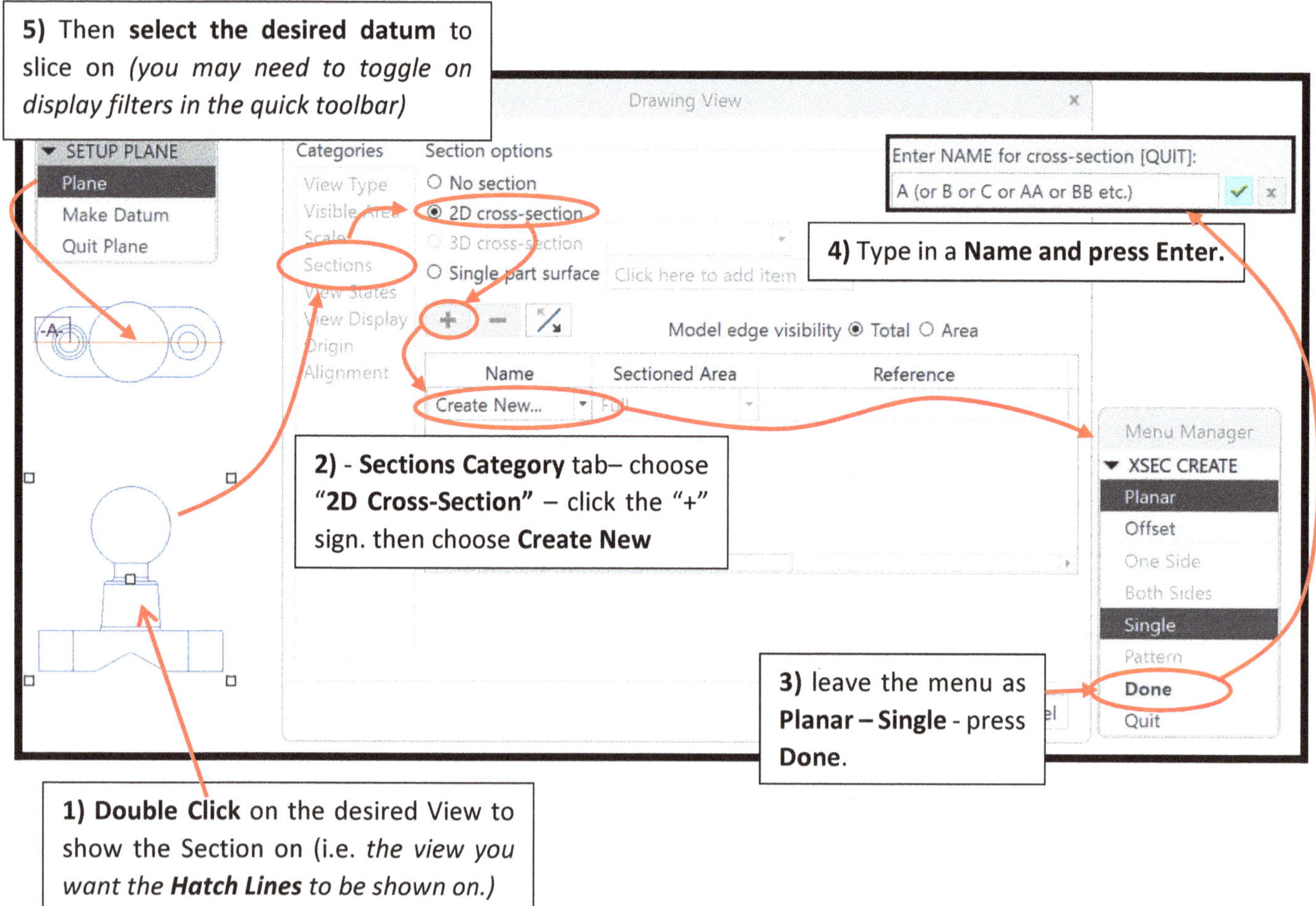

Cross-Section Problems?

- ***Does your Cross-Section have a*** *red X **by the name?** Press the **Minus sign** on the section list to remove the failed section, or toggle on No Section if you want to remove the cross section from the view. Then create a new one, but you will need to use a slightly different name (i.e. AA or B)*

- ***Need to Rotate a view by angles to match the key that already has a Cross-Section on it?*** *The section must first be toggled off or removed, as the Cross Section will prevent the view orientation from being rotated.*

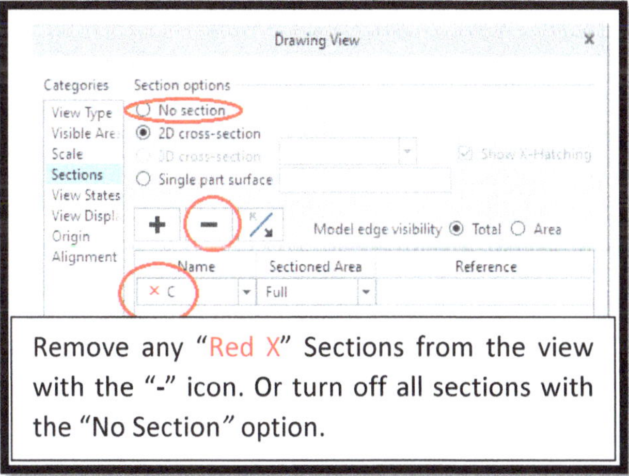

Remove any *"Red X"* Sections from the view with the *"-"* icon. Or turn off all sections with the *"No Section"* option.

Step 8b – Adjust the Cross-Section Hatch Line Spacing & Add Placement Arrows

Now you must **adjust the spacing** of the Hatch Lines as well as **Add Arrows** to show where the section datum is located.

1) Adjust the Hatch Line Spacing:

- **Layout Tab – LMB in a blank area** of the screen – **double click LMB on the section Hatch lines** (not the drawing view, but the actual hatch lines) to bring up the Hatch menu – set a **Scale value** as per the Lesson Key.

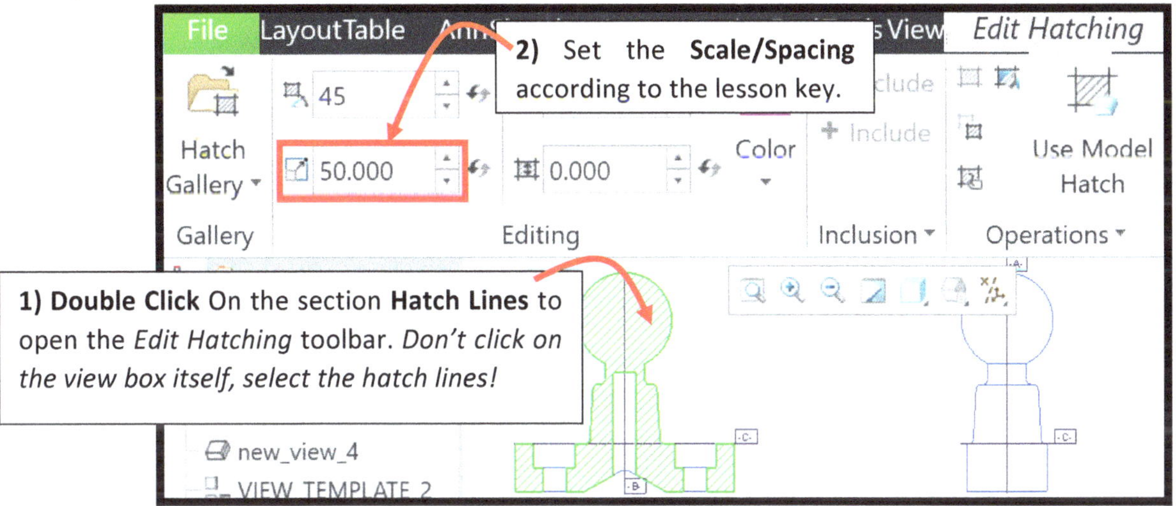

2) Add Placement Arrows for the Cross Section:

- With the **Hatch lines still selected & the *Edit Hatching* toolbar open – go to the Layout Tab -** click the **"Arrows"**** tool from the toolbar– click on the desired view** to show the Section Placement Line & Arrows on (Top View for this example) – the arrows will appear on that view – press **MMB** or **Escape** to exit the *Add Arrow* tool.

**** You only get 1 chance to place the cross-section placement Arrows!** *Once Arrows are placed for a given Cross-Section you cannot use that tool again for it. You will need to Create a New Section to use the Arrows again.*

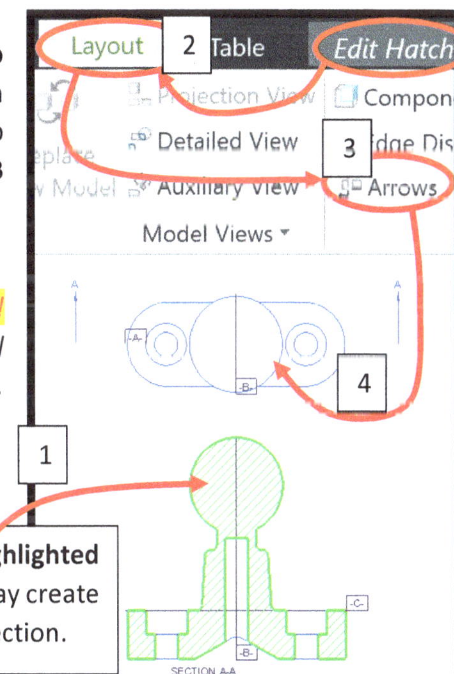

1) **The Hatch Lines must still be selected/highlighted** with the *Edit Hatching* **toolbar still open** or it may create arrows for the Projection view instead of the Section.

Step 9 – Clean up or "Peel-Back" the Datum Tag Lines *(starting in Lesson 4)*

Datum Lines often cover the dotted centerlines. You must "Peel Back" the Datum Tag lines so that they do not "touch" the drawing views. Note- this is only possible on 'Set' geometric tolerance datums, not the Display Filter Datums which can be shown or hidden in the Quick Toolbar.

- You must be on the **Annotate Tab** – then **LMB click on a datum line on a view** – with the datum selected you will see the **white boxes on the Tag Lines – drag the white boxes away from the view to** adjust the length and position of the Datum Lines. **Repeat for each Datum on each view**.

Note – it does not matter if your Datum Name/Tag is positioned differently than the key, or in a different location, as long as the Datum lines are not touching the view edges.

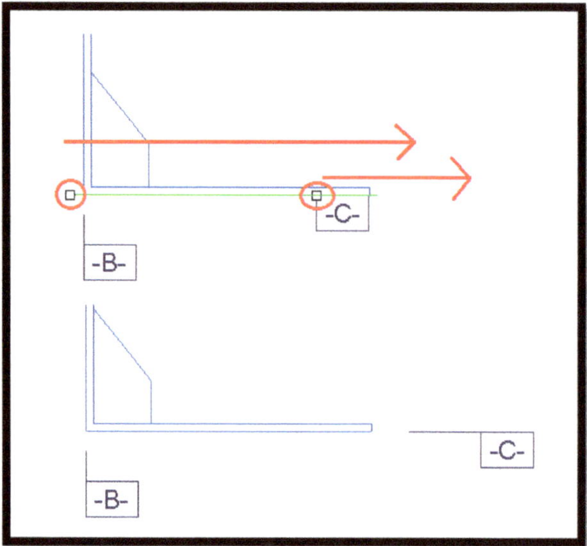

While on the Annotate Tab: Click on a Datum line to select the Datum. The white boxes or "drag handles" will appear – drag both of them away from the view edges.

You can only drag datum tags of "Set" datums, not Display Filter datums or datums you did not Set in the Properties of the datum in the Part Model.

Continued on Next Page:

Step 10 - Drawing Notes & Axis/Centerlines *(starting in Lesson 3)*

Add **Annotations** such as centerlines and notes that may be required on the drawings. Note that being on the correct tab (i.e. Annotate vs. Layout tab) is important as it will allow or limit what types of operations can be completed.

1) Select the **Annotate Tab** in the top toolbar.
2) Select the **Show Model Annotations** tool
3) In the menu, select the desired category for the type of annotations you are showing (Axis, Notes, etc.)

For Centerlines: Choose the far-right **tab icon ("*Model Datums*")** which will show the centerlines/Axis of any revolves, holes, or circles.

- While on the Model Datums tab – LMB on the **"View Box"** that surrounds the desired view to add all entities to the menu (if you try to choose the part model itself it may not populate all options for the view)– **checkmark all items to show them** – press **Apply**. Repeat to add annotations to the other views by selecting the view box for the Top or Right views. **Press OK or Cancel** to close out the tool when finished

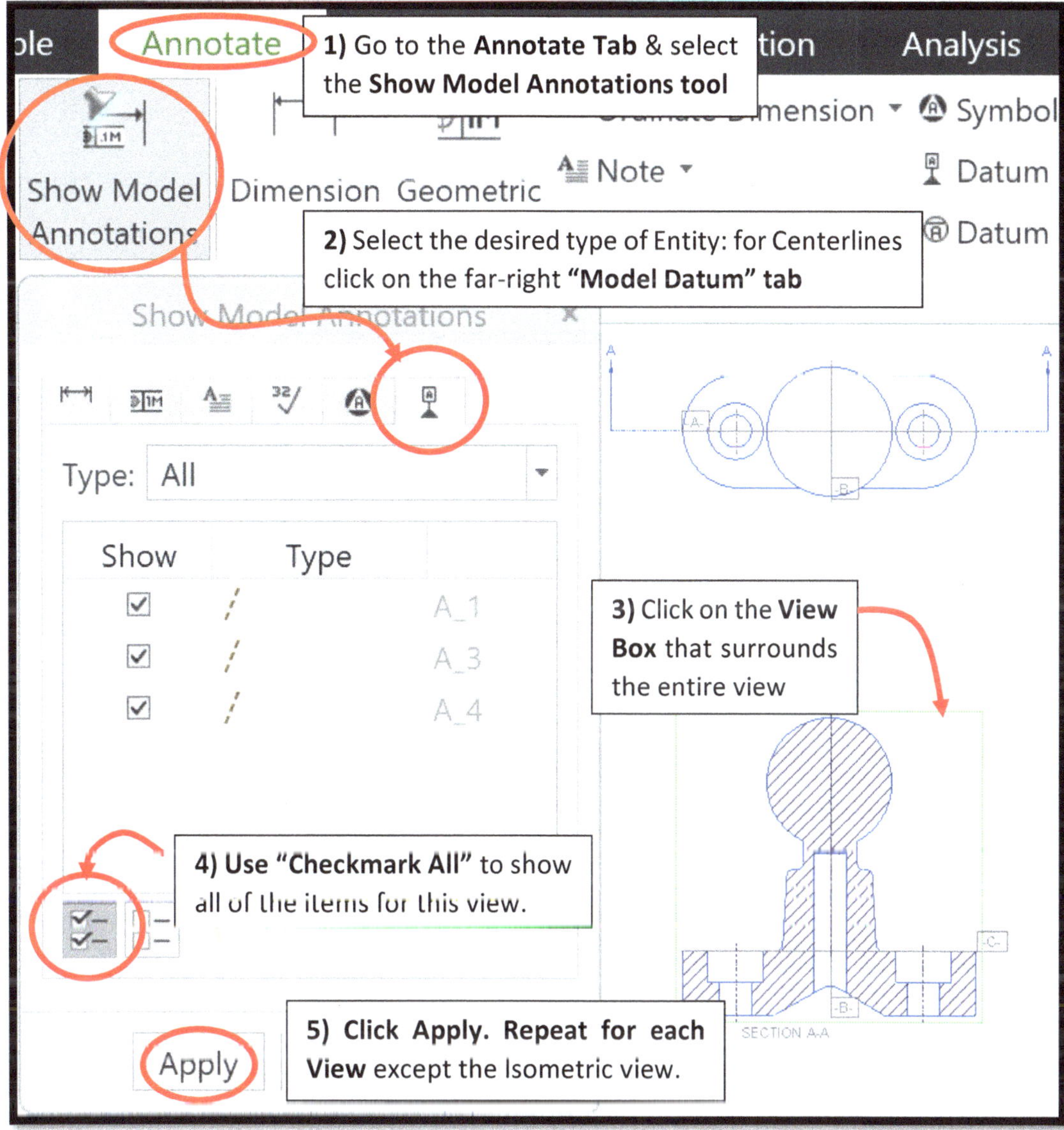

Do not show centerlines on the Isometric / General view!

For Hole Notes: Notes can also be shown using the Show Model Annotations tool for features that have pre-built notes, such as Standard Holes, Cosmetic Threads, etc. That process is covered in the *Detail Drawing Steps (#7)*.

End of the *Basic Drawing Steps*

"Detail" Drawing Steps

A detail drawing will have more details about the model than the basic drawings, such as dimensions, notes, additional views or sheets, and ASME style datums. These details are important to communicate the design and fabrication requirements for a part model.

Add the following steps to those in the ***Basic Drawing Steps*** as needed. Not all steps will be required on every drawing.

Overview of Steps:

1) Set the Datums to ASME Style (gtol_datums set to *std_asme*), if datums are shown.
2) Erase or Un-erase datums from a Drawing (*if needed*)
3) Create or Show Dimensions
4) Flip Diameter Dimensions to the single arrow style
5) Adjust the Decimal Places of Dimensions
6) Add Text to a Dimension, and add details for Hole Position and Spacing *(starting in Lesson 6)*
7) Show Pre-Generated Notes (for Standard/Threaded Holes) *(starting in Lesson 4)*
     7.b) Edit Attachment of Notes to clean up the arrow pointer position
8) Create your own General Notes or Leader Notes

Step 1 – Set the Geometric Tolerance Datums to ASME style

If datums are shown on a drawing, then they should be adjusted to the *American Society of Mechanical Engineering* (ASME) setting. This will not affect the datums shown from the quick toolbar Display Filter, only datums that were **Set** Renamed in the part model properties of each datum.

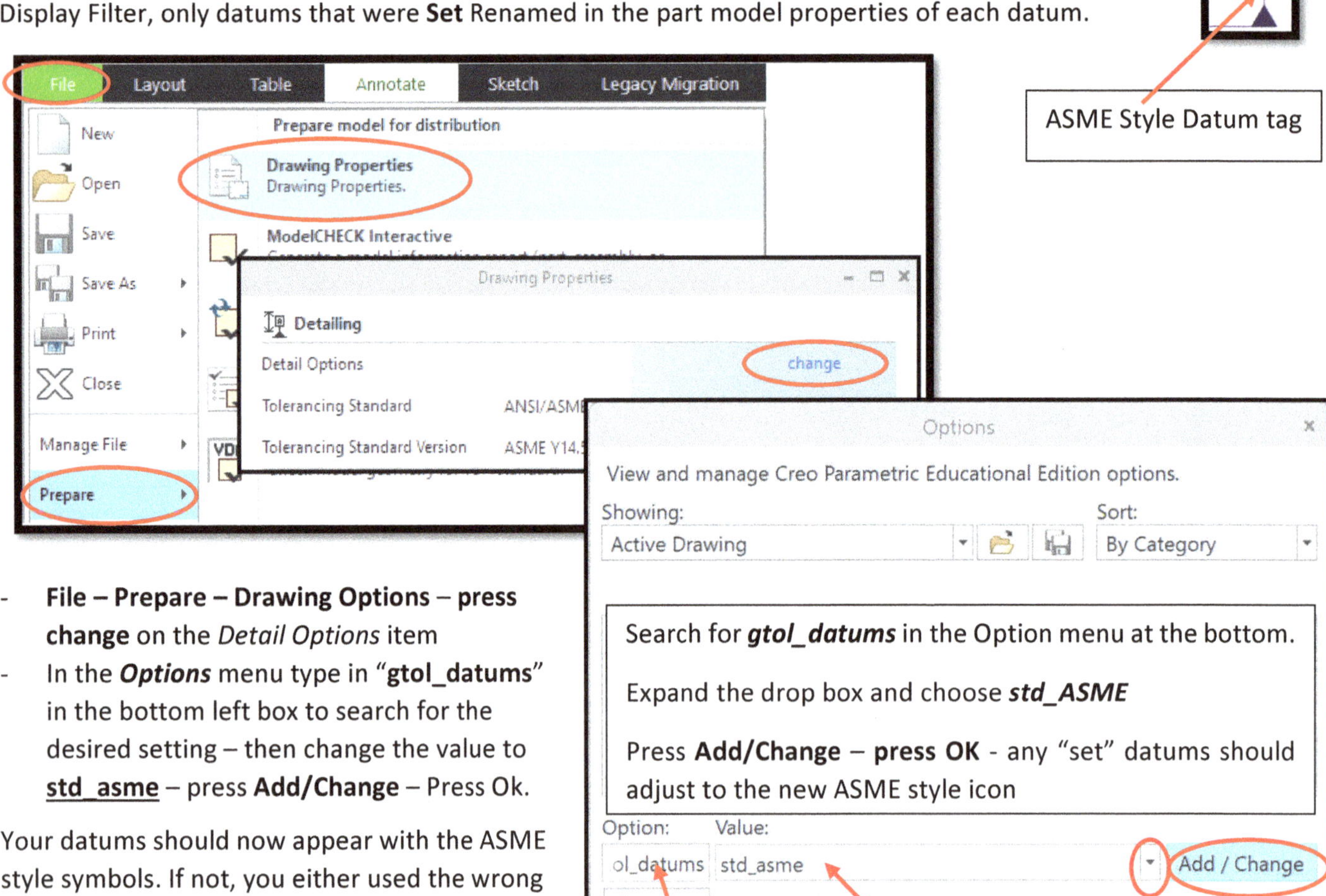

- **File – Prepare – Drawing Options – press change** on the *Detail Options* item
- In the ***Options*** menu type in "**gtol_datums**" in the bottom left box to search for the desired setting – then change the value to **std_asme** – press **Add/Change** – Press Ok.

Your datums should now appear with the ASME style symbols. If not, you either used the wrong setting or the datums were not "Set" in Part mode under the datums properties during the renaming process.

Step 2 - Erase & Un-Erase Datums from a Drawing (if needed):

If you have datums that you do not want included on a drawing you can **Erase** them from the drawing view. This is a better option that deleting them out, as erasing can be undone using un-erase. This only works on Set Datums, not the Display Filter datums that can be toggled off from the quick toolbar.

-   While on the Annotate tab - **select a datum** from the drawing or you can select them from the Drawing Tree – then **choose the Eraser icon** in the pop-up menu. The same thing can be done to Un-erase a datum.

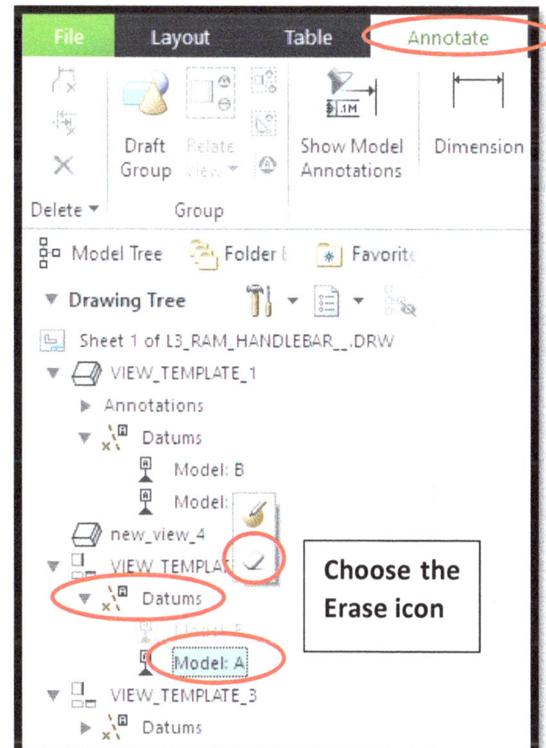

You can also hide all Part Datums all at once from a Model or Drawing using the **Layer Tool** (covered in Lesson 11).

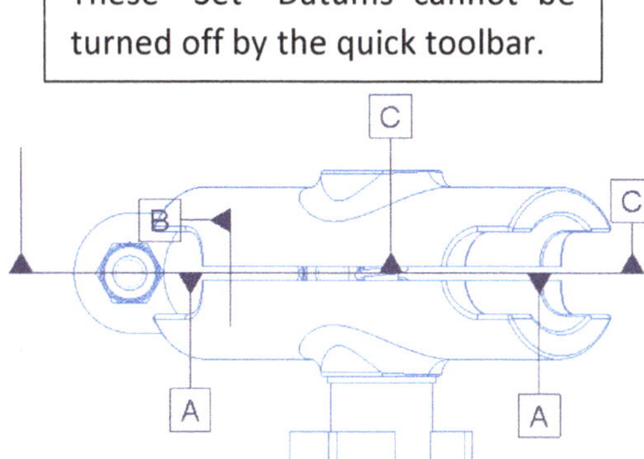

These "Set" Datums cannot be turned off by the quick toolbar.

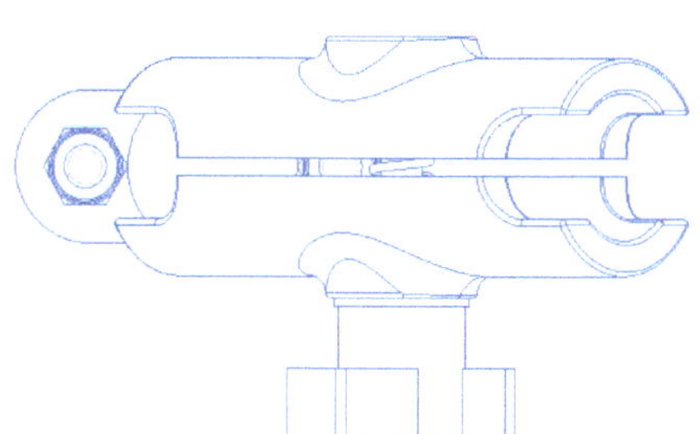

Example – This assembly drawings has too many part datums showing. You can hide these to make a cleaner drawing if the datums are not needed. Why are some needed and some not? It depends on the lesson Key and if you want Geometric Tolerance datums. For your own projects, typically you would not set datums unless you are setting up geometric tolerances that must be met during manufacturing.

Step 3 - Show or Create Dimensions

There are two methods to add dimensions to a drawing view – Creating new dimensions in place using the Dimension tool, or showing Driving dimensions (those used to define the model features) using Show Model Annotations. Depending on the geometry, you may need to use a combination of both options to show the required dimensions on a drawing.

☐ Option 1 - Create the dimensions using the Dimension tool.

- On the **Annotate Tab – Dimension Tool – hold Control on the keyboard and click the desired entities to measure from - MMB to place the dimension measurement.**

Reference Selection Type Options:

- *Select by Entity* (click on any edge or point).
- *Intersection Point* (hold Control and select two intersecting edges to get 1 reference point at that intersection), then switch back to Entity selection and hole Control to get the last measurement reference point.
- *Midpoint* (click a line or edge to get the midpoint of it)
- *Surface* (to grab a center of sphere).

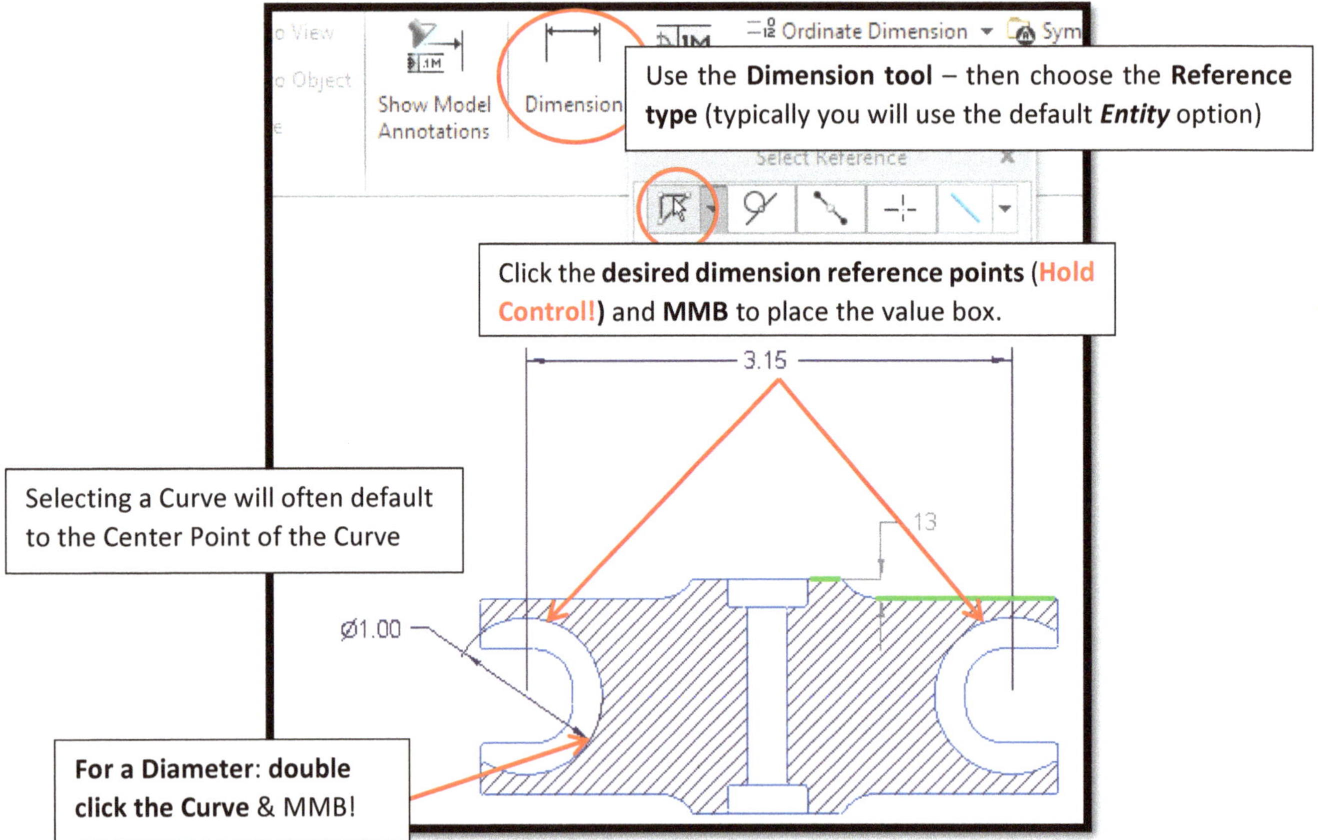

Tip: Where you press the MMB to place the dimension may affect what type of Dimension you get as well; CREO will try to interpret if you want a Vertical, Horizontal, or a Slanted measurement between your reference points. If you want a Slanted dimension that follow the slope along an angled edge, MMB closer to the line.

Note: Some edges will not allow you to measure them to place a dimension, such as face ends edges on a Blend.

☐ Option 2 – Show Dimensions using the Show Model Annotations

Use the **Show Model Annotations tool** to show the <u>Driving</u> dimensions for each drawing view as needed. *Driving dimensions are those that were setup during part modeling and sketching to define the model.*

- On the **Annotate Tab** - **Show Model Annotations Tool**- select the **Dimension category tab** – select the <u>View Box</u> that surrounds the desired view – **check mark** the needed dimensions to match the key – **press Ok**.
  Tip: It may be difficult to find the correct dimensions; choose a few that look possible, press OK, then reposition those dimensions to view them better and Erase (do not Delete) the ones that are not needed.

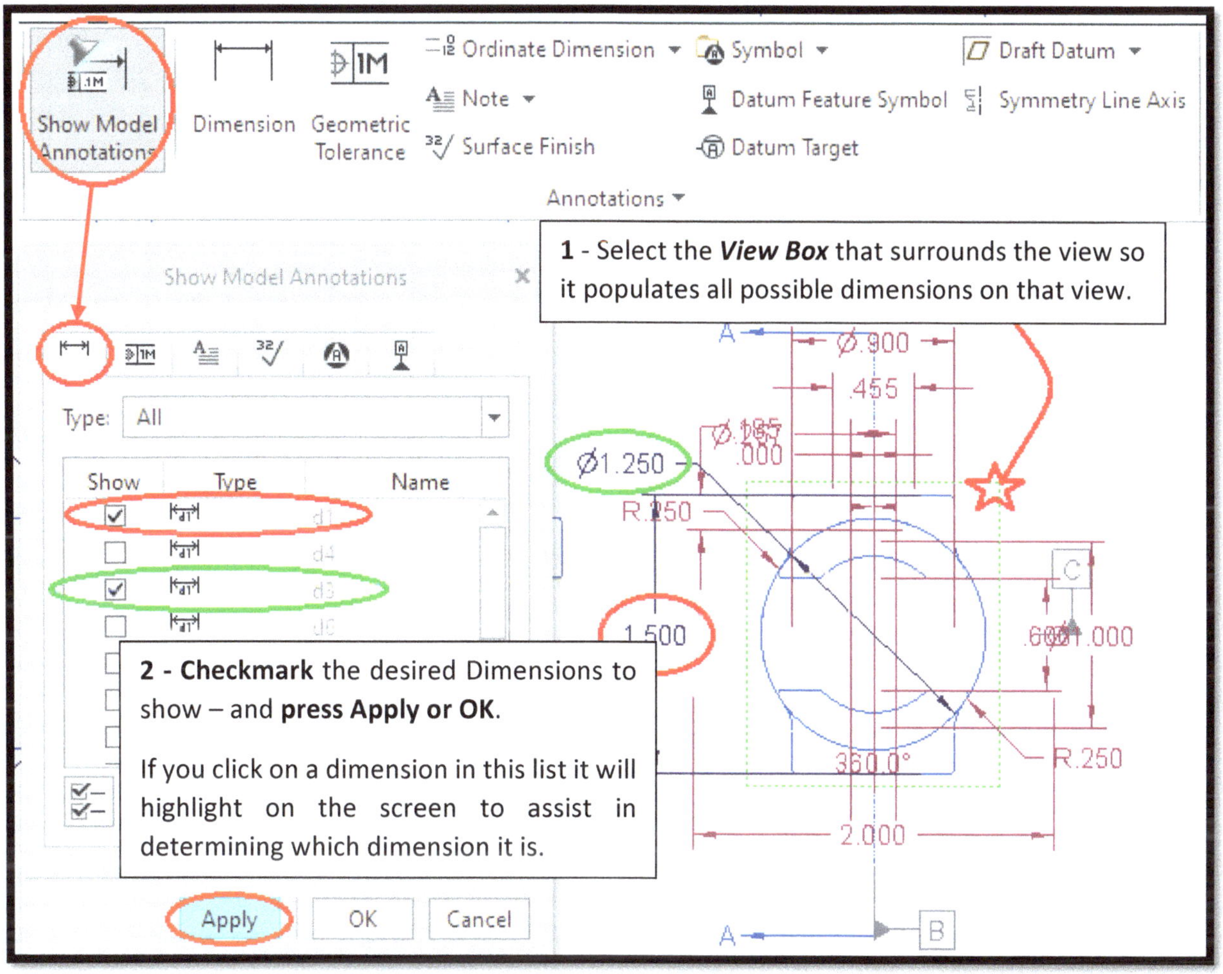

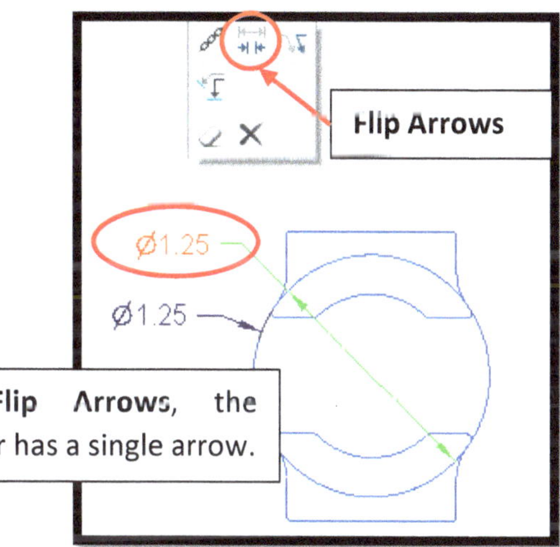

Step 4 - Flip Arrows on Diameter Dimensions

"Flip" Diameter Dimensions to have a single Arrow style, instead of 2 arrows which cross the center of the circle

- **Click on the Dimension** – choose the **Flip Arrows icon** to toggle the arrow style. **Repeat for each Diameter dimension.**

Step 5 – Adjust the Decimal Places as needed (for Tolerance)

The decimal places on dimensions need to be adjusted to the desired manufacturing tolerance as per the Tolerance guidance in the UND template box. You can select and change all dimensions at once, or change only the few that need a different decimal place setting.

- **Left click on the desired dimension** (or make a drag selection box over many dimensions) - the *Dimension* toolbar should appear at the top – change the **Precision value** to the desired number of decimal places.
- **Repeat as necessary** to match the drawing key or your desired tolerance.

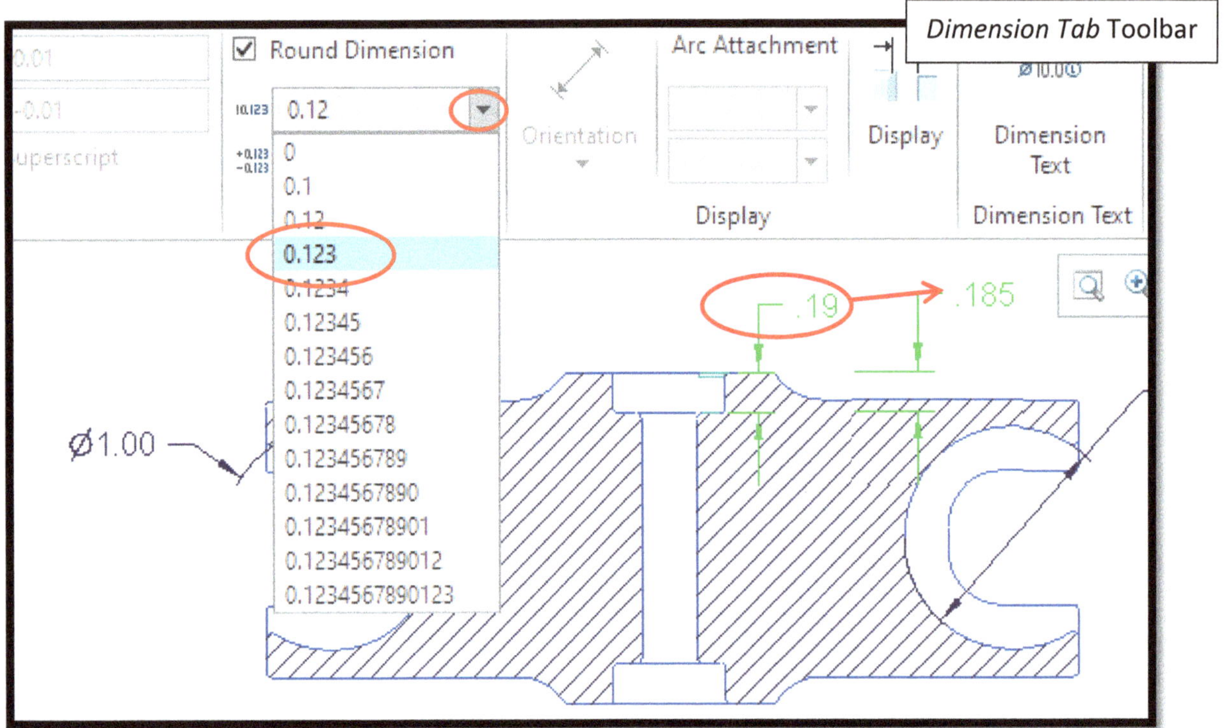

Tip: Sometimes it is useful to show more decimal places to show a standard size, such using 1/8" thick material which is very common, you might show the dimension as 0.125 instead of the rounded 0.13 value.

Step 6 – Add Dimension Text (*starting in Lesson 6*)

If you need to **Add Text** to a dimension to add more details, such as the Quantity of a series of simple holes, you can do so using the **Dimension Text tool**.

Holes require a **placement dimension** (*X & Y from an edge*), the **Pattern spacing dimension**, as well as the **Hole Quantity**.

These dimensions detail the placement of a hole from the edge as well as the spacing of the Hole pattern.

Click on the desired **hole diameter dimension** – use the **Dimension Text** tool – **type in your text** for the quantity of the hole, leaving the **@D placeholder** in place.

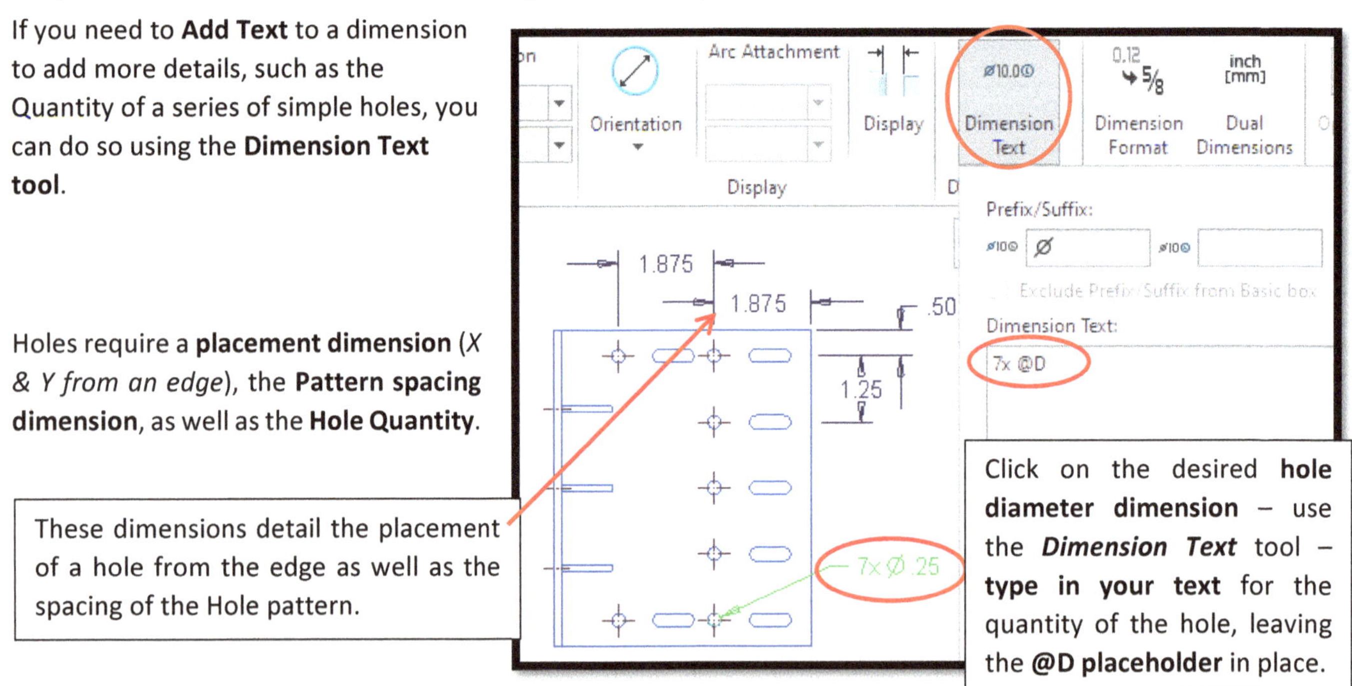

Step 7 – Show Pre-Generated Notes (*Standard Hole Notes*)

If you have features that have pre-generated Notes, like Standard Threaded Holes or Cosmetic Threads, you can show those Notes using the Show Model Annotations tool.

For a Standard Hole note you should show the Note for one of a series of Holes, not on every hole. The Note should automatically display the Tap or Drill Size, Thread Size, Threaded Length, Depth of the Hole, and Quantity of Holes.

- On the **Annotate Tab - Show Model Annotations** tool – use the **Note tab** - click the ***View Box*** of the desired view – **checkmark** any desired Notes - **press OK**.

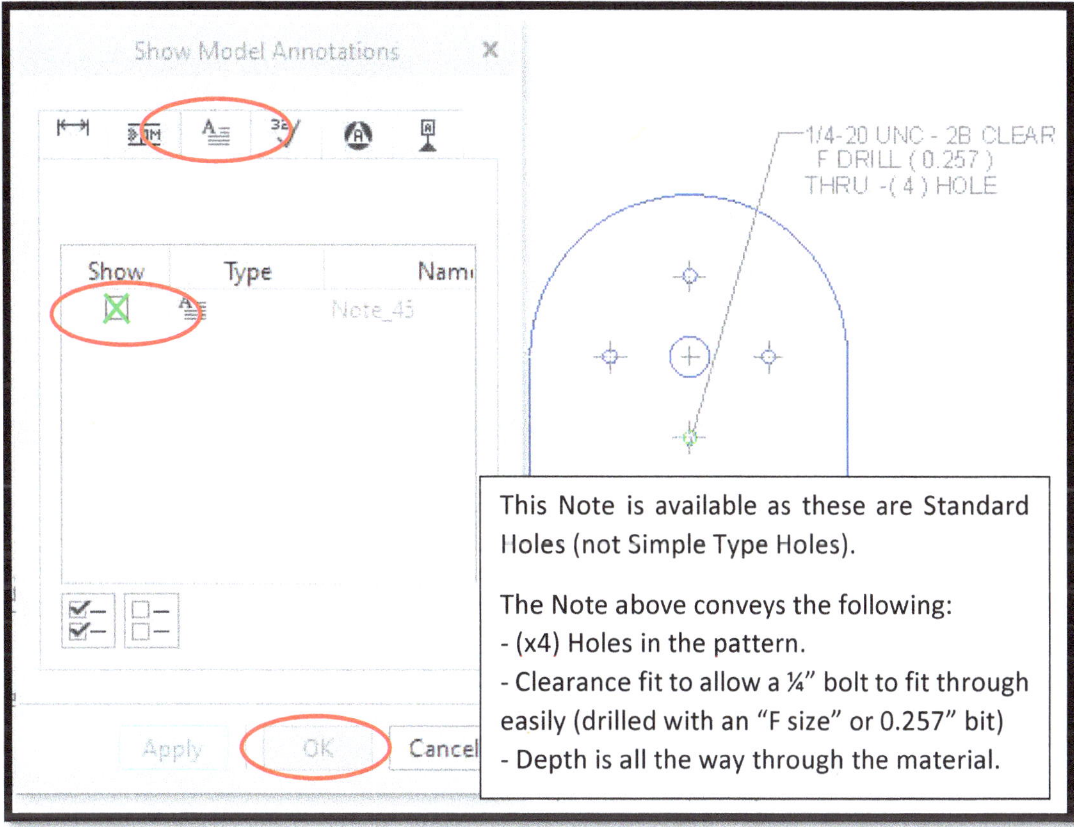

Tip: If you show a note on the wrong view it will not appear in the menu again to show on the other view. You must instead select the Note on the screen – in the pop-up menu select **Move Item to view** – **LMB** on a new view.

If you accidently delete or erase a note, you may be able to Show or Unerase the note by looking on the Annotate Tab in the far left of the screen at the View Template list. Expand the drop arrow under each view and look to see if there are any hidden annotations you can click on and show again.

Step 7b - Edit Attachment to Clean up the Reference of Standard Hole Notes

Sometimes the note arrow will show up on a hole that has the arrow crossing over other details. You can choose a new reference to have the note point at to make a cleaner drawing using the **Edit Attachment tool**.

- **Annotate tab** – use the **Attachment tool** – select the desired **Note** – press Yes in the pop-up menu – use the **Change Ref tool** in the next pop-up menu – **select the new edge** to point at.

Note – this will allow you to choose any edge as a reference, so choose carefully so it is still pointing at a hole from that pattern. If you need to point back at the original feature you can hold RMB on the note – use the **Restore 3D dependencies** option to reset it.

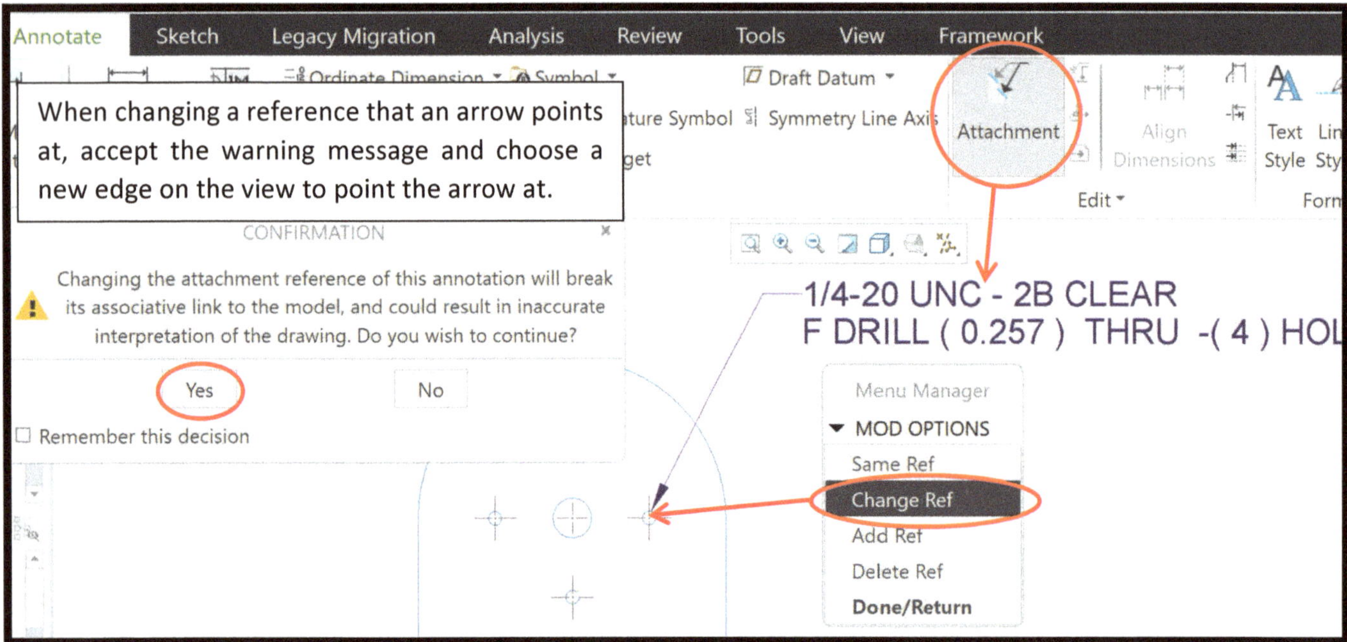

Step 8 – Create your own General *(unattached)* or Leader *(arrow)* Notes

You can also add your own Notes to a drawing, using an **Unattached Note** (i.e. "All Rounds are 0.05 unless otherwise Noted") or a **Leader Note** that has an arrow that points at a specific feature (i.e. "Apply wood glue to this edge")

- Use the **Annotate Tab – Note tool** drop arrow - and **choose Unattached or Leader note**. For a leader note: select the edge to point the arrow at – MMB to place the note text box position - type in your text as desired.

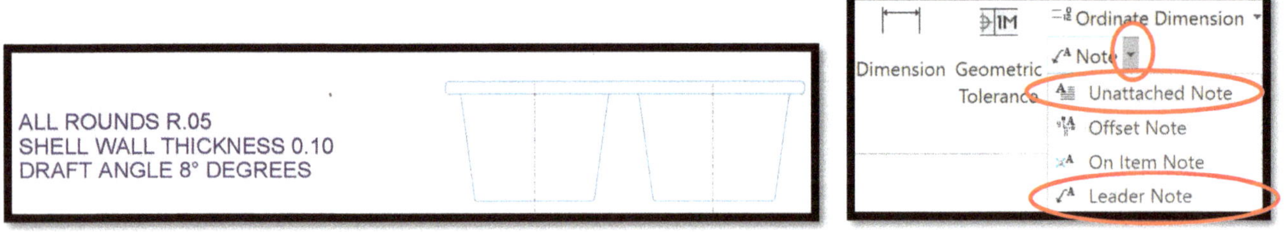

End of the Detail Drawing Steps

Assembly Drawing Steps

Assembly Drawings **have different requirements** than the Detail Part Drawings that are created for individual parts. The focus of an Assembly Drawing is on the connection and placement of components. Only information relevant to the Assembly itself should be present on an Assembly drawing, such as the placement dimensions, overall assembly size dimensions, clearance between components, and of course a Bill of Materials (BOM) Table, Balloon Notes, and Exploded views to identify components that make up the Assembly.

Note – "Feature" dimensions about features on a single part should <u>not</u> be shown! Hole sizes, slot dimensions, part lengths are all found on the detail part drawings, **<u>not</u>** the Assembly drawing. A part detail drawing will have more details about the model than the basic drawings, such as dimensions, notes, additional views or sheets, and ASME style datums. These details are important to communicate the design and fabrication requirements for a part model.

Add the following steps to those in the ***Basic Drawing Steps & Detail Drawing Steps*** as needed. Not all steps will be required on every assembly drawing.

Overview of Assembly Drawings Steps:

0) Setup the drawing views, orientation, etc. to show the desired views.
1) Create the custom Isometric orientation and Exploded View state. (starting with Lesson 11)
2) Show the Exploded View on the Drawing (starting with Lesson 11)
3) Add the Bill of Materials (BOM) Tables & Balloon notes (Starting in Lesson 10)
4) Add Assembly dimensions & notes (Starting in Lesson 11)
5) Add an Assembly Cross-Section (if useful) (Starting in Lesson 11)

Step 1 - Create and Show the Exploded View

Create an exploded view state while the Assembly is open that can alter be shown on a drawing. Follow the steps carefully, as being in the wrong tab or missing a step can cause issues!

Step 1 – 2 - Rotate the model and create a custom **MYISO** Orientation that will be shown on the Drawing. Use this view to show the Exploded Part positions on so you know it will look OK on the drawing.
- **View Tab- Manage Views –View Manager – <u>Orient Tab</u> – New – "MYISO" – press Enter – press Close.**

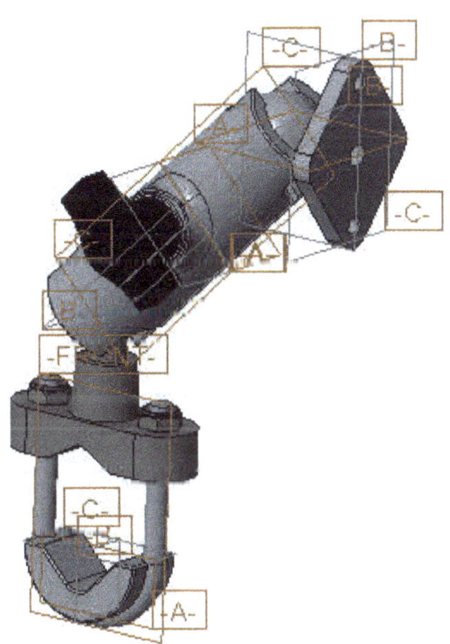
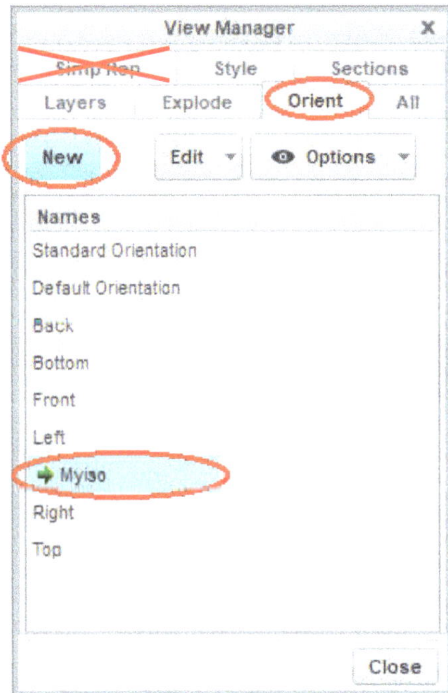

Tip: *A common mistake is you make the isometric view in the "Simp Rep" tab… if you do that you will need to set it back to Master rep, delete out the one you added from the list, and then do this step again under the correct Orient Tab!*

Step 1 – 3 - In the View Manager go to the **Explode Tab**.

    - **Click New** to create your own Exploded View State and name it "MYEXPLODE".

    - Press **Edit - Edit Position** to bring up the explode toolbar.

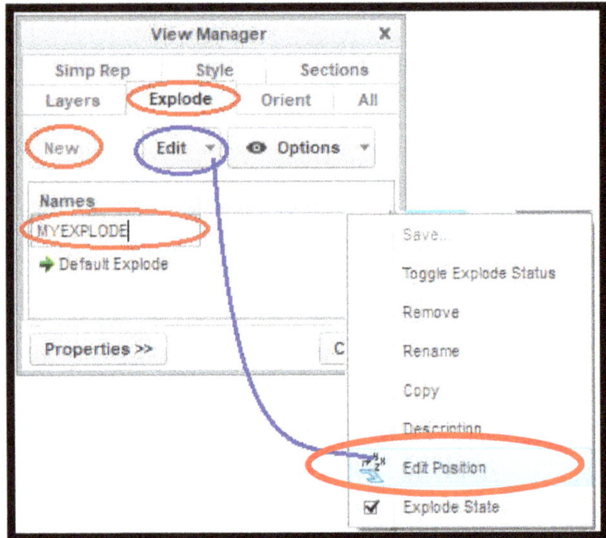

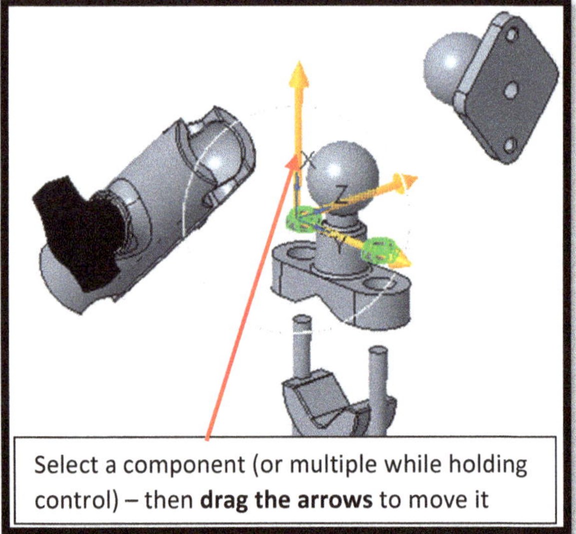

Step 1 – 4 - In the **Explode Toolbar** you can click on components and then move them to their "Exploded" position by dragging the desired movement arrows on the graphics window. You can move groups of components together if you hold control while selecting them, or even select entire subassemblies by clicking on them in the Model Tree.

1) Move the components in a linear fashion to their exploded positions so it is clear how the components assemble.
2) Move sub-assemblies together as a single unit (*except Lesson 11, they were all exploded individually to better view the assembly for grading*).
3) After moving the pieces to their desired position, **press the checkmark** in the top toolbar.
4) **REQUIRED:** You must then press **Edit - Save** in the **Explode Tab Menu** or it will not save the exploded positions!

Step 1 – 5 - Toggle On/Off the Exploded State. The exploded state can be toggled on or off on the model screen as desired, toggling it off does not delete the state from being shown on the drawing.

    - **View manager - Explode Tab – Edit** – uncheck the **Explode State** option.

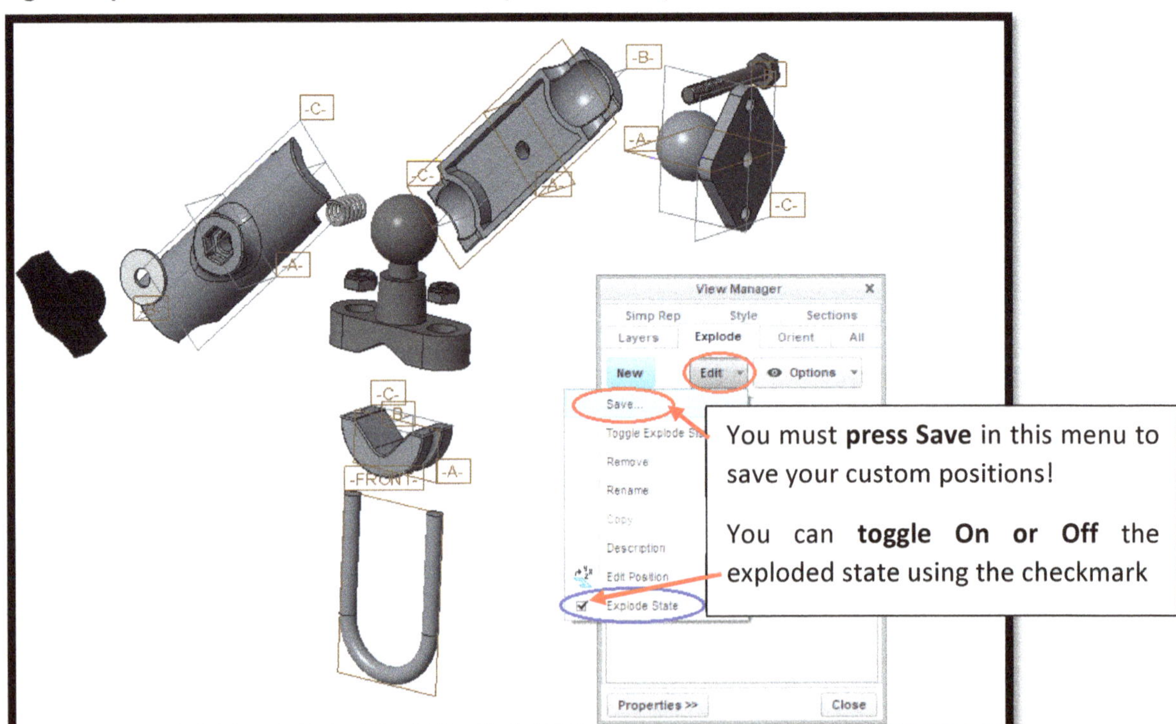

Step 2 – Show the Exploded View on the Drawing

Often you will want to show the Exploded view on its own sheet so it can be scaled up and have more room on the drawing page.

1)   On the drawing sheet tab area (bottom left corner) press the + symbol to insert a Sheet #2.

2)   On that sheet, use the **General View tool** to add your custom **MyIso view** to the drawing

3)   In the **View States** category - **checkmark** the "Explode Components in View" option

4)   Change the drop box menu from the "Default" explode to the **MyExplode** that you created previously

5)   Setup a **Custom Scale** for this view (under the **Scale category**) so it fits the sheet.

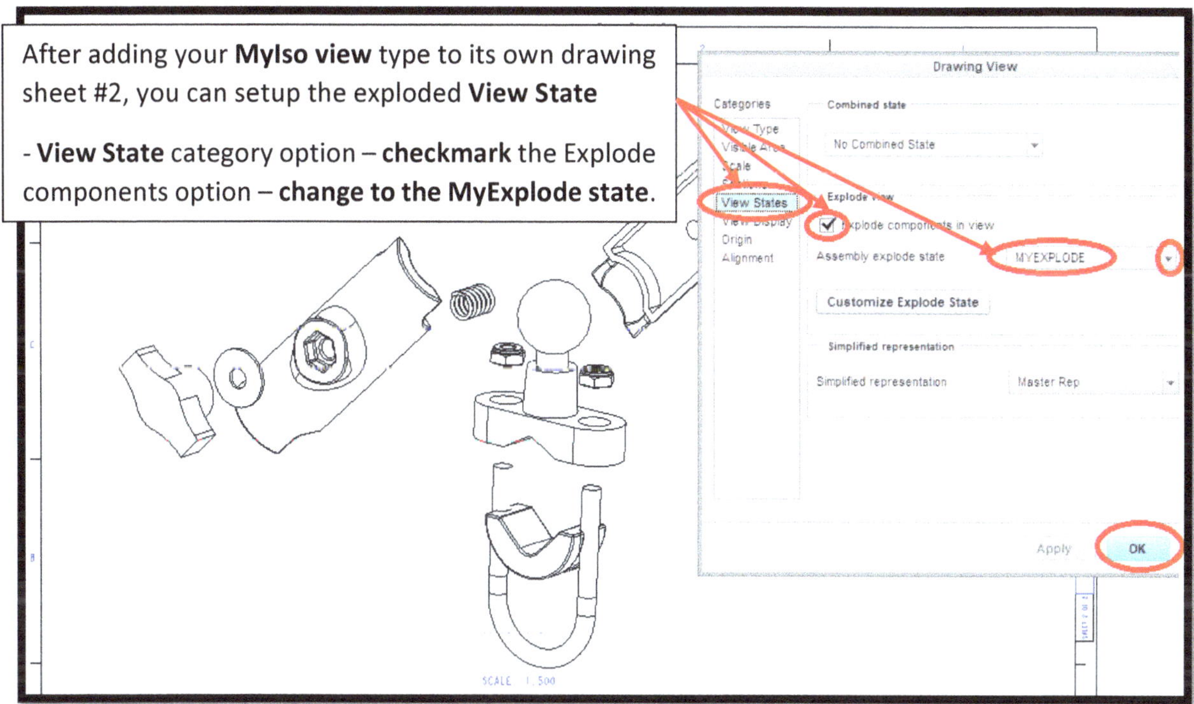

If your exploded view does not look correct it could be due to some common mistakes.

-   Make sure you are showing the *MyExplode*, not the *Default* explode
    Check that the view is set to the **MyIso view type**, so it is angled correctly to view the explode position properly.
-   Go back to the Assembly model – open the View Manager – Explode Tab - toggle on the Exploded view state   if it does not look correct or doesn't explode you can **Edit – Edit Position** to move the components again. Make sure after you set the position you click **Edit - Save** from that menu, or the positions will not save.
-   Also make sure you are viewing the MyIso view orientation when creating the exploded positions. If you rotate the model while setting the positions it may not look good when viewed from the MyIso angle of the drawing.

Tip: Having an exploded view is great to show how the components are assembled. On your own projects, **you may also want to consider having a second Isometric view that is not exploded**, often on the main sheet like was done on part drawings.

Step 3a – Add the Bill of Materials

A Bill of Material (BOM) Table file is provided by the instructor. Make sure this has been downloaded and saved to your working directory. This Table file is setup to list all "main" level parts or sub-assemblies in the model tree along with their specified material and quantity. Parts within a sub-assembly should not be found by the BOM table.

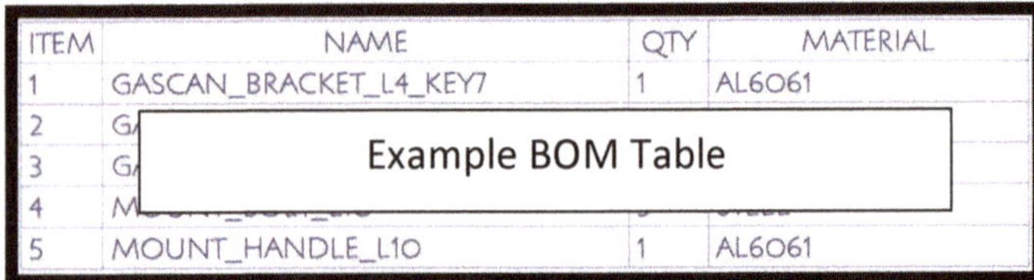

6)   **Table Tab** – choose the **"Table from File"** option – navigate to your Working Directory - **choose the desired "UND_BOM_table" file*** – LMB to place it** in the upper right corner of the Drawing (or where it fits).

***Note -** There are **two table files provided**; one that auto fills in the material and one that keeps the material cell blank. If you are diligent about setting the material of all your parts you can use the autofill version, otherwise use the "blank" table file and type in the material cell-by-cell while on the Table tab. The Autofill table will not let you modify the Cells.

Step 3b - Add the Balloon Notes

Add Balloon Notes to the drawing to show which part corresponds to an item in the Bill of Materials Table. **You must have a BOM table in place before you can add Balloons**. Note – typically you would add these to the Exploded view.

- **Table Tab – Create Balloons – Create by All** (or By View)– it may ask you to select the region in the bottom left menu (select the BOM Table cells you added in earlier) – click the desired View (if adding "by View". The Balloon notes should appear on the drawing.

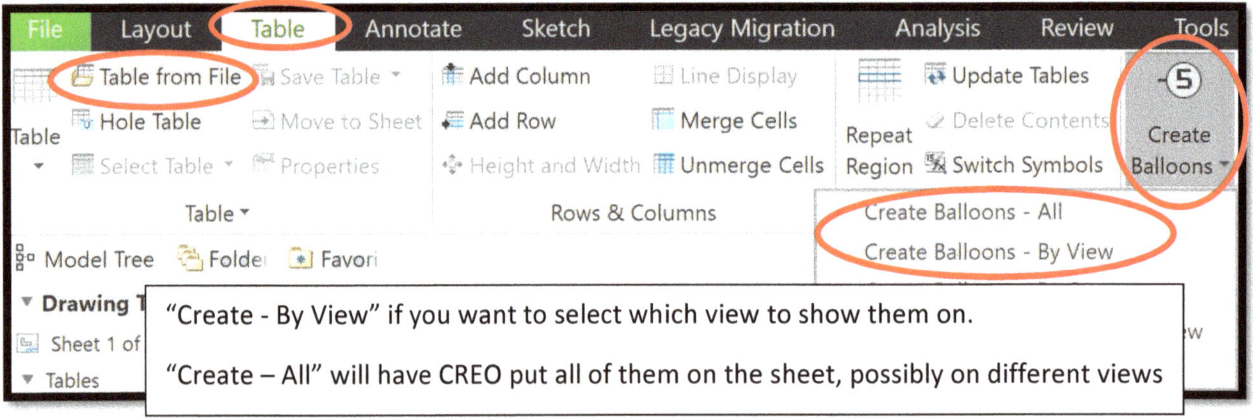

Step 3c - Clean up the Balloon Notes

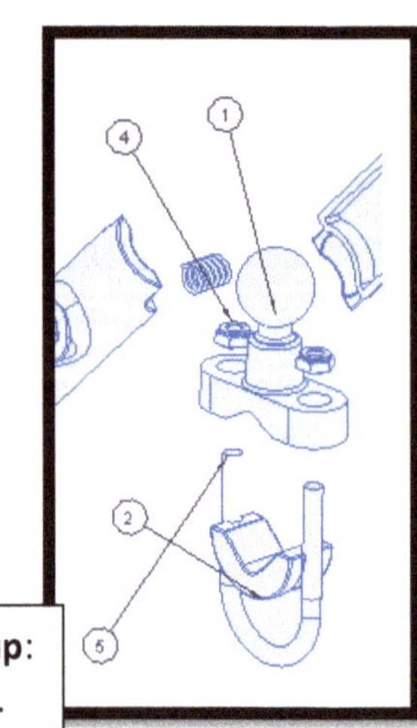

Move the balloon notes to other views and new positions as needed so they are clear what they are pointing at and do not cross over each other. Use "Move item to view" to have a balloon point at a different view, then use "Edit Attachment" to change which edge of the component the arrow is pointing at (only if needed).

1)   **LMB to select a Balloon – hold RMB- "Move Item to View" – LMB on a new view** to show it on that view.
2)   **LMB to select a Balloon – hold RMB – "Edit Attachment" – LMB on a new reference edge** for the arrow.

This example needs some cleaning up: Move the balloons to better positions.

Step 4 – Add Assembly Dimensions and Notes

Use the **Dimension tool** or **Show Model annotations tool** to create or show any desired Assembly dimensions. Focus the dimensions on the position and placement of components, clearance between parts that are important for the design, and an overall Length, Width, Height to provide sense of size of the assembly.

Important Tip: You should **not** be dimensioning the part features like hole diameters, length of a part, width of a feature, etc. as all of those dimensions should be located on the Detail Part drawing for the individual parts.

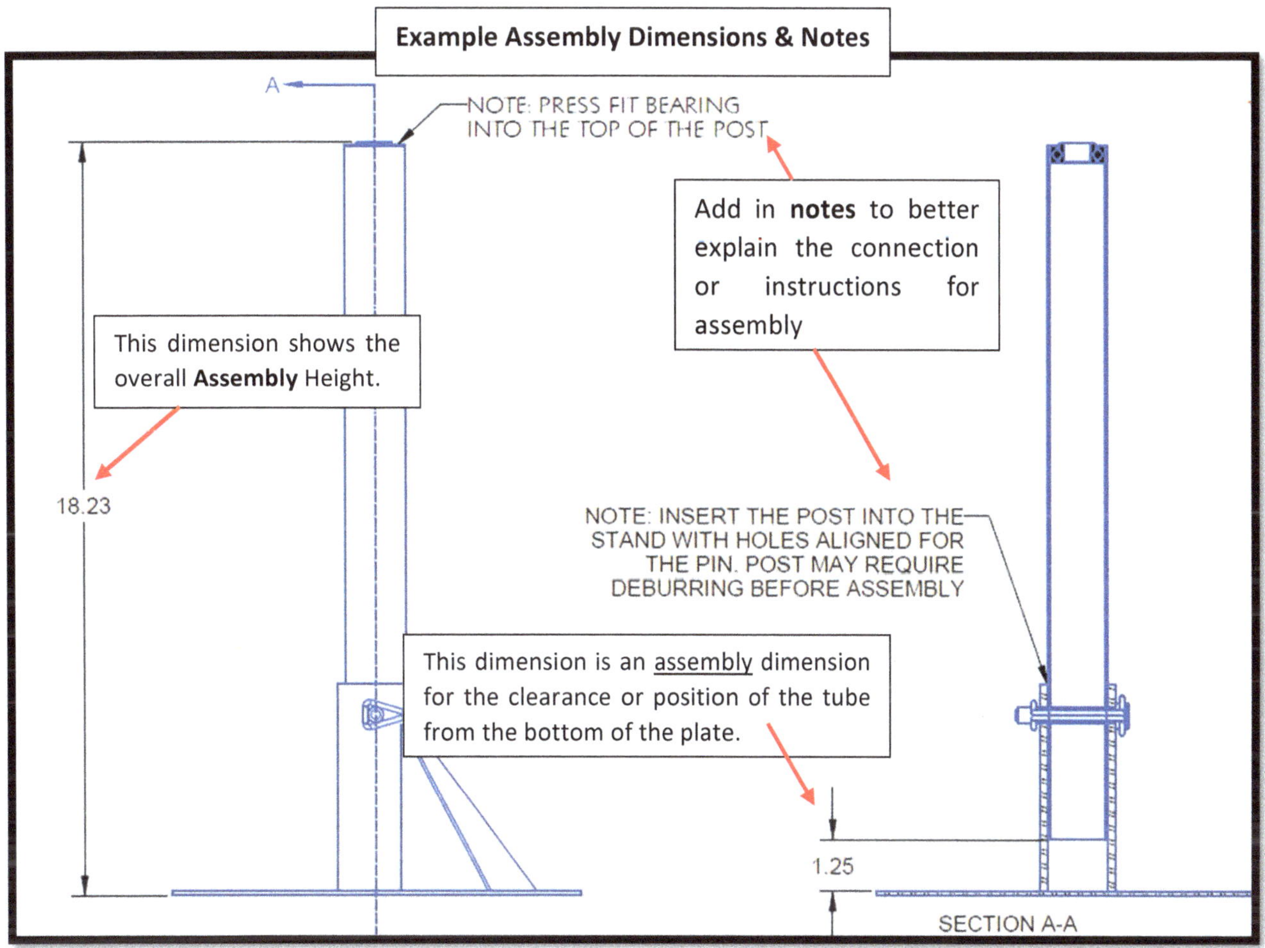

Use the **Note Tool** (either **Leader Note type** (with an arrow) or an **Unattached Note** type) to add in extra information, such as how parts are connected if it is not clear that they are bolted together (i.e. *"Apply wood glue and use #8 2" Construction Screws spaced Every 4 inches"*).

Step 5 – Create an Assembly Cross-section

If useful for your drawing, you may want to show an Assembly Cross-Section to better show how parts are connected and fit together internally. Similar to creating a Section on a part but create and use an Assembly Datum for the plane.

Step 5 – 1 - Create a new Assembly Datum (ADTM1) on the model that will be used to place a Cross-Section.

- Go back to the Assembly Model window – use the **Plane Tool** – Select the **ASM_FRONT datum** in the Model Tree as the reference – set the **offset translation to 0.00 – press OK**.

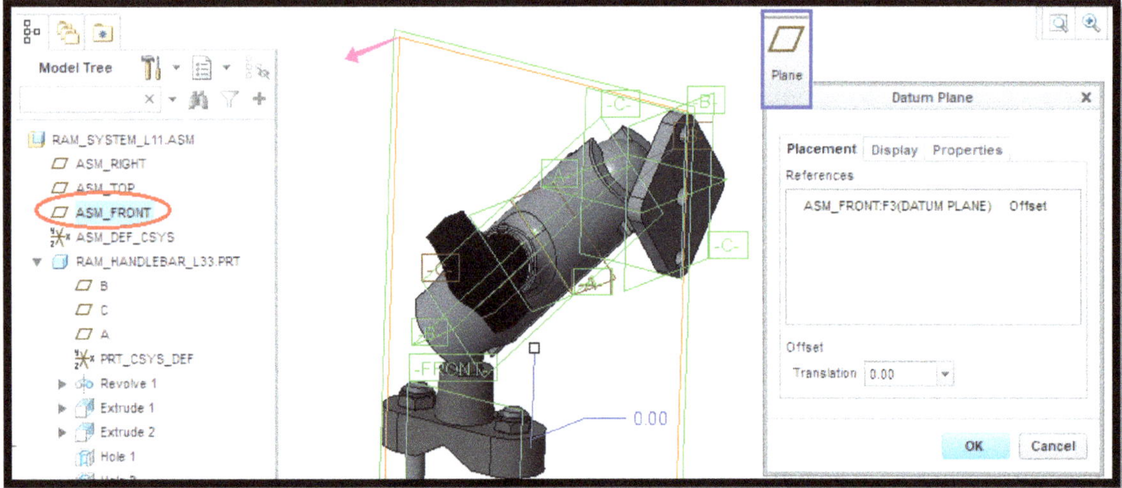

Step 5 – 2 - Follow the steps to **create a new Cross-Section** (*Step 8 of the <u>Basic</u> Drawing Steps*). When requested to choose the section plane, **select ADTM1** or another other Assembly datum from the Model Tree to create the Section. You **must use an Assembly Datum** so it interacts with all components on that section plane.

Step 5 – 3 - Adjust the hatch line settings for each component on the cross-section to differentiate the components.

- **Double-click on the hatch lines** (*while on the Layout tab*) to bring up the ***Edit Hatching*** top toolbar – click on the hatch lines for a chosen Component on the section – choose the desired settings in the *Hatching Toolbar* such as **Spacing & Angle, Include or Exclude**, and the option for "**Use a Solid Hatch**" to match the key as needed.

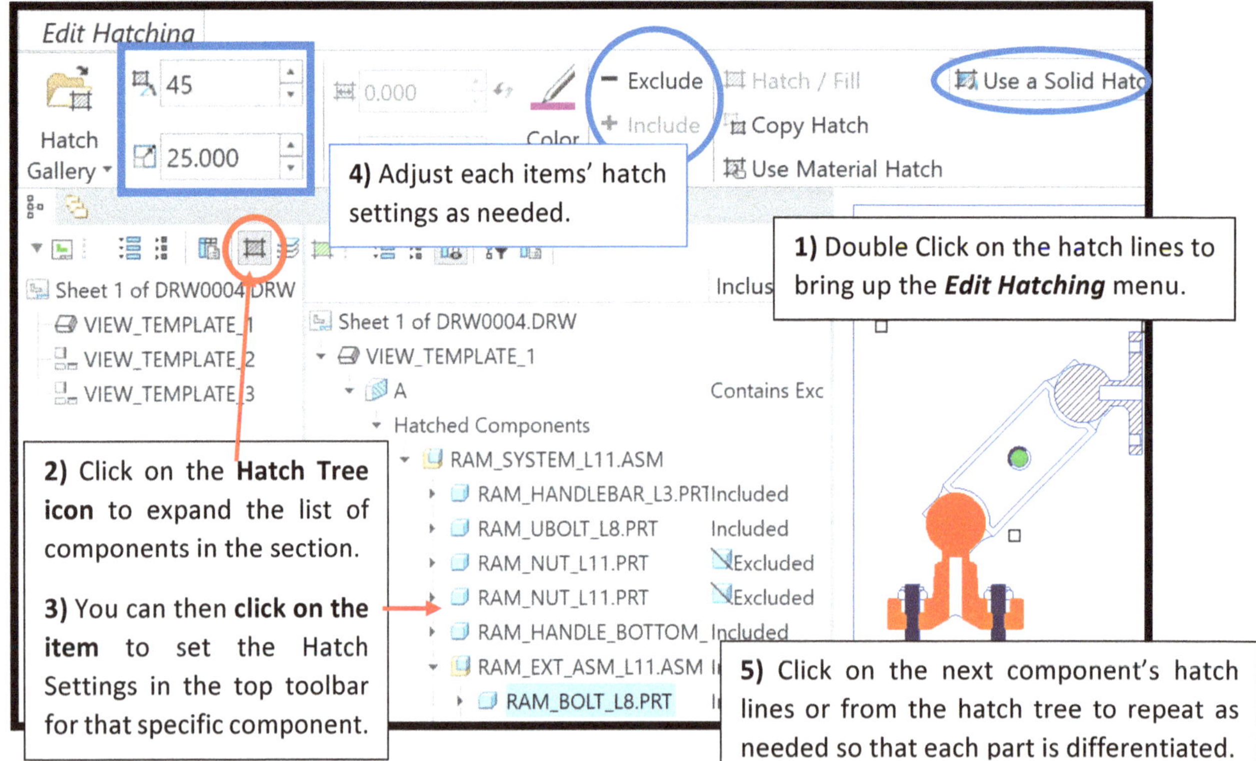

Step 5 – 4 – "Add Arrows" for the Section Placement (*Layout Tab*). Refer to *Basic Drawings Steps - Step 8b #2 for help.*

"Class Standards" Drawing Checklist

Part Drawings

Drawings should be created to class standards with proper details and dimensions similar to what was done in the lesson assignments (refer to the lesson keys for examples). Use these checklists when working on your own drawings.

☐	**Proper dimensions & clean Dimension layout** (Limit crossing of dimensions, show enough but no more than about 15 -20 dimensions on a drawing (use your judgement)

☐	**Show Centerlines on Holes** (Show Model Annotations Tools) and **dimension the placement of holes to the centerlines**. (i.e. distance from X & Y edges to location of a hole, and a note or dimension for the hole spacing for additional copies)

☐	Show a **note for any Standard holes or Threaded surfaces**

☐	**Datums, if shown, should be ASME Style** (gtol_datums). *Datums are not required to be shown.*

☐	**Use Notes** to add more detail about features (Spacing of a pattern, Shell thickness etc.)

☐	Use **Detail Views** to detail & dimension smaller features that are important so they can be zoomed in.

☐	**View Display** – typically should be No Hidden, Tangents None (Except leave Tangents Default on the Iso View)

☐	**Cross-Sections** should be added as needed to detail inner geometry: Add a Cross-section, **Add Arrows** for its placement, and reduce the **hatch spacing**.

☐	**UND Template**: should show proper part name, drawing name, material, and your name.

☐	**Views:** Symmetric parts may only require two or three views, while others you may want to add an additional Projection view or Detail View. Make sure the views are Aligned. Use additional views as needed to dimension features, such as some components may require a bottom/left/or back view to detail an extrude or hole.

☐	**Dimension Tolerance & Decimal Places**: Adjust the dimension decimal places as needed (should not be X.XXX unless that tolerance is required, or if it is a fractional like 1/8th then showing 0.125 would be OK)

☐	**Diameter Dimensions: Flip Style** to be single arrow style, and use **Diameters** (double click) instead of Radius (single click) to detail holes. Radius would be OK for rounds, fillets, etc. but not Holes or Cylinders.

Assembly & Sub-Assembly Drawings

The goal of an Assembly drawing is to detail which components are used and how they connect together. Details for specific parts (size of hole, feature dimensions, etc.) are <u>not included</u> in the assembly drawing as they are detailed in the part drawings. Include Placement dimensions, clearance or gaps that need to be maintained, etc.

☐	**BOM (bill of materials table)** must be shown. Parts would have material listed (if Model Properties were set), while the Assemblies material would be blank.

☐	**Placement Dimensions** (distance a shaft is slid through a bearing, etc.) or other important dimensions for the assembly **(Overall Length, Width, Height).**

☐	Do not include "Part Feature" dimensions! Hole sizes, edge dimensions, etc. of part features should not be dimensioned on an assembly drawing; the part drawings have that information.

☐	**Exploded View** on an Isometric (MyIso) View, to show the different components, typically on a 2nd sheet. Explode components as they are listed in the BOM table; so only "explode" the items with a balloon; if the item is actually a sub-assembly you would move that entire sub-assembly as a unit, not individual components of that unit.

☐	**Balloon Notes** – list our which components are the ones shown in the BOM table. Typically shown on the exploded view.

☐	**UND Format Template** – for material, either leave blank or "SEE BOM"

☐	**Notes** – use **Notes to detail specific assembly requirements as needed**. Example: "Apply Wood Glue & Use 1" Wood Screws every 4 inches along edges" or "Rotor Diameter = 18.75")

☐	**Assembly Cross-Sections** can be added to show better details of the interior of an assembly (make a Cross-section on an Assembly Datum, adjust the spacing/scale & of each hatch, & add arrows for the placement.

End of this Checklist

Creating a Custom Isometric View

1) **Rotate the model** on the screen to <u>approximate</u> the **Isometric View** on the drawing key.
2) In the Top Toolbar go to the **View Tab** – select **Manage Views** – in the View Manager change to the **Orient** tab.
3) **In the Orient Tab** select **New** – type in the name for the new view **"MYISO"** – press Enter.

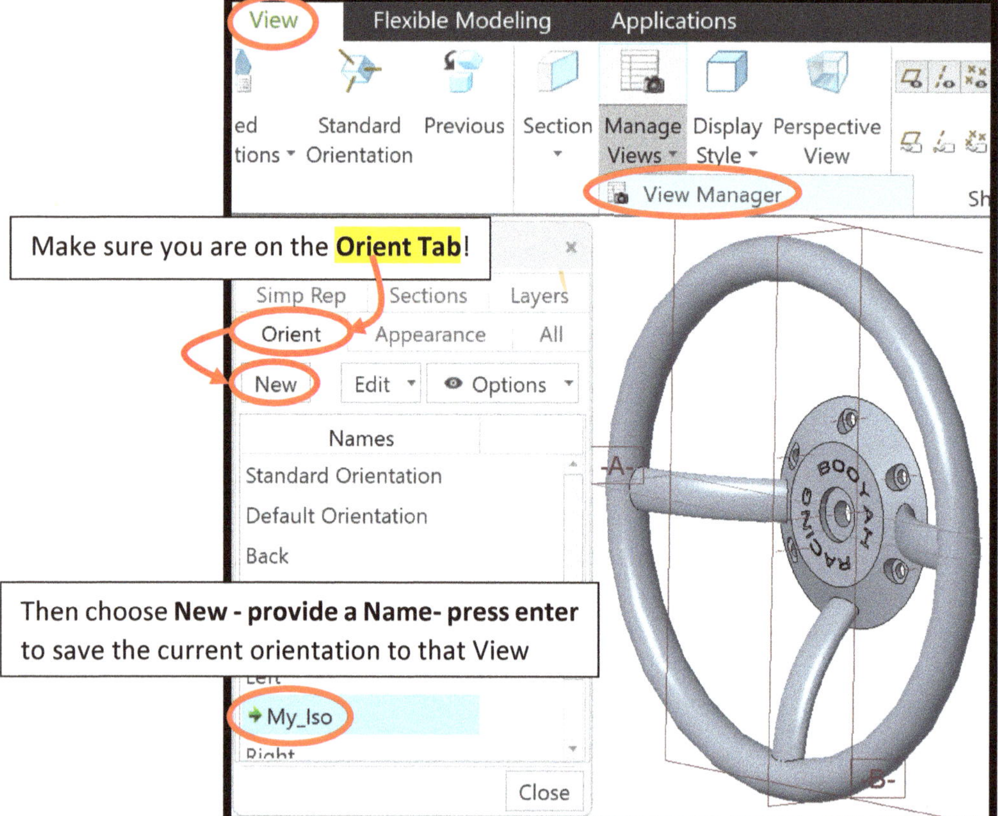

4) **<u>Test out the new view</u>**: Rotate the model in the graphics window – in the quick toolbar select the **View Orientation** tool – select the new **MY_ISO view**. The view should rotate to your custom Isometric view.

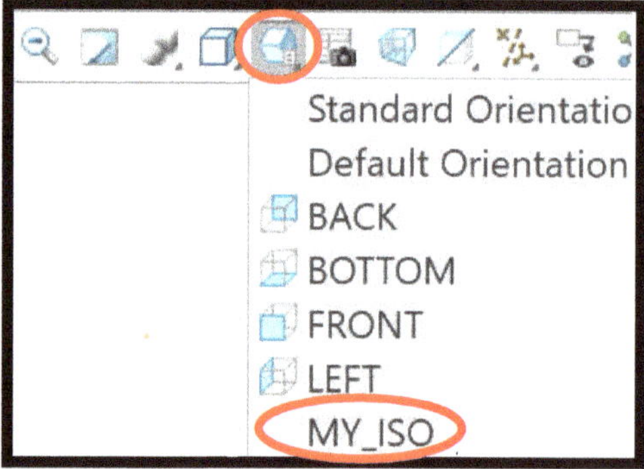

Note – If the view does not show your Isometric view, you likely did not make the view while on the **Orient Tab!** You will need to fix your error by changing the tab you did make the view on (usually the Simple Rep tab) back to its default setting (i.e. back to Master rep setting) and then create the view under the correct Orient Tab as described above.

Note – When inserting the **General View** on a drawing you can choose the view type as the **MY_ISO** custom view state you created above. If you do not see the *My_Iso* as a view state option on the drawing, you likely made the custom view state while on the "Simple Rep" tab instead of the "Orient tab" (see note above). Also, sometimes students have saved a backup of the model and then have their drawing using a version of their model that they did not setup the custom View state on.

End of this Topic

Creating an Auxiliary Drawing View

Part models may often have a surface that is angled out of plane of the typical Top, Right, Front views, and showing this plane perpendicular to the screen will allow dimensions and other details to be shown for any features on that plane. Use the steps below to learn how to create an Auxiliary View on a drawing.

1) With a drawing setup, use the Layout tab – **Auxiliary View** tool to add the new view.
2) **Select an edge** or **Datum** (from a side view) to use as the auxiliary plane reference. In the example below, the top angled line from the front view was selected as the reference. On some models, or if there is a Cross-section present, you may need to create a new datum plane (using the Plane tool on the part model) coincident with the angled surface that can then be selected when adding the Auxiliary tool.
3) **Click to place the new Auxiliary View**. This new view will be constrained to be projected at an angle from the selected surface, but will rotate the plane towards the screen.
4) Then use the new view to **add dimensions or other details** to better detail hole locations or other details that occur in that plane.

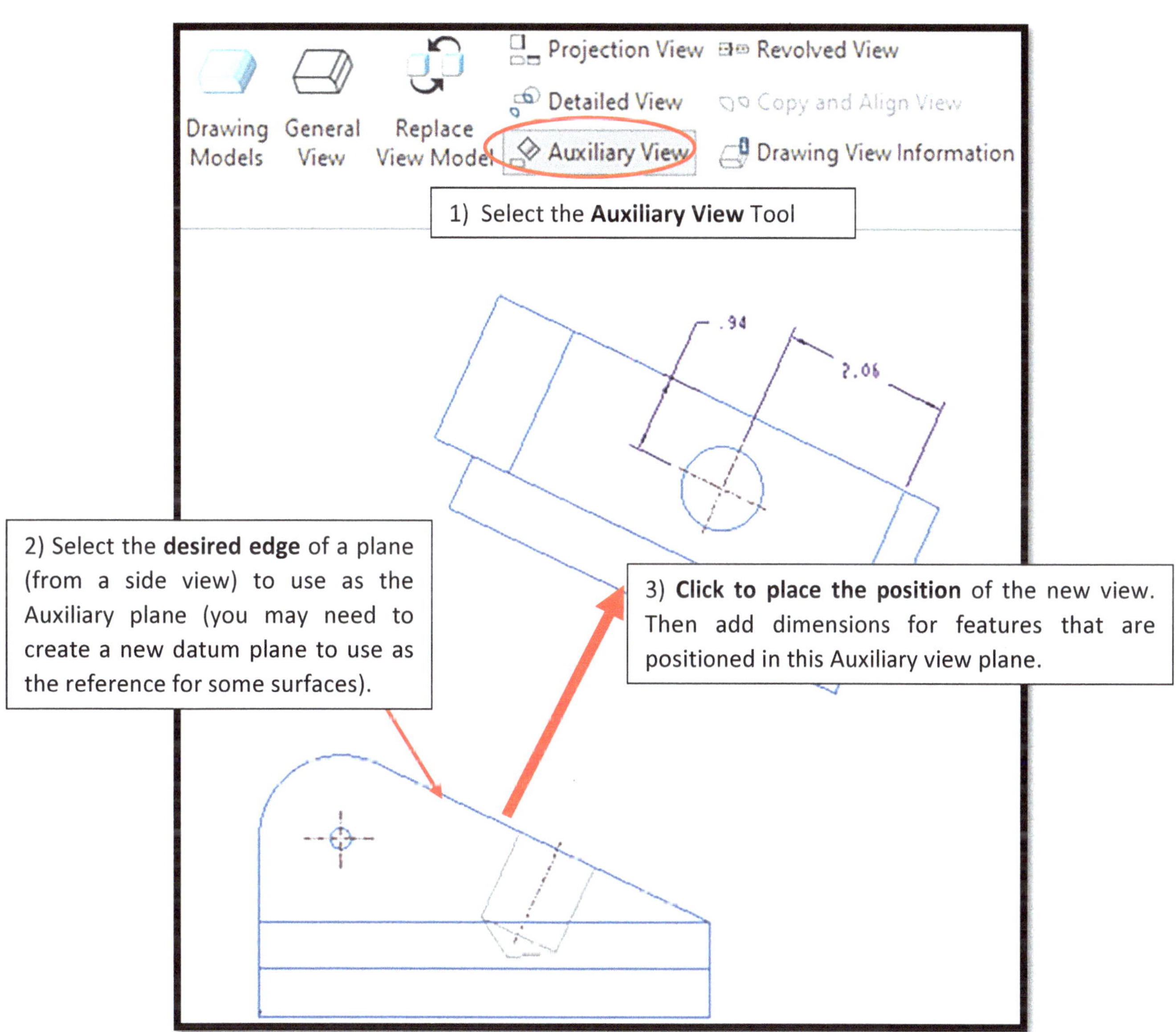

End of this Topic

Recovering from a Crash (Trail Files)

Computer crashes happen and they can be especially frustrating when you realize there is no Auto-Save in CREO (this is to prevent unwanted design changes from happening when you are examining a model). But there is a way to recover your modeling work using a **Trail file**. By default, CREO will record a trail of all the steps and commands you have completed during your current session. This can be "replayed" to repeat all the steps you have done and get back to what you were doing. Note – if you have a very long session it will redo all of your steps, it may take a while.

Step 1 - Locate your Trail.txt file and edit it. It is likely stored to some default location like C:/Users/Public/Public Documents, or your working directory. You can use the Search button in the bottom windows toolbar and search for "trail", then RMB on it and choose **Open Folder Location** when it finds a file. Look for the file with the accurate date and time of the crash, as you will likely have multiple trail files from past sessions of using CREO.

Step 2 - After locating the file copy and paste it to your Desktop

Step 3 – Edit the Trail file to remove any exit command, as well as change it to a **non-versioned Text file type**.

- **RMB** on the trail file – **Open With** - choose **Notepad** or similar – scroll to the bottom and look for an Exit Dialogue command – if present, change it to no to prevent CREO from closing when it runs the trail file.
- **Use File - Save As – rename the file as a "My_trail.txt"** so that it becomes a text file instead of a CREO file.
  - *You can double check the file type by RMB on the file – Properties – and it should list a Text file type instead of CREO Version file type. You may need to adjust your windows "view" settings in File explorer to view file extensions so you can change the type to .txt when you rename it.*

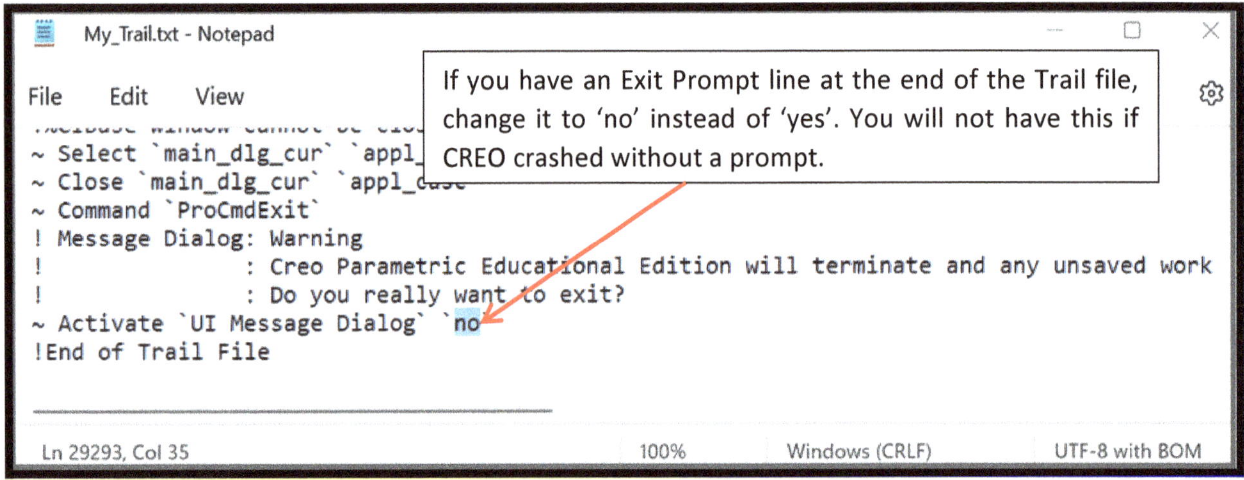

Step 4 - Load the text file into CREO: Open CREO Parametric – in the Top Toolbar choose **Play Trail File** – find your edited or renamed trail file. The screen will then go through all the commands in the Trail file, and if the command to close was set to "no", the program should stay open.

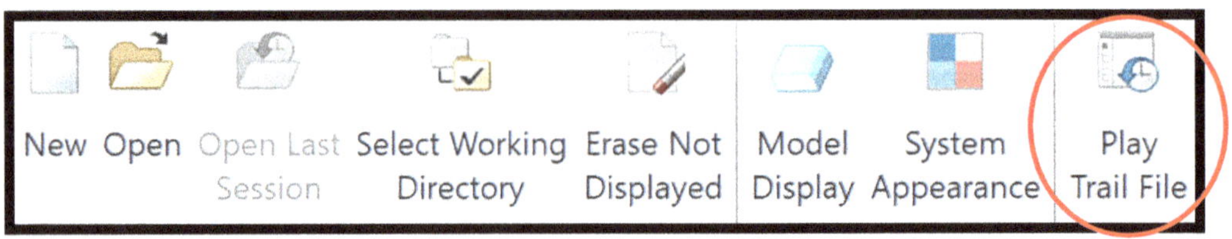

If this step does not work, look in the bottom left corner. If it says "Version numbers not allowed" you likely need to save the trail file as a .txt file as shown above

Set your Directory and Save your file. You can then continue as normal where you left off.

End of this Topic

Setting the Model Units

By default, CREO will be setup for inch units of measurements ("*inch-lbm-seconds*"). You can easily setup other units of measurement for the current part such as cm or mm. When changing the units, you will be asked if you want to "interpret" or "convert" the measurements of the model that may already have been placed. Note the difference between the options further below.

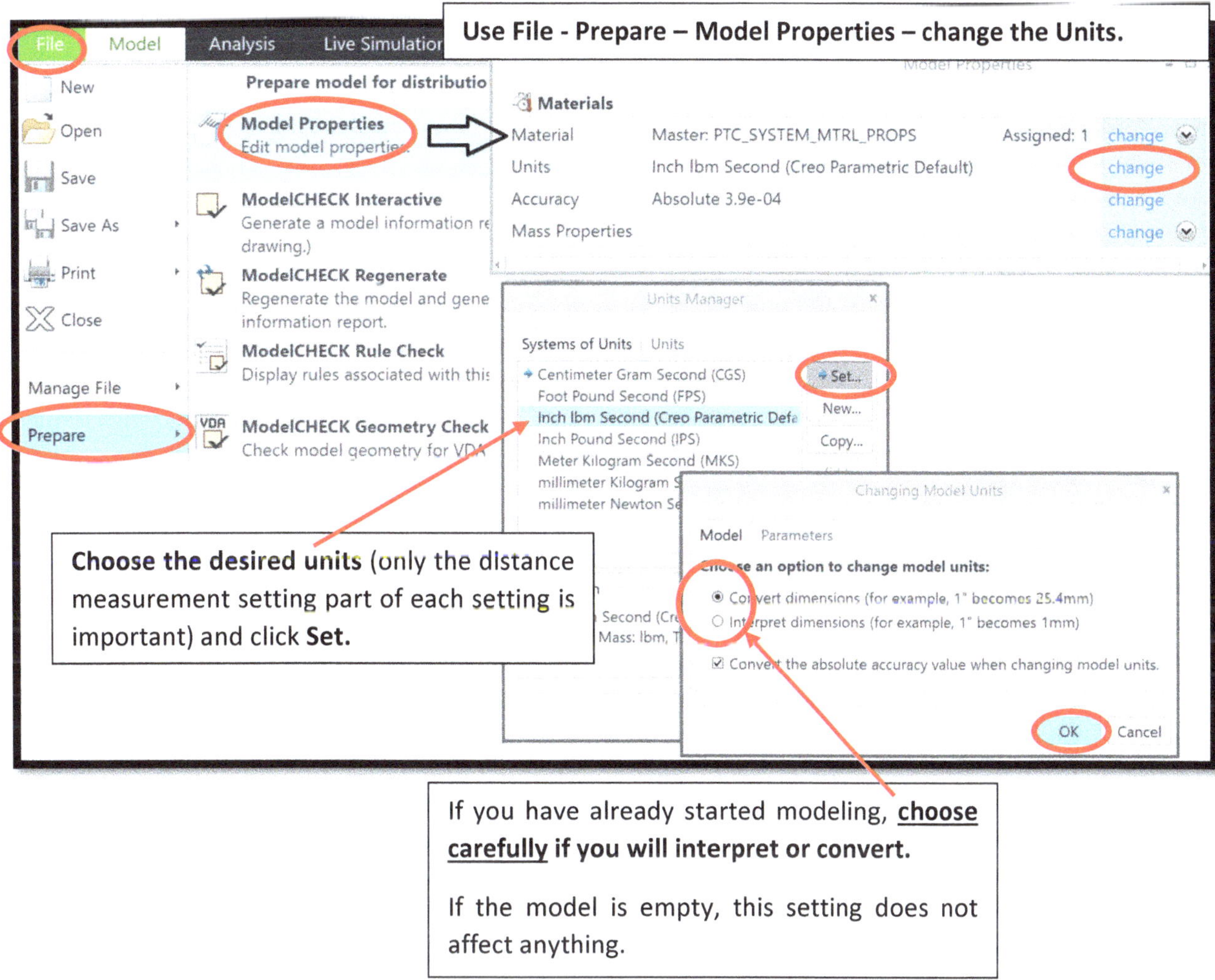

Interpret the Unit Change: Use this setting if you do **not** want to add a conversion factor to any dimensions or values that are currently on the model. A value of 10 inches will be interpreted to be 10 cm or 10mm depending on the units chosen, without applying a conversion factor.

Converting the Units: Use this setting if the model was created properly with one unit of measurement, and you desire to convert all values. A value of 10 inches will become (10 in * 2.54 cm /in = 25.4 cm).

If you are always working in a specific set of units, you can web search how to adjust the "config file" to default to a specific unit setting on models and drawings.

End of this Topic

Save, Save As, Save a Copy, and Renaming a File

Save your model often! Pressing **Save** will save the model to the <u>current </u>directory. This will be either the default directory (bad!), your correctly specified ME101 Working Directory (good!), or the location that you last used in this session of CREO.

> **Example:** You set your working directory to your U: drive, but later opened a CREO part file you had moved to the desktop. Since the last location used was different from your set Working Directory, CREO will likely save the file to that "last used" location unless you set the Working Directory again.

> **Double Check**: Using the Windows **File Explorer**, navigate to your working directory and see if your saved part file is present. You may see the same file multiple times – every time you press "Save", CREO saves an iteration of the same file (GasCan.prt.1, GasCan.prt.2, etc.). You can remove the old versions if desired

<u>**Saving a Part vs. Saving a Sketch:**</u> If you press Save while a sketch is active, CREO will try to save the sketch file (.sec) only and will not save your Part model. Make sure to finish or cancel any sketch and feature you are working on so that when you press Save it will save the model file as desired.

<u>**Can't Save? Model Not Regenerated?**</u> When you are working on Assemblies, often you will need to **Regenerate** the model before CREO will allow you to save. Press Control + G to regenerate which forces the model to be rebuilt and allowing you to save.

<u>**Save A Backup:**</u> If you **do not** see your recently saved file in your ME101 Working Directory, you can always reset the working directory (File - Manage Session –Set Working Directory) or use **File- Save As – Save a Backup** to save a copy of the current file to a different location as a backup.

<u>**Save a Copy:**</u> allows you to pick a new location to save a copy to but you **must** provide a new filename (i.e. GasCanV2). Note this is slightly different from "**Save a Backup**" which makes a copy to a new location but with the same name. Typically, Save a Copy would only be used if you wanted 2 versions on a model to test out some design options while still having the original model untouched.

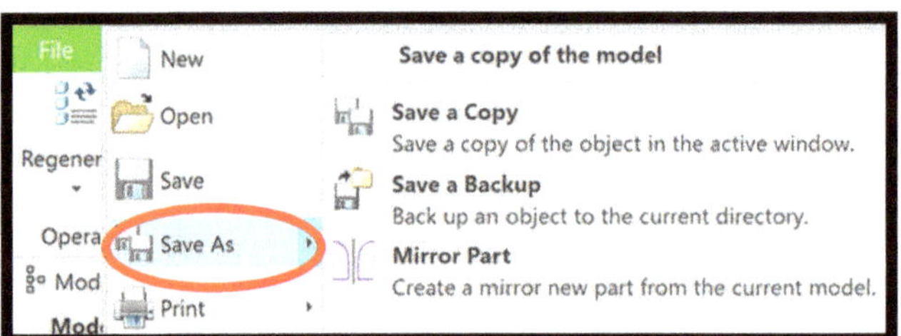

<u>**Renaming a File:**</u> Did you leave your part model name as *PRT0001* when you started the model? While the model is open in CREO you can use **File - Manage File- Rename** to name the file appropriately. This is the proper way to do so, instead of renaming the file through Windows Explorer which won't be effective until you close CREO since the old file name would still be in the current session's memory.

Be careful when renaming a file, as any other CREO files (drawings, assemblies) that are looking for the original name will not be able to find it!

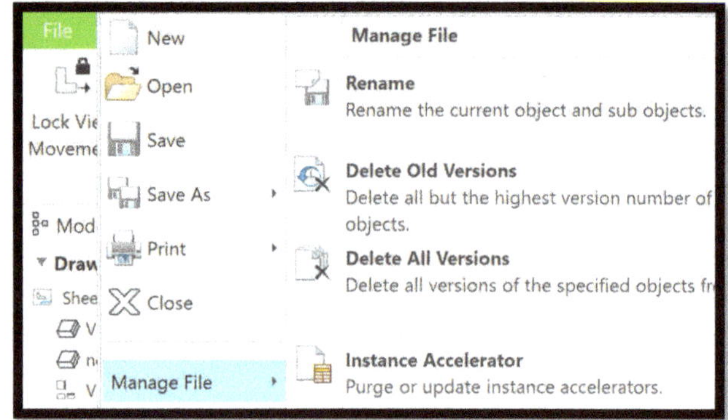

End of this Topic

Using the Measuring Tool & Mass Properties

Measuring Tool: A useful tool in CREO is to use the Measuring Tool to check lengths, distances, diameters of holes, or clearance of parts on the fly.

While in Modeling mode, use the **Analysis tab – Measure Tool –** and choose which type of measurement you want. The Diameter and the Distance measurement options are shown below.

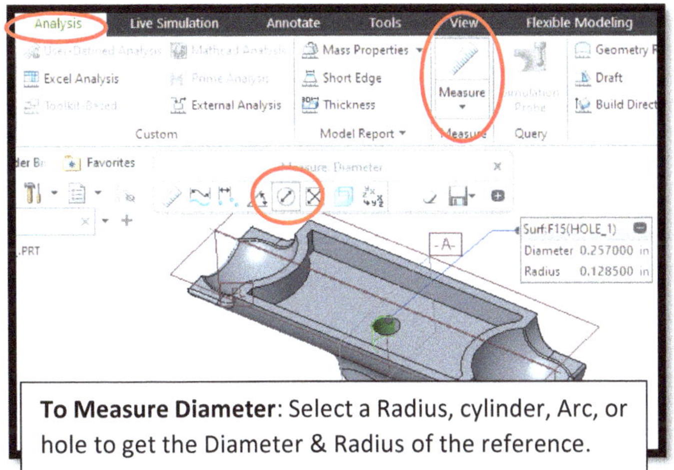

To Measure Diameter: Select a Radius, cylinder, Arc, or hole to get the Diameter & Radius of the reference.

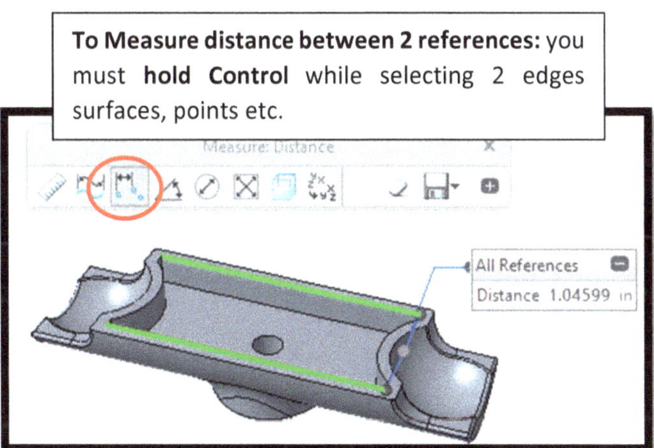

Weight & Center of Gravity: use the **Mass Properties** tool (Analysis Tab) to have CREO calculate the weight and Center of Gravity position (Distance from the X, Y, Z axis of the chosen coordinate system (CSYS)).

Note: this is only accurate if components have had their Model Properties set. If Parts within an assembly do not have material properties set, you can set the Model Properties of an Assembly so all components will use the same property (doing so overrides any individual part model properties for the Mass Analysis calculation.)

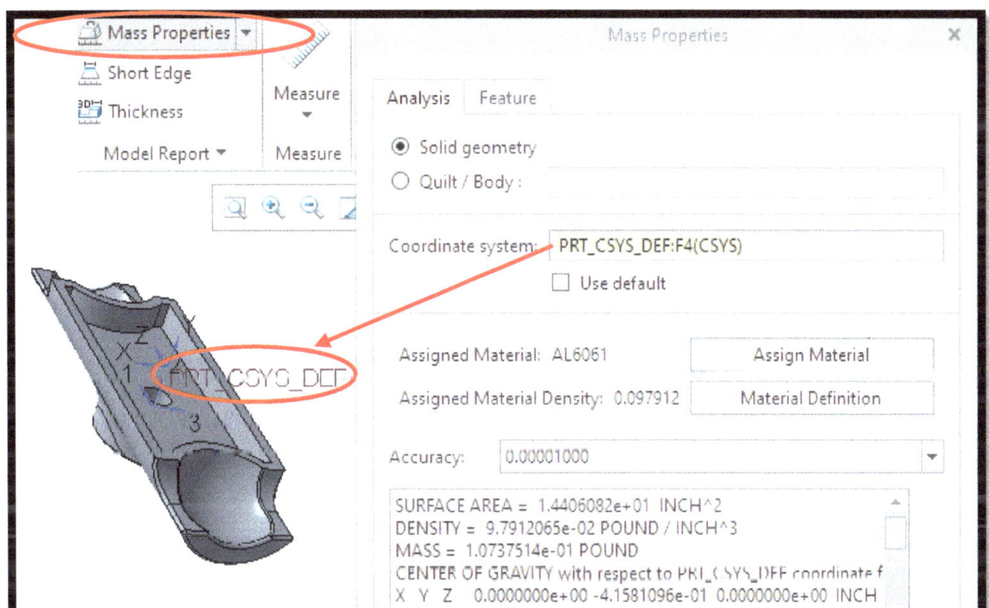

LMB to choose the desired CSYS. Then the Mass and Centroids will be listed in exponential notation. This example weights 0.0107 lb*, and has a center of gravity centered about the X&Z Axis, but offset from the Y by about -0.04158". *Note that mass given is pound-mass, which happens to be the same as a pound-force (assuming on Earth/Standard Gravity), you do not need to convert it to get a weight on Earth.

End of this Topic

Importing a Points File or X, Y, Z Coordinates

Engineers will sometimes need to import a series of coordinates into a CAD model for use as a trajectory or to approximate a shape or curve (such as an Airfoil cross-section). You can import a Points file, fit a curve or sketch between them, then use that in modeling. Brief Steps are shown below.

- Start a new Part File – in the Model Tab – expand the drop box of the Points tool – select Offset Coordinate System – in the menu click the Reference box and LMB on the desired CSYS from the graphics window.

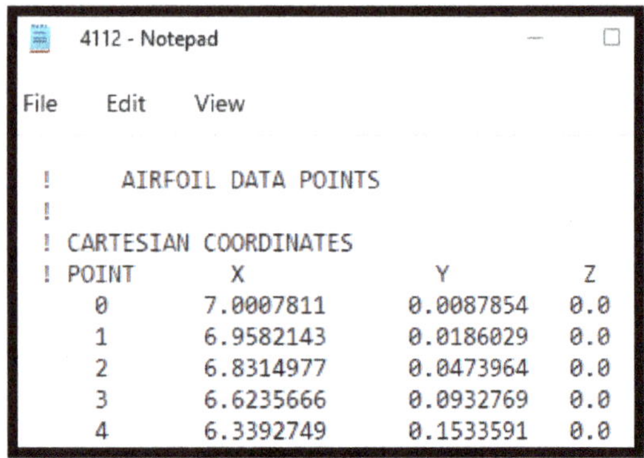
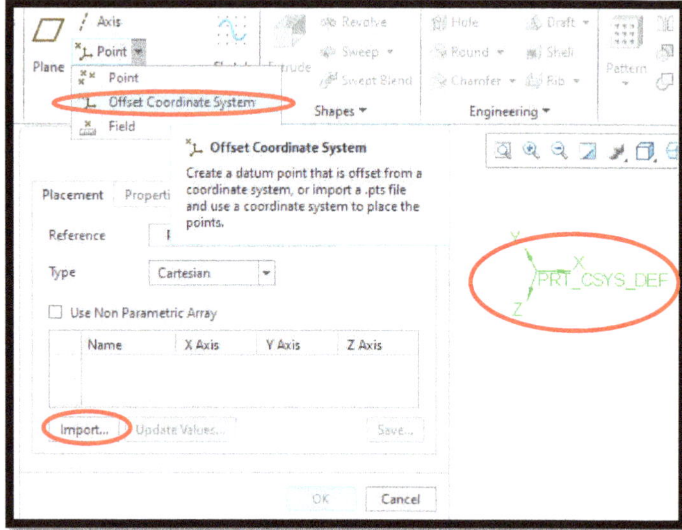

- Then choose **Import**. Browse for your **Points file** (if you have a text file, rename it or change the file extension so it is a .pts file). The Data should be populated if it is delimited consistently into columns for the Point #, X, Y, & Z values. **Press OK** if the table looks correct in CREO.
- You should see the Points on the screen. While still selected on the points, expand the **Datum Menu drop box – choose Curve – Curve through Points**.
- The curve should be added between the points without needing to change any options.

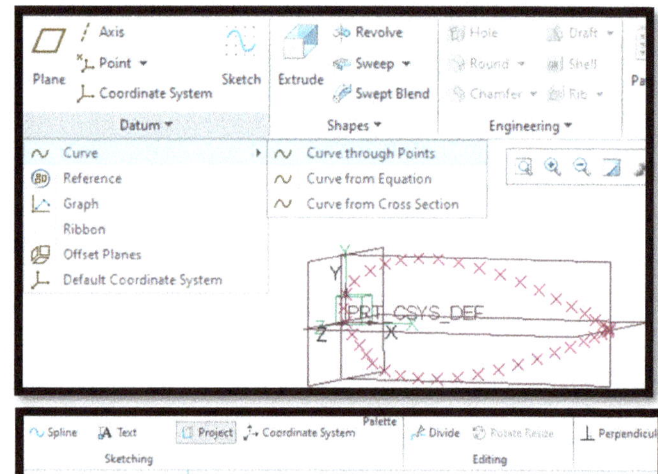

Once you have a curve, you can use it as a trajectory (sweep), or can turn it into a sketch or solid feature.

To convert to a sketch: Use the Sketch tool – use **the Project Tool** – project the Curve as the closed-loop. Add entities as needed to complete the closed-loop. You could even save this sketch to have available in the **Palette Tool** in other sketches. That can be useful as then you can deploy the saved sketch and resize it as desired using the Scaling Factor of the Palette Tool settings.

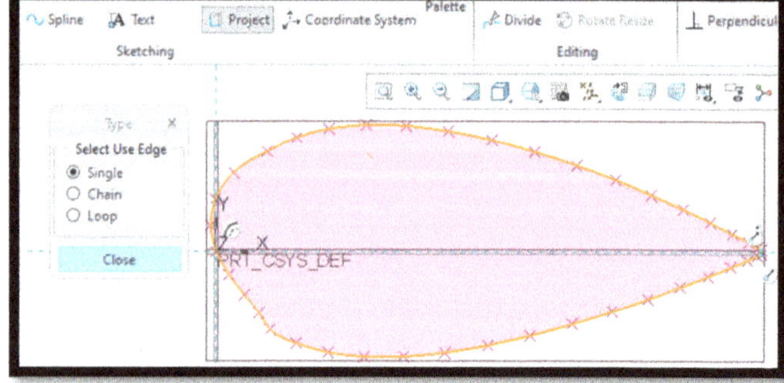

Tip: Make sure you have a Points file with standard delimiting between columns (tab or space), and be aware you cannot easily change the size of a feature made with Points (as the points would need to be edited). Using the palette Tool can get around this if it should be scaled, but the shape will not be adjustable.

End of this Topic

Setting Colors of the Model & Renderings

You can adjust the color and texture of models to better identify components on an assembly, make a model more realistic, or to prepare a Rendering image for a cover page or presentation.

Adjusting the Color of a model: Use the **Appearance Gallery Tool** (View tab) and select a desired Color Setting.

>   **- View Tab - Appearances Tool** – click on a **Libraries option – browse for a metal library** (or other type if you want) – **select the color/texture** you want to use - a small **selection menu** box will appear in the upper right corner of your screen – **select the Part File name** from the Model tree – **press OK** in the small selection menu. This will change all the model material to the selected color.

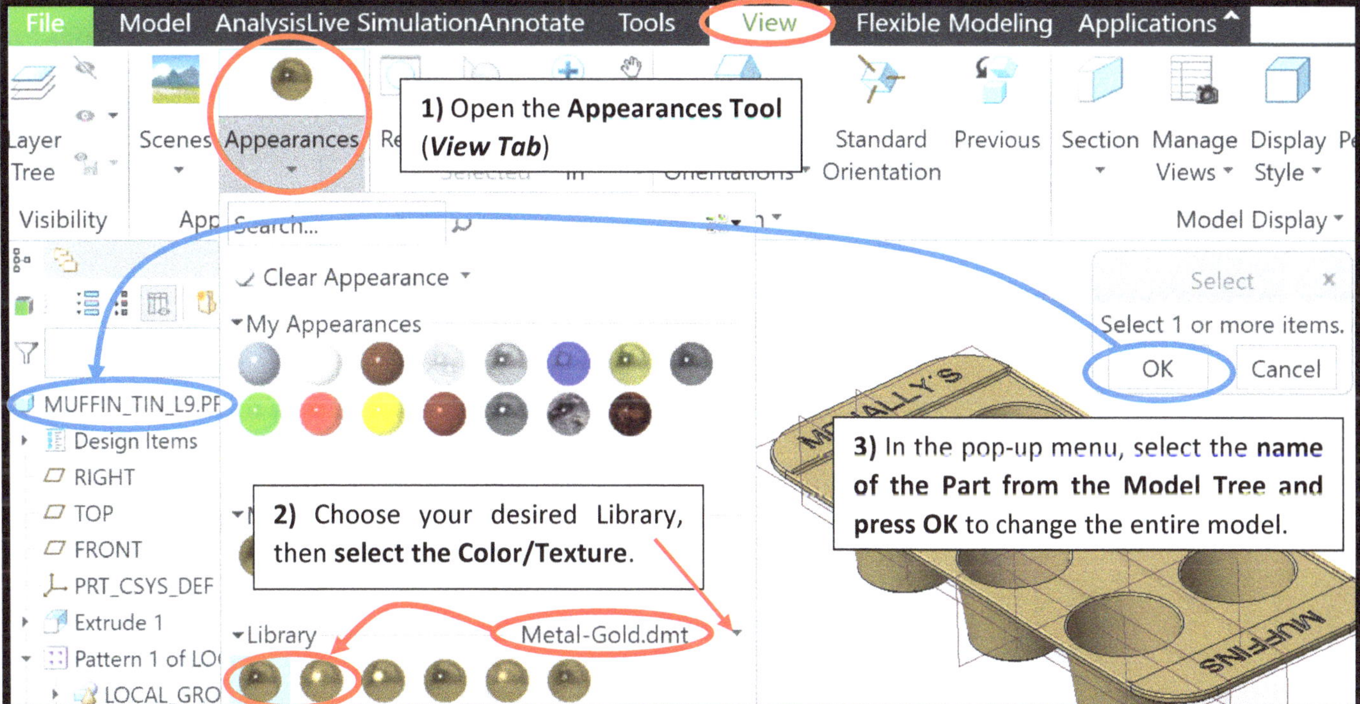

Rendering a high-resolution image: Some graphics cards may not work properly for rendering, so if you have issues you can toggle off "Real Time Rendering". Also, some student home versions of CREO may not have the render studio installed.

- **Applications Tab** in the top toolbar – **Render Studio** – **Scenes Tool** – **double click on a Scene** to add it as the background.
- Use the **Scenes tool – click on Edit Scenes – click on the Environment tab** to adjust the **"Floor Plane"** as needed: depending on your part model you can choose a different plane (Front, Bottom, Right, etc.) to orient the scene to your part model. **Press Close.**
- Toggle on **Perspective View** in the top toolbar to add depth to the model graphics window.
- **Press Render – Browse to choose a location** to output the image to and setup a max time of rendering.
- Use the **Screenshot Tool** (*or windows Snipping Tool*) to take a screenshot to use in marketing your muffin tin.

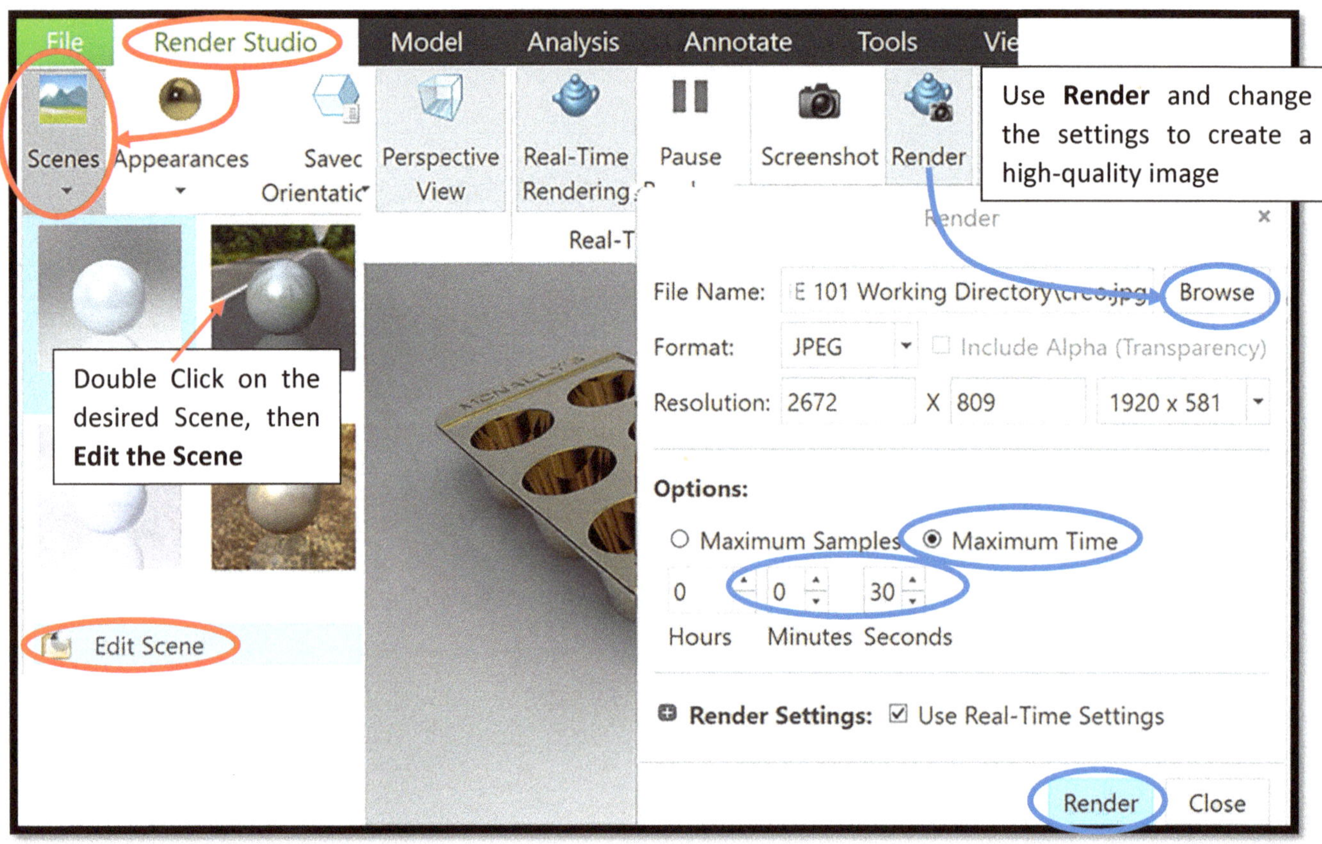

End of this Topic

Importing ServoCity/McMaster Step Files in CREO

STEP files are 3D model files that contains 3D data in a format universally recognized by CAD and 3D printing software (similar to a .PDF vs a .DOC file for documents). These files contain the geometry and size of a model but lack the parametric features that allow it to be modified or useful in further design work on the component. While STEP files are the most common filetype you can also use these same steps for other CAD files (Inventor, Solidworks, .STL).

Overview of the steps:

1) Locate the desired part from the vendor, then download and extract the STEP file (if provided by the supplier).
2) Start the **import process** in CREO
3) Check the model for errors, such as surface modeling or validation errors, that may cause issues with the model.
4) Specify the model Material, adjust the colors, and add Axis or Datums to use as assembly constraint references.
5) As an alternative to importing a model, consider modeling your own simplified Model in CREO yourself.
6) Download the vendor 2D Drawing PDF, or if needed, setup your own custom Front view to create a Detail Drawing showing fabrication modifications of the vendor components.

1) Sourcing Vendor Supplied Component CAD Models

McMaster.com has many of their parts available as STEP file formats that can be downloaded and imported. Servocity.com, a supplier of robotic parts, also provides STEP files for most of their components (mounts, adapters, etc., except most motors seem to be missing) and even a library zip file of all of them you can download at once.

Note – you normally should not import "Stock Material" that your models will be machined from during fabrication, such as chunks of aluminum, angle iron, tubing etc. Only import finished products or those that will have only minor modifications (i.e. cut to length), as imported files are not suitable to edit features and lack drawing view templates.

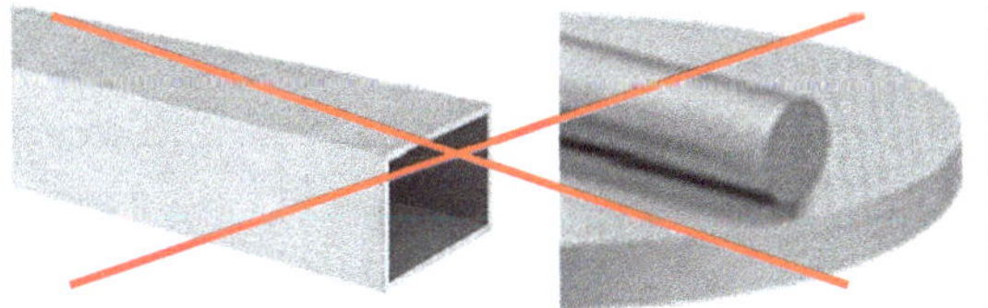

Do not import "raw material" CAD models that will be machined to model your own parts! Instead, model from scratch starting with a new Part file. Some exceptions may be for complex shapes like "80-20" tubing or box tube that needs holes added.

Download & Extract the vendor STEP file.

You can use www.McMaster.com to source bolts, bearings, etc. and download their CAD file in the desired format.

- **Navigate the catalog** and find the desired part, **click on the part #** to expand the details – then change the CAD file download to **"3D Step File"** – **press Download** – Save to a known location and **extract or unzip the file**.
- You can also download the **2D PDF Detail** drawing from the CAD list to include in your project drawing package.

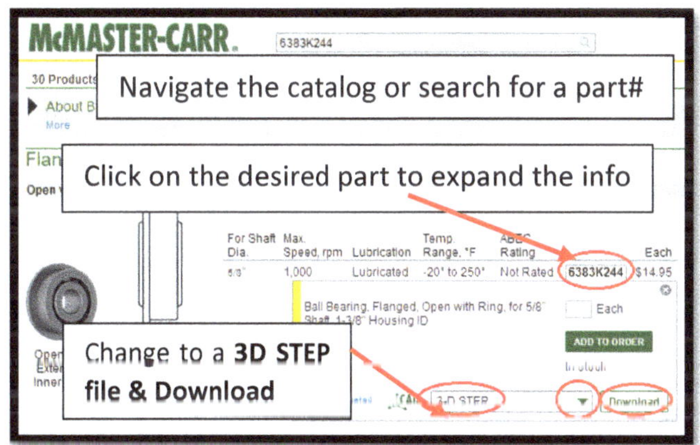

Another website for robotic projects is to use www.ServoCity.com to find the desired components.

- On the desired product page look for the **Downloads area** – download the **STEP file** (if available) and extract and unzip to a known location. Not all of their components have CAD files available (most motors do not), so for those components you should model a simplified representative model yourself.
- You can also download the entire library of CAD models from them from the bottom menu "Tech Resources" link – **STEP File Library**. These files must be unzipped or extracted before they can be imported into CREO.

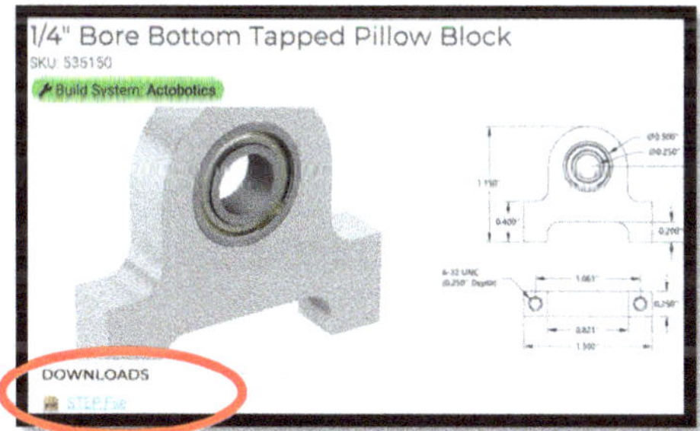

2) Import the Step File into CREO

2 – 1 - In CREO use **File – Open – change the File type to "Step" or "All Files"** - locate the downloaded STEP File - **press Import.** *By default, CREO only looks for CREO files so you must change the Type setting to show ".STEP" or "All Files". Make sure the files have been extracted if they are compressed when downloaded from the vendor.*

2 – 2 - Select either the **Part file** or **Assembly file option**. Single objects, like a bolt, must be imported as a Part or it may have issues. Some objects, like bearings, should behave as a single unit in the model so you should choose to import them as Part files, even though it is really an assembly model. **Tip:** Some files may work better if you choose the Assembly file option, it depends on the specifics of the model (*see example on next page Step 3-B*)

2 – 3 - Provide a name for the new file, typically you will want to keep the Part# or name from the vendor but may want to add in some descriptive info such as that is a bearing or bolt or similar (i.e. *638K244_BallBearing*).

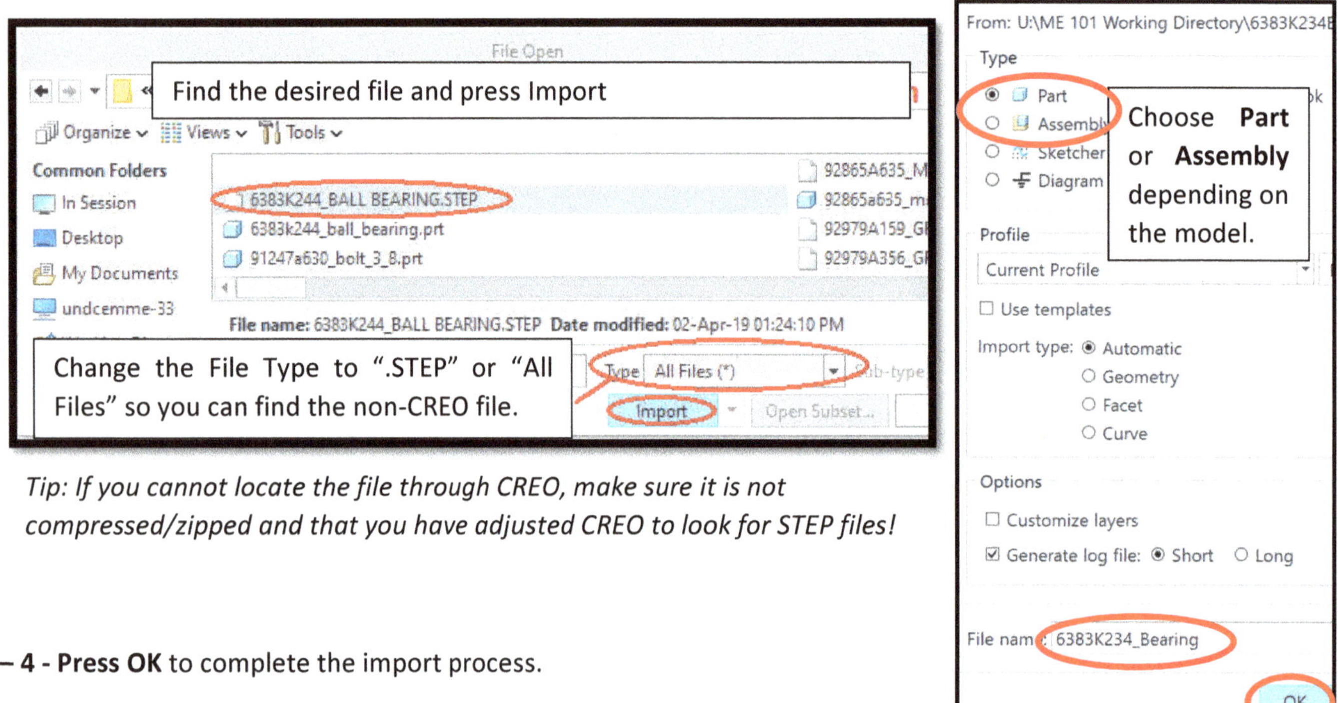

Tip: If you cannot locate the file through CREO, make sure it is not compressed/zipped and that you have adjusted CREO to look for STEP files!

2 – 4 - Press OK to complete the import process.

The part model appears on the screen. You may get an error message about the names of some features in the new model being modified; disregard these errors and press close to continue on. You may also get some warnings about some **Import Validation errors**, those are usually resolved by following the instructions on the next page and choosing a different file type when importing the model.

If you get errors about the names conflicting when you try to save an imported file, you may have to work through the menu that asks for new suffixes on the model until it is acceptable (can happen when importing SolidWorks files).

Continued on Next Page:

3) Check for Model Import Errors:

After importing you may notice some errors in the model. Often, choosing the incorrect Part or Assembly **file type** can cause an error. Other times, it is the model that is provided by the vendor being made poorly making it mostly unusable for use in CAD. If you get a Model Validation error, it is likely due to one of the reasons below.

A) Purple Color (Surface Model) Error:
A purple colored model usually indicates the model is a hollow Surface, not a solid material. If you notice this purple color on your model, repeat the **import process as a Part instead of an Assembly (or vice versa)** to see if it comes in properly. Some models will come in purple no matter the import setting (common with some McMaster bolts). **If so, it is suggested not to use them**. Find an alternative part # & model, or create your own simplified representation model (discussed in Step 5 later on). If you do decide to use a purple surface model it will show up as "Wireframe" on all drawings that include it, cluttering up the assembly drawings (see example in "C" below).

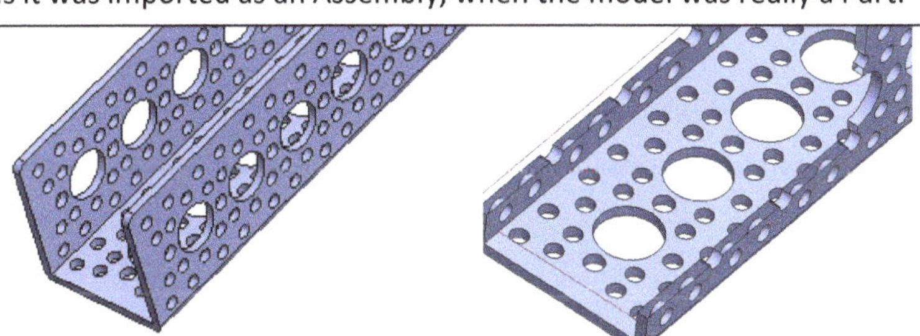

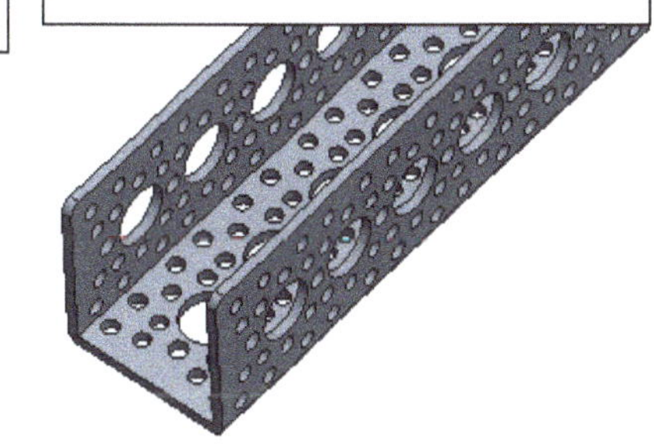

B) Disconnected Components Error:
Likewise, an Assembly incorrectly imported as a Part may have some issues as well. The motor mount shown in the image is really an Assembly, as the screws are a separate component in the same STEP file. If this model is imported as Part the screw will not be positioned properly. When imported as an Assembly, the screws are placed correctly in the hole.

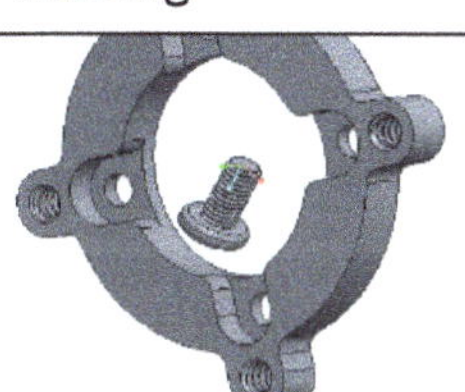

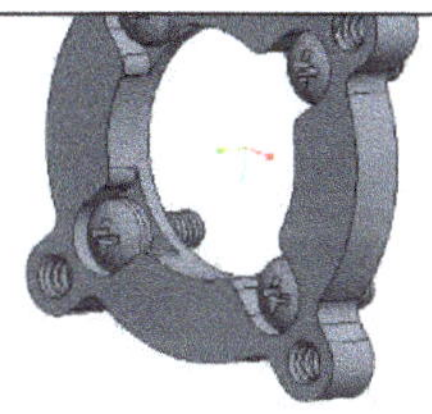

C) Partially Purple Surface Models:
Some models may import with just an accessory screw as a surface. One solution for this scenario would be to **expand the Model Tree** – find the **Part** that is worth keeping – **RMB – Open** to open it in its own window, and work with that part file which would not include the purple screw. You could also delete the screw out from the assembly, but may need to set the option to ignore some validation warnings (click on the items in Red in the model tree – RMB - Import Validation – Disable Ignore Validation.)

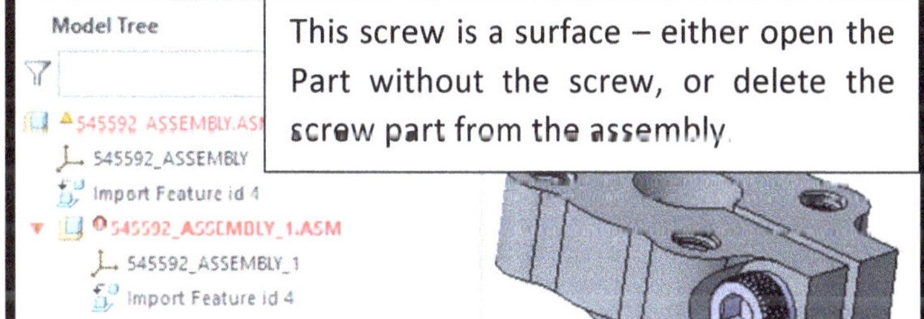

D) Wireframe lines showing up on drawings

The purple surface models, discussed above, will show up as a purple wireframe line on drawings that use component and it cannot be fixed. Either accept the fact of messy lines being on the drawing or find/create a new model for that part.

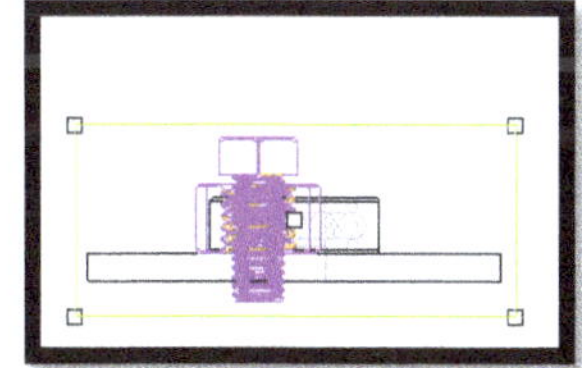

4) Setup Axis, Color, & Model Material Properties:

After importing models into CREO you will need to **create Datum Axis** *(and Datum Planes, if needed)* to use as Assembly Constraint references. The imported models will not have Axis on holes, default datums, default Drawing views, etc.

4 - 1 - Set the **material properties (File – Prepare - Model Properties)** so that any future assembly BOM table using this model will show the material properly, and any weight analysis will be more accurate.

4 - 2 - Set the **color** as desired (**View tab - Color Appearance tool).**

4 - 3 - Use the **Axis tool** from the top toolbar to add any axis you may find useful for assembly constraints later on – LMB to select the outer edge of a hole, cylinder, or arc as a reference – **press OK**. The Axis is now added to the part Model. You can add datums if desired using the Plane tool.

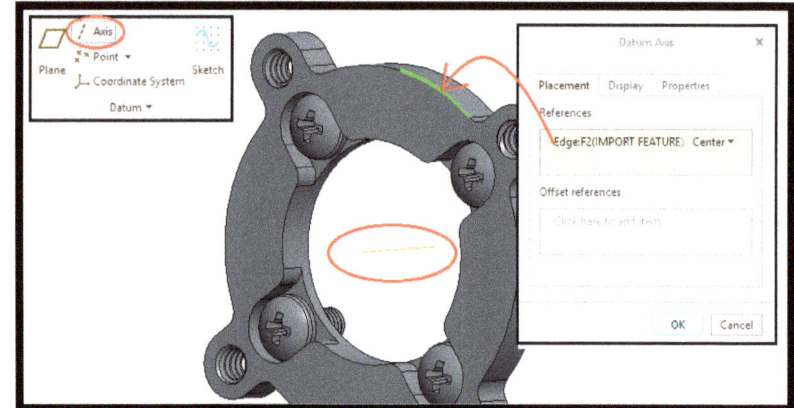

4 - 4 - If desired, you can also add Datums to the model using the Plane Tool. Typically, you would only need to do this for parts that don't have flat surfaces that could be used for assembly constraint references instead, or if you want to setup Drawing Views.

- **Datum Plane Tool –** choose your reference(s) as desired to setup a datum plane and the options (through, offset, etc.).

- **Example: To setup a Datum plane using an Axis as a Reference:** You can use an Axis as a primary reference, then while <u>holding Control</u> on the keyboard select a vertex along an arc to act as the refence that the plane will run through. Or, on this part there is a small vertical flat surface near the screws, you could use the Center Axis and the vertical surface as references, then set the Offset angle as desired to make this datum.

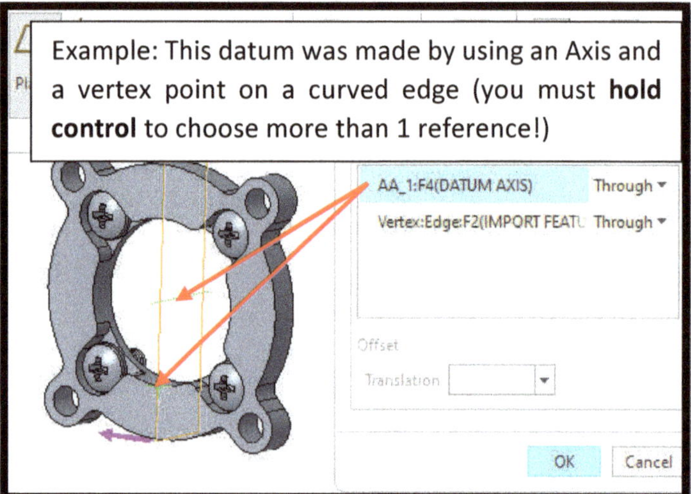

4 - 5 - Press Save to save the CREO model (make sure it is saved to your working directory), and use that from now on instead of the vendor Step file. Using the CREO file in your assemblies usually works better than using the original STEP file directly without doing an import. If you imported an Assembly file, note that you will need to keep all the individual Part files that make up the imported assembly within the same directory, as each assembly likely has many part files created with it.

5) Simple Representation Models

Sometimes it is easier to create a "simple representation" of a model that you create yourself instead of using the more detailed STEP/CAD files from a vendor. A Threaded Rod can be downloaded easily but the threaded surface may have CAD defects such as "Stitching" or "Surface" errors, or be created as Surface which can cause problems if you need to modify the part for your design.

A simple representation is a quick and easy model you create to show the **critical dimensions and shape** of a component, without adding in the fine details such as threading, tapping, notes, surface effects, curvatures, etc. These are often easier to work with than the complex CAD files of the finished product, while still showing fit, clearance, and function of the component in the assembly.

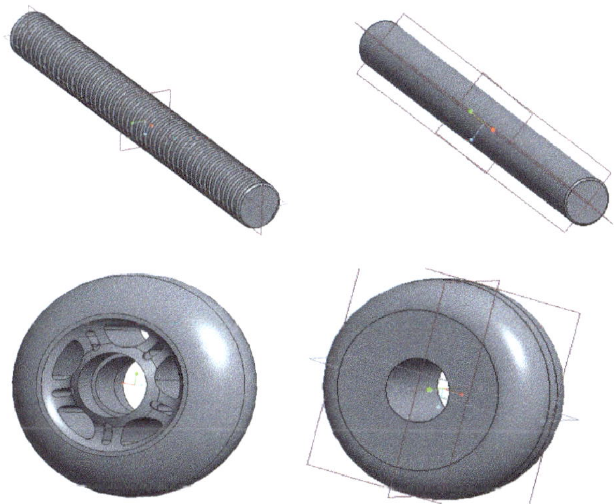

Threaded Rod Example: The simplified model does not need to show threads. The user could model a simple cylinder of the correct OD & Length, then add a Cosmetic Thread for the thread info, if desired. Fit and function is still maintained for the assembly, and a note could be added to an assembly to show it is threaded.

Robot Wheel Example: The ServoCity STEP file is highly detailed, but a simple wheel with correct ØOD, ØID, & thickness could be modeled easily. Add a Rounded edge and it looks good enough to show fit and function.

6) Creating Drawings of Imported Vendor Models

By default, imported CAD models do **not** have the Front, Top, Right datums and drawing views built in the model. If you try to create a drawing the template will be empty or show the views of the CREO "placeholder" model.

Typically, **you do not need to create Drawings of vendor products** as you are not manufacturing them. Instead, download the vendor detail drawing for your project records (McMaster provides a **2D PDF** download option, while ServoCity only shows on image of the dimensions on the product detail page that can be screenshot and saved.)

If you are making modification to a vendor supplied part (adding holes, milling out slots, etc.), you will need to follow the steps on the next page to create **Custom View States** and create a drawing that shows details about the fabrication modifications.

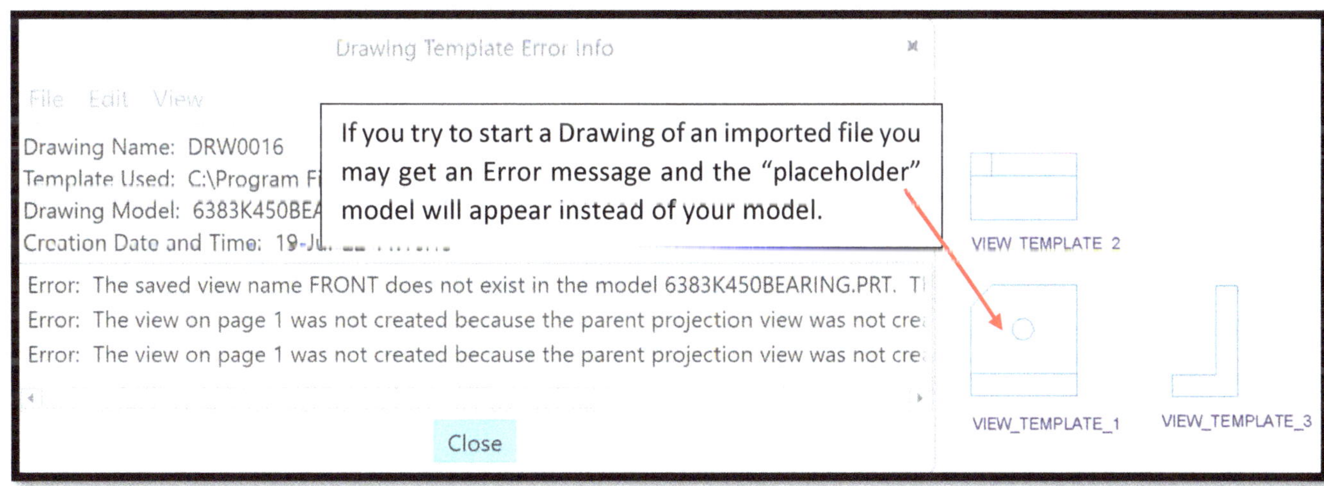

6 - 1 - Set up a custom Front View with "View Manager" to create a new view state.

- **Model Tab - Manage Views – View Manager – Orient tab – New** – enter in **"MYFRONT"** and press Enter – use the **Edit** tool from the view manager menu **– Edit Definition –** in the **Reference One Box LMB to** select the **planar surface*** or datum to use as the Front View – then **LMB to select a planar surface** to act as the secondary reference (perpendicular to the "Front" to act as a "Top" plane). **Press OK** to accept the shown view as the "Front View" state or press Reset to retry.

*Note: If your part does not have any planar surfaces then you may need to create some using the **Datum Plane** tool or extruding to add/remove some material to create a flat surface.

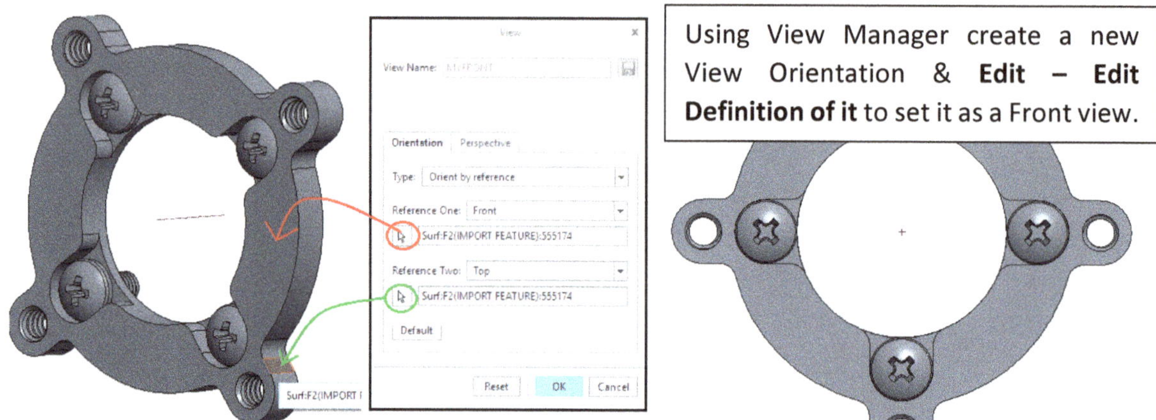

6 - 2 - Test out if the view works: In the **View Manager - Orient tab** – double click on Standard Orientation – double click on the "MyFront" – the model will readjust to on the screen which should be perpendicular to the screen.

6 - 3 - Create a new drawing with an **empty** template, and insert a general view using the MyFront view state.

- **File – New - Drawing –** provide a filename – **OK – choose Empty** as the template – **OK –** select **Insert General View** – place the view on the screen – **select "MYFRONT" from the View Type List – OK.**

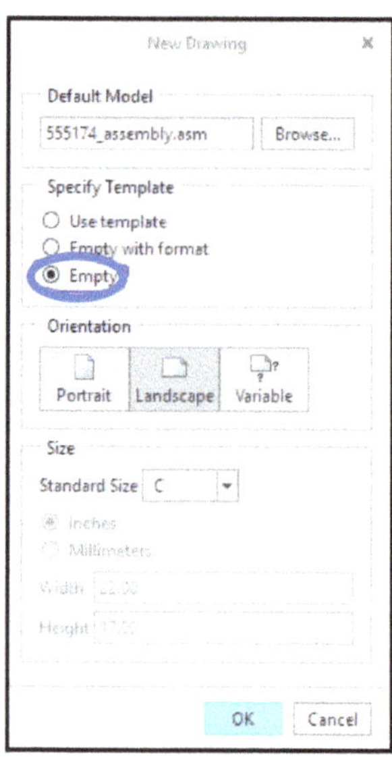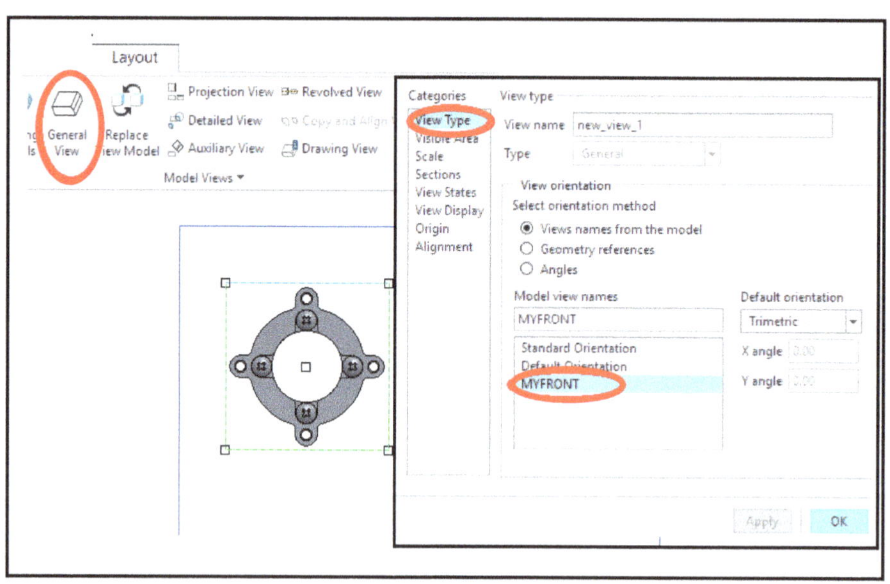

6 - 4 - Project the other views: Use the **Projection View** tool from the top toolbar (Layout Tab) – select the Front view to act as the "Parent View" – LMB to click to the right to place the Right View. Repeat to create the Top View.

6 - 5 - Continue with the drawing as needed to detail the fabrication changes to the model. Include some details about the part itself (A note specifying the Vendor part#, etc.) on the drawing.

End of This Topic

Creating a DXF file for CNC Milling

At some point you may need to have a CAD Model fabricated using 2D CNC Milling operations. The steps below show how to create a **DXF file** type that can be imported by software used by a 2D CNC/Router/Plasma table for **two-dimensional** profiling or cutting out shapes in a sheet material.

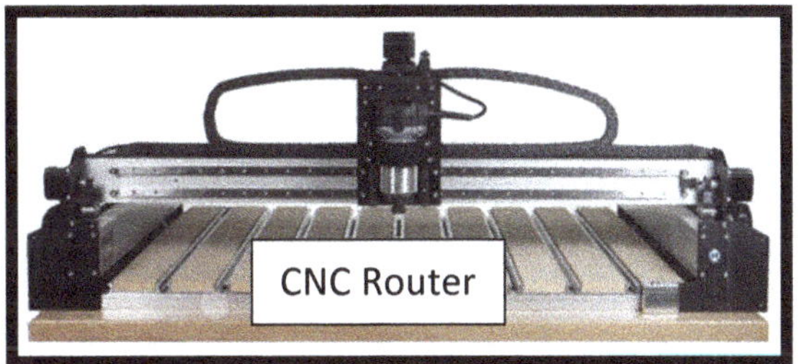

It is possible to have 3D parts CNC milled but that is not covered here as typically you will need to send the 3D CAD file as a STEP, IGES, STL or similar for the fabricator to import into their CAM (computer aided manufacturing) software.

The Goal of creating a DXF file is to provide a 2D layout that the CAM software can use to choose the edges, curves, etc. that will be used to setup toolpaths in the CNC software to mill or drill shapes.

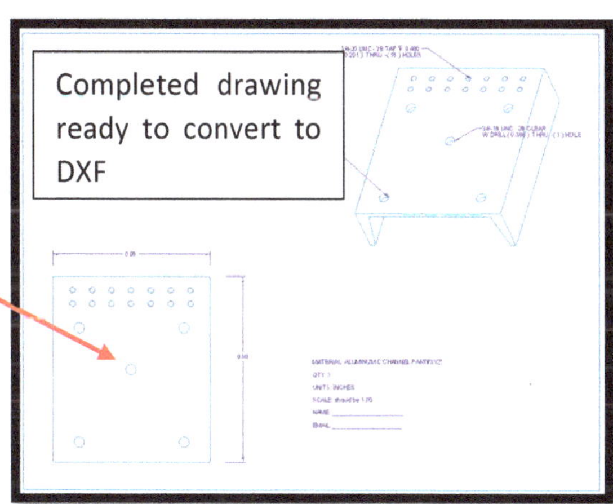

This front view without dimensions or other lines touching it is used to allow the CNC operator to choose which edges, curves, lines will be used to setup the toolpaths. The other view is for reference.

Step 1 – After completing your model in CREO a new Drawing must be started to be the basis of the 2D geometry needed in the DXF.

- File – New – Drawing – Provide a Name – Press OK. Browse if needed for the Part Model – do not change the Template settings – **Press OK**.

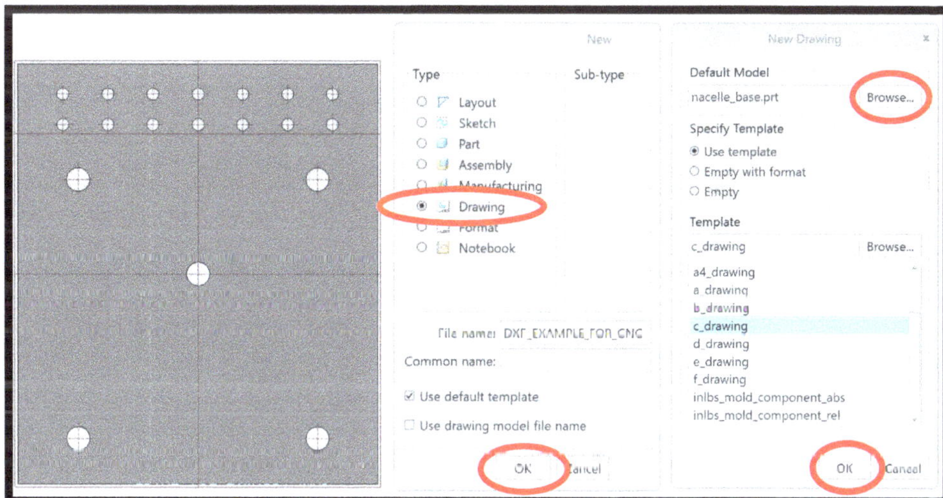

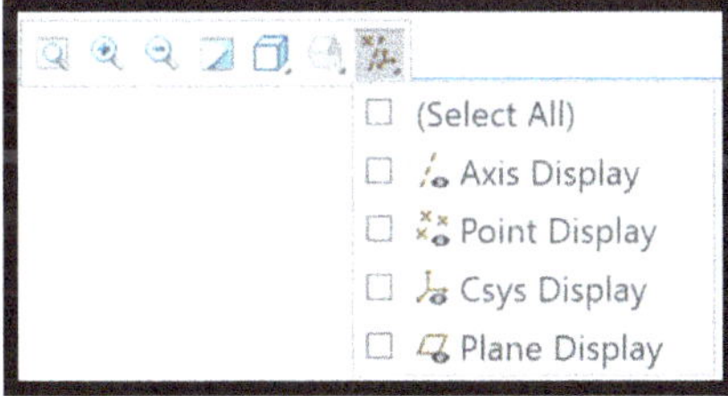

Step 2 – Turn off the Display Filters to remove the brown Datum Lines, Axis, & Coordinate entities from the views.

Step 2 – The DXF must show the Planar surface that will be operated on by the equipment (not a side surface). Rotate the Parent View (bottom left view) to a different **View Orientation** or **By Angles** <u>if needed</u> to show the desired plane.

- Double Click on the Parent View – View Type category – choose the View or Rotate by angles as needed – Apply/OK.

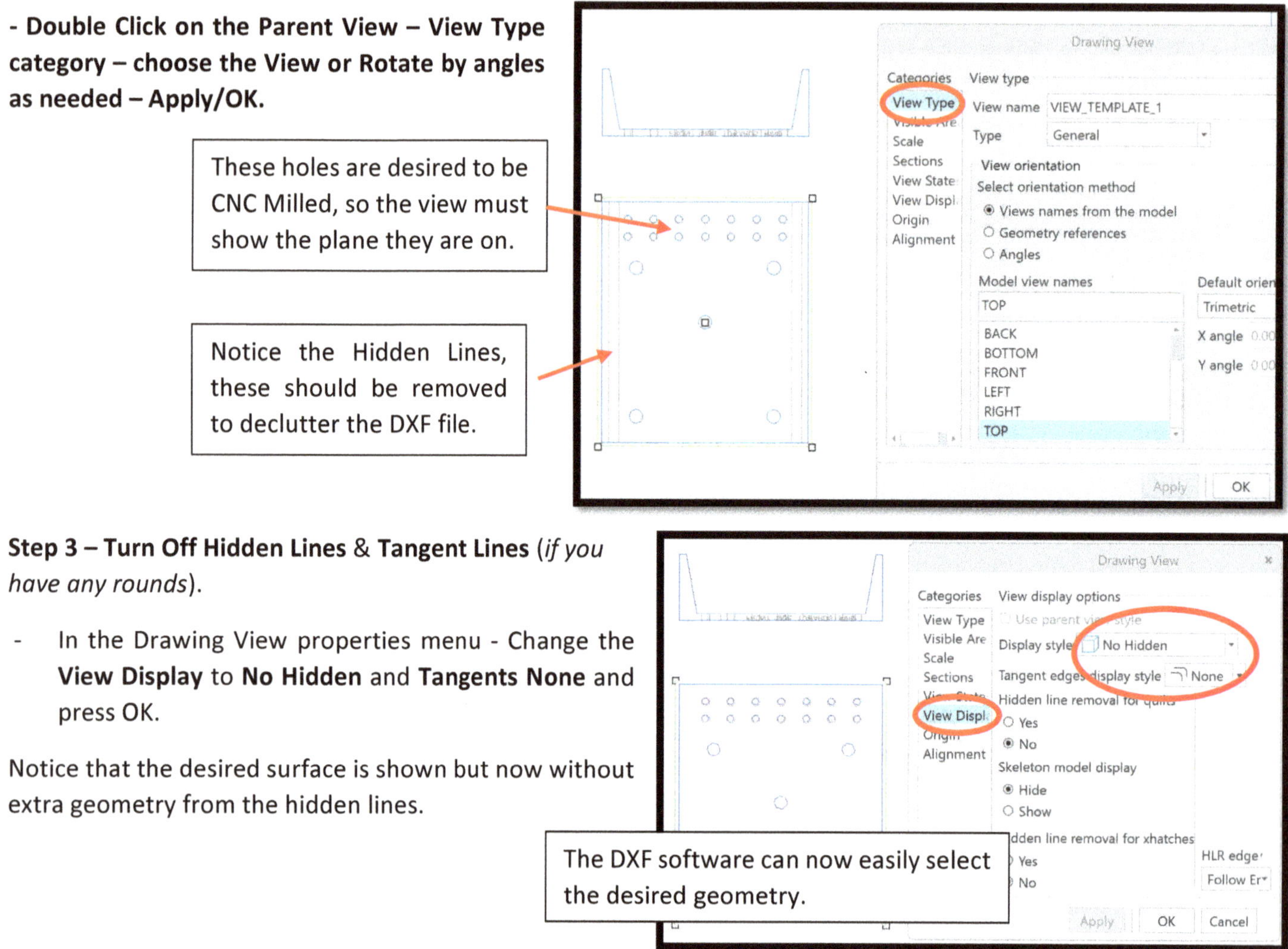

Step 3 – Turn Off Hidden Lines & **Tangent Lines** (*if you have any rounds*).

- In the Drawing View properties menu - Change the **View Display** to **No Hidden** and **Tangents None** and press OK.

Notice that the desired surface is shown but now without extra geometry from the hidden lines.

Now prepare the Drawing for use as a DXF. The operator of the CNC will use a CNC software such as CARBIDE, VCARVE, SURFCAM, etc. to select any entities like edge lines and hole curves on the drawing and then setup the details and toolpaths about how it should be cut (depth of cut, # of passes, speed, bit size, should the bit follow the inside, outside, or middle of the lines, etc.).

To make it easy for the correct geometry edges to be selected properly, the DXF should not have a lot of extra hidden lines, other geometry, Datum lines, dimension lines that they need to pick through to choose what will be cut.

Step 4 – Set the Scale to 1.00 in the bottom left corner. By using a scale of 1.00 the DXF will not need to be scaled.

Step 5 – Also make sure there are not any standard hole notes, datums, etc. on the main view. Do not include the UND template (not needed as this is not a detail drawing).

Step 6 – Toggle Off "Lock View Movement" in the top toolbar, then move the drawing view to the bottom left corner without touching the edge of the drawing.

Step 7 – At this point the drawing is ready to be saved as a DXF. **But to assist the operator you should either:**

A) **Save a Copy** to add Dimensions/Notes on, save as a PDF, and share along with the minimal detailed DXF.
> **or**

B) **Add info to this drawing but do NOT add anything that touches the lines of the Main View!**
- o Isometric view, General Note (Annotate Tab) about Scale, Units, Qty, Material, Contact Info
- o Include a few dimensions for the Overall Size (That do not touch the Model lines!)

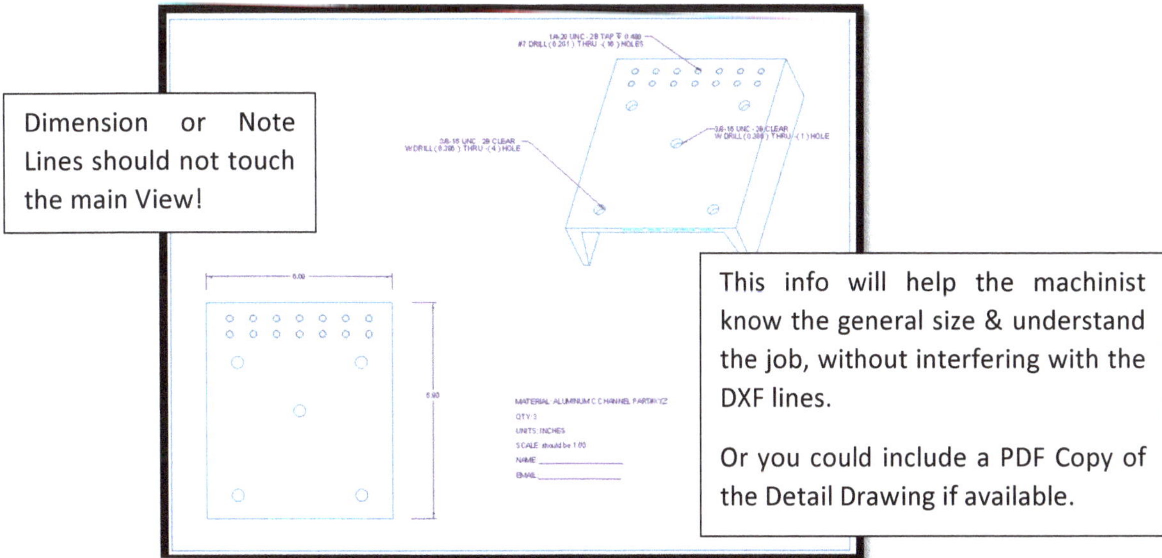

Step 8 – Now Export as a DXF file.

- File- Save a Copy – Choose the Type as "DXF", then type in a Name. The next menu may appear asking for export options. Leave these as the default settings.

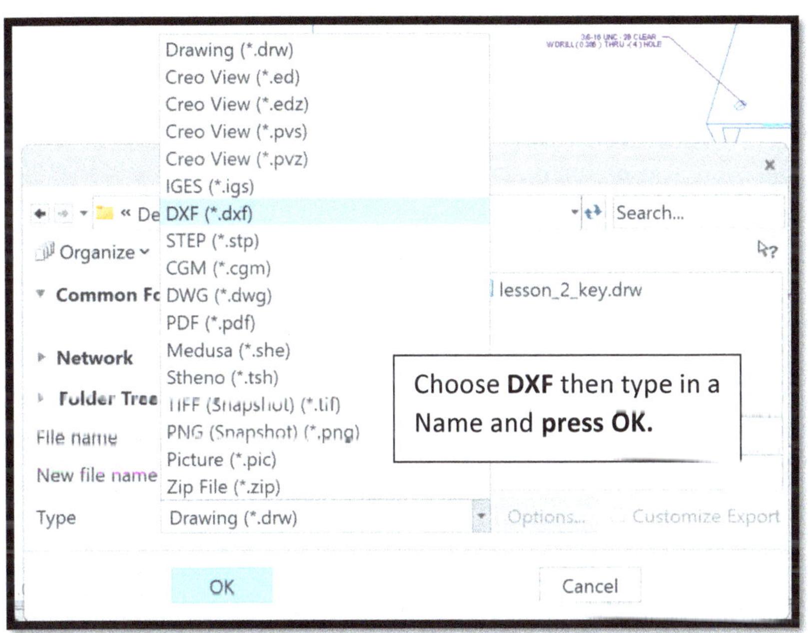

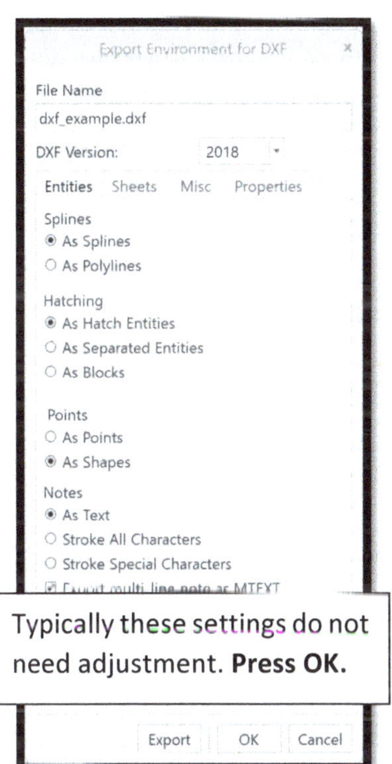

At the very bottom of the screen you will see confirmation the DXF was created.

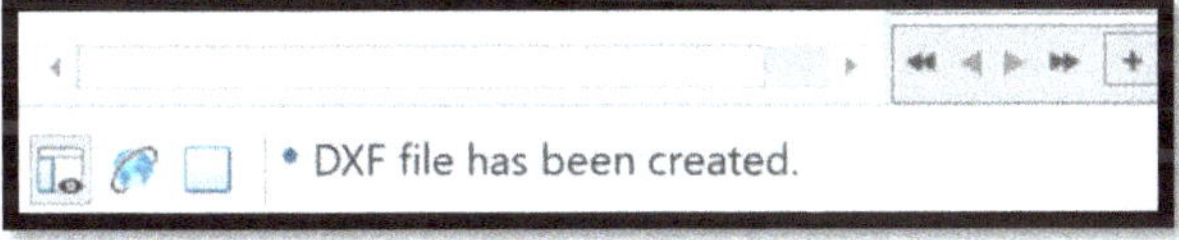

Step 9 – You now have a **.DXF** you can share or load into the CNC machine software in your working directory (ignore the Log file). Also Save As – Quick Export to create a PDF to print & available for discussion with the machinist.

Creating a File for 3D Printing

After creating a CAD model, you may need to prepare a copy of the model as a new file type to prepare it for 3D printing. Some 3D Printers, like the Stratasys printers used in UND Engineering, are able to open a CREO Part or Assembly file directly and do not need any special file processing.

Other services online, in libraries, etc. may need to have a more universal file type shared, such as a Stereolithography (.STL) format file. An STL file approximates a model by creating a mesh of triangular points. If the Mesh isn't adjusted to be smaller when exporting the file then the curves and edges of the model can turn out blocky. Also, STL files do not include any metadata like units so you must specify to the producer the units used in the model so they can scale the model properly.

 - **When using CREO and 3D Printing at UND**, all you need to do is **share a copy of the CREO Part file with the instructor** and **rename it** according to the naming convention (i.e. *Yourname_PartName_UnitsUsed*). You do not need to follow the steps below when sharing CREO files with the instructor to 3D print unless otherwise directed.

- **When using other 3D printing services** (personal printers, Chester Fritz Library, etc.), you will likely need to save a copy as a .STL file, with **custom export settings** using the steps below.

Creating an STL File for 3D Printing

1) Save your CREO file as normal. Then **Save a Copy** as an STL with <u>**custom Export Settings**</u>.
- **File – Save a Copy –** change file type to Stereolithography (.STL) and provide a name.
- Use the file naming convention of *"Yourname_PartName_UnitsUsed"*
2) **You <u>must checkmark</u> "*Customize Export*"** to control the STL mesh settings to prevent a blocky model.
3) **Press OK.** The Export setting will popup – change the settings following the tips in the image below to reduce the chord height and angle control. Usually, you only need to change the Chord Height to "0" so it will default to a minimum value. On some larger models, you may also want to change the Angle Control to 1 (more accurate), or adjust the Step Size lower (will drastically increase the file size).

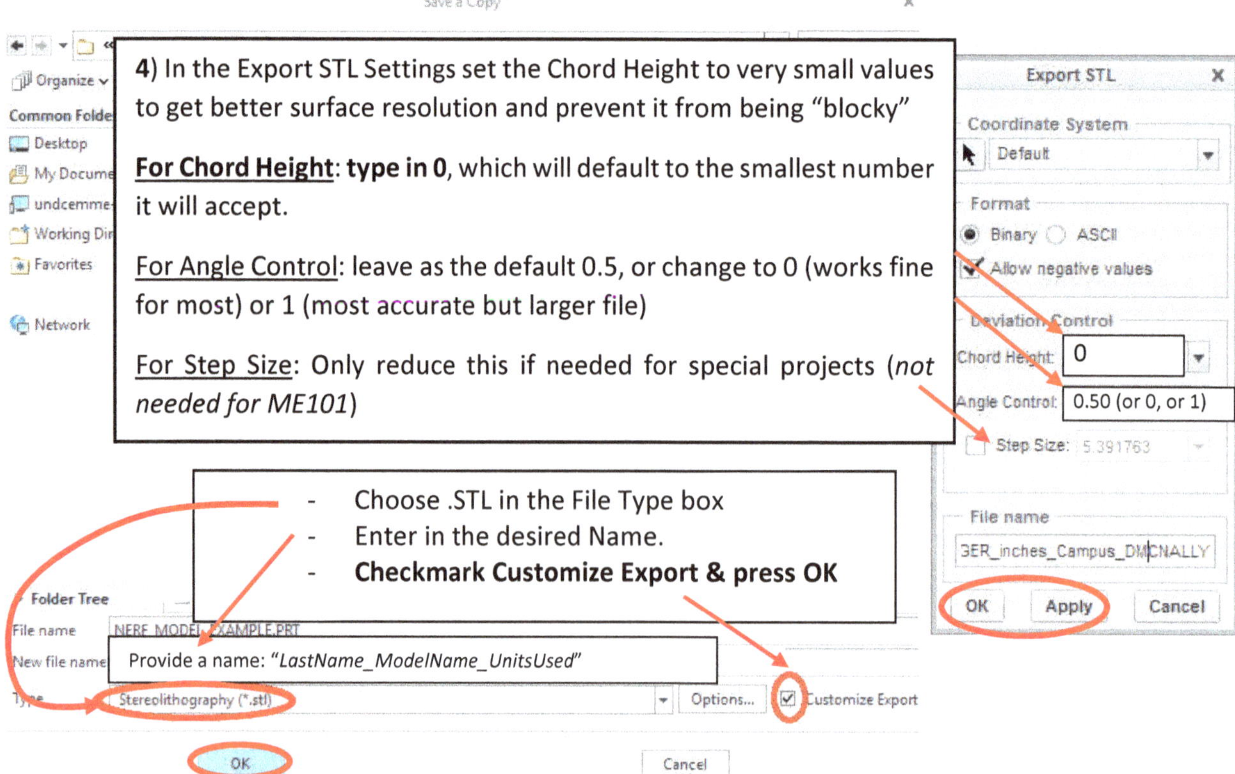

4) **Press Apply** after setting the Export settings. You will see the mesh triangles and lines on the screen. If they are not close together make sure you adjusted the chord height to the minimum, and you may want to adjust the step size down (try halving it). Reducing the Step Size will increase the file size, possibly making it unsuitable for emailing.

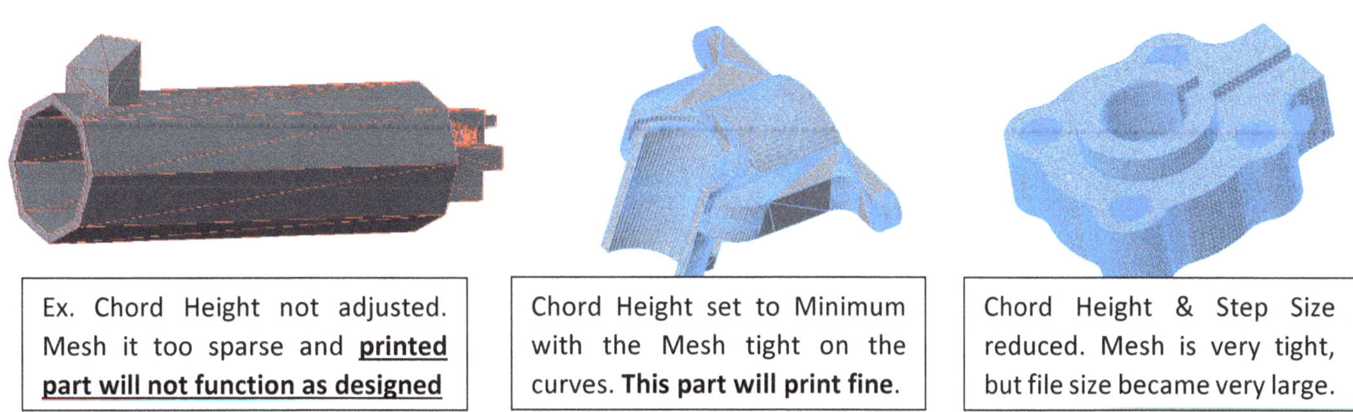

| Ex. Chord Height not adjusted. Mesh it too sparse and **printed part will not function as designed** | Chord Height set to Minimum with the Mesh tight on the curves. **This part will print fine**. | Chord Height & Step Size reduced. Mesh is very tight, but file size became very large. |

5) **Press OK** to accept the STL export settings which will create the new file in your current directory.
6) Use Windows File Explorer to find this file and share with your 3D printing service (if large, it may need to be zipped.) Keep in mind that the STL file does not include units so you must notify the printer operator which units the model uses, and perhaps provide a measurement of the overall length so they can double check the scale.

Specifying 3D Print Infill Setting and Orientation *(if desired for some projects)*

If desired for your print, you may want to request a specific print orientation and infill or density setting. Most 3D prints will work fine by having the instructor or printer orient the file to reduce print time, cost, and support material usage. Rarely does a part actually need to be printed solid, as most of the stiffness and strength of a 3D print comes from the geometry on smaller parts.

If you have a special requirement for a component you may want to consider a different infill setting. The Stratasys 3D printers are setup with only a few standard options, shown below. The overall density depends on the geometry and processing that the software performs to prep the file for printing.

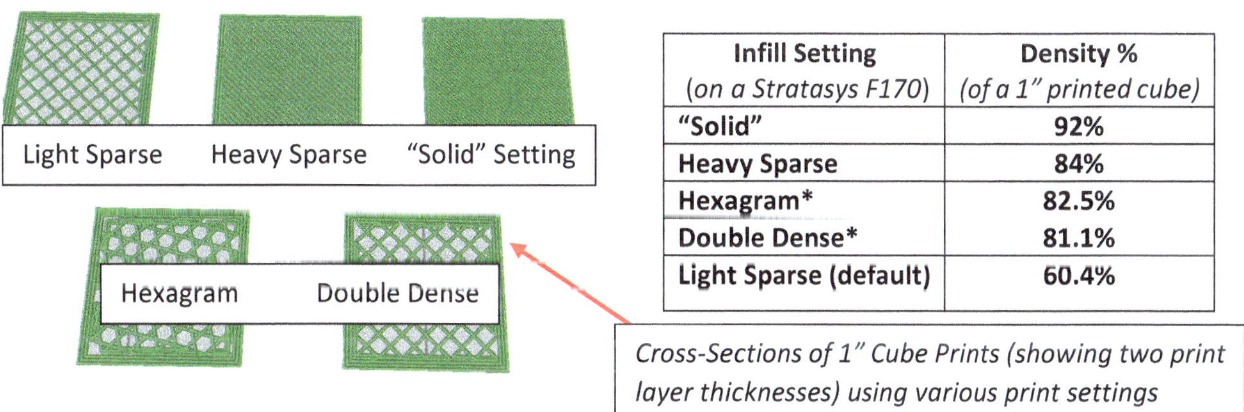

Infill Setting (on a Stratasys F170)	Density % (of a 1" printed cube)
"Solid"	92%
Heavy Sparse	84%
Hexagram*	82.5%
Double Dense*	81.1%
Light Sparse (default)	60.4%

Cross-Sections of 1" Cube Prints (showing two print layer thicknesses) using various print settings

*At UND, only the newer F170 series Stratasys printers have additional options such as hexagram (hex shaped infill) or Double Dense (thicker walls). These may not be available on all prints depending on 3D printer availability.

Tip: Always double check that the file you are sharing for 3D printing is the correct and most recent file, that the units are setup in the model properties as desired, and that you have been clear about how many copies you need printed. You may consider opening the file in CREO and making sure it looks accurate before printing (if using an STL, change the "File - Open type" setting to All Files so you can find it).

End of This Topic

www.ingramcontent.com/pod-product-compliance
Lightning Source LLC
Chambersburg PA
CBHW040140110726
48005CB00018B/2594